THEORY OF MATRIX STRUCTURAL ANALYSIS

J. S. PRZEMIENIECKI

Institute Senior Dean and Dean of Engineering
Air Force Institute of Technology

DOVER PUBLICATIONS, INC.
New York

To Fiona, Anita and Chris

Copyright © 1968, 1985 by J. S. Przemieniecki.
All rights reserved under Pan American and International Copyright Conventions.

Published in Canada by General Publishing Company, Ltd., 30 Lesmill Road, Don Mills, Toronto, Ontario.
Published in the United Kingdom by Constable and Company, Ltd., 10 Orange Street, London WC2H 7EG.

This Dover edition, first published in 1985, is an unabridged and corrected republication of the work originally published by the McGraw-Hill Book Company, New York, 1968.

Manufactured in the United States of America
Dover Publications, Inc., 31 East 2nd Street, Mineola, N.Y. 11501

Library of Congress Cataloging in Publication Data

Przemieniecki, J. S.
Theory of matrix structural analysis.

Reprint. Originally published: New York : McGraw-Hill, 1968.
Bibliography: p.
Includes index.
1. Structures, Theory of—Matrix methods. 2. Matrices. I. Title.
TA642.P77 1985 624.1'71 85-6818
ISBN 0-486-64948-2 (pbk.)

PREFACE

The matrix methods of structural analysis developed for use on modern digital computers, universally accepted in structural design, provide a means for rapid and accurate analysis of complex structures under both static and dynamic loading conditions.

The matrix methods are based on the concept of replacing the actual continuous structure by an equivalent model made up from discrete structural elements having known elastic and inertial properties expressible in matrix form. The matrices representing these properties are considered as building blocks which, when fitted together in accordance with a set of rules derived from the theory of elasticity, provide the static and dynamic properties of the actual structure.

In this text the general theory of matrix structural analysis is presented. The following fundamental principles and theorems and their applications to matrix theory are discussed: principles of virtual displacements and virtual forces, Castigliano's theorems, minimum-strain-energy theorem, minimum-complementary-strain-energy theorem, and the unit-displacement and unit-load theorems. The matrix displacement and force methods of analysis are presented together with the elastic, thermal, and inertial properties of the most commonly used structural elements. Matrix formulation of dynamic analysis of structures, calculation of vibration frequencies and modes, and dynamic response of undamped and damped structural systems are included. Furthermore, structural synthesis, nonlinear effects due to large deflections, inelasticity, creep, and buckling are also discussed.

The examples illustrating the various applications of the theory of matrix structural analysis have been chosen so that a slide rule is sufficient to carry out the numerical calculations. For the benefit of the reader who may be unfamiliar with the matrix algebra, Appendix A discusses the matrix operations and their applications to structural analysis. Appendix B gives an extensive bibliography on matrix methods of structural analysis.

This book originated as lecture notes prepared for a graduate course in Matrix Structural Analysis, taught by the author at the Air Force Institute of Technology and at The Ohio State University. The book is intended for both the graduate student and the structural engineer who wish to study modern methods of structural analysis; it should also be valuable as a reference source for the practicing structural engineer.

Dr. Peter J. Torvik, Associate Professor of Mechanics, Air Force Institute of Technology, and Walter J. Mykytow, Assistant for Research and Technology, Vehicle Dynamics Division, Air Force Flight Dynamics Laboratory, carefully read the manuscript and made many valuable suggestions for improving the contents. Their contributions are gratefully acknowledged. Wholehearted thanks are also extended to Sharon Coates for her great patience and cooperation in typing the entire manuscript.

J. S. PRZEMIENIECKI

CONTENTS

CHAPTER 11

INERTIA PROPERTIES OF STRUCTURAL ELEMENTS 288

CHAPTER 12

VIBRATIONS OF ELASTIC SYSTEMS 310

CHAPTER 13

DYNAMIC RESPONSE OF ELASTIC SYSTEMS 341

CHAPTER 1
MATRIX METHODS

1.1 INTRODUCTION

Recent advances in structural technology have required greater accuracy and speed in the analysis of structural systems. This is particularly true in aerospace applications, where great technological advances have been made in the development of efficient lightweight structures for reliable and safe operation in severe environments. The structural design for these applications requires consideration of the interaction of aerodynamic, inertial, elastic, and thermal forces. The environmental parameters used in aerospace design calculations now include not only the aerodynamic pressures and temperature distributions but also the previous load and temperature history in order to account for plastic flow, creep, and strain hardening. Furthermore, geometric nonlinearities must also be considered in order to predict structural instabilities and determine large deflections. It is therefore not surprising that new methods have been developed for the analysis of the complex structural configurations and designs used in aerospace engineering. In other fields of structural engineering, too, more refined methods of analysis have been developed. Just to mention a few examples, in nuclear-reactor structures many challenging problems for the structures engineer call for special methods of analysis; in architecture new structural-design concepts require reliable and accurate methods; and in ship construction accurate methods are necessary for greater strength and efficiency.

The requirement of accuracy in analysis has been brought about by a need for demonstrating structural safety. Consequently, accurate methods of analysis had to be developed since the conventional methods, although perfectly satisfactory when used on simple structures, have been found inadequate when applied to complex structures. Another reason why greater accuracy is required results from the need to establish the fatigue strength level of structures; therefore, it is necessary to employ methods of analysis capable of predicting accurately any stress concentrations so that we may avoid structural fatigue failures.

The requirement of speed, on the other hand, is imposed by the need of having comprehensive information on the structure sufficiently early in the design cycle so that any structural modifications deemed necessary can be incorporated before the final design is decided upon and the structure enters into the production stages. Furthermore, in order to achieve the most efficient design a large number of different structural configurations may have to be analyzed rapidly before a particular configuration is selected for detailed study.

The methods of analysis which meet the requirements mentioned above use matrix algebra, which is ideally suited for automatic computation on high-speed digital computers. Numerous papers on the subject have been published, but it is comparatively recently that the scope and power of matrix methods have been brought out by the formulation of general matrix equations for the analysis of complex structures. In these methods the digital computer is used not only for the solution of simultaneous equations but also for the whole process of structural analysis from the initial input data to the final output, which represents stress and force distributions, deflections, influence coefficients, characteristic frequencies, and mode shapes.

Matrix methods are based on the concept of replacing the actual continuous structure by a mathematical model made up from structural elements of finite size (also referred to as discrete elements) having known elastic and inertial properties that can be expressed in matrix form. The matrices representing these properties are considered as building blocks, which, when fitted together according to a set of rules derived from the theory of elasticity, provide the static and dynamic properties of the actual structural system. In order to put matrix methods in the correct perspective, it is important to emphasize the relationship between matrix methods and classical methods as used in the theory of deformations in continuous media. In the classical theory we are concerned with the deformational behavior on the macroscopic scale without regard to the size or shape of the particles confined within the prescribed boundary of the structure. In the matrix methods particles are of finite size and have a specified shape. Such finite-sized particles are referred to as the *structural elements*, and they are specified somewhat arbitrarily by the analyst in the process of defining the mathematical model of the continuous structure. The properties of each element are calculated, using the theory of continuous elastic media, while the analysis of the entire structure is carried out for the assembly of the individual structural elements. When the size of the elements is decreased, the deformational behavior of the mathematical model converges to that of the continuous structure.

Matrix methods represent the most powerful design tool in structural engineering. Matrix structural-analysis programs for digital computers are now available which can be applied to general types of built-up structures. Not only can these programs be used for routine stress and deflection analysis of complex structures, but they can also be employed very effectively for studies in applied elasticity.

Although this text deals primarily with matrix methods of structural analysis of aircraft and space-vehicle structures, it should be recognized that these methods are also applicable to other types of structures. The general theory for the matrix methods is developed here on the basis of the algebraic symbolism of the various matrix operations; however, the computer programming and computational procedures for the high-speed electronic computers are not discussed. The basic theory of matrix algebra necessary for understanding matrix structural analysis is presented in Appendix A with a view to convenient reference rather than as an exhaustive treatment of the subject. For rigorous proofs of the various theorems in matrix algebra and further details of the theory of matrices, standard textbooks on the subject should be consulted.

1.2 DESIGN ITERATIONS

The primary function of any structure is to support and transfer externally applied loads to the reaction points while at the same time being subjected to some specified constraints and a known temperature distribution. In civil engineering the reaction points are those points on the structure which are attached to a rigid foundation. On a flight-vehicle structure the concept of reaction points is not required, and the points can now be chosen somewhat arbitrarily.

The structures designer is therefore concerned mainly with the *analysis* of known structural configurations which are subjected to known distributions of static or dynamic loads, displacements, and temperatures. From his point of view, however, what is really required is not the analysis but structural *synthesis* leading to the most efficient design (optimum design) for the specified load and temperature environment. Consequently, the ultimate objective in structural design should not be the analysis of a given structural configuration but the automated generation of a structure, i.e., structural synthesis, which will satisfy the specified design criteria.

In general, structural synthesis applied to aerospace structures requires selection of configuration, member sizes, and materials. At present, however, it is not economically feasible to consider all parameters. For this reason, in developing synthesis methods attention has been focused mainly on the variation of member sizes to achieve minimum weight subject to restrictions on the stresses, deflections, and stability. Naturally, in any structural synthesis all design conditions must be considered. Some significant progress has already been made in the development of structural-synthesis methods. This has been prompted by the accomplishments in the fields of computer technology, structural theory, and operations research, all of which could be amalgamated and developed into automated design procedures. Synthesis computer programs are now available for relatively simple structures, but one can foresee that in the near future these programs will be extended to the synthesis of large structural configurations.

In present-day structural designs the structure is designed initially on the basis of experience with similar types of structures, using perhaps some simple analytical calculations, then the structure is analyzed in detail by numerical methods, and subsequently the structure is modified by the designer after examination of the numerical results. The modified structure is then reanalyzed, the analysis examined, and the structure modified again, and so on, until a satisfactory structural design is obtained. Each design cycle may introduce some feedback on the applied loading if dynamic and aeroelastic conditions are also considered. This is due to the dynamic-loading dependence on the mass and elastic distributions and to the aerodynamic-loading dependence on elastic deformations of the structure.

To focus attention on the design criteria, the subsequent discussion will be restricted to structures for aerospace applications; however, the general conclusions are equally applicable to other types of structures. The complexity of structural configurations used on modern supersonic aircraft and aerospace vehicles is illustrated in Figs. 1.1 and 1.2 with perspective cutaway views of the structure of the XB-70 supersonic aircraft and the Titan III launch vehicle. These structures are typical of modern methods of construction for aerospace applications. The magnitude of the task facing the structures designer can be appreciated only if we consider the many design criteria which must all be satisfied when designing these structures. The structural design criteria are related mainly to two characteristics of the structure, structural strength and structural stiffness. The design criteria must specify the required strength to

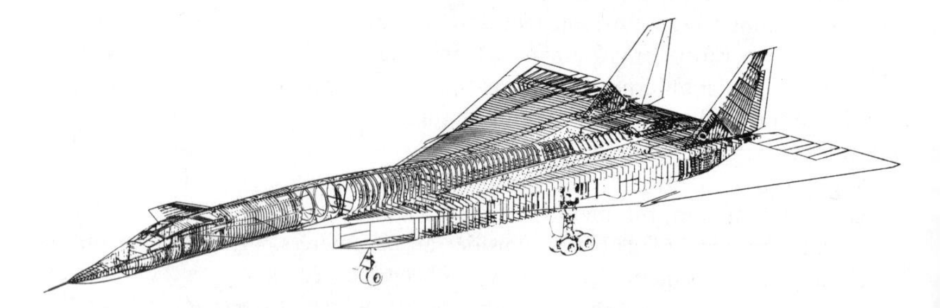

FIG. 1.1 Structural details of the XB-70 supersonic aircraft. (*North American Aviation Company, Inc.*)

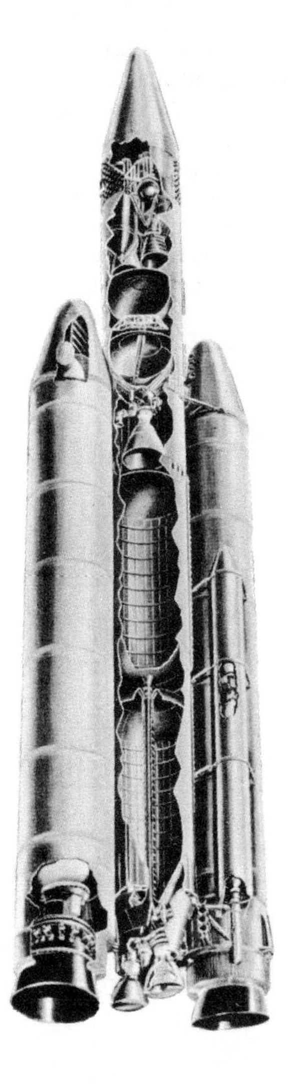

FIG. 1.2 Structural details of the Titan III launch vehicle. (*Martin Company*)

ensure structural integrity under any loading and environment to which the structure may be subjected in service and the required stiffness necessary to prevent such adverse aeroelastic effects as flutter, divergence, and reversal of controls. Whether or not these criteria are satisfied in a specific design is usually verified through detailed stress and aeroelastic analysis. Naturally, experimental verification of design criteria is also used extensively.

The structural design criteria are formed on the basis of aircraft performance and the aerodynamic characteristics in terms of maneuvers and other conditions, e.g., aerodynamic heating, appropriate to the intended use of the structure, and they are specified as the so-called design conditions. Before the actual structural analysis can be started, it is necessary to calculate the loading systems due to the dynamic loads, pressures, and temperatures for each design condition. Once the design conditions and the corresponding loading systems have been formulated, the structural and aeroelastic analyses of the structure can be performed provided its elastic properties and mass distributions are known.

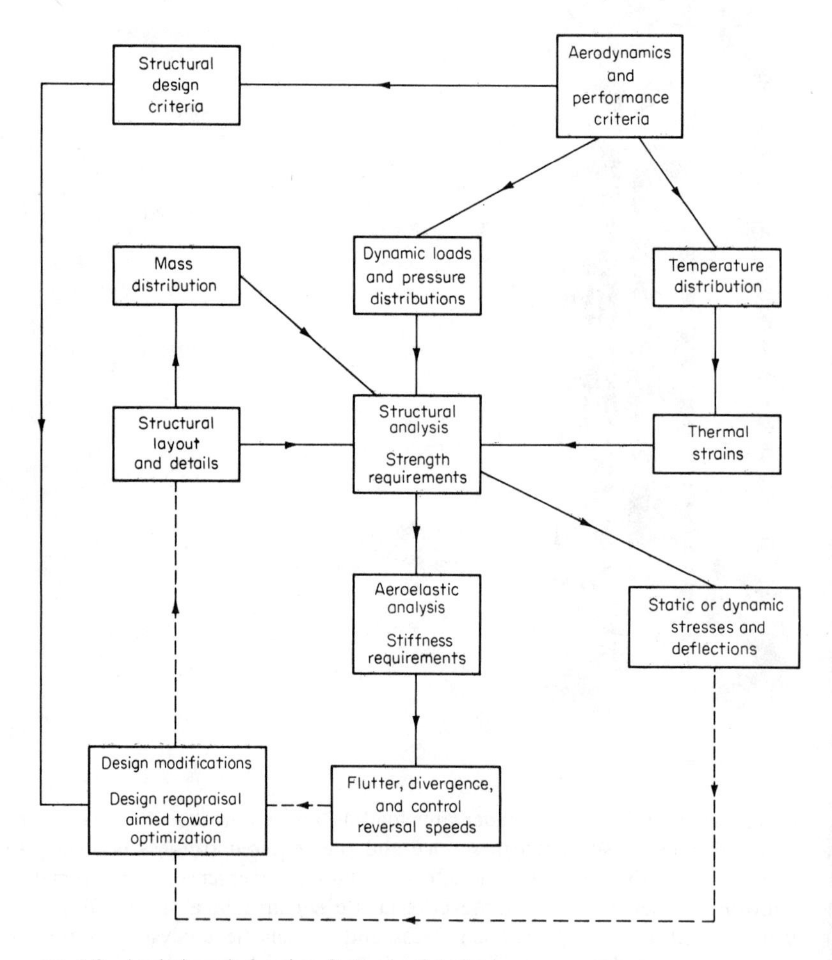

FIG. 1.3 Analysis cycle for aircraft structural design.

The structural analysis gives stress distributions which can be compared with the maximum allowable stresses, and if the stress levels are unsatisfactory (either too high or too low), structural modifications are necessary to achieve the optimum structural design. This usually implies a minimum-weight structure, although in the optimization process economic aspects may also have a decisive influence in selecting materials or methods of construction. It should also be mentioned that factors of safety are used in establishing design conditions. These factors are necessary because of the possibility that (1) the loads in service exceed the design values and (2) the structure is actually less strong than determined by the design calculations. Similarly, the aeroelastic analysis must demonstrate adequate margins of safety in terms of structural stiffness for the specified performance and environment to avoid adverse aeroelastic phenomena and ensure flight safety.

The structural modifications deemed necessary for reasons of strength or stiffness may be so extensive as to require another complete cycle of structural and aeroelastic analysis. In fact, it is not uncommon to have several design iterations before achieving a satisfactory design which meets the required criteria of strength and stiffness. A typical structural-analysis cycle in aircraft design is presented in Fig. 1.3, where some of the main steps in the analysis are indicated. The dotted lines represent the feedback of design information, which is evaluated against the specified requirements (design criteria) so that any necessary modifications in the structural layout or structural details can be introduced in each design cycle.

1.3 METHODS OF ANALYSIS

Methods of structural analysis can be divided into two groups (see Fig. 1.4), analytical methods and numerical methods. The limitations imposed by the analytical methods are well known. Only in special cases are closed-form solutions possible. Approximate solutions can be found for some simple structural configurations, but, in general, for complex structures analytical methods cannot be used, and numerical methods must invariably be employed. The numerical methods of structural analysis can be subdivided into two types, (1) numerical solutions of differential equations for displacements or stresses and (2) matrix methods based on discrete-element idealization.

In the first type the equations of elasticity are solved for a particular structural configuration, either by finite-difference techniques or by direct numerical integration. In this approach the analysis is based on a mathematical approximation of differential equations. Practical limitations, however, restrict the application of these methods to simple structures. Although the various operations in the finite-difference or numerical-integration techniques could be cast into matrix notation and the matrix algebra applied to the solution of the governing equations for the unknowns, these techniques are generally not

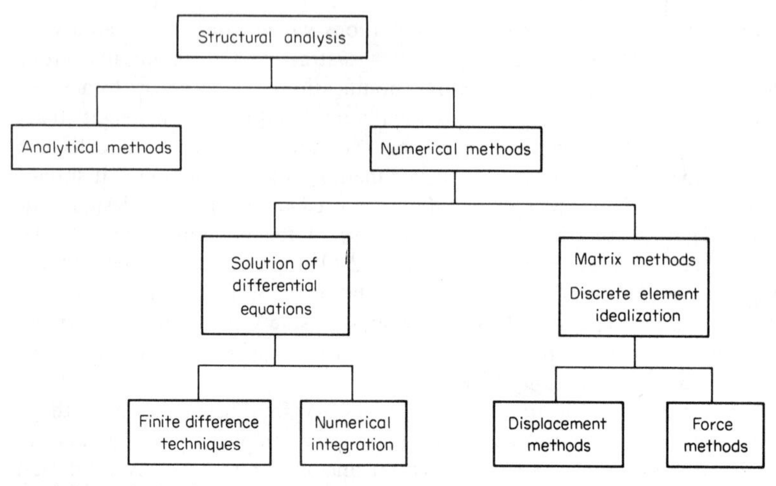

FIG. 1.4 Methods of structural analysis.

described as matrix methods since matrices are not essential in formulating the analysis.

In the second type the complete structural theory is developed *ab initio* in matrix algebra, through all stages in the analysis. The structure is first idealized into an assembly of discrete structural elements with assumed form of displacement or stress distribution, and the complete solution is then obtained by combining these individual approximate displacement or stress distributions in a manner which satisfies the force-equilibrium and displacement compatibility at the junctions of these elements. Methods based on this approach appear to be suitable for the analysis of complex structures. These methods involve appreciable quantities of linear algebra, which must be organized into a systematic sequence of operations, and to this end the use of matrix algebra is a convenient method of defining the various processes involved in the analysis without the necessity of writing out the complete equations in full. Furthermore, the formulation of the analysis in matrix algebra is ideally suited for subsequent solution on the digital computer, and it also allows an easy and systematic compilation of the required data.

Two complementary matrix methods of formulation of any structural problem are possible: (1) the displacement method (stiffness method), where displacements are chosen as unknowns, and (2) the force method (flexibility method), where forces are unknowns. In both these methods the analysis can be thought of as a systematic combination of individual unassembled structural elements into an assembled structure in which the conditions of equilibrium and compatibility are satisfied.

1.4 AREAS OF STRUCTURAL ANALYSIS

Structural analysis deals essentially with the determination of stress and displacement distributions under prescribed loads, temperatures, and constraints, both under static and dynamic conditions. Numerous other areas, however, must also be explored through detailed analysis in order to ensure structural integrity and efficiency. The main areas of investigation in structural design are summarized below:

stress distribution
displacement distribution
structural stability
thermoelasticity (thermal stresses and displacements)
plasticity
creep
creep buckling
vibration frequencies
normal modes of vibration
aeroelasticity, e.g., flutter, divergence
aerothermoelasticity, e.g., loss of stiffness due to aerodynamic heating
dynamic response, e.g., due to gust loading
stress concentrations
fatigue and crack propagation, including sonic fatigue
optimization of structural configurations

CHAPTER 2 BASIC EQUATIONS OF ELASTICITY

For an analytical determination of the distribution of static or dynamic displacements and stresses in a structure under prescribed external loading and temperature, we must obtain a solution to the basic equations of the theory of elasticity, satisfying the imposed boundary conditions on forces and/or displacements. Similarly, in the matrix methods of structural analysis we must also use the basic equations of elasticity. These equations are listed below, with the number of equations for a general three-dimensional structure in parentheses:

strain-displacement equations	(6)
stress-strain equations	(6)
equations of equilibrium (or motion)	(3)

Thus there are fifteen equations available to obtain solutions for fifteen unknown variables, three displacements, six stresses, and six strains. For two-dimensional problems we have eight equations with two displacements, three stresses, and three strains. Additional equations pertain to the continuity of strains and displacements (compatibility equations) and to the boundary conditions on forces and/or displacements.

To provide a ready reference for the development of the general theory of matrix structural analysis, all the basic equations of the theory of elasticity are summarized in this chapter, and when convenient, they are also presented in matrix form.

2.1 STRAIN-DISPLACEMENT EQUATIONS

The deformed shape of an elastic structure under a given system of loads and temperature distribution can be described completely by the three displacements

$$
\begin{aligned}
u_x &= u_x(x,y,z) \\
u_y &= u_y(x,y,z) \\
u_z &= u_z(x,y,z)
\end{aligned}
\tag{2.1}
$$

The vectors representing these three displacements at a point in the structure are mutually orthogonal, and their positive directions correspond to the positive directions of the coordinate axes. In general, all three displacements are represented as functions of x, y, and z. The strains in the deformed structure can be expressed as partial derivatives of the displacements u_x, u_y, and u_z. For small deformations the strain-displacement relations are linear, and the strain components are given by (see, for example, Timoshenko and Goodier)*

$$
e_{xx} = \frac{\partial u_x}{\partial x} \qquad e_{yy} = \frac{\partial u_y}{\partial y} \qquad e_{zz} = \frac{\partial u_z}{\partial z} \tag{2.2a}
$$

$$
\begin{aligned}
e_{xy} = e_{yx} &= \frac{\partial u_y}{\partial x} + \frac{\partial u_x}{\partial y} \\
e_{yz} = e_{zy} &= \frac{\partial u_z}{\partial y} + \frac{\partial u_y}{\partial z} \\
e_{zx} = e_{xz} &= \frac{\partial u_x}{\partial z} + \frac{\partial u_z}{\partial x}
\end{aligned}
\tag{2.2b}
$$

where e_{xx}, e_{yy}, and e_{zz} represent normal strains, while e_{xy}, e_{yz}, and e_{zx} represent shearing strains. Some textbooks on elasticity define the shearing strains with a factor $\frac{1}{2}$ at the right of Eqs. (2.2*b*). Although such a definition allows the use of a single expression for both the normal and shearing strains in tensor notation, this has no particular advantage in matrix structural analysis. From Eqs. (2.2*b*) it follows that the symmetry relationship

$$
e_{ij} = e_{ji} \qquad i,j = x, y, z \tag{2.3}
$$

is valid for all shearing strains, and therefore a total of only six strain components is required to describe strain states in three-dimensional elasticity problems.

To derive the strain-displacement equations (2.2) we shall consider a small rectangular element *ABCD* in the *xy* plane within an elastic body, as shown in Fig. 2.1. If the body undergoes a deformation, the undeformed element *ABCD* moves to *A′B′C′D′*. We observe here that the element has two basic geometric

* General references are listed in Appendix B.

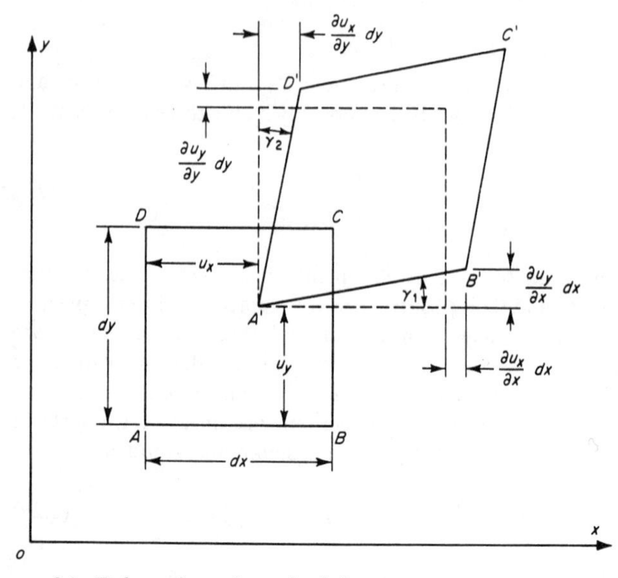

FIG. 2.1 Deformations of a strained element.

deformations, change in length and angular distortion. The change in length of AB is $(\partial u_x/\partial x)\,dx$, and if we define the normal strain as the ratio of the change of length over the original length, it follows that the normal strain in the x direction is $\partial u_x/\partial x$. Similarly it can be shown that the normal strains in the y and z directions are given by the derivatives $\partial u_y/\partial y$ and $\partial u_z/\partial z$. The angular distortion on the element can be determined in terms of the angles γ_1 and γ_2 shown in Fig. 2.1. It is clear that for small deformations $\gamma_1 = \partial u_y/\partial x$ and $\gamma_2 = \partial u_x/\partial y$. If the shearing strain e_{xy} in the xy plane is defined as the total angular deformation, i.e., sum of the angles γ_1 and γ_2, it follows that this shearing-strain component is given by $\partial u_y/\partial x + \partial u_x/\partial y$. The other two shearing-strain components can be obtained by considering angular deformations in the yz and zx planes.

2.2 STRESS-STRAIN EQUATIONS

THREE-DIMENSIONAL STRESS DISTRIBUTIONS

Since the determination of thermal stresses plays an important part in the design of structures operating at elevated temperatures, the stress-strain equations must include the effects of temperature. To explain how temperature modifies the three-dimensional isothermal stress-strain equations, we shall consider a small element in the elastic body subjected to a temperature change T.

If the length of this element is dl, then under the action of the temperature change T the element will expand to a new length $(1 + \alpha T)\,dl$, where α is the coefficient of thermal expansion. For isotropic and homogeneous materials this coefficient is independent of the direction and position of the element but may depend on the temperature.

Attention will subsequently be confined to isotropic bodies, for which thermal expansions are the same in all directions. This means that an infinitely small unrestrained parallelepiped in an isotropic body subjected to a change in temperature will experience only a uniform expansion without any angular distortions, and the parallelepiped will retain its rectangular shape. Thus, the thermal strains (thermal dilatations) in an unrestrained element may be expressed as

$$e_{T_{ij}} = \delta_{ij}\alpha T \qquad i, j = x, y, z \tag{2.4}$$

where δ_{ij} is the Kronecker delta, given by

$$\delta_{ij} = \begin{cases} 1 & \text{when } i = j \\ 0 & \text{when } i \neq j \end{cases} \tag{2.5}$$

Equation (2.4) expresses the fact that in isotropic bodies the temperature change T produces only normal thermal strains while the shearing thermal strains are all equal to zero; that is,

$$e_{T_{ij}} = 0 \qquad \text{for } i \neq j \tag{2.6}$$

Imagine now an elastic isotropic body made up of a number of small rectangular parallelepiped elements of equal size which fit together to form a continuous body. If the temperature of the body is increased uniformly and no external restraints are applied on the body boundaries, each element will expand freely by an equal amount in all directions. Since all elements are of equal size, they will still fit to form a continuous body, although slightly expanded, and no thermal stresses will be induced. If, however, the temperature increase is not uniform, each element will expand by a different amount, proportional to its own temperature, and the resulting expanded elements will no longer fit together to form a continuous body; consequently, elastic strains must be induced so that each element will restrain the distortions of its neighboring elements and the continuity of displacements on the distorted body will be preserved.

The total strains at each point of a heated body can therefore be thought of as consisting of two parts; the first part is the thermal strain $e_{T_{ij}}$ due to the uniform thermal expansion, and the second part is the elastic strain ϵ_{ij} which is required to maintain the displacement continuity of the body subjected to a nonuniform temperature distribution. If at the same time the body is subjected to a system of external loads, ϵ_{ij} will also include strains arising from such loads. Now since the strains e_{ij} in Sec. 2.1 were derived from the total displacements due to a system of loads and temperature distribution, they represent the total strains, and they can be expressed as the sum of the elastic strains ϵ_{ij} and the thermal strains $e_{T_{ij}}$. Hence

$$\begin{aligned} e_{ij} &= \epsilon_{ij} + e_{T_{ij}} \\ &= \epsilon_{ij} + \alpha T \delta_{ij} \end{aligned} \tag{2.7}$$

The elastic strains ϵ_{ij} are related to the stresses by means of the usual Hooke's law for linear isothermal elasticity

$$\begin{aligned} \epsilon_{xx} &= \frac{1}{E}[\sigma_{xx} - \nu(\sigma_{yy} + \sigma_{zz})] \\ \epsilon_{yy} &= \frac{1}{E}[\sigma_{yy} - \nu(\sigma_{zz} + \sigma_{xx})] \\ \epsilon_{zz} &= \frac{1}{E}[\sigma_{zz} - \nu(\sigma_{xx} + \sigma_{yy})] \\ \epsilon_{xy} &= 2\frac{1+\nu}{E}\sigma_{xy} \\ \epsilon_{yz} &= 2\frac{1+\nu}{E}\sigma_{yz} \\ \epsilon_{zx} &= 2\frac{1+\nu}{E}\sigma_{zx} \end{aligned} \tag{2.8}$$

where E denotes Young's modulus and ν is Poisson's ratio.

Substituting Eqs. (2.8) into (2.7) gives

$$\begin{aligned} e_{xx} &= \frac{1}{E}[\sigma_{xx} - \nu(\sigma_{yy} + \sigma_{zz})] + \alpha T \\ e_{yy} &= \frac{1}{E}[\sigma_{yy} - \nu(\sigma_{zz} + \sigma_{xx})] + \alpha T \\ e_{zz} &= \frac{1}{E}[\sigma_{zz} - \nu(\sigma_{xx} + \sigma_{yy})] + \alpha T \\ e_{xy} &= 2\frac{1+\nu}{E}\sigma_{xy} \\ e_{yz} &= 2\frac{1+\nu}{E}\sigma_{yz} \\ e_{zx} &= 2\frac{1+\nu}{E}\sigma_{zx} \end{aligned} \tag{2.9}$$

Equations (2.9) represent the three-dimensional Hooke's law generalized for thermal effects. These equations can be solved for the stresses σ_{ij}, and the following stress-strain relationships are then obtained:

$$
\begin{aligned}
\sigma_{xx} &= \frac{E}{(1+\nu)(1-2\nu)}[(1-\nu)e_{xx} + \nu(e_{yy} + e_{zz})] - \frac{E\alpha T}{1-2\nu} \\
\sigma_{yy} &= \frac{E}{(1+\nu)(1-2\nu)}[(1-\nu)e_{yy} + \nu(e_{zz} + e_{xx})] - \frac{E\alpha T}{1-2\nu} \\
\sigma_{zz} &= \frac{E}{(1+\nu)(1-2\nu)}[(1-\nu)e_{zz} + \nu(e_{xx} + e_{yy})] - \frac{E\alpha T}{1-2\nu} \\
\sigma_{xy} &= \frac{E}{2(1+\nu)} e_{xy} \\
\sigma_{yz} &= \frac{E}{2(1+\nu)} e_{yz} \\
\sigma_{zx} &= \frac{E}{2(1+\nu)} e_{zx}
\end{aligned}
\tag{2.10}
$$

Equations (2.9) and (2.10) can be written in matrix form as

$$
\begin{bmatrix} e_{xx} \\ e_{yy} \\ e_{zz} \\ e_{xy} \\ e_{yz} \\ e_{zx} \end{bmatrix}
= \frac{1}{E}
\begin{bmatrix}
1 & -\nu & -\nu & 0 & 0 & 0 \\
-\nu & 1 & -\nu & 0 & 0 & 0 \\
-\nu & -\nu & 1 & 0 & 0 & 0 \\
0 & 0 & 0 & 2(1+\nu) & 0 & 0 \\
0 & 0 & 0 & 0 & 2(1+\nu) & 0 \\
0 & 0 & 0 & 0 & 0 & 2(1+\nu)
\end{bmatrix}
\begin{bmatrix} \sigma_{xx} \\ \sigma_{yy} \\ \sigma_{zz} \\ \sigma_{xy} \\ \sigma_{yz} \\ \sigma_{zx} \end{bmatrix}
+ \alpha T \begin{bmatrix} 1 \\ 1 \\ 1 \\ 0 \\ 0 \\ 0 \end{bmatrix}
\tag{2.11}
$$

and

$$
\begin{bmatrix} \sigma_{xx} \\ \sigma_{yy} \\ \sigma_{zz} \\ \sigma_{xy} \\ \sigma_{yz} \\ \sigma_{zx} \end{bmatrix}
= \frac{E}{(1+\nu)(1-2\nu)}
\begin{bmatrix}
1-\nu & \nu & \nu & 0 & 0 & 0 \\
\nu & 1-\nu & \nu & 0 & 0 & 0 \\
\nu & \nu & 1-\nu & 0 & 0 & 0 \\
0 & 0 & 0 & \frac{1-2\nu}{2} & 0 & 0 \\
0 & 0 & 0 & 0 & \frac{1-2\nu}{2} & 0 \\
0 & 0 & 0 & 0 & 0 & \frac{1-2\nu}{2}
\end{bmatrix}
\begin{bmatrix} e_{xx} \\ e_{yy} \\ e_{zz} \\ e_{xy} \\ e_{yz} \\ e_{zx} \end{bmatrix}
- \frac{E\alpha T}{1-2\nu} \begin{bmatrix} 1 \\ 1 \\ 1 \\ 0 \\ 0 \\ 0 \end{bmatrix}
\tag{2.12}
$$

It should be noted that in all previous equations the shearing stress–strain relationships are expressed in terms of Young's modulus E and Poisson's ratio

ν. If necessary, the shear modulus G can be introduced into these equations using

$$G = \frac{E}{2(1+\nu)} \tag{2.13}$$

Equation (2.12) can be expressed symbolically as

$$\boldsymbol{\sigma} = \boldsymbol{\varkappa}\mathbf{e} + \alpha T \boldsymbol{\varkappa}_T \tag{2.14}$$

where

$$\boldsymbol{\sigma} = \{\sigma_{xx} \quad \sigma_{yy} \quad \sigma_{zz} \quad \sigma_{xy} \quad \sigma_{yz} \quad \sigma_{zx}\} \tag{2.15}$$

$$\mathbf{e} = \{e_{xx} \quad e_{yy} \quad e_{zz} \quad e_{xy} \quad e_{yz} \quad e_{zx}\} \tag{2.16}$$

$$\boldsymbol{\varkappa}_T = \frac{E}{1-2\nu}\{-1 \quad -1 \quad -1 \quad 0 \quad 0 \quad 0\} \tag{2.17}$$

$$\boldsymbol{\varkappa} = \frac{E}{(1+\nu)(1-2\nu)}\left[\begin{array}{ccc|ccc} 1-\nu & \nu & \nu & & & \\ \nu & 1-\nu & \nu & & \mathbf{0} & \\ \nu & \nu & 1-\nu & & & \\ \hline & & & \frac{1-2\nu}{2} & 0 & 0 \\ & \mathbf{0} & & 0 & \frac{1-2\nu}{2} & 0 \\ & & & 0 & 0 & \frac{1-2\nu}{2} \end{array}\right] \tag{2.18}$$

The braces used in Eqs. (2.15) to (2.17) represent column matrices written horizontally to save space. The term $\alpha T \boldsymbol{\varkappa}_T$ in Eq. (2.14) can be interpreted physically as the matrix of stresses necessary to suppress thermal expansion so that $\mathbf{e} = \mathbf{0}$.

When we premultiply Eq. (2.14) by $\boldsymbol{\varkappa}^{-1}$ and solve for $\mathbf{e}$, it follows that

$$\begin{aligned} \mathbf{e} &= \boldsymbol{\varkappa}^{-1}\boldsymbol{\sigma} - \alpha T \boldsymbol{\varkappa}^{-1}\boldsymbol{\varkappa}_T \\ &= \boldsymbol{\phi}\boldsymbol{\sigma} + \mathbf{e}_T \end{aligned} \tag{2.19}$$

where

$$\boldsymbol{\phi} = \boldsymbol{\varkappa}^{-1} = \frac{1}{E}\begin{bmatrix} 1 & -\nu & -\nu & 0 & 0 & 0 \\ -\nu & 1 & -\nu & 0 & 0 & 0 \\ -\nu & -\nu & 1 & 0 & 0 & 0 \\ 0 & 0 & 0 & 2(1+\nu) & 0 & 0 \\ 0 & 0 & 0 & 0 & 2(1+\nu) & 0 \\ 0 & 0 & 0 & 0 & 0 & 2(1+\nu) \end{bmatrix} \tag{2.20}$$

and $$\mathbf{e}_T = -\alpha T \boldsymbol{\varkappa}^{-1} \boldsymbol{\varkappa}_T$$
$$= \alpha T\{1 \quad 1 \quad 1 \quad 0 \quad 0 \quad 0\} \tag{2.21}$$

Equation (2.19) is, of course, the matrix representation of the strain-stress relationships given previously by Eqs. (2.9).

TWO-DIMENSIONAL STRESS DISTRIBUTIONS

There are two types of two-dimensional stress distributions, plane-stress and plane-strain distributions. The first type is used for thin flat plates loaded in the plane of the plate, while the second is used for elongated bodies of constant cross section subjected to uniform loading.

PLANE STRESS The plane-stress distribution is based on the assumption that

$$\sigma_{zz} = \sigma_{zx} = \sigma_{zy} = 0 \tag{2.22}$$

where the z direction represents the direction perpendicular to the plane of the plate, and that no stress components vary through the plate thickness. Although these assumptions violate some of the compatibility conditions, they are sufficiently accurate for practical applications if the plate is thin.

Using Eqs. (2.22), we can reduce the three-dimensional Hooke's law represented by Eq. (2.12) to

$$\begin{bmatrix} \sigma_{xx} \\ \sigma_{yy} \\ \sigma_{xy} \end{bmatrix} = \frac{E}{1-\nu^2} \begin{bmatrix} 1 & \nu & 0 \\ \nu & 1 & 0 \\ 0 & 0 & \dfrac{1-\nu}{2} \end{bmatrix} \begin{bmatrix} e_{xx} \\ e_{yy} \\ e_{xy} \end{bmatrix} - \frac{E\alpha T}{1-\nu} \begin{bmatrix} 1 \\ 1 \\ 0 \end{bmatrix} \tag{2.23}$$

which in matrix notation can be presented as

$$\boldsymbol{\sigma} = \boldsymbol{\varkappa}\mathbf{e} + \alpha T \boldsymbol{\varkappa}_T \tag{2.24}$$

where

$$\boldsymbol{\sigma} = \{\sigma_{xx} \quad \sigma_{yy} \quad \sigma_{xy}\} \tag{2.25}$$

$$\mathbf{e} = \{e_{xx} \quad e_{yy} \quad e_{xy}\} \tag{2.26}$$

$$\boldsymbol{\varkappa}_T = \frac{E}{1-\nu}\{-1 \quad -1 \quad 0\} \tag{2.27}$$

$$\boldsymbol{\varkappa} = \frac{E}{1-\nu^2} \begin{bmatrix} 1 & \nu & 0 \\ \nu & 1 & 0 \\ 0 & 0 & \dfrac{1-\nu}{2} \end{bmatrix} \tag{2.28}$$

The matrix form of Eq. (2.24) is identical to that for the three-dimensional stress distributions. Although the same symbols have been introduced here for both the three- and two-dimensional cases, no confusion will arise, since in the actual analysis it will always be clear which type of stress-strain relationship should be used. The strains can also be expressed in terms of the stresses, and therefore solving Eq. (2.23) for the strains e_{xx}, e_{yy}, and e_{xy} gives

$$\begin{bmatrix} e_{xx} \\ e_{yy} \\ e_{xy} \end{bmatrix} = \frac{1}{E}\begin{bmatrix} 1 & -\nu & 0 \\ -\nu & 1 & 0 \\ 0 & 0 & 2(1+\nu) \end{bmatrix}\begin{bmatrix} \sigma_{xx} \\ \sigma_{yy} \\ \sigma_{xy} \end{bmatrix} + \alpha T\begin{bmatrix} 1 \\ 1 \\ 0 \end{bmatrix} \tag{2.29}$$

Furthermore, it follows from Eqs. (2.22) and (2.11) that

$$\begin{aligned} e_{zz} &= \frac{-\nu}{E}(\sigma_{xx} + \sigma_{yy}) + \alpha T \\ &= \frac{-\nu}{1-\nu}(e_{xx} + e_{yy}) + \frac{1+\nu}{1-\nu}\alpha T \end{aligned} \tag{2.30}$$

and $$e_{yz} = e_{zx} = 0 \tag{2.31}$$

Equation (2.30) indicates that the normal strain e_{zz} is linearly dependent on the strains e_{xx} and e_{yy}, and for this reason it has not been included in the matrix equation (2.29).

The strain-stress equation for plane-stress problems can therefore be represented symbolically as

$$\mathbf{e} = \boldsymbol{\phi}\boldsymbol{\sigma} + \mathbf{e}_T \tag{2.32}$$

where $$\boldsymbol{\phi} = \frac{1}{E}\begin{bmatrix} 1 & -\nu & 0 \\ -\nu & 1 & 0 \\ 0 & 0 & 2(1+\nu) \end{bmatrix} \tag{2.33}$$

and $$\mathbf{e}_T = \alpha T\{1 \quad 1 \quad 0\} \tag{2.34}$$

PLANE STRAIN The plane-strain distribution is based on the assumption that

$$\frac{\partial}{\partial z} = u_z = 0 \tag{2.35}$$

where z represents now the lengthwise direction of an elastic elongated body of constant cross section subjected to uniform loading. With the above assumption, it follows then immediately from Eqs. (2.2) that

$$e_{zz} = e_{zx} = e_{zy} = 0 \tag{2.36}$$

When Eq. (2.36) is used, the three-dimensional Hooke's law represented by Eq. (2.12) reduces to

$$\begin{bmatrix} \sigma_{xx} \\ \sigma_{yy} \\ \sigma_{xy} \end{bmatrix} = \frac{E}{(1+\nu)(1-2\nu)} \begin{bmatrix} 1-\nu & \nu & 0 \\ \nu & 1-\nu & 0 \\ 0 & 0 & \dfrac{1-2\nu}{2} \end{bmatrix} \begin{bmatrix} e_{xx} \\ e_{yy} \\ e_{xy} \end{bmatrix} - \frac{E\alpha T}{1-2\nu} \begin{bmatrix} 1 \\ 1 \\ 0 \end{bmatrix} \tag{2.37}$$

$$\sigma_{zz} = \nu(\sigma_{xx} + \sigma_{yy}) - E\alpha T \tag{2.38}$$

$$\sigma_{yz} = \sigma_{zx} = 0 \tag{2.39}$$

Equation (2.38) implies that the normal stress σ_{zz} in the case of plane-strain distribution is linearly dependent on the normal stresses σ_{xx} and σ_{yy}. For this reason the stress component σ_{zz} is not included in the matrix stress–strain equation (2.37). Equation (2.37) can be expressed symbolically as

$$\boldsymbol{\sigma} = \boldsymbol{\varkappa}\mathbf{e} + \alpha T \boldsymbol{\varkappa}_T \tag{2.40}$$

where

$$\boldsymbol{\sigma} = \{\sigma_{xx} \quad \sigma_{yy} \quad \sigma_{xy}\} \tag{2.41}$$

$$\mathbf{e} = \{e_{xx} \quad e_{yy} \quad e_{xy}\} \tag{2.42}$$

$$\boldsymbol{\varkappa}_T = \frac{E}{1-2\nu}\{-1 \quad -1 \quad 0\} \tag{2.43}$$

$$\boldsymbol{\varkappa} = \frac{E}{(1+\nu)(1-2\nu)} \begin{bmatrix} 1-\nu & \nu & 0 \\ \nu & 1-\nu & 0 \\ 0 & 0 & \dfrac{1-2\nu}{2} \end{bmatrix} \tag{2.44}$$

Solving Eqs. (2.37) for the strains e_{xx}, e_{yy}, and e_{zz} gives the matrix equation

$$\begin{bmatrix} e_{xx} \\ e_{yy} \\ e_{xy} \end{bmatrix} = \frac{1+\nu}{E} \begin{bmatrix} 1-\nu & -\nu & 0 \\ -\nu & 1-\nu & 0 \\ 0 & 0 & 2 \end{bmatrix} \begin{bmatrix} \sigma_{xx} \\ \sigma_{yy} \\ \sigma_{xy} \end{bmatrix} + (1+\nu)\alpha T \begin{bmatrix} 1 \\ 1 \\ 0 \end{bmatrix} \tag{2.45}$$

which may be expressed symbolically as

$$\mathbf{e} = \boldsymbol{\phi}\boldsymbol{\sigma} + \mathbf{e}_T \tag{2.46}$$

where for this case

$$\boldsymbol{\phi} = \frac{1+\nu}{E}\begin{bmatrix} 1-\nu & -\nu & 0 \\ -\nu & 1-\nu & 0 \\ 0 & 0 & 2 \end{bmatrix} \tag{2.47}$$

and $$\mathbf{e}_T = (1+\nu)\alpha T\{1 \quad 1 \quad 0\} \tag{2.48}$$

ONE-DIMENSIONAL STRESS DISTRIBUTIONS

If all stress components are zero except for the normal stress σ_{xx}, the Hooke's law generalized for thermal effects takes a particularly simple form

$$\sigma_{xx} = Ee_{xx} - E\alpha T \tag{2.49}$$

and $$e_{xx} = \frac{1}{E}\sigma_{xx} + \alpha T \tag{2.50}$$

Equation (2.49) can be written symbolically as

$$\boldsymbol{\sigma} = \boldsymbol{\varkappa}\mathbf{e} + \alpha T\boldsymbol{\varkappa}_T \tag{2.51}$$

where

$$\boldsymbol{\sigma} = \sigma_{xx} \tag{2.52}$$

$$\mathbf{e} = e_{xx} \tag{2.53}$$

$$\boldsymbol{\varkappa} = E \tag{2.54}$$

$$\boldsymbol{\varkappa}_T = -E \tag{2.55}$$

2.3 STRESS-STRAIN EQUATIONS FOR INITIAL STRAINS

If any initial strains $e_{I_{ij}}$ are present, e.g., those due to lack of fit when the elastic structure was assembled from its component parts, the total strains, including the thermal strains $e_{T_{ij}}$, must be expressed as

$$e_{ij} = \epsilon_{ij} + e_{T_{ij}} + e_{I_{ij}} \tag{2.56}$$

where ϵ_{ij} represents the elastic strains required to maintain continuity of displacements due to external loading and thermal and initial strains. The elastic strains are related to the stresses through Hooke's law, and hence it follows immediately that the total strains for three-dimensional distributions are given by

$$\mathbf{e} = \boldsymbol{\phi}\boldsymbol{\sigma} + \mathbf{e}_T + \mathbf{e}_I \tag{2.57}$$

where $$\mathbf{e}_I = \{e_{I_{xx}} \quad e_{I_{yy}} \quad e_{I_{zz}} \quad e_{I_{xy}} \quad e_{I_{yz}} \quad e_{I_{zx}}\} \tag{2.58}$$

represents the column matrix of initial strains. When Eq. (2.57) is solved for the stresses $\boldsymbol{\sigma}$, it follows that

$$\begin{aligned}\boldsymbol{\sigma} &= \boldsymbol{\varkappa}\mathbf{e} - \boldsymbol{\varkappa}\mathbf{e}_T - \boldsymbol{\varkappa}\mathbf{e}_I \\ &= \boldsymbol{\varkappa}\mathbf{e} + \alpha T\boldsymbol{\varkappa}_T - \boldsymbol{\varkappa}\mathbf{e}_I \end{aligned} \tag{2.59}$$

The concept of initial strains has also been applied in the analysis of structures with cutouts. In this technique fictitious elements with some unknown initial strains are introduced into the cutout regions so that the uniformity of the pattern of equations is not disturbed by the cutouts. The analysis of the structure with the cutouts filled in is then carried out, and the unknown initial strains are determined from the condition of zero stress in the fictitious elements. Thus the fictitious elements, which were substituted in the place of missing elements, can be removed from the structure, and the resulting stress distribution corresponds to that in a structure with the cutouts present.

2.4 EQUATIONS OF EQUILIBRIUM

Equations of internal equilibrium relating the nine stress components (three normal stresses and six shearing stresses) are derived by considering equilibrium of moments and forces acting on a small rectangular parallelepiped (see Fig. 2.2). Taking first moments about the x, y, and z axes, respectively, we can

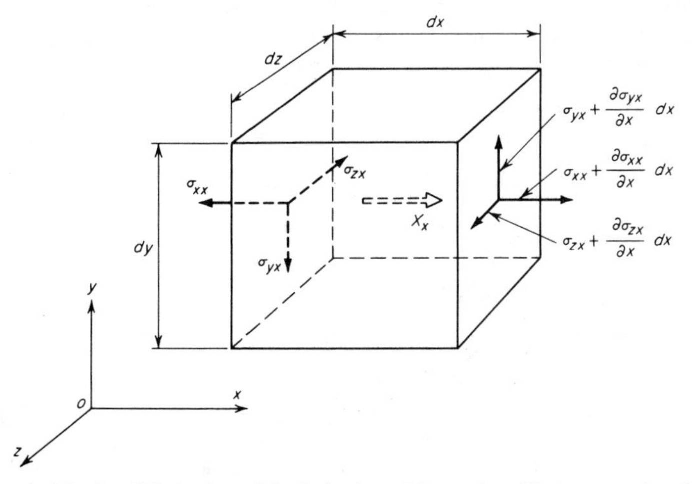

FIG. 2.2 Parallelepiped used in derivation of internal-equilibrium equations; only the stress components on a typical pair of faces are shown.

show that in the absence of body moments

$$\sigma_{ij} = \sigma_{ji} \tag{2.60}$$

Resolving forces in the x, y, and z directions, we obtain three partial differential equations

$$\begin{aligned}
\frac{\partial \sigma_{xx}}{\partial x} + \frac{\partial \sigma_{xy}}{\partial y} + \frac{\partial \sigma_{xz}}{\partial z} + X_x &= 0 \\
\frac{\partial \sigma_{yx}}{\partial x} + \frac{\partial \sigma_{yy}}{\partial y} + \frac{\partial \sigma_{yz}}{\partial z} + X_y &= 0 \\
\frac{\partial \sigma_{zx}}{\partial x} + \frac{\partial \sigma_{zy}}{\partial y} + \frac{\partial \sigma_{zz}}{\partial z} + X_z &= 0
\end{aligned} \tag{2.61}$$

where X_x, X_y, and X_z represent the body forces in the x, y, and z directions, respectively.

Equation (2.61) must be satisfied at all points of the body. The stresses σ_{ij} vary throughout the body, and at its surface they must be in equilibrium with the external forces applied on the surface. When the component of the surface force in the ith direction is denoted by Φ_i, it can be shown that the consideration of equilibrium at the surface leads to the following equations:

$$\begin{aligned}
l\sigma_{xx} + m\sigma_{xy} + n\sigma_{xz} &= \Phi_x \\
l\sigma_{yx} + m\sigma_{yy} + n\sigma_{yz} &= \Phi_y \\
l\sigma_{zx} + m\sigma_{zy} + n\sigma_{zz} &= \Phi_z
\end{aligned} \tag{2.62}$$

where l, m, and n represent the direction cosines for an outward-drawn normal at the surface. These equations are obtained by resolving forces acting on a small element at the surface of the body, as shown in Fig. 2.3. In the particular case of the plane-stress distribution Eqs. (2.62) reduce to

$$l\sigma_{xx} + m\sigma_{xy} = \Phi_x \qquad l\sigma_{yx} + m\sigma_{yy} = \Phi_y \tag{2.63}$$

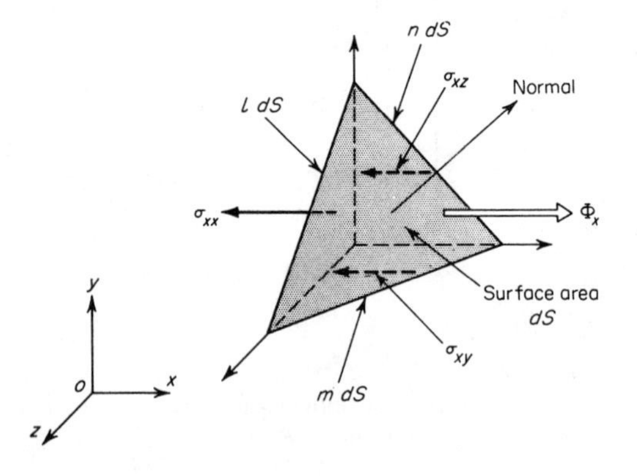

FIG. 2.3 Equilibrium at the surface; only the stress and surface components in the x direction are shown.

The surface forces Φ_i and the body forces X_i must also satisfy the equations of overall equilibrium; i.e., all external forces, including reactive forces, must constitute a self-equilibrating load system. If the external load system consists of a set of concentrated loads P_i and concentrated moments M_i, in addition to Φ_i and X_i, the following six equations must be satisfied

$$\begin{aligned}
&\int_S \Phi_x \, dS + \int_V X_x \, dV + \Sigma P_x = 0 \\
&\int_S \Phi_y \, dS + \int_V X_y \, dV + \Sigma P_y = 0 \\
&\int_S \Phi_z \, dS + \int_V X_z \, dV + \Sigma P_z = 0
\end{aligned} \tag{2.64}$$

$$\begin{aligned}
&\int_S (\Phi_z y - \Phi_y z) \, dS + \int_V (X_z y - X_y z) \, dV + \Sigma M_x = 0 \\
&\int_S (\Phi_x z - \Phi_z x) \, dS + \int_V (X_x z - X_z x) \, dV + \Sigma M_y = 0 \\
&\int_S (\Phi_y x - \Phi_x y) \, dS + \int_V (X_y x - X_x y) \, dV + \Sigma M_z = 0
\end{aligned} \tag{2.65}$$

Equations (2.64) represent the condition that the sum of all applied loads in the x, y, and z directions, respectively, must be equal to zero, while Eqs. (2.65) represent the condition of zero moment about the x, y, and z axes, respectively.

2.5 COMPATIBILITY EQUATIONS

The strains and displacements in an elastic body must vary continuously, and this imposes the condition of continuity on the derivatives of displacements and strains. Consequently, the displacements u_i in Eqs. (2.2) can be eliminated, and the following six equations of compatibility are obtained:

$$\begin{aligned}
&\frac{\partial^2 e_{xx}}{\partial y^2} + \frac{\partial^2 e_{yy}}{\partial x^2} = \frac{\partial^2 e_{xy}}{\partial x \, \partial y} \\
&\frac{\partial^2 e_{yy}}{\partial z^2} + \frac{\partial^2 e_{zz}}{\partial y^2} = \frac{\partial^2 e_{yz}}{\partial y \, \partial z} \\
&\frac{\partial^2 e_{zz}}{\partial x^2} + \frac{\partial^2 e_{xx}}{\partial z^2} = \frac{\partial^2 e_{zx}}{\partial z \, \partial x} \\
&\frac{\partial^2 e_{xx}}{\partial y \, \partial z} = \frac{1}{2} \frac{\partial}{\partial x}\left(- \frac{\partial e_{yz}}{\partial x} + \frac{\partial e_{zx}}{\partial y} + \frac{\partial e_{xy}}{\partial z} \right) \\
&\frac{\partial^2 e_{yy}}{\partial z \, \partial x} = \frac{1}{2} \frac{\partial}{\partial y}\left(\frac{\partial e_{yz}}{\partial x} - \frac{\partial e_{zx}}{\partial y} + \frac{\partial e_{xy}}{\partial z} \right) \\
&\frac{\partial^2 e_{zz}}{\partial x \, \partial y} = \frac{1}{2} \frac{\partial}{\partial z}\left(\frac{\partial e_{yz}}{\partial x} + \frac{\partial e_{zx}}{\partial y} - \frac{\partial e_{xy}}{\partial z} \right)
\end{aligned} \tag{2.66}$$

For two-dimensional plane-stress problems, the six equations of compatibility reduce to only one equation

$$\frac{\partial^2 e_{xx}}{\partial y^2} + \frac{\partial^2 e_{yy}}{\partial x^2} = \frac{\partial^2 e_{xy}}{\partial x \, \partial y} \tag{2.67}$$

For multiply connected elastic bodies, such as plates with holes, additional equations are required to ensure single-valuedness of the solution. These additional equations are provided by the Cesàro integrals (see, for example, Boley and Weiner). In matrix methods of structural analysis, however, we do not use the compatibility equations or the Cesàro integrals. The fundamental equations of matrix structural theory require the use of displacements, and only when the displacements are not used in elasticity problems are the compatibility relations and the Cesàro integrals needed.

CHAPTER 3
ENERGY THEOREMS

3.1 INTRODUCTION

Exact solution of the differential equations of elasticity for complex structures presents a formidable analytical problem which can be solved in closed form only in special cases. Although the basic equations of elasticity have been known since the beginning of the last century, it was not until the introduction of the concept of strain energy toward the latter half of the last century that big strides in the development of structural-analysis methods were made possible.

Early methods of structural analysis dealt mainly with the stresses and deformations in trusses, and interest was then centered mainly on statically determinate structures, for which the repeated application of equilibrium equations at the joints was sufficient to determine completely the internal-force distribution and hence the displacements. For structures which were statically indeterminate (redundant), equations of internal-load equilibrium were insufficient to determine the distribution of internal forces, and it was therefore realized that additional equations were needed. Navier pointed out in 1827 that the whole problem could be simply solved by considering the displacements at the joints, instead of the forces. In these terms there are always as many equations available as there are unknown displacements; however, even on simple structures, this method leads to a very large number of simultaneous equations for the unknown displacements, and it is therefore not at all surprising that displacement methods of analysis found only a very limited application before the introduction of electronic digital computers.

Further progress in structural analysis was not possible until Castigliano enunciated his strain-energy theorems in 1873. He stated that if U_i is the internal energy (strain energy) stored in the structure due to a given system of loads and if $X_1, X_2, \ldots$ are the forces in the redundant members, these forces can be determined from the linear simultaneous equations

$$\frac{\partial U_i}{\partial X_1} = \frac{\partial U_i}{\partial X_2} = \cdots = 0 \tag{3.1}$$

These equations are, in fact, the deflection-compatibility equations supplementing the force-equilibrium equations that were inadequate in number to determine all the internal forces. Other energy theorems developed by Castigliano dealt with the determination of external loads and displacements using the concept of strain energy.

The first significant step forward in the development of strain-energy methods of structural analysis since the publication of Castigliano's work is due to Engesser. Castigliano assumed that the displacements are linear functions of external loads, but there are naturally many instances where this assumption is not valid. Castigliano's theory, as presented originally, indicated that the displacements could be calculated from the partial derivatives of the strain energy with respect to the corresponding external forces, that is,

$$\frac{\partial U_i}{\partial P_r} = U_r \tag{3.2}$$

where P_r is an external force and U_r is the displacement in the direction of P_r. In 1889 Engesser introduced the concept of the complementary strain energy U_i^* and showed that derivatives of U_i^* with respect to external forces always give displacements, even if the load-displacement relationships are nonlinear. Thus the correct form of Eq. (3.2) should have been

$$\frac{\partial U_i^*}{\partial P_r} = U_r \tag{3.3}$$

Similarly, in Eqs. (3.1) for generality Castigliano should have dealt not with the strain energy U_i but with the complementary strain energy U_i^*, and the equations should have been written as

$$\frac{\partial U_i^*}{\partial X_1} = \frac{\partial U_i^*}{\partial X_2} = \cdots = 0 \tag{3.4}$$

The complementary energy U_i^* has no direct physical meaning and therefore is to be regarded only as a quantity formally defined by the appropriate equation.

Engesser's work received very little attention, since the main preoccupation of structural engineers at that time was with linear structures, for which the differences between energy and complementary energy disappear, and his work was not followed up until 1941, when Westergaard developed Engesser's basic idea further.

Although the various energy theorems formed the basis for analysis of redundant structures, very little progress was made toward fundamental understanding of the underlying principles. Only in more recent years has the whole approach to structural analysis based on energy methods been put on a more rational basis. It has been demonstrated[10]* that all energy theorems can be derived directly from two complementary energy principles:

1. The principle of virtual work (or virtual displacements)
2. The principle of complementary virtual work (or virtual forces)

These two principles form the basis of any strain-energy approach to structural analysis. The derivation of these two principles using the concept of virtual displacements and forces is discussed in subsequent sections.

The energy principles presented here will be restricted to small strains and displacements so that strain-displacement relationships can be expressed by linear equations; such displacements and the corresponding strains are obviously additive. Furthermore, a nonlinear elastic stress-strain relationship will be admitted unless otherwise stated.

3.2 WORK AND COMPLEMENTARY WORK; STRAIN ENERGY AND COMPLEMENTARY STRAIN ENERGY

Consider a force-displacement diagram, as shown in Fig. 3.1, which for generality is taken to be elastically nonlinear. The area W under the force-displacement curve (shaded horizontally) is obviously equal to the work done by the external force P in moving through the displacement u, and for linear systems this area is given by

$$W = \tfrac{1}{2}Pu \tag{3.5}$$

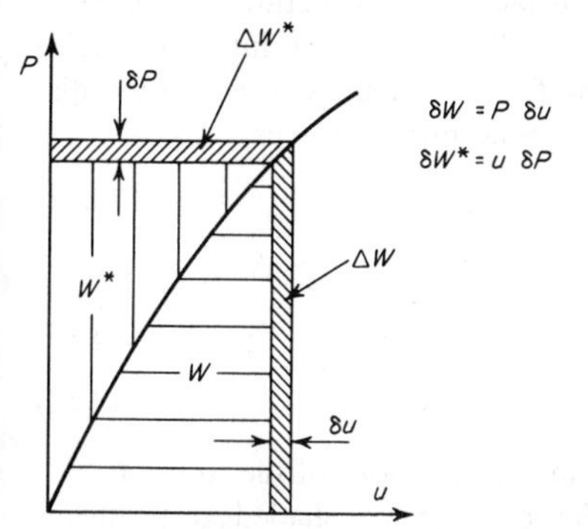

$\delta W = P\,\delta u$

$\delta W^* = u\,\delta P$

FIG. 3.1 Work and complementary work.

* Numbers refer to works listed in Appendix B.

In a linear system if displacement u is increased to $u + \delta u$, the corresponding increment in W becomes

$$\Delta W = P\,\delta u + \tfrac{1}{2}\delta P\,\delta u \tag{3.6}$$

For nonlinear systems terms of orders higher than $\delta P\,\delta u$ would also be present in ΔW, but detailed discussion of the higher-order terms will not be required, since only first variations will be considered in the derivation of strain-energy theorems. Only for stability analysis must the second variation of W also be included.

In a three-dimensional structure subjected to a system of surface forces Φ_i and body forces X_i if the displacements are increased from u_i to $u_i + \delta u_i$ while the temperature and also any initial strains are kept constant, the corresponding increment in work is given by

$$\begin{aligned}\Delta W &= \int_v \mathbf{X}^T\,\delta\mathbf{u}\,dV + \int_s \mathbf{\Phi}^T\,\delta\mathbf{u}\,dS + \text{terms of higher order} \\ &= \delta W + \tfrac{1}{2}\delta^2 W + \cdots \end{aligned} \tag{3.7}$$

where $$\delta W = \int_v \mathbf{X}^T\,\delta\mathbf{u}\,dV + \int_s \mathbf{\Phi}^T\,\delta\mathbf{u}\,dS \tag{3.8}$$

represents the first variation and $\delta^2 W$ is the second variation in W. The remaining symbols in Eq. (3.7) are defined by

$$\delta\mathbf{u} = \{\delta u_x \quad \delta u_y \quad \delta u_z\} \tag{3.9}$$

$$\mathbf{X} = \{X_x \quad X_y \quad X_z\} \tag{3.10}$$

$$\mathbf{\Phi} = \{\Phi_x \quad \Phi_y \quad \Phi_z\} \tag{3.11}$$

The first integral in (3.8) represents the work done by the body forces $\mathbf{X}$, while the second integral is the work done by the surface forces $\mathbf{\Phi}$. The integrals are evaluated over the whole volume and surface of the structure, respectively. If any concentrated forces are applied externally, the surface integral in (3.8) will also include a sum of products of the forces and the corresponding displacement variations. For example, if only concentrated forces

$$\mathbf{P} = \{P_1 \quad P_2 \quad \cdots \quad P_n\} \tag{3.12}$$

are applied to the structure, then

$$\delta W = \mathbf{P}^T\,\delta\mathbf{U} \tag{3.13}$$

where $$\delta\mathbf{U} = \{\delta U_1 \quad \delta U_2 \quad \cdots \quad \delta U_n\} \tag{3.14}$$

represents the variations of displacements in the directions of the forces $\mathbf{P}$.

The area to the left of the force-displacement curve, shaded vertically in Fig. 3.1, is defined as the complementary work W^*, since it can be regarded as the complementary area within the rectangle Pu. A perusal of Fig. 3.1 shows that for linear elasticity $W = W^*$, but even then it is still useful to differentiate between work and complementary work. If now the body forces and surface forces in a three-dimensional elastic structure are increased from $\mathbf{X}$ to $\mathbf{X} + \delta\mathbf{X}$ and from $\mathbf{\Phi}$ to $\mathbf{\Phi} + \delta\mathbf{\Phi}$, respectively, the corresponding increment in the complementary work is given by

$$\Delta W^* = \int_v \mathbf{u}^T\, \delta\mathbf{X}\, dV + \int_s \mathbf{u}^T\, \delta\mathbf{\Phi}\, dS + \text{terms of higher order}$$
$$= \delta W^* + \tfrac{1}{2}\delta^2 W^* + \cdots \tag{3.15}$$

where
$$\delta W^* = \int_v \mathbf{u}^T\, \delta\mathbf{X}\, dV + \int_s \mathbf{u}^T\, \delta\mathbf{\Phi}\, dS \tag{3.16}$$

represents the first variation and $\delta^2 W^*$ is the second variation in W^*. Other symbols used are defined by

$$\mathbf{u} = \{u_x \quad u_y \quad u_z\} \tag{3.17}$$

$$\delta\mathbf{X} = \{\delta X_x \quad \delta X_y \quad \delta X_z\} \tag{3.18}$$

$$\delta\mathbf{\Phi} = \{\delta\Phi_x \quad \delta\Phi_y \quad \delta\Phi_z\} \tag{3.19}$$

If only concentrated forces are applied, then

$$\delta W^* = \mathbf{U}^T\, \delta\mathbf{P} \tag{3.20}$$

where
$$\mathbf{U} = \{U_1 \quad U_2 \quad \cdots \quad U_n\} \tag{3.21}$$

and
$$\delta\mathbf{P} = \{\delta P_1 \quad \delta P_2 \quad \cdots \quad \delta P_n\} \tag{3.22}$$

The stress-strain relationship will also be assumed to be given by a nonlinear elastic law. It can easily be demonstrated that the area under the stress–elastic-strain curve, shown shaded horizontally in Fig. 3.2, represents the density of

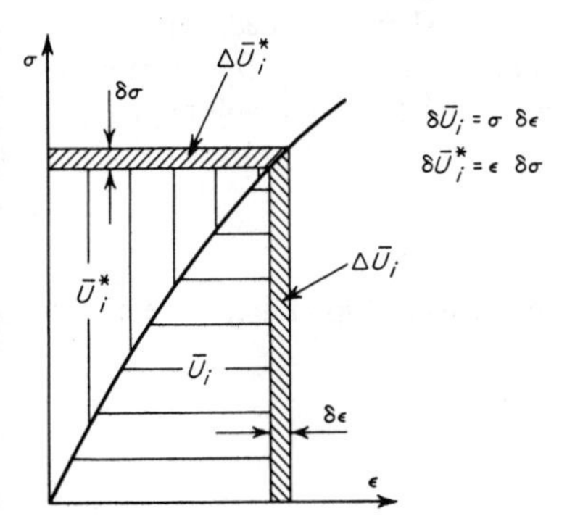

FIG. 3.2 Strain-energy and complementary-strain-energy densities.

strain energy $\bar{U}_i$, which may be measured in pound-inches per cubic inch. The elastic strain energy U_i stored in the structure can be obtained by integrating the strain-energy density $\bar{U}_i$ over the whole volume of the structure. Hence

$$U_i = \int_v \bar{U}_i \, dV \tag{3.23}$$

If now the displacements are increased from $\mathbf{u}$ to $\mathbf{u} + \delta\mathbf{u}$, there will be an accompanying increase in strains from $\boldsymbol{\epsilon}$ to $\boldsymbol{\epsilon} + \delta\boldsymbol{\epsilon}$, and the corresponding increment in the strain-energy density will be given by

$$\begin{aligned} \Delta\bar{U}_i &= \boldsymbol{\sigma}^T \, \delta\boldsymbol{\epsilon} + \text{terms of higher order} \\ &= \delta\bar{U}_i + \tfrac{1}{2}\delta^2\bar{U}_i + \cdots \end{aligned} \tag{3.24}$$

where

$$\delta\bar{U}_i = \boldsymbol{\sigma}^T \, \delta\boldsymbol{\epsilon} \tag{3.25}$$

$$\boldsymbol{\sigma} = \{\sigma_{xx} \quad \sigma_{yy} \quad \sigma_{zz} \quad \sigma_{xy} \quad \sigma_{yz} \quad \sigma_{zx}\} \tag{3.26}$$

$$\delta\boldsymbol{\epsilon} = \{\delta\epsilon_{xx} \quad \delta\epsilon_{yy} \quad \delta\epsilon_{zz} \quad \delta\epsilon_{xy} \quad \delta\epsilon_{yz} \quad \delta\epsilon_{zx}\} \tag{3.27}$$

From Eqs. (3.23) and (3.25) it follows therefore that

$$\delta U_i = \int_v \delta\bar{U}_i \, dV = \int_v \boldsymbol{\sigma}^T \, \delta\boldsymbol{\epsilon} \, dV \tag{3.28}$$

The derivation of Eq. (3.25) follows immediately if we observe that strain components $\delta\epsilon_{xx}, \ldots$ are all independent variables so that in order to form $\delta\bar{U}_i$, which is a scalar quantity, individual contributions such as $\sigma_{xx}\,\delta\epsilon_{xx}, \ldots$ must be summed, and this can be obtained most conveniently from the matrix product $\boldsymbol{\sigma}^T \, \delta\boldsymbol{\epsilon}$.

The area to the left of the stress-strain curve, shown shaded vertically in Fig. 3.2, represents the density of complementary strain energy $\bar{U}_i^*$, from which the complementary strain energy can be calculated using the volume integral

$$U_i^* = \int_v \bar{U}_i^* \, dV \tag{3.29}$$

If the stresses are increased from $\boldsymbol{\sigma}$ to $\boldsymbol{\sigma} + \delta\boldsymbol{\sigma}$, the corresponding increment in the density of complementary strain energy is given by

$$\begin{aligned} \Delta\bar{U}_i^* &= \boldsymbol{\epsilon}^T \, \delta\boldsymbol{\sigma} + \text{terms of higher order} \\ &= \delta\bar{U}_i^* + \tfrac{1}{2}\delta^2\bar{U}_i^* + \cdots \end{aligned} \tag{3.30}$$

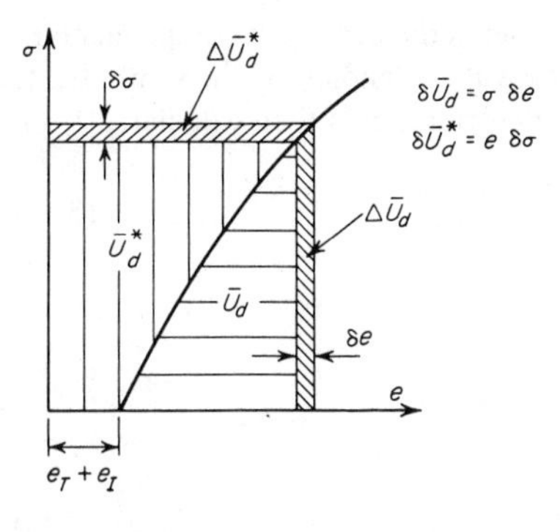

FIG. 3.3 Strain-energy and complementary-strain-energy densities of total deformation.

where $$\delta \bar{U}_i^* = \boldsymbol{\epsilon}^T \, \delta \boldsymbol{\sigma} \tag{3.31}$$

$$\boldsymbol{\epsilon} = \{\epsilon_{xx} \quad \epsilon_{yy} \quad \epsilon_{zz} \quad \epsilon_{xy} \quad \epsilon_{yz} \quad \epsilon_{zx}\} \tag{3.32}$$

$$\delta\boldsymbol{\sigma} = \{\delta\sigma_{xx} \quad \delta\sigma_{yy} \quad \delta\sigma_{zz} \quad \delta\sigma_{xy} \quad \delta\sigma_{yz} \quad \delta\sigma_{zx}\} \tag{3.33}$$

Hence from Eqs. (3.29) and (3.31) it follows that

$$\delta U_i^* = \int_v \delta \bar{U}_i^* \, dV = \int_v \boldsymbol{\epsilon}^T \, \delta\boldsymbol{\sigma} \, dV \tag{3.34}$$

If initial strains are present, or if the structure is subjected to a nonuniform temperature distribution, the stress–total-strain diagram must be of the form shown in Fig. 3.3. It should be noted that for linear elasticity the stress-strain relationship in Fig. 3.3 would be reduced to a straight line displaced to the right relative to the origin, depending on the amount of the initial strain and/or thermal strain. The horizontally shaded area represents the density of strain energy of total deformation $\bar{U}_d$, from which the strain energy of total deformation can be calculated using the volume integral

$$U_d = \int_v \bar{U}_d \, dV \tag{3.35}$$

When the total strains are increased from $\mathbf{e}$ to $\mathbf{e} + \delta\mathbf{e}$, we have

$$\Delta \bar{U}_d = \delta \bar{U}_d + \tfrac{1}{2}\delta^2 \bar{U}_d + \cdots \tag{3.36}$$

$$\delta \bar{U}_d = \boldsymbol{\sigma}^T \, \delta\mathbf{e} \tag{3.37}$$

$$\delta U_d = \int_v \boldsymbol{\sigma}^T \, \delta\mathbf{e} \, dV \tag{3.38}$$

$$\delta\mathbf{e} = \{\delta e_{xx} \quad \delta e_{yy} \quad \delta e_{zz} \quad \delta e_{xy} \quad \delta e_{yz} \quad \delta e_{zx}\} \tag{3.39}$$

The vertically shaded area in Fig. 3.3 represents the density of complementary strain energy of total deformation; this area will be denoted by the symbol $\bar{U}_d^*$. The complementary strain energy of total deformation is then calculated from

$$U_d^* = \int_v \bar{U}_d^* dV \tag{3.40}$$

When the stresses are increased from $\boldsymbol{\sigma}$ to $\boldsymbol{\sigma} + \delta\boldsymbol{\sigma}$, we have

$$\Delta\bar{U}_d^* = \delta\bar{U}_d^* + \tfrac{1}{2}\delta^2\bar{U}_d^* + \cdots \tag{3.41}$$

$$\delta\bar{U}_d^* = \mathbf{e}^T\,\delta\boldsymbol{\sigma} \tag{3.42}$$

$$\delta U_d^* = \int_v \mathbf{e}^T\,\delta\boldsymbol{\sigma}\,dV \tag{3.43}$$

$$\mathbf{e} = \{e_{xx} \quad e_{yy} \quad e_{zz} \quad e_{xy} \quad e_{yz} \quad e_{zx}\} \tag{3.44}$$

Since the total strains $\mathbf{e}$ are expressed as (see Chap. 2)

$$\mathbf{e} = \boldsymbol{\epsilon} + \mathbf{e}_T + \mathbf{e}_I \tag{3.45}$$

the first variation of the complementary strain energy of total deformation becomes

$$\begin{aligned} \delta U_d^* &= \int_v \boldsymbol{\epsilon}^T\,\delta\boldsymbol{\sigma}\,dV + \int_v \mathbf{e}_T{}^T\,\delta\boldsymbol{\sigma}\,dV + \int_v \mathbf{e}_I{}^T\,\delta\boldsymbol{\sigma}\,dV \\ &= \delta U_i^* + \int_v \alpha T\,\delta s\,dV + \int_v \mathbf{e}_I{}^T\,\delta\boldsymbol{\sigma}\,dV \end{aligned} \tag{3.46}$$

where $\qquad \delta s = \delta\sigma_{xx} + \delta\sigma_{yy} + \delta\sigma_{zz} \qquad (3.47)$

Before leaving the subject of energy and complementary energy it may be interesting to mention that two similar complementary functions are used in thermodynamics; the free-energy function A of von Helmholtz and the energy function G of Gibbs.

3.3 GREEN'S IDENTITY

In deriving the fundamental energy principles volume integrals of the type

$$I = \int_v \frac{\partial\Phi}{\partial i}\frac{\partial\psi}{\partial i}\,dV \qquad i = x, y, z \tag{3.48}$$

must be transformed into a surface and a volume integral; the functions Φ and ψ can be any functions of x, y, and z provided they are continuous within the specified volume for integration. The necessary transformation will be carried out through integration by parts, and then it will be shown that it actually leads to Green's first identity.

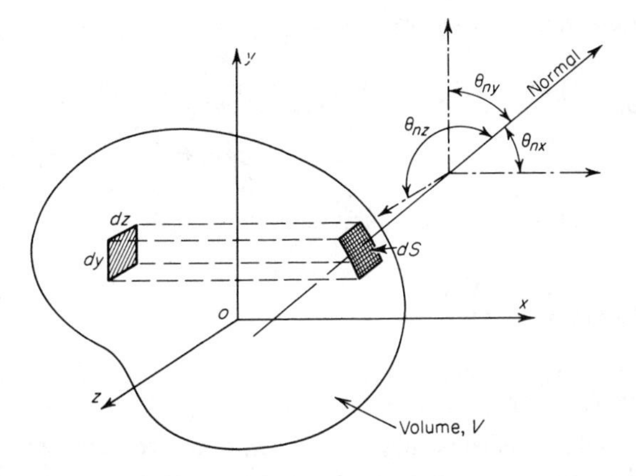

FIG. 3.4 Projection of the surface area dS onto the yz plane.

Consider a small surface area dS whose projection on the yz plane is a rectangle $dy\,dz$ (see Fig. 3.4). It is clear that the projections of dS onto the three coordinate planes are related to dS by the following equations:

$$\begin{aligned} dy\,dz &= dS \cos \theta_{nx} = l\,dS \\ dz\,dx &= dS \cos \theta_{ny} = m\,dS \\ dx\,dy &= dS \cos \theta_{nz} = n\,dS \end{aligned} \tag{3.49}$$

where l, m, and n denote the direction cosines of the outward normal at the surface. Upon substituting $i = x$ in Eq. (3.48) and using the first equation in (3.49) it follows that

$$\begin{aligned} \iiint \frac{\partial \Phi}{\partial x}\frac{\partial \psi}{\partial x}\,dx\,dy\,dz &= \int_s \int_x \frac{\partial \Phi}{\partial x}\frac{\partial \psi}{\partial x}\,l\,dS\,dx \\ &= \int_s \Phi \frac{\partial \psi}{\partial x}\,l\,dS - \int_s \int_x \Phi \frac{\partial^2 \psi}{\partial x^2}\,l\,dS\,dx \\ &= \int_s \Phi \frac{\partial \psi}{\partial x}\,l\,dS - \int_v \Phi \frac{\partial^2 \psi}{\partial x^2}\,dV \end{aligned} \tag{3.50}$$

When $i = y$ and z is substituted in Eq. (3.48), two more similar relationships can be obtained. Hence

$$\int_v \frac{\partial \Phi}{\partial y}\frac{\partial \psi}{\partial y}\,dV = \int_s \Phi \frac{\partial \psi}{\partial y}\,m\,dS - \int_v \Phi \frac{\partial^2 \psi}{\partial y^2}\,dV \tag{3.51}$$

and

$$\int_v \frac{\partial \Phi}{\partial z}\frac{\partial \psi}{\partial z}\,dV = \int_s \Phi \frac{\partial \psi}{\partial z}\,n\,dS - \int_v \Phi \frac{\partial^2 \psi}{\partial z^2}\,dV \tag{3.52}$$

If these three identities are added together, we obtain

$$\int_v \left(\frac{\partial \Phi}{\partial x}\frac{\partial \psi}{\partial x} + \frac{\partial \Phi}{\partial y}\frac{\partial \psi}{\partial y} + \frac{\partial \Phi}{\partial z}\frac{\partial \psi}{\partial z}\right) dV = \int_s \Phi\left(l\frac{\partial \psi}{\partial x} + m\frac{\partial \psi}{\partial y} + n\frac{\partial \psi}{\partial z}\right) dS - \int_v \Phi\, \nabla^2\psi\, dV \quad (3.53)$$

which represents the standard form of Green's first identity.

3.4 ENERGY THEOREMS BASED ON THE PRINCIPLE OF VIRTUAL WORK

THE PRINCIPLE OF VIRTUAL WORK (THE PRINCIPLE OF VIRTUAL DISPLACEMENTS)

In Sec. 3.2 the variations in displacements $\delta\mathbf{u}$ were assumed to be accompanied by corresponding variations in stresses and forces, so that all increments in work and strain energy were derived from two neighboring equilibrium states in which the strains $\delta\boldsymbol{\epsilon}$ derived from the displacements $\delta\mathbf{u}$ satisfied both the equations of equilibrium and compatibility. It should be noticed, however, that the first variations δW and δU_i are independent of the δP's and $\delta\sigma$'s. Thus for the purpose of finding δW and δU_i, the forces and stresses in the structure can be assumed to remain constant while the displacements are varied from $\mathbf{u}$ to $\mathbf{u} + \delta\mathbf{u}$. It follows therefore that the displacements $\delta\mathbf{u}$ must give rise to strains which satisfy the equations of compatibility but not necessarily the equations of equilibrium expressed in terms of strain components. This means that the displacements $\delta\mathbf{u}$ can be any infinitesimal displacements as long as they are geometrically possible; they must be continuous in the interior (within the structure boundaries) and must satisfy any kinematic boundary conditions which may be imposed on the actual displacements $\mathbf{u}$; for example, zero displacements and slopes at the built-in end of a cantilever beam. These infinitesimal displacements are referred to as *virtual displacements*, a term borrowed from rigid-body mechanics. In the subsequent analysis the variations δW and δU_i derived from the virtual displacements $\delta\mathbf{u}$ will be called *virtual work* and *virtual strain energy*, respectively.

If the three equations of internal equilibrium, Eqs. (2.61), are multiplied by the virtual displacements δu_x, δu_y, and δu_z, respectively, and integrated over the whole volume of the structure, the addition of the resulting three integrals leads to

$$\int_v \left(\frac{\partial \sigma_{xx}}{\partial x} + \frac{\partial \sigma_{xy}}{\partial y} + \frac{\partial \sigma_{xz}}{\partial z} + X_x\right) \delta u_x\, dV + \int_v \left(\frac{\partial \sigma_{yx}}{\partial x} + \frac{\partial \sigma_{yy}}{\partial y} + \frac{\partial \sigma_{yz}}{\partial z} + X_y\right) \delta u_y\, dV + \int_v \left(\frac{\partial \sigma_{zx}}{\partial x} + \frac{\partial \sigma_{zy}}{\partial y} + \frac{\partial \sigma_{zz}}{\partial z} + X_z\right) \delta u_z\, dV = 0 \quad (3.54)$$

Applying now Green's identities (3.50) to (3.52) to Eq. (3.54), we have

$$\int_s \sigma_{xx} l\,\delta u_x\,dS - \int_v \sigma_{xx}\frac{\partial \delta u_x}{\partial x}\,dV + \int_s \sigma_{xy} m\,\delta u_x\,dS - \int_v \sigma_{xy}\frac{\partial \delta u_x}{\partial y}\,dV + \int_s \sigma_{xz} n\,\delta u_x\,dS$$
$$-\int_v \sigma_{xz}\frac{\partial \delta u_x}{\partial z}\,dV + \int_v X_x\,\delta u_x\,dV + \int_s \sigma_{yx} l\,\delta u_y\,dS - \int_v \sigma_{yx}\frac{\partial \delta u_y}{\partial x}\,dV + \int_s \sigma_{yy} m\,\delta u_y\,dS$$
$$-\int_v \sigma_{yy}\frac{\partial \delta u_y}{\partial y}\,dV + \int_s \sigma_{yz} n\,\delta u_y\,dS - \int_v \sigma_{yz}\frac{\partial \delta u_y}{\partial z}\,dV + \int_v X_y\,\delta u_y\,dV + \int_s \sigma_{zx} l\,\delta u_z\,dS$$
$$-\int_v \sigma_{zx}\frac{\partial \delta u_z}{\partial x}\,dV + \int_s \sigma_{zy} m\,\delta u_z\,dS - \int_v \sigma_{zy}\frac{\partial \delta u_z}{\partial y}\,dV + \int_s \sigma_{zz} n\,\delta u_z\,dS$$
$$-\int_v \sigma_{zz}\frac{\partial \delta u_z}{\partial z}\,dV + \int_v X_z\,\delta u_z\,dV = 0 \quad (3.55)$$

Since the virtual displacements δu_x, δu_y, and δu_z are continuous, $\partial \delta u_x/\partial x = \delta(\partial u_x/\partial x)$, etc., and since $\sigma_{xy} = \sigma_{yx}$, etc., Eq. (3.55) can be simplified to

$$\int_s [(\sigma_{xx} l + \sigma_{xy} m + \sigma_{xz} n)\,\delta u_x + (\sigma_{yx} l + \sigma_{yy} m + \sigma_{yz} n)\,\delta u_y$$
$$+ (\sigma_{zx} l + \sigma_{zy} m + \sigma_{zz} n)\,\delta u_z]\,dS + \int_v (X_x\,\delta u_x + X_y\,\delta u_y + X_z\,\delta u_z)\,dV$$
$$= \int_v \Big[\sigma_{xx}\,\delta\Big(\frac{\partial u_x}{\partial x}\Big) + \sigma_{yy}\,\delta\Big(\frac{\partial u_y}{\partial y}\Big) + \sigma_{zz}\,\delta\Big(\frac{\partial u_z}{\partial z}\Big) + \sigma_{xy}\,\delta\Big(\frac{\partial u_x}{\partial y} + \frac{\partial u_y}{\partial x}\Big)$$
$$+ \sigma_{yz}\,\delta\Big(\frac{\partial u_y}{\partial z} + \frac{\partial u_z}{\partial y}\Big) + \sigma_{zx}\,\delta\Big(\frac{\partial u_z}{\partial x} + \frac{\partial u_x}{\partial z}\Big)\Big]\,dV \quad (3.56)$$

Using now the strain-displacement relationships (2.2) for the strain increments and equations of stress equilibrium on the surface (2.62), and assuming that $\delta \mathbf{e}_T = \delta \mathbf{e}_I = \mathbf{0}$, we obtain from (3.56)

$$\int_s \mathbf{\Phi}^T\,\delta\mathbf{u}\,dS + \int_v \mathbf{X}^T\,\delta\mathbf{u}\,dV = \int_v \boldsymbol{\sigma}^T\,\delta\boldsymbol{\epsilon}\,dV \quad (3.57)$$

Hence, from Eqs. (3.8), (3.28), and (3.57), it follows that

$$\delta W = \delta U_i \quad (3.58)$$

Equation (3.58) represents the principle of virtual work, which states that an elastic structure is in equilibrium under a given system of loads and temperature distribution if for any virtual displacement $\delta\mathbf{u}$ from a compatible state of deformation $\mathbf{u}$ the virtual work is equal to the virtual strain energy. An alternative name for this principle is the principle of virtual displacements. Equation (3.58) can also be used for large deflections provided that when the strain energy U_i is calculated, nonlinear (large-deformation) strain-displacements relations are used and the equilibrium equations are considered for the deformed structure.[10]

THE PRINCIPLE OF A STATIONARY VALUE OF TOTAL POTENTIAL ENERGY

The principle of virtual work can be expressed in a more concise form as

$$\delta_\epsilon U = \delta(U_i + U_e) = 0 \tag{3.59}$$

where $\delta U_e = -\delta W$ (3.60)

and U_e is the potential of external forces. The subscript ϵ has been added to the variation symbol δ to emphasize that only elastic strains and displacements are to be varied. In calculating the potential of external forces U_e it should be observed that all displacements are treated as variables wherever the corresponding forces are specified. The quantity

$$U = U_i + U_e \tag{3.61}$$

is the total potential energy of the system provided the potential energy of the external forces in the unstressed condition is taken as zero. Thus Eq. (3.59) may be described as the principle of stationary value of total potential energy. Since only displacements, and hence strains, are varied, this principle could alternatively be expressed by the statement that "of all compatible displacements satisfying given boundary conditions, those which satisfy the equilibrium conditions make the total potential energy U assume a stationary value."

The stationary value of U is always a minimum, and therefore a structure under a system of external loads and temperature distribution represents a stable system. The proof that U is a minimum for the exact system of stresses when compatible strain and displacement variations are considered has been given by Biezeno and Grammel.

CASTIGLIANO'S THEOREM (PART I)

Consider a structure having some specified temperature distribution and subjected to a system of external forces $P_1, P_2, \ldots, P_r, \ldots, P_n$, and then apply only one virtual displacement δU_r in the direction of the load P_r while keeping temperatures constant. The virtual work

$$\delta W = P_r \, \delta U_r \tag{3.62}$$

and from the principle of virtual work it immediately follows that

$$\delta U_i = P_r \, \delta U_r \tag{3.63}$$

Hence, in the limit

$$\left(\frac{\partial U_i}{\partial U_r}\right)_{T=\text{const}} = P_r \tag{3.64}$$

which is the well-known Castigliano's theorem (part I). Equation (3.64) can also be used for large deflections provided the strain energy U_i is calculated using large deformation strains (see Chap. 15). Matrix formulation of this theorem for linear problems is discussed in Sec. 5.3.

If instead of one virtual displacement δU_r we introduce a virtual rotation $\delta\theta_r$ in the direction of a concentrated moment M_r, then Castigliano's theorem becomes

$$\left(\frac{\partial U_i}{\partial \theta_r}\right)_{T=\text{const}} = M_r \tag{3.65}$$

THE THEOREM OF MINIMUM STRAIN ENERGY

If in a strained structure only such virtual displacements are selected which are zero in the direction of the applied forces, then

$$\delta\boldsymbol{\epsilon}_r = \underline{\boldsymbol{\epsilon}}_r\,\delta U_r \tag{3.69}$$

and therefore Eq. (3.58) reduces to

$$\delta_\epsilon U_i = 0 \tag{3.67}$$

where the subscript ϵ has been added to indicate that only strains and displacements are varied. If the second variation of the strain energy is considered, it can be shown that $\delta_\epsilon U_i = 0$ represents in fact the condition for minimum strain energy. For this reason, therefore, Eq. (3.67) is referred to as the theorem of minimum strain energy.

THE UNIT-DISPLACEMENT THEOREM[10]

This theorem is used to determine the force P_r necessary to maintain equilibrium in a structure for which the distribution of true stresses is known. Let the known stresses be given by a stress matrix $\boldsymbol{\sigma}$. If a virtual displacement δU_r is applied at the point of application and in the direction of P_r so that virtual strains $\delta\boldsymbol{\epsilon}_r$ are produced in the structure, then when the virtual work is equated to the virtual strain energy, the following equation is obtained:

$$P_r\,\delta U_r = \int_v \boldsymbol{\sigma}^T\,\delta\boldsymbol{\epsilon}_r\,dV \tag{3.68}$$

In a linearly elastic structure $\delta\boldsymbol{\epsilon}_r$ is proportional to δU_r; that is,

$$\delta\boldsymbol{\epsilon}_r = \underline{\boldsymbol{\epsilon}}_r\,\delta U_r \tag{3.69}$$

where $\underline{\boldsymbol{\epsilon}}_r$ represents compatible strain distribution due to a unit displacement ($\delta U_r = 1$) applied in the direction of P_r. The strains $\underline{\boldsymbol{\epsilon}}_r$ must be compatible only with themselves and with the imposed unit displacement. Substituting Eq. (3.69) into (3.68) and canceling out the factor δU_r gives

$$P_r = \int_v \boldsymbol{\sigma}^T\,\underline{\boldsymbol{\epsilon}}_r\,dV \tag{3.70}$$

Equation (3.70) represents the unit-displacement theorem,[10] which states that the force necessary, at the point r in a structure, to maintain equilibrium under a specified stress distribution (which can also be derived from a specified displacement distribution) is equal to the integral over the volume of the structure of true stresses $\boldsymbol{\sigma}$ multiplied by strains $\underline{\boldsymbol{\epsilon}}_r$ compatible with a unit displacement corresponding to the required force. This theorem is restricted to linear elasticity in view of the assumption made in Eq. (3.69).

The unit-displacement theorem can be used very effectively for the calculation of stiffness properties of structural elements used in the matrix methods of structural analysis. Matrix formulation of the unit-displacement theorem and its application to the displacement method of analysis are discussed at some length in Sec. 5.2.

3.5 ENERGY THEOREMS BASED ON THE PRINCIPLE OF COMPLEMENTARY VIRTUAL WORK

THE PRINCIPLE OF COMPLEMENTARY VIRTUAL WORK (THE PRINCIPLE OF VIRTUAL FORCES)

The variations δW^* and δU_d^* are independent of δu's and δe's. Thus, for the purpose of finding δW^* and δU_d^*, the displacements and total strains can be assumed to remain constant while the forces and stresses are varied. It follows therefore that the virtual stresses $\delta\boldsymbol{\sigma}$ and virtual body forces $\delta\mathbf{X}$ must satisfy the equations of internal stress equilibrium, and the virtual surface forces $\delta\boldsymbol{\Phi}$ must satisfy equations for boundary equilibrium, but the strains derived from the virtual stresses need not necessarily satisfy the equations of compatibility. This also implies that the virtual stresses and forces, that is, $\delta\boldsymbol{\sigma}$, $\delta\mathbf{X}$, and $\delta\boldsymbol{\Phi}$, can be any infinitesimal stresses or forces as long as they are statically possible; they must satisfy all equations of equilibrium throughout the whole structure. In the subsequent analysis, the increments δW^* and δU_d^* derived from the virtual forces will be referred to as *complementary virtual work* and *complementary virtual strain energy of total deformation*, respectively.

If the three equations of internal stress equilibrium for the virtual stresses $\delta\boldsymbol{\sigma}$ and the virtual body forces $\delta\mathbf{X}$ are multiplied by the actual displacements u_x, u_y, and u_z, respectively, and integrated over the whole volume of the structure, the addition of the resulting three integrals leads to the following equation:

$$\begin{aligned}\int_v \left(\frac{\partial\delta\sigma_{xx}}{\partial x} + \frac{\partial\delta\sigma_{xy}}{\partial y} + \frac{\partial\delta\sigma_{xz}}{\partial z} + \delta X_x\right) u_x\, dV \\ + \int_v \left(\frac{\partial\delta\sigma_{yx}}{\partial x} + \frac{\partial\delta\sigma_{yy}}{\partial y} + \frac{\partial\delta\sigma_{yz}}{\partial z} + \delta X_y\right) u_y\, dV \\ + \int_v \left(\frac{\partial\delta\sigma_{zx}}{\partial x} + \frac{\partial\delta\sigma_{zy}}{\partial y} + \frac{\partial\delta\sigma_{zz}}{\partial z} + \delta X_z\right) u_z\, dV = 0 \qquad (3.71)\end{aligned}$$

Applying Green's identities (3.50), (3.51), and (3.52) to Eq. (3.71), we obtain

$$\int_s \delta\sigma_{xx}\, lu_x\, dS - \int_v \delta\sigma_{xx} \frac{\partial u_x}{\partial x}\, dV + \int_s \delta\sigma_{xy}\, mu_x\, dS - \int_v \delta\sigma_{xy} \frac{\partial u_x}{\partial y}\, dV$$
$$+ \int_s \delta\sigma_{xz}\, nu_x\, dS - \int_v \delta\sigma_{xz} \frac{\partial u_x}{\partial z}\, dV + \int_v \delta X_x\, u_x\, dV$$
$$+ \int_s \delta\sigma_{yx}\, lu_y\, dS - \int_v \delta\sigma_{yx} \frac{\partial u_y}{\partial x}\, dV + \int_s \delta\sigma_{yy}\, mu_y\, dS - \int_v \delta\sigma_{yy} \frac{\partial u_y}{\partial y}\, dV$$
$$+ \int_s \delta\sigma_{yz}\, nu_y\, dS - \int_v \delta\sigma_{yz} \frac{\partial u_y}{\partial z}\, dV + \int_v \delta X_y\, u_y\, dV$$
$$+ \int_s \delta\sigma_{zx}\, lu_z\, dS - \int_v \delta\sigma_{zx} \frac{\partial u_z}{\partial x}\, dV + \int_s \delta\sigma_{zy}\, mu_z\, dS - \int_v \delta\sigma_{zy} \frac{\partial u_z}{\partial y}\, dV$$
$$+ \int_s \delta\sigma_{zz}\, nu_z\, dS - \int_v \delta\sigma_{zz} \frac{\partial u_z}{\partial z}\, dV + \int_v \delta X_z\, u_z\, dV = 0 \tag{3.72}$$

which can be rearranged into

$$\int_s [(\delta\sigma_{xx}\, l + \delta\sigma_{xy}\, m + \delta\sigma_{xz}\, n)u_x + (\delta\sigma_{yx}\, l + \delta\sigma_{yy}\, m + \delta\sigma_{yz}\, n)u_y$$
$$+ (\delta\sigma_{zx}\, l + \delta\sigma_{zy}\, m + \delta\sigma_{zz}\, n)u_z]\, dS$$
$$+ \int_v (\delta X_x\, u_x + \delta X_y\, u_y + \delta X_z\, u_z)\, dV$$
$$= \int_v \left[\delta\sigma_{xx} \frac{\partial u_x}{\partial x} + \delta\sigma_{yy} \frac{\partial u_y}{\partial y} + \delta\sigma_{zz} \frac{\partial u_z}{\partial z} + \delta\sigma_{xy}\left(\frac{\partial u_x}{\partial y} + \frac{\partial u_y}{\partial x}\right)\right.$$
$$\left. + \delta\sigma_{yz}\left(\frac{\partial u_y}{\partial z} + \frac{\partial u_z}{\partial y}\right) + \delta\sigma_{zx}\left(\frac{\partial u_z}{\partial x} + \frac{\partial u_x}{\partial z}\right)\right] dV \tag{3.73}$$

Noting that on the surface

$$\begin{aligned} \delta\sigma_{xx}\, l + \delta\sigma_{xy}\, m + \delta\sigma_{xz}\, n &= \delta\Phi_x \\ \delta\sigma_{yx}\, l + \delta\sigma_{yy}\, m + \delta\sigma_{yz}\, n &= \delta\Phi_y \\ \delta\sigma_{zx}\, l + \delta\sigma_{zy}\, m + \delta\sigma_{zz}\, n &= \delta\Phi_z \end{aligned} \tag{3.74}$$

and using strain-displacement relations (2.2), we obtain from Eq. (3.73)

$$\int_s \mathbf{u}^T\, \delta\mathbf{\Phi}\, dS + \int_v \mathbf{u}^T\, \delta\mathbf{X}\, dV = \int_v \mathbf{e}^T\, \delta\boldsymbol{\sigma}\, dV \tag{3.75}$$

Hence from Eqs. (3.16), (3.43), (3.46), and (3.75), it follows that

$$\begin{aligned} \delta W^* &= \delta U_d^* \\ &= \delta U_i^* + \int_v \alpha T\, \delta s\, dV + \int_v \mathbf{e}_I{}^T\, \delta\boldsymbol{\sigma}\, dV \end{aligned} \tag{3.76}$$

where $\quad \delta s = \delta\sigma_{xx} + \delta\sigma_{yy} + \delta\sigma_{zz} \quad$ (3.47)

Equation (3.76) represents the principle of complementary virtual work, which states that an elastic structure is in a compatible state of deformation under a given system of loads and temperature distribution if for any virtual stresses and forces $\delta\boldsymbol{\sigma}$, $\delta\boldsymbol{\Phi}$, $\delta\mathbf{X}$ away from the equilibrium state of stress the virtual complementary work is equal to the virtual complementary strain energy of total deformation. An alternative name for this principle is the principle of virtual forces. Since the linear strain-displacement equations (2.2) were used to derive Eq. (3.76), the principle of complementary work as stated here can be applied only to small deflections.

THE PRINCIPLE OF A STATIONARY VALUE OF TOTAL COMPLEMENTARY POTENTIAL ENERGY

The principle of complementary virtual work can also be expressed in a more concise form as

$$\delta_\sigma U^* = \delta(U_d^* + U_e^*) = 0 \tag{3.77}$$

where $$\delta U_e^* = -\delta W^* \tag{3.78}$$

and U_e^* is the complementary potential of external forces. The subscript σ has been added to emphasize that only the stresses and forces are to be varied. In calculating the complementary potential of external forces U_e^*, it should be observed that all forces are treated as variables wherever the corresponding displacements are specified. The quantity

$$U^* = U_d^* + U_e^* \tag{3.79}$$

is the total complementary potential energy of the system. Thus Eq. (3.77) may be described as the principle of a stationary value of total complementary potential energy. This principle may also be interpreted as a statement that "of all statically equivalent, i.e., satisfying equations of equilibrium, stress states satisfying given boundary conditions on the stresses, those which satisfy the compatibility equations make the total complementary potential energy U^* assume a stationary value." It can be demonstrated that this stationary value is a minimum.

CASTIGLIANO'S THEOREM (PART II)

If only one virtual force δP_r is applied in the direction of the displacement U_r in a structure subjected to a system of external loads and temperature distribution, then

$$\delta W^* = U_r\,\delta P_r \tag{3.80}$$

From the principle of complementary virtual work it follows that

$$\delta U_d^* = U_r\,\delta P_r \tag{3.81}$$

which in the limit becomes

$$\left(\frac{\partial U_d^*}{\partial P_r}\right)_{T=\text{const}} = U_r \tag{3.82}$$

Equation (3.82) represents the well-known Castigliano's theorem (part II), generalized for thermal and initial strains. Matrix formulation of this theorem is discussed in Sec. 7.4.

If instead of virtual force δP_r we introduce a virtual concentrated moment δM_r, Castigliano's theorem (part II) becomes

$$\left(\frac{\partial U_d^*}{\partial M_r}\right)_{T=\text{const}} = \theta_r \tag{3.83}$$

where θ_r is the rotation in the direction of δM_r.

THE THEOREM OF MINIMUM COMPLEMENTARY STRAIN ENERGY OF TOTAL DEFORMATION

If in a strained structure no external virtual forces are introduced but only virtual stresses which are zero on the boundaries, the complementary virtual work is

$$\delta W^* = 0 \tag{3.84}$$

and, from the principle of complementary virtual work the complementary virtual strain energy of total deformation

$$\delta_\sigma U_d^* = 0 \tag{3.85}$$

where the subscript σ has been added to indicate that only stresses are varied. If the second variation in the complementary strain energy of total deformation is considered, it can be shown that $\delta_\sigma U^* = 0$ represents the condition for minimum U^*. Since the virtual stresses are kept zero on the boundaries, this theorem can be applied only to redundant structures, for which variation of internal stresses is possible.

THE UNIT-LOAD THEOREM[10]

This theorem is used to determine the displacement U_r in a structure for which the distribution of true total strains is known. These strains will be denoted by the strain matrix $\mathbf{e}$. If a virtual force δP_r is applied in the direction of U_r so that virtual stresses $\delta\boldsymbol{\sigma}_r$ are produced in the structure, then the following equation is obtained by equating the complementary virtual work to the complementary virtual strain energy of total deformation:

$$U_r\,\delta P_r = \int_v \mathbf{e}^T\,\delta\boldsymbol{\sigma}_r\,dV \tag{3.86}$$

In a linearly elastic structure $\delta\boldsymbol{\sigma}_r$ is proportional to δP_r, that is,

$$\delta\boldsymbol{\sigma}_r = \bar{\boldsymbol{\sigma}}_r \, \delta P_r \tag{3.87}$$

where $\bar{\boldsymbol{\sigma}}_r$ represents statically equivalent stress distribution due to a unit load ($\delta P_r = 1$) applied in the direction U_r. The stresses $\bar{\boldsymbol{\sigma}}_r$ must be in equilibrium only with themselves and with the imposed unit load. Substituting Eq. (3.87) into (3.86) and canceling out the virtual load δP_r, we obtain

$$U_r = \int_v \mathbf{e}^T \bar{\boldsymbol{\sigma}}_r \, dV \tag{3.88}$$

Equation (3.88) represents the unit-load theorem,[10] which states that the displacement U_r in a structure under any system of loads and temperature distribution is equal to the integral over the volume of the structure of the true strains $\mathbf{e}$, which include both the initial and thermal strains, multiplied by the stresses $\bar{\boldsymbol{\sigma}}_r$ equivalent to a unit load corresponding with the required displacement.

The unit-load theorem is particularly useful in calculating deflections on redundant structures and in determining flexibility matrices as demonstrated in Secs. 7.3 and 8.1. Furthermore, this theorem forms the basis of the matrix force method of structural analysis of redundant structures. It should be noted that since $\bar{\boldsymbol{\sigma}}_r$ must satisfy only the equations of equilibrium, the $\bar{\boldsymbol{\sigma}}_r$ distributions can be determined in the simplest statically determinate systems. Clearly if $\bar{\boldsymbol{\sigma}}_r$ can be chosen in such a way that it is equal to zero over much of the structure, the integration in Eq. (3.88) can be considerably simplified. In a statically determinate system $\bar{\boldsymbol{\sigma}}_r$ naturally represents the true stress distribution for a unit load applied in the direction of U_r.

3.6 CLAPEYRON'S THEOREM

For linearly elastic structures

$$\begin{aligned} \delta U_i = \delta U_i^* &= \int_v \boldsymbol{\sigma}^T \, \delta\boldsymbol{\epsilon} \, dV \\ &= \int_v \boldsymbol{\epsilon}^T \, \delta\boldsymbol{\sigma} \, dV \end{aligned} \tag{3.89}$$

and

$$\begin{aligned} \delta W = \delta W^* &= \int_v \mathbf{X}^T \, \delta\mathbf{u} \, dV + \int_s \boldsymbol{\Phi}^T \, \delta\mathbf{u} \, dS \\ &= \int_v \mathbf{u}^T \, \delta\mathbf{X} \, dV + \int_s \mathbf{u}^T \, \delta\boldsymbol{\Phi} \, dS \end{aligned} \tag{3.90}$$

The above equations can be deduced from Figs. 3.1 and 3.2 if we note that the force-displacement and stress-strain diagrams for linear elasticity are represented by straight lines. Assuming that body forces, surface forces, and the temperature distribution are all increased from zero to their final values, we have that

$$U_i = U_i^* = \frac{1}{2}\int_v \boldsymbol{\sigma}^T \boldsymbol{\epsilon}\, dV$$

$$= \frac{1}{2}\int_v \boldsymbol{\epsilon}^T \boldsymbol{\sigma}\, dV \tag{3.91}$$

and $$W = W^* = \frac{1}{2}\int_v \mathbf{X}^T \mathbf{u}\, dV + \frac{1}{2}\int_s \boldsymbol{\Phi}^T \mathbf{u}\, dS$$

$$= \frac{1}{2}\int_v \mathbf{u}^T \mathbf{X}\, dV + \frac{1}{2}\int_s \mathbf{u}^T \boldsymbol{\Phi}\, dS \tag{3.92}$$

where all matrices refer to their final values. Equations (3.91) and (3.92) are sometimes referred to as Clapeyron's theorem, and they are used for calculating strain energy and work in linearly elastic structures.

The complementary strain energy of total deformation for linear elasticity can be deduced from Fig. 3.3. Hence

$$U_d^* = \frac{1}{2}\int_v \boldsymbol{\sigma}^T \boldsymbol{\epsilon}\, dV + \int_v s\alpha T\, dV + \int_v \boldsymbol{\sigma}^T \mathbf{e}_I\, dV \tag{3.93}$$

where $$s = \sigma_{xx} + \sigma_{yy} + \sigma_{zz} \tag{3.94}$$

The first term on the right of Eq. (3.93) represents the complementary strain energy, while the second and third terms represent the respective contributions due to thermal and initial strains.

3.7 BETTI'S THEOREM

We shall now consider a linearly elastic structure subjected to two force systems represented by the matrices $\mathbf{P}_{\mathrm{I}}$ and $\mathbf{P}_{\mathrm{II}}$, respectively. The displacements due to system $\mathbf{P}_{\mathrm{I}}$ alone will be represented by $\mathbf{U}_{\mathrm{I}}$, and those due to $\mathbf{P}_{\mathrm{II}}$ alone by $\mathbf{U}_{\mathrm{II}}$. If the system $\mathbf{P}_{\mathrm{I}}$ is applied first followed by the system $\mathbf{P}_{\mathrm{II}}$, the work done by external forces is given by

$$W_{\mathrm{I,II}} = \tfrac{1}{2}\mathbf{P}_{\mathrm{I}}{}^T\mathbf{U}_{\mathrm{I}} + \tfrac{1}{2}\mathbf{P}_{\mathrm{II}}{}^T\mathbf{U}_{\mathrm{II}} + \mathbf{P}_{\mathrm{I}}{}^T\mathbf{U}_{\mathrm{II}} \tag{3.95}$$

where the subscript I, II with W indicates the sequence of application of the load systems. If the sequence is reversed, the work done by external forces becomes

$$W_{\mathrm{II,I}} = \tfrac{1}{2}\mathbf{P}_{\mathrm{II}}{}^T\mathbf{U}_{\mathrm{II}} + \tfrac{1}{2}\mathbf{P}_{\mathrm{I}}{}^T\mathbf{U}_{\mathrm{I}} + \mathbf{P}_{\mathrm{II}}{}^T\mathbf{U}_{\mathrm{I}} \tag{3.96}$$

In both instances work is stored as elastic strain energy U_i, and the amount of energy so stored must be the same, since the final deformed configuration in a linear system must be independent of the sequence of load application. Therefore

$$U_i = W_{\mathrm{I,II}} = W_{\mathrm{II,I}} \tag{3.97}$$

from which it follows that

$$\mathbf{P}_{\mathrm{I}}{}^T\mathbf{U}_{\mathrm{II}} = \mathbf{P}_{\mathrm{II}}{}^T\mathbf{U}_{\mathrm{I}} \tag{3.98}$$

Equation (3.98) is usually referred to as the reciprocal theorem of Betti, which states that "the work done by the system of forces $\mathbf{P}_{\mathrm{I}}$ over the displacements $\mathbf{U}_{\mathrm{II}}$ is equal to the work done by the system of forces $\mathbf{P}_{\mathrm{II}}$ over the displacements $\mathbf{U}_{\mathrm{I}}$, where $\mathbf{U}_{\mathrm{I}}$ and $\mathbf{U}_{\mathrm{II}}$ are the displacements due to $\mathbf{P}_{\mathrm{I}}$ and $\mathbf{P}_{\mathrm{II}}$, respectively."

3.8 MAXWELL'S RECIPROCAL THEOREM

If systems of forces $\mathbf{P}_{\mathrm{I}}$ and $\mathbf{P}_{\mathrm{II}}$ used in Betti's theorem consist each of one force, for example, P_1 and P_2, applied at two different locations, then Eq. (3.98) becomes

$$[P_1 \quad 0]\begin{bmatrix} f_{12}P_2 \\ f_{22}P_2 \end{bmatrix} = [0 \quad P_2]\begin{bmatrix} f_{11}P_1 \\ f_{21}P_1 \end{bmatrix} \tag{3.99}$$

where $f_{i,j}$ is the displacement in the ith direction due to a unit force in the jth direction. After multiplying out matrices in Eq. (3.99) it follows that

$$f_{12} = f_{21} \tag{3.100}$$

or, in general,

$$f_{i,j} = f_{j,i} \tag{3.100a}$$

The result expressed by (3.100*a*) is the well-known Maxwell reciprocal theorem. It represents the reciprocal relationship for the influence coefficients which form the elements of flexibility matrices. Thus all flexibility matrices, and hence stiffness matrices, must be symmetric for linear structures.

3.9 SUMMARY OF ENERGY THEOREMS AND DEFINITIONS

A summary of definitions used in the energy theorems of Chap. 3 is presented in Table 3.1. In addition, a convenient summary of energy theorems is provided in Table 3.2.

TABLE 3.1 SUMMARY OF DEFINITIONS

Symbol	Definition	Calculated from
W	Work	$W = \int P\,du$
W^*	Complementary work	$W^* = \int u\,dP$
$\bar{U}_i$	Density of strain energy	$\bar{U}_i = \int \sigma\,d\epsilon$
$\bar{U}_d$	Density of strain energy of total deformation	$\bar{U}_d = \int \sigma\,de$
$\bar{U}_i^*$	Density of complementary strain energy	$\bar{U}_i^* = \int \epsilon\,d\sigma$
$\bar{U}_d^*$	Density of complementary strain energy of total deformation	$\bar{U}_d^* = \int e\,d\sigma$
U_i	Strain energy	$U_i = \int \bar{U}_i\,dV$
U_d	Strain energy of total deformation	$U_d = \int \bar{U}_d\,dV$
U_i^*	Complementary strain energy	$U_i^* = \int \bar{U}_i^*\,dV$
U_d^*	Complementary strain energy of total deformation	$U_d^* = \int \bar{U}_d^*\,dV$
U_e	Potential of external forces	$\delta U_e = -\delta W$
U_e^*	Complementary potential of external forces	$\delta U_e^* = -\delta W^*$
U	Potential energy	$U = U_i + U_e$
U^*	Complementary potential energy	$U^* = U_i^* + U_e^*$

TABLE 3.2 SUMMARY OF ENERGY THEOREMS

Theorems based on the principle of virtual work (virtual-displacements principle) $\delta U_i = \delta W$	Theorems based on the principle of complementary virtual work (virtual-forces principle) $\delta U_d^* = \delta W^*$
The principle of a stationary value of total potential energy (principle of minimum potential energy) $\delta_\epsilon U = \delta(U_i + U_e) = 0$	The principle of a stationary value of total complementary potential energy (principle of minimum complementary potential energy) $\delta_\sigma U^* = \delta(U_d^* + U_e^*) = 0$
Castigliano's theorem (part I) $\left(\dfrac{\partial U_i}{\partial U_r}\right)_{T=\text{const}} = P_r$	Castigliano's theorem (part II) $\left(\dfrac{\partial U_d^*}{\partial P_r}\right)_{T=\text{const}} = U_r$
The theorem of minimum strain energy $\delta_\epsilon U_i = 0$	The theorem of minimum complementary strain energy of total deformation (applicable only to redundant structures) $\delta_\sigma U_d^* = 0$
The unit-displacement theorem $P_r = \int_v \boldsymbol{\sigma}^T \underline{\boldsymbol{\epsilon}}_r\,dV$	The unit-load theorem $U_r = \int_v \mathbf{e}^T \bar{\boldsymbol{\sigma}}_r\,dV$

PROBLEMS

3.1 Derive a general formula for calculating strain energy in terms of stresses for plane-stress problems.

3.2 Derive a general formula for calculating complementary strain energy of total deformation in terms of stresses for plane-stress problems.

3.3 A uniform bar of cross-sectional area A is held between two rigid walls as shown in Fig. 3.5. Young's modulus of the material is E, and the coefficient of expansion is α. Determine the complementary strain energy of total deformation when the bar is subjected to a temperature increase T. What is the strain energy stored in the bar? Comment on your results.

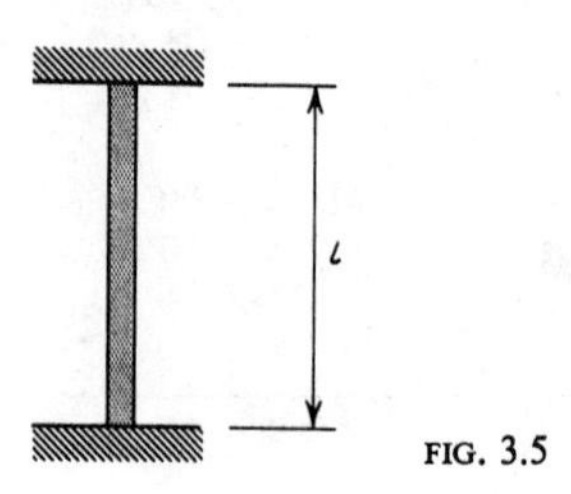

FIG. 3.5

3.4 Using Green's identities, show that

$$\int_v \mathbf{e}^T \boldsymbol{\sigma}^* \, dV = 0$$

where $\mathbf{e} = \{e_{xx} \quad e_{yy} \quad e_{zz} \quad e_{xy} \quad e_{yz} \quad e_{zx}\}$

represents total strains due to a system of external forces applied to a structure and

$$\boldsymbol{\sigma}^* = \{\sigma^*_{xx} \quad \sigma^*_{yy} \quad \sigma^*_{zz} \quad \sigma^*_{xy} \quad \sigma^*_{yz} \quad \sigma^*_{zx}\}$$

represents any self-equilibrating stress system in the same structure. (*Note:* A self-equilibrating stress system is one in which all surface and body forces are equal to zero.)

3.5 A framework consisting of a single joint connected by three pin-jointed bars attached to a rigid foundation is loaded by forces P_1 and P_2 (see Fig. 3.6). The cross-sectional areas of the bars and the framework geometry are as shown. Using the principle of virtual work, determine the internal forces in the three bars.

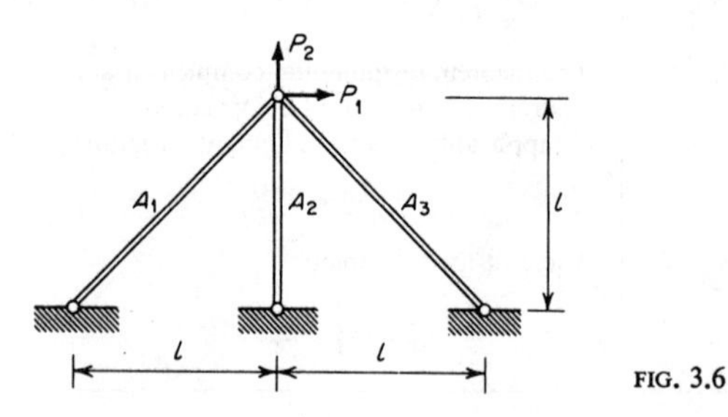

FIG. 3.6

3.6 Show, using the principle of complementary virtual work, that if w is the distributed loading on a uniform cantilever beam (see Fig. 3.7) with constant temperature, then the area swept by the beam when undergoing deflections due to w is given by $\partial U_i^*/\partial w$.

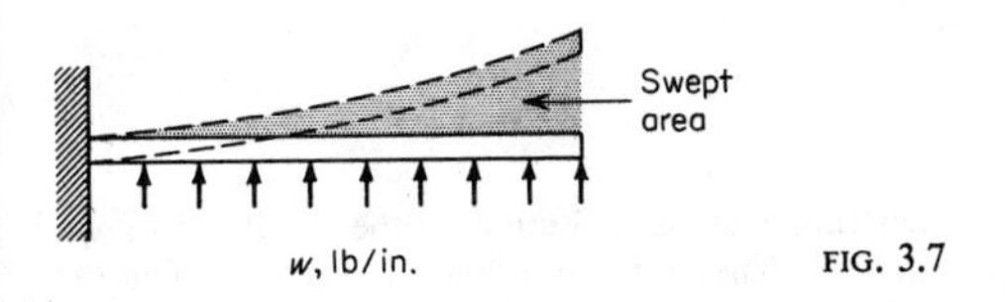

FIG. 3.7

3.7 Using the principle of complementary virtual work shown that if m is the distributed moment on a uniform cantilever beam (see Fig. 3.8) with constant temperature, the tip deflection is given by $\partial U_i^*/\partial m$.

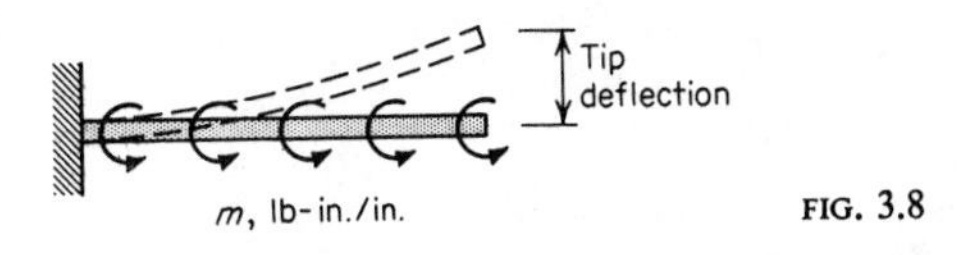

FIG. 3.8

3.8 The beam element shown in Fig. 3.9 is subjected to applied moments M_1 and M_2 and a temperature distribution T, which is constant along the length. The applied moments are equilibrated by the shear reactions at the two ends. Show that if the shear deformations are taken into account, the rotation θ_1 in the direction of M_1 is given by

$$\theta_1 = \frac{l}{12EI}(4 + \Phi)M_1 + \frac{l}{12EI}(2 - \Phi)M_2 - \frac{1}{2}\frac{\alpha l}{I}\int_A Ty\, dA$$

where l = beam length

EI = flexural stiffness (constant)

Φ = $12EI/GA_s l^2$-shear parameter

A_s = cross-sectional area effective in shear

G = shear modulus

T = temperature (function of y only)

y = distance from neutral axis of the beam

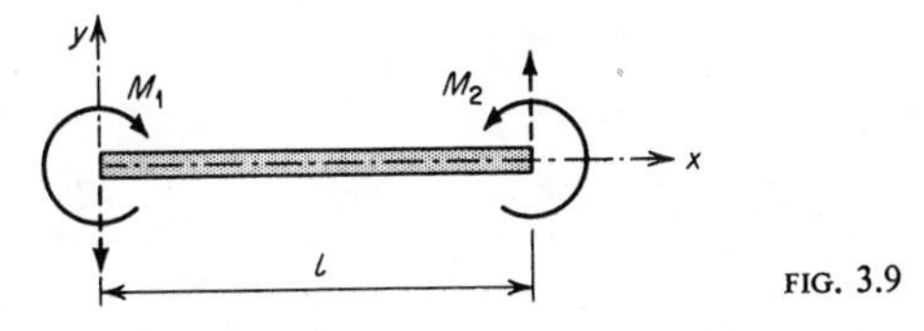

FIG. 3.9

3.9 Calculate the tip deflection of a uniform cantilever beam subjected to uniform loading w lb/in., as shown in Fig. 3.10. The effects of shear deformations must be included. The flexural stiffness is EI, the shear modulus is G, and the effective cross-sectional area in shear is A_s.

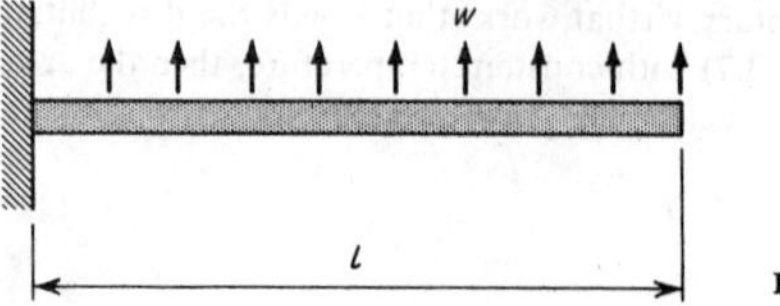

FIG. 3.10

3.10 Apply the unit-load theorem to determine lateral deflection on the cantilever shown in Fig. 3.11 at a distance b from the built-in end. The cantilever is loaded by a force W at the tip. Effects of shear deformations must be included. The stiffness and geometrical data are the same as in Prob. 3.9.

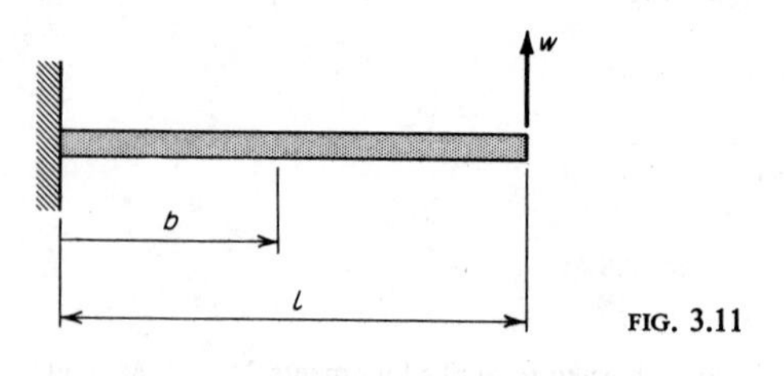

FIG. 3.11

CHAPTER 4
STRUCTURAL IDEALIZATION

The most important step in matrix structural analysis is the formulation of a discrete-element mathematical model equivalent to the actual continuous structure. This model is necessary in order to have a system with a finite number of degrees of freedom upon which matrix algebra operations can be performed. The formulation of such a model, usually referred to as structural idealization, is accomplished essentially by equating energies of the continuous and discrete element systems. For some types of structures this energy equivalence leads to the exact representation by the discrete systems; for others we are forced to use approximate representations. Structures which are already made up from elements with discrete attachments, such as trusses or rigid jointed frames, present no difficulty in the formulation of their discrete models. If the elements are made up with fictitious boundaries and attachments, exact discrete-element representations are not possible, and we must resort to the use of assumed stress or displacement distributions within the elements. The assumed distributions must be such that when the size of elements is decreased, the matrix solutions for the stresses and displacements must tend to the exact values for the continuous system. In this chapter the underlying principles of structural idealization are discussed. Also the most commonly used idealized structural elements are described, with special emphasis on those elements whose elastic and inertia matrix properties are derived in subsequent chapters.

4.1 STRUCTURAL IDEALIZATION

Most engineering structures consist of an assembly of different *structural elements* connected together either by discrete or by continuous attachments.

The size of the elements may vary; at the one end of the scale a single element may represent a stringer extending over one frame bay in a fuselage shell, while on the other end such large components as a complete wing structure may also be regarded as elements. In the latter case, however, because of the size and complexity of the structure it is customary to refer to such elements as *substructures*.

If the structural elements on the actual structure are connected together by discrete joints, the necessary interaction between individual unassembled elements is introduced as joint forces or displacements. Pin-jointed trusses or rigidly connected frames are typical examples of structures made up from components with discrete joints. The interaction forces between the various elements in a framework structure are represented by discrete-joint forces (shear and axial forces), bending moments, and torques. For these cases the elastic characteristics of structural elements can be determined accurately using elementary theories of bending and torsion, particularly for elements which are long compared with their cross-sectional dimensions. If the assembled framework structure is statically determinate, the equations of statics, i.e., the equations of equilibrium in terms of forces, are sufficient to determine all the joint forces, bending moments, and torques. For statically indeterminate (redundant) structures the equations of equilibrium are insufficient in number to determine all the unknown forces, and therefore these equations must be supplemented by the equations of compatibility. Alternatively, the equations of equilibrium can be reformulated in terms of displacements, in which case there will always be a sufficient number of equations to determine the unknown displacements (deflections and rotations).

In the case of a framework structure the transition from the differential equations of continuum mechanics (elasticity) into a set of algebraic (matrix) operations can be easily interpreted. The differential equations for elasticity for each structural element can be solved initially in terms of the boundary values. This gives the element force-displacement relationship or element displacement-force relationship, depending on the method of solution. The satisfaction of boundary conditions on adjacent elements leads then directly to a set of algebraic equations which are used to determine the unknown force and/or displacement boundary values. Thus it is clear that structures comprising elements with discrete attachments (see Fig. 4.1) lend themselves ideally to a structural analysis based on algebraic operations on matrix equations in which the element forces and displacements may be identified as specific matrices.

The discrete-element analysis of structural elements which are continuously attached presents some difficulties. For such elements there is no longer one-to-one correspondence between the element forces used in the matrix analysis and the forces in the actual structure, and consequently the analyst must exercise great care in interpreting the results.

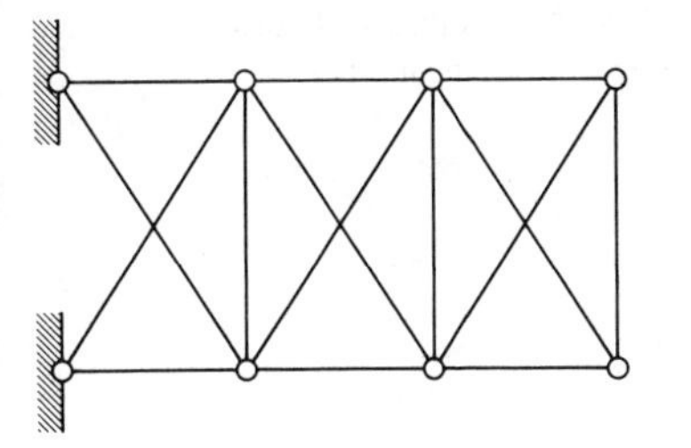

FIG. 4.1 Example of a discrete-attachment structure: pin-jointed truss.

Continuously attached structural elements are used in many applications; e.g., aircraft or missile skin panels are attached to other components by bolts or rivets or are welded to them. Also, in order to obtain a detailed stress distribution in a large panel attached continuously to the surrounding structure it may be necessary to subdivide each panel into a number of smaller panels by a system of grid lines, the intersections of which are referred to as *nodes* or *nodal points*. The smaller panels can then be regarded as structural elements attached continuously to other surrounding panels (see Fig. 4.2). Furthermore, a somewhat similar idealization can be used for analysis of solid bodies, where the body may be regarded as an assembly of solid tetrahedra, and each tetrahedron is treated as a structural element.

In continuously attached panels the main difficulty lies in the physical interpretation of the element forces, since the mechanism of modeling the continuum mechanics by discrete-element techniques is by no means a simple procedure. First, the displacements of node points are interpreted as the actual displacements of the corresponding points on the structure. Second, the varying stress field in the element must be replaced by an equivalent set of element discrete forces. These element forces, unlike the case of structural elements with discrete attachments, have no physical counterpart in the actual structure, and therefore they must be regarded only as fictitious forces introduced by the structural idealization. Figure 4.3 shows a thin-walled shell idealized into an aggregate of triangular panels. In this case the element forces would be represented by fictitious concentrated forces and moments at the nodal points formed by the intersection of the grid lines outlining the triangular plate elements. Another example is shown in Fig. 4.4. A sweptwing root

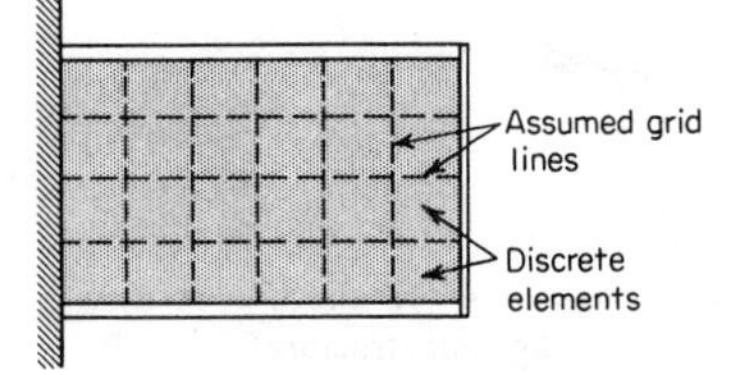

FIG. 4.2 Example of a continuous-attachment structure: two-flange spar.

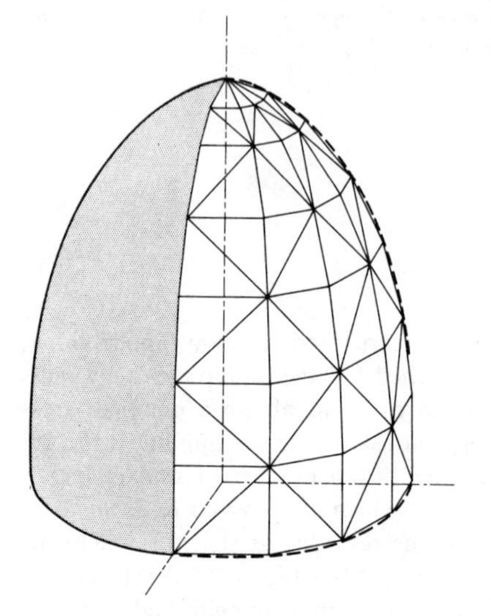

FIG. 4.3 Thin-walled shell idealized into an aggregate of triangular panels.

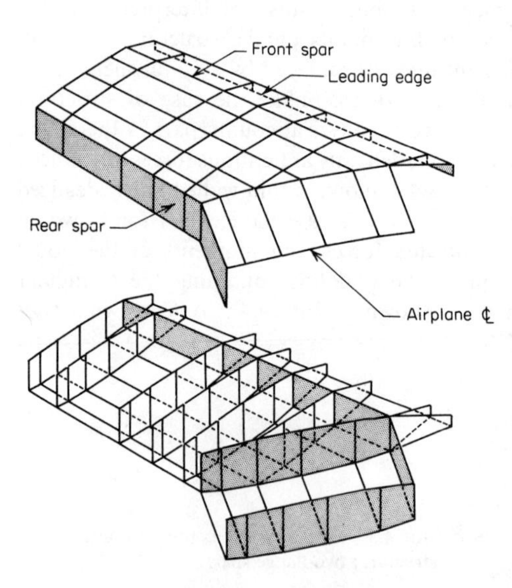

FIG. 4.4 Idealization of a swept-wing root structure.

structure is idealized into panel elements representing the skins, ribs, and spar webs. Although not shown, the actual idealization would also include any reinforcing members, such as spar caps, rib-to-skin attachment angles, etc. Here fictitious concentrated forces in the plane of each panel and acting at points of spar and web intersections would be used as the element forces.

It is evident from the previous discussion that the structural idealization is simply a process whereby a complex structure is reduced into an assembly of discrete structural elements. The elastic and inertia properties of these elements must first be established before we can proceed with the static or dynamic analyses of the system.

4.2 ENERGY EQUIVALENCE

The basis for determining elastic and inertia properties of the idealized structural element is the equivalence of strain and kinetic energies of the actual continuous element and its equivalent discrete model. To illustrate this method we shall consider a rectangular plate element formed by the intersection of grid lines in a flat panel subjected to some general loading, as shown in Fig. 4.5. We assume first that the element properties will be specified for twelve deflections and eight rotations. For convenience, both the deflections and rotations will be referred to as element displacements. The idealized element has a stress distribution carried over from the surrounding structure and, in addition, is also subjected to surface forces $\mathbf{\Phi}$, body forces $\mathbf{X}$, and temperature T.

The first step in determining the properties of this idealized element is to assume that the interior displacements $\mathbf{u} = \{u_x \quad u_y \quad u_z\}$ are expressible in terms

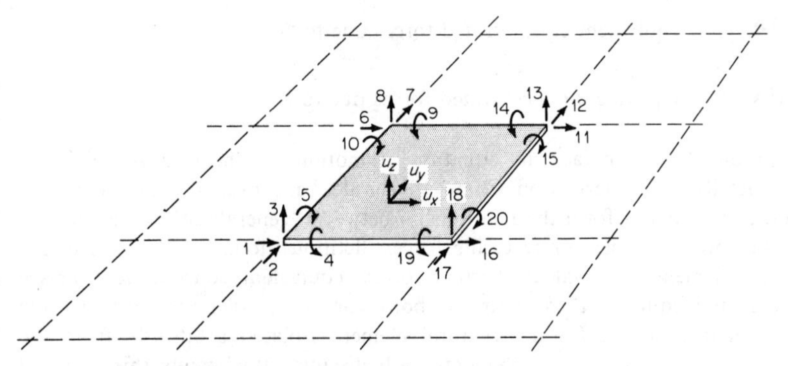

FIG. 4.5 Typical idealized element in a flat plate.

of the discrete displacements $\mathbf{U} = \{U_1 \quad U_2 \quad \cdots \quad U_{20}\}$ by the matrix equation

$$\mathbf{u} = \mathbf{aU} \tag{4.1}$$

where $\mathbf{a} = \mathbf{a}(x,y,z)$ is a function of the position coordinates. Clearly, such a relationship can be valid only for static stress distributions and for elements attached to the surrounding structure by means of discrete joints. Only then will the interior displacements be proportional to $\mathbf{U}$, and $\mathbf{a}$ will be a unique matrix. For continuously attached elements, as in Fig. 4.5, the matrix $\mathbf{a}$ is not unique, since it depends on the global stress distribution in the panel. Furthermore, for general dynamic problems the matrix $\mathbf{a}$ does not exist, even for elements with discrete attachments. This means that, in general, we can expect only approximate expressions for the matrix $\mathbf{a}$.

The total strains $\mathbf{e}$ can always be obtained by differentiation of Eq. (4.1) leading to the matrix equation

$$\mathbf{e} = \mathbf{bU} \tag{4.2}$$

Using virtual-work and d'Alembert's principles, virtual work can be equated to the virtual strain energy, and the following equation is obtained for the element:

$$\mathbf{m}\ddot{\mathbf{U}} + \mathbf{kU} = \mathbf{P} - \mathbf{Q} + \int_s \mathbf{a}^T\mathbf{\Phi}\, dS + \int_v \mathbf{a}^T\mathbf{X}\, dV \tag{4.3}$$

where $\mathbf{m} = \int_v \rho\mathbf{a}^T\mathbf{a}\, dV =$ mass matrix

$\mathbf{k} = \int_v \mathbf{b}^T\boldsymbol{\varkappa}\mathbf{b}\, dV =$ stiffness matrix

$\mathbf{P} = \{P_1 \quad P_2 \quad \cdots\} =$ element concentrated forces

$\mathbf{Q} = \int_v \mathbf{b}^T\boldsymbol{\varkappa}_T\alpha T\, dV =$ equivalent thermal forces

$\int_s \mathbf{a}^T\mathbf{\Phi}\, dS =$ equivalent concentrated forces due to $\mathbf{\Phi}$

$\int_v \mathbf{a}^T\mathbf{X}\, dV =$ equivalent concentrated forces due to $\mathbf{X}$

Equation (4.3) is in fact the equation of motion for the idealized element. For details of its derivation Chap. 10 should be consulted. This equation indicates in matrix form the relationship between "generalized" forces such as inertia forces, stiffness or restoring forces, fictitious nodal forces from neighboring elements, equivalent thermal forces, equivalent concentrated surface forces, and equivalent concentrated body forces. It therefore provides the mathematical model for a structural element isolated arbitrarily from the continuous structure. For elements with discrete attachments this idealized model is exact for static loading; however, for all other cases only approximate representations are possible.

For elements with discrete attachments the matrix **a** is found from the solution of the boundary-value elasticity problem for specified boundary displacements. For continuously attached elements we usually determine **a** by assuming the form of the displacement distribution. Specifically, a linear combination of different continuous functions with arbitrary multiplying constants is selected for the displacements. If the number of constants is equal to the number of discrete displacements, these constants can be determined directly in terms of the element displacements; however, if this number is greater than the number of displacements, we must use the condition of minimum potential energy to derive additional equations for the constants[253] (see Sec. 5.13).

When selecting the displacement functions it is desirable, although not necessary, to ensure compatibility of deflections and slopes on the boundaries of adjacent elements. This requirement is satisfied provided the distribution of deflections and slopes on a boundary depends only on the discrete displacements at the end points defining that particular boundary.

The matrix **a** can also be determined from an assumed stress distribution, but this involves integration of the strain-displacement equations to obtain the required displacement relations. In the force method of analysis the flexibility properties of elements can also be calculated directly from an assumed stress distribution satisfying the equation of stress equilibrium. Alternatively, flexibility properties are obtained by inverting stiffness matrices.

The matrix displacement and force methods yield identical results if the assumptions in deriving the discrete-element properties are identical in both methods. Different assumptions in deriving the element properties may also be used in the displacement and force methods of analysis, so that the two different results bracket the true results by providing upper and lower bounds. For example, an equilibrium noncompatible stress field in elements would overestimate the overall structural flexibility, while a compatible nonequilibrium field would underestimate the flexibility.

As already mentioned before, the assumed displacement functions must be continuous and should preferably satisfy compatibility of deflections and slopes on the boundaries. The satisfaction of the stress-equilibrium equation is also desirable. The choice of the best assumed deflection form is not an easy one. Melosh[226] has proposed a criterion for the monotonic convergence of the discrete-element solution to be used for selecting element properties. It appears, however, that such a criterion may perhaps be too restrictive. Bazeley et al.[35] postulated that the only requirement on the assumed functions is that they be able to represent rigid-body translations and rotations and constant strain distributions. More work is still required in order to select the best deflection and/or stress distributions for the calculation of properties of the idealized elements.

4.3 STRUCTURAL ELEMENTS

BAR ELEMENT

The pin-jointed bar is the simplest structural element. It requires two element forces for the displacement method (DM) and only one element force for the force method (FM). This element has no bending stiffness, and it carries only a one-dimensional stress distribution. In Fig. 4.6 and in all subsequent figures in this section the element forces for the FM are indicated by solid arrows while the corresponding reaction forces are shown as dashed-line arrows.

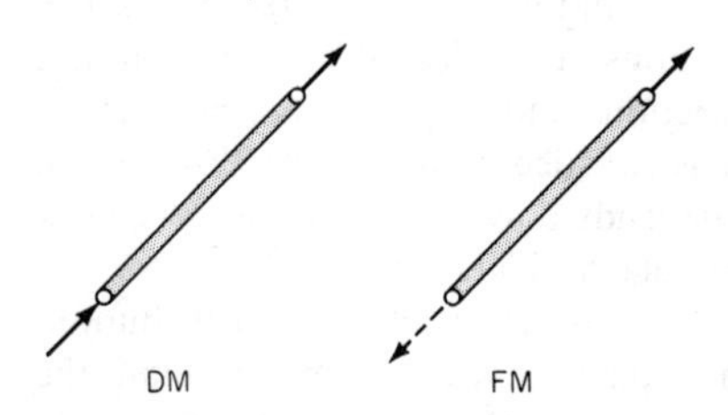

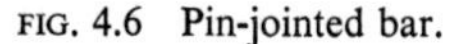
FIG. 4.6 Pin-jointed bar.

BEAM ELEMENT

For the DM we require four shear forces, two axial forces, four bending moments, and two torsional moments. All forces are acting at the ends of the beam. For the FM we require two shear forces, one axial force, two bending moments and one torsional moment (see Fig. 4.7). Other choices of reactions and element forces are possible, and it is largely the matter of personal preference which particular system is used.

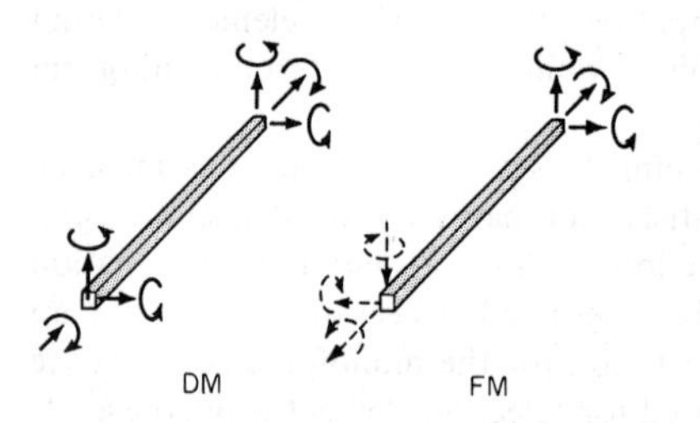

FIG. 4.7 Beam element.

TRIANGULAR PLATE (IN-PLANE FORCES)

The in-plane and bending deformations are uncoupled for small deflections, and consequently elastic properties can be evaluated separately for the in-plane and out-of-plane forces. In the DM we use two forces at each vertex of the triangle, while in the FM it is preferable to select three sets of edge forces, as

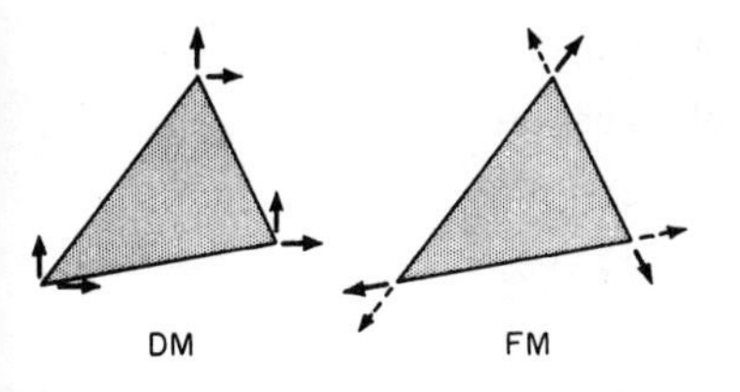

FIG. 4.8 Triangular plate (in-plane forces).

indicated in Fig. 4.8. Other choices of element forces in the FM are also possible.

RECTANGULAR PLATE (IN-PLANE FORCES)

No. of element forces in DM = 8
No. of element forces in FM = 5
As in the case of triangular plate elements, it is preferable to use edge-force systems for the FM. Four edge-force systems plus a diagonal system may be used.

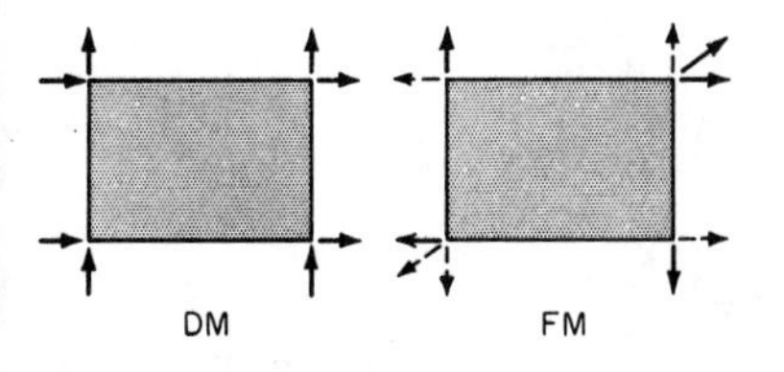

FIG. 4.9 Rectangular plate (in-plane forces).

TETRAHEDRON

No. of element forces in DM = 12
No. of element forces in FM = 6
Here again the edge-force systems may be used in the FM.

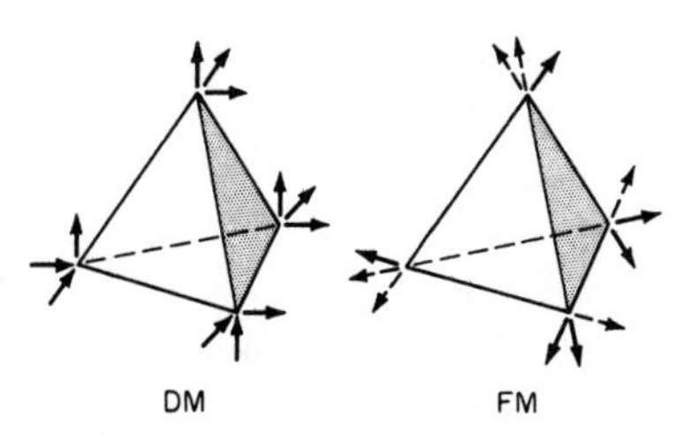

FIG. 4.10 Solid tetrahedron.

TRIANGULAR PLATE (BENDING FORCES)

Considering only the bending forces (out-of-plane forces) we have
No. of element forces in DM = 9
No. of element forces in FM = 6

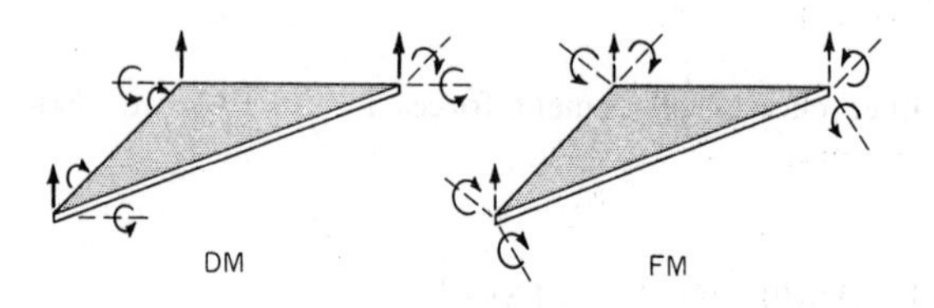

FIG. 4.11 Triangular plate (bending forces).

RECTANGULAR PLATE (BENDING FORCES)

For the bending forces only we have
No. of element forces in DM = 12
No. of element forces in FM = 9

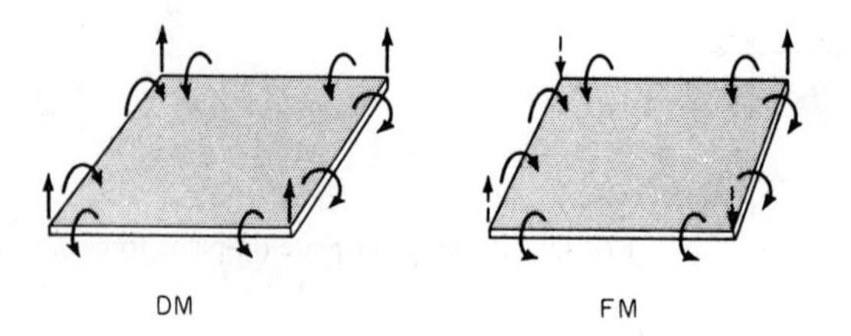

FIG. 4.12 Rectangular plate (bending forces).

HRENNIKOFF'S PLATE MODEL

For rectangular plates in bending Hrennikoff[143] introduced an idealization based on four edge beams and two diagonal beams having cross-sectional properties derived from the plate stiffness. Application of this idealized element in matrix methods has been limited.

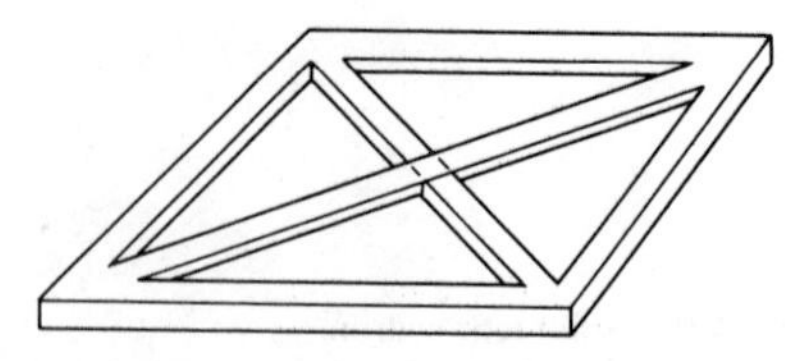

FIG. 4.13 Hrennikoff's model for plate in bending.

CONSTANT-SHEAR-FLOW PANEL

CONTINUOUS ATTACHMENTS: The constant-shear-flow-panel idealization has been used extensively in the stress analysis of stressed-skin aircraft structures. It generally gives excellent results provided the effects of Poisson's ratio can be neglected.
No. of element forces in DM = 4
No. of element forces in FM = 1

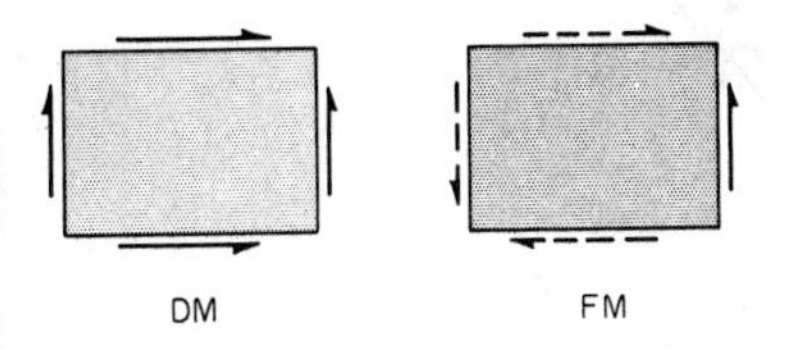

FIG. 4.14 Constant-shear-flow panel: continuous attachments.

MIDPOINT ATTACHMENTS: The constant shear flows can be replaced by shear forces applied at the midpoints of the panel sides. This idealization can then be used in conjunction with the pin-jointed bar elements to represent normal stress capability of the panel.
No. of element forces in DM = 4
No. of element forces in FM = 1

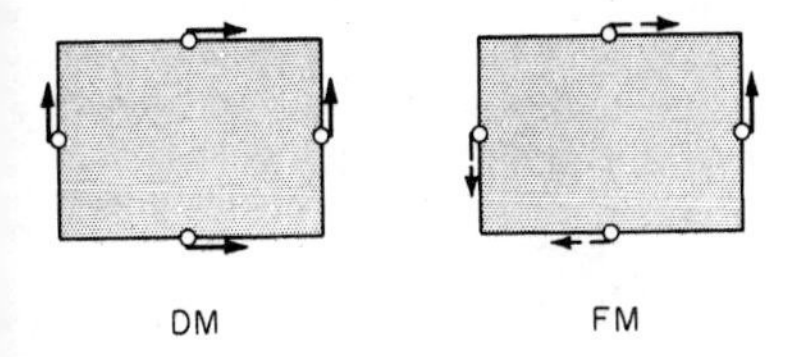

FIG. 4.15 Constant-shear-flow panel; midpoint attachments.

AXIAL-FORCE ELEMENT

This element is used in conjunction with the constant-shear-flow panel to represent the normal stress capability of the panel.
No. of element forces in DM = 3
No. of element forces in FM = 2

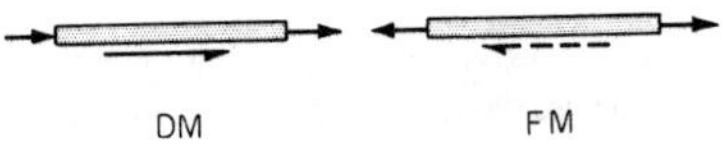

DM FM

FIG. 4.16 Axial-force member.

AXISYMMETRICAL SHELL ELEMENT

The element consists of a truncated cone with distributed edge forces and moments as the element forces. Tangential forces are also included.
No. of element forces in DM = 8
No. of element forces in FM = 6

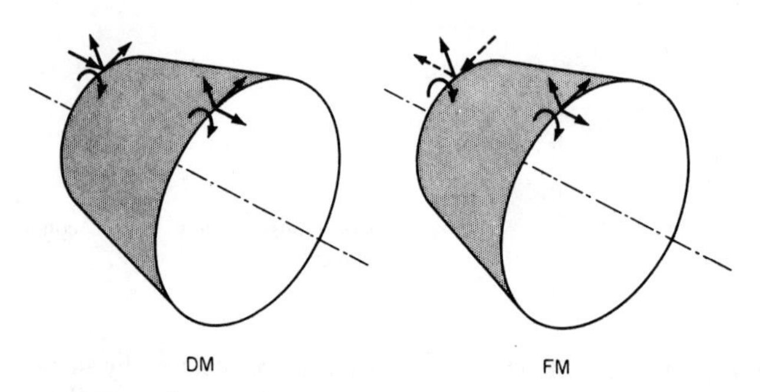

FIG. 4.17 Axisymmetrical shell element; axisymmetrical loading including torsion.

AXISYMMETRICAL TRIANGULAR RING

This element is used in the analysis of axisymmetrical solid rings. Tangential forces are not included.
No. of element forces in DM = 6
No. of element forces in FM = 5

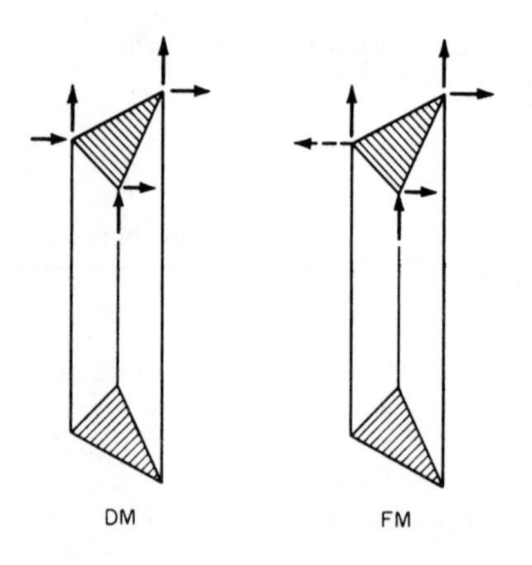

FIG. 4.18 Axisymmetrical ring element; axisymmetrical loading excluding torsion.

CHAPTER 5 STIFFNESS PROPERTIES OF STRUCTURAL ELEMENTS

We have seen in the previous chapter that in order to use matrix methods of structural analysis, the structure is first idealized into an assembly of structural elements and that these elements are attached to the adjacent elements at node points, which may be either the actual joints or fictitious points obtained by the intersecting grid lines. To determine stiffness characteristics of the entire assembled structure, which are required in the analysis, we must find stiffness properties of individual unassembled elements. In this chapter stiffness properties are developed for the following elements: pin-jointed bars, beams, triangular plates, rectangular plates, quadrilateral plates, and solid tetrahedra. Stiffness properties of the linearly varying axial-load element and constant-shear-flow panel are presented in Chap. 6. For other types of elements the general principles discussed in this chapter may be used to derive the required stiffness properties.

5.1 METHODS OF DETERMINING ELEMENT FORCE-DISPLACEMENT RELATIONSHIPS

The fundamental step in the application of the matrix displacement method is the determination of the stiffness characteristics of structural elements into which the structure is idealized for the purpose of the analysis. A number of alternative methods are available for the calculation of force-displacement relationships describing the stiffness characteristics of structural elements, and the choice of a particular method depends mainly on the type of element.

The following methods can be used:

1. Unit-displacement theorem
2. Castigliano's theorem (part I)
3. Solution of differential equations for the element displacements
4. Inversion of the displacement-force relationships

Of these methods, the first one, the application of the unit-displacement theorem, is undoubtedly the most convenient since it leads directly to the required matrix equation relating element forces to their corresponding displacements. In the second method, based on a direct application of Castigliano's theorem (part I), the strain energy is first calculated in terms of element displacements and temperature and then is differentiated with respect to a selected displacement. The differentiation is repeated for each element displacement in turn, and this generates a complete set of force-displacement equations, which can then be formulated in matrix notation. In the third method the solutions of the differential equations for displacements are used to derive the required stiffness relationships. Naturally the application of this method is limited to structural elements for which solutions for displacements are available. In the fourth method, the equations for the displacement-force relationships are determined first, and then these equations are inverted to find force-displacement relationships; however, the force-displacement relationships thus obtained must subsequently be modified to include the rigid-body degrees of freedom. This method is discussed in Chap. 6.

5.2 DETERMINATION OF ELEMENT STIFFNESS PROPERTIES BY THE UNIT-DISPLACEMENT THEOREM

We shall consider an elastic element (Fig. 5.1) subjected to a set of n forces

$$\mathbf{S} = \{S_1 \quad S_2 \quad \cdots \quad S_i \quad S_j \quad \cdots \quad S_n\} \tag{5.1}$$

and some specified temperature distribution

$$T = T(x,y,z) \tag{5.2}$$

The displacements corresponding to the forces $\mathbf{S}$ will be denoted by the column matrix

$$\mathbf{u} = \{u_1 \quad u_2 \quad \cdots \quad u_i \quad u_j \quad \cdots \quad u_n\} \tag{5.3}$$

To determine a typical force S_i we can use the unit-displacement theorem. Hence

$$S_i = \int_v \underline{\boldsymbol{\epsilon}}_i^T \boldsymbol{\sigma} \, dV \tag{5.4}$$

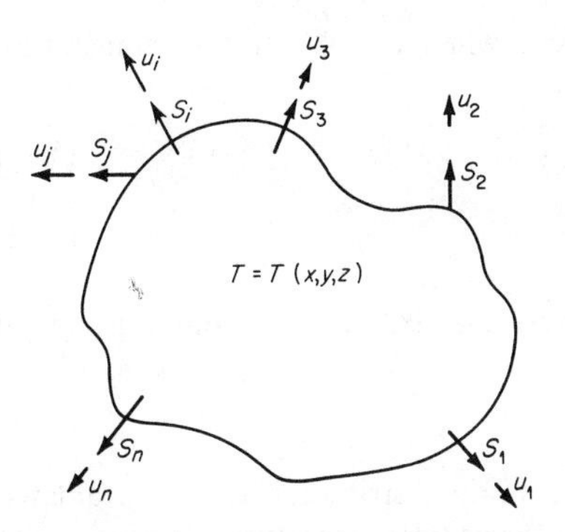

FIG. 5.1 Elastic body or structure subjected to static loading.

where $\underline{\boldsymbol{\epsilon}}_i$ represents the matrix of compatible strains due to a unit displacement in the direction of S_i and $\boldsymbol{\sigma}$ is the exact stress matrix due to the applied forces **S** and the temperature distribution T. The unit displacements can be applied in turn at all points where the forces are impressed, and hence

$$\mathbf{S} = \int_v \underline{\boldsymbol{\epsilon}}^T \boldsymbol{\sigma}\, dV \tag{5.5}$$

where $$\underline{\boldsymbol{\epsilon}} = [\underline{\boldsymbol{\epsilon}}_1 \quad \underline{\boldsymbol{\epsilon}}_2 \quad \cdots \quad \underline{\boldsymbol{\epsilon}}_i \quad \underline{\boldsymbol{\epsilon}}_j \quad \cdots \quad \underline{\boldsymbol{\epsilon}}_n] \tag{5.6}$$

Since we are dealing with a linear system, the total strains **e** must be expressed by the relationship

$$\mathbf{e} = \mathbf{bu} \tag{5.7}$$

where **b** represents a matrix of the *exact* strains due to unit displacements **u.** Substituting Eq. (5.7) into (2.14), we have

$$\boldsymbol{\sigma} = \boldsymbol{\varkappa}\mathbf{bu} + \boldsymbol{\varkappa}_T \alpha T \tag{5.8}$$

Hence from Eqs. (5.5) and (5.8) the element force-displacement relationship becomes

$$\mathbf{S} = \int_v \underline{\boldsymbol{\epsilon}}^T \boldsymbol{\varkappa}\mathbf{b}\, dV\, \mathbf{u} + \int_v \underline{\boldsymbol{\epsilon}}^T \boldsymbol{\varkappa}_T \alpha T\, dV \tag{5.9}$$

or $$\mathbf{S} = \mathbf{ku} + \mathbf{Q} \tag{5.10}$$

where $$\mathbf{k} = \int_v \underline{\boldsymbol{\epsilon}}^T \boldsymbol{\varkappa}\mathbf{b}\, dV \tag{5.11}$$

represents the element stiffness matrix and

$$\mathbf{Q} = \int_v \underline{\boldsymbol{\epsilon}}^T \boldsymbol{\varkappa}_T \alpha T\, dV \tag{5.12}$$

represents thermal forces on the element when $\mathbf{u} = \mathbf{0}$. If the temperature throughout the element is constant, then

$$\mathbf{Q} = \mathbf{h}\alpha T \tag{5.13}$$

where $$\mathbf{h} = \int_v \underline{\boldsymbol{\epsilon}}^T \boldsymbol{\varkappa}_T \, dV \tag{5.14}$$

may be described as the thermal stiffness matrix. Hence from Eqs. (5.10) and (5.13) we have

$$\mathbf{S} = \mathbf{k}\mathbf{u} + \mathbf{h}\alpha T \tag{5.15}$$

The matrix $\underline{\boldsymbol{\epsilon}}$ representing compatible strain distribution can be evaluated without any appreciable difficulties, even for complex structural elements. On the other hand, evaluation of the matrix $\mathbf{b}$, representing exact strain distributions, is often exceedingly difficult, if not impossible. In cases for which no exact strain distribution can be found approximate procedures must be used. This requires determination of approximate functional relationships between strains and displacements. Naturally, the degree of approximation then depends on the extent to which the equations of equilibrium and compatibility are satisfied. One possible approach is to select the matrix $\mathbf{b}$ in such a way that it will satisfy only the equations of compatibility. Denoting this approximate matrix by $\underline{\mathbf{b}}$, and noting that $\underline{\boldsymbol{\epsilon}} = \underline{\mathbf{b}}$, we obtain from Eq. (5.11)

$$\mathbf{k} \simeq \int_v \underline{\mathbf{b}}^T \boldsymbol{\varkappa} \underline{\mathbf{b}} \, dV \tag{5.16}$$

For convenience in subsequent analysis Eq. (5.14) may be rewritten as

$$\mathbf{h} = \int_v \underline{\mathbf{b}}^T \boldsymbol{\varkappa}_T \, dV \tag{5.17}$$

A typical application of Eqs. (5.16) and (5.17) will be illustrated on the pin-jointed bar shown in Fig. 5.2. It can be shown that the displacement u_x

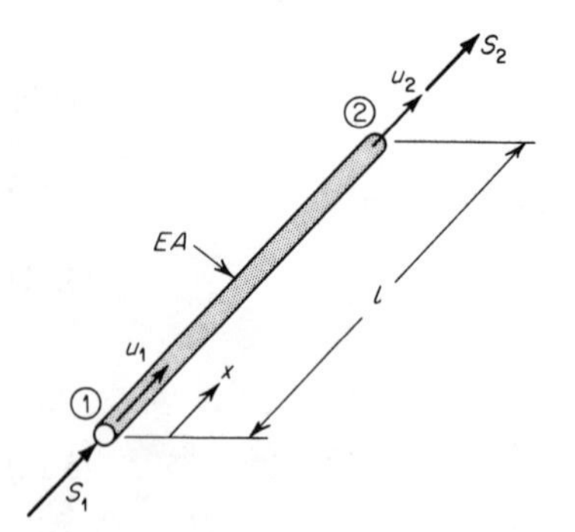

FIG. 5.2 Pin-jointed bar element.

along the longitudinal axis of the bar is given by

$$u_x = u_1 + (u_2 - u_1)\frac{x}{l} \tag{5.18}$$

where u_1 and u_2 are the end displacements at $x = 0$ and $x = l$, respectively. From Eq. (5.18) it follows that

$$\begin{aligned} e_{xx} &= \frac{\partial u_x}{\partial x} = \frac{1}{l}(u_2 - u_1) \\ &= \frac{1}{l}[-1 \quad 1]\begin{bmatrix} u_1 \\ u_2 \end{bmatrix} \end{aligned} \tag{5.19}$$

It should be noted that in this case the strain distribution given by Eq. (5.19) is not only compatible but also exact. Hence

$$\underline{\mathbf{b}} = \mathbf{b} = \frac{1}{l}[-1 \quad 1] \tag{5.20}$$

Since the pin-jointed bar is a one-dimensional element for which

$$\boldsymbol{\varkappa} = E \tag{2.54}$$

$$\boldsymbol{\varkappa}_T = -E \tag{2.55}$$

the stiffness matrix $\mathbf{k}$, determined from Eq. (5.16), becomes

$$\begin{aligned} \mathbf{k} &= \int_0^l \frac{1}{l}\begin{bmatrix} -1 \\ 1 \end{bmatrix}\frac{E}{l}[-1 \quad 1]A\,dx \\ &= \frac{AE}{l}\begin{bmatrix} 1 & -1 \\ -1 & 1 \end{bmatrix} \end{aligned} \tag{5.21}$$

where A represents the cross-sectional area of the bar. Similarly, substituting Eqs. (5.20) and (2.55) into (5.17) gives

$$\mathbf{h} = -\int_0^l \frac{1}{l}\begin{bmatrix} -1 \\ 1 \end{bmatrix} EA\,dx = EA\begin{bmatrix} 1 \\ -1 \end{bmatrix} \tag{5.22}$$

The complete force-displacement relationship can now be written as

$$\begin{bmatrix} S_1 \\ S_2 \end{bmatrix} = \frac{AE}{l}\begin{bmatrix} 1 & -1 \\ -1 & 1 \end{bmatrix}\begin{bmatrix} u_1 \\ u_2 \end{bmatrix} + AE\alpha T\begin{bmatrix} 1 \\ -1 \end{bmatrix} \tag{5.23}$$

If $u_1 = u_2 = 0$ and the element is subjected to a temperature change T, then $S_1 = AE\alpha T$, and $S_2 = -AE\alpha T$, which according to the sign convention in Fig. 5.2 implies that both forces are compressive.

5.3 APPLICATION OF CASTIGLIANO'S THEOREM (PART I) TO DERIVE STIFFNESS PROPERTIES

Applying Castigliano's theorem (part I), as given by Eq. (3.64), to the structural element in Fig. 5.1, we have

$$S_i = \left(\frac{\partial U_i}{\partial u_i}\right)_{T=\text{const}} \tag{5.24}$$

By varying the subscript i from 1 to n we obtain a set of n equations relating the element forces S_i to their corresponding displacements u_i. Symbolically this can be represented by

$$\mathbf{S} = \left(\frac{\partial U_i}{\partial \mathbf{u}}\right)_{T=\text{const}} \tag{5.25}$$

where the strain energy U_i must of course be expressed in terms of $\mathbf{u}$.

Using now the results of Chap. 3, we have

$$\begin{aligned} U_i &= \frac{1}{2}\int_v \boldsymbol{\epsilon}^T \boldsymbol{\sigma}\, dV \\ &= \frac{1}{2}\int_v (\mathbf{e}^T - \mathbf{e}_T{}^T)(\boldsymbol{\varkappa}\mathbf{e} + \boldsymbol{\varkappa}_T \alpha T)\, dV \\ &= \frac{1}{2}\int_v (\mathbf{e}^T\boldsymbol{\varkappa}\mathbf{e} + \mathbf{e}^T\boldsymbol{\varkappa}_T\alpha T - \mathbf{e}_T{}^T\boldsymbol{\varkappa}\mathbf{e} - \mathbf{e}_T{}^T\boldsymbol{\varkappa}_T\alpha T)\, dV \end{aligned} \tag{5.26}$$

From Hooke's law

$$\boldsymbol{\sigma} = \boldsymbol{\varkappa}\mathbf{e} + \boldsymbol{\varkappa}_T \alpha T$$

it follows that

$$\begin{aligned} \mathbf{e} &= \boldsymbol{\varkappa}^{-1}\boldsymbol{\sigma} - \boldsymbol{\varkappa}^{-1}\boldsymbol{\varkappa}_T\alpha T \\ &= \boldsymbol{\epsilon} + \mathbf{e}_T \end{aligned}$$

Hence $\quad \boldsymbol{\varkappa}\mathbf{e}_T = -\boldsymbol{\varkappa}_T \alpha T \qquad (5.27)$

Substituting now Eq. (5.27) into (5.26) and noting that U_i is a 1×1 matrix, we obtain

$$U_i = \frac{1}{2}\int_v (\mathbf{e}^T\boldsymbol{\varkappa}\mathbf{e} + 2\mathbf{e}^T\boldsymbol{\varkappa}_T\alpha T + \mathbf{e}_T{}^T\boldsymbol{\varkappa}\mathbf{e}_T)\, dV \tag{5.28}$$

When we use the strain-displacement relationship

$$\mathbf{e} = \mathbf{b}\mathbf{u} \tag{5.7}$$

the strain-energy expression becomes

$$U_i = \frac{1}{2}\int_v \mathbf{u}^T\mathbf{b}^T\boldsymbol{\varkappa}\mathbf{b}\mathbf{u}\, dV + \int_v \mathbf{u}^T\mathbf{b}^T\boldsymbol{\varkappa}_T\alpha T\, dV + \frac{1}{2}\int_v \mathbf{e}_T{}^T\boldsymbol{\varkappa}\mathbf{e}_T\, dV \tag{5.29}$$

Hence

$$\mathbf{S} = \left(\frac{\partial U_i}{\partial \mathbf{u}}\right)_{T=\text{const}} = \int_v \mathbf{b}^T \boldsymbol{\varkappa} \mathbf{b} \, dV \, \mathbf{u} + \int_v \mathbf{b}^T \boldsymbol{\varkappa}_T \alpha T \, dV \tag{5.30}$$

which agrees with the result obtained by the unit-displacement theorem when $\mathbf{b} = \underline{\mathbf{b}}$.

5.4 TRANSFORMATION OF COORDINATE AXES: λ MATRICES

In order to determine the stiffness property of the complete structure, a common datum must be established for all unassembled structural elements so that all the displacements and their corresponding forces will be referred to a common coordinate system. The choice of such a datum is arbitrary, and in practice it is best selected to correspond to the coordinate system used on the engineering drawings, from which coordinates of different points on the structure can easily be found.

Since the stiffness matrices **k** and **h** are initially calculated in local coordinates, suitably oriented to minimize the computing effort, it is necessary to introduce transformation matrices changing the frame of reference from a local to a datum coordinate system. The first step in deriving such a transformation is to obtain a matrix relationship between the element displacements **u** in the local system and the element displacement $\bar{\mathbf{u}}$ in the datum system. This relationship is expressed by the matrix equation

$$\mathbf{u} = \boldsymbol{\lambda} \bar{\mathbf{u}} \tag{5.31}$$

where $\boldsymbol{\lambda}$ is a matrix of coefficients obtained by resolving datum displacements in the directions of local coordinates. It will be shown later that the elements of $\boldsymbol{\lambda}$ are obtained from the direction cosines of angles between the local and datum coordinate systems.

If virtual displacements $\delta\bar{\mathbf{u}}$ are introduced on an element, then, from (5.31),

$$\delta\mathbf{u} = \boldsymbol{\lambda} \, \delta\bar{\mathbf{u}} \tag{5.32}$$

Since the resulting virtual work (a scalar quantity) must obviously be independent of the coordinate system, it follows that

$$\delta\bar{\mathbf{u}}^T \bar{\mathbf{S}} = \delta\mathbf{u}^T \mathbf{S} \tag{5.33}$$

where $\bar{\mathbf{S}}$ refers to forces in datum system corresponding to the displacements $\bar{\mathbf{u}}$. Substituting Eq. (5.32) into (5.33) gives

$$\delta\bar{\mathbf{u}}^T (\bar{\mathbf{S}} - \boldsymbol{\lambda}^T \mathbf{S}) = \mathbf{0} \tag{5.34}$$

and since $\delta\bar{\mathbf{u}}$'s are arbitrary, we must have that

$$\bar{\mathbf{S}} - \boldsymbol{\lambda}^T \mathbf{S} = \mathbf{0} \tag{5.35}$$

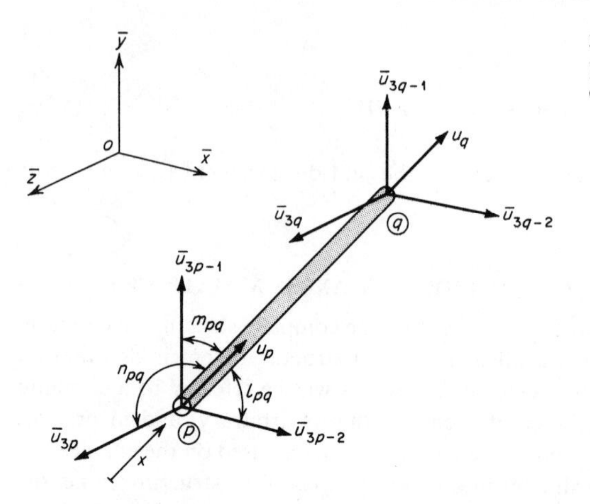

FIG. 5.3 Pin-jointed bar displacements in local and datum coordinate systems.

By use of Eqs. (5.15), (5.31), and (5.35) the following element force-displacement equation is obtained in datum system:

$$\bar{\mathbf{S}} = \bar{\mathbf{k}}\bar{\mathbf{u}} + \bar{\mathbf{h}}\alpha T = \bar{\mathbf{k}}\bar{\mathbf{u}} + \bar{\mathbf{Q}} \tag{5.36}$$

where

$$\bar{\mathbf{k}} = \boldsymbol{\lambda}^T\mathbf{k}\boldsymbol{\lambda} \tag{5.37}$$

$$\bar{\mathbf{h}} = \boldsymbol{\lambda}^T\mathbf{h} \tag{5.38}$$

$$\bar{\mathbf{Q}} = \boldsymbol{\lambda}^T\mathbf{h}\alpha T = \boldsymbol{\lambda}^T\mathbf{Q} \tag{5.39}$$

The formulation of the transformation matrix $\boldsymbol{\lambda}$ will be illustrated for a pin-jointed bar element orientated arbitrarily in space, as shown in Fig. 5.3. The displacements in local coordinates can be related to those in datum coordinates by the equations

$$\begin{aligned} u_p &= l_{pq}\bar{u}_{3p-2} + m_{pq}\bar{u}_{3p-1} + n_{pq}\bar{u}_{3p} \\ u_q &= l_{pq}\bar{u}_{3q-2} + m_{pq}\bar{u}_{3q-1} + n_{pq}\bar{u}_{3q} \end{aligned} \tag{5.40}$$

where l_{pq}, m_{pq}, and n_{pq} represent direction cosines of angles between the line pq and ox, oy, and oz directions, respectively. Equations (5.40) can be arranged in matrix notation as

$$\begin{bmatrix} u_p \\ u_q \end{bmatrix} = \begin{bmatrix} l_{pq} & m_{pq} & n_{pq} & 0 & 0 & 0 \\ 0 & 0 & 0 & l_{pq} & m_{pq} & n_{pq} \end{bmatrix} \begin{bmatrix} \bar{u}_{3p-2} \\ \bar{u}_{3p-1} \\ \bar{u}_{3p} \\ \bar{u}_{3q-2} \\ \bar{u}_{3q-1} \\ \bar{u}_{3q} \end{bmatrix} \tag{5.41}$$

Hence the transformation matrix $\boldsymbol{\lambda}$ is given by

$$\boldsymbol{\lambda} = \begin{bmatrix} l_{pq} & m_{pq} & n_{pq} & 0 & 0 & 0 \\ 0 & 0 & 0 & l_{pq} & m_{pq} & n_{pq} \end{bmatrix} \tag{5.42}$$

Substitution of Eq. (5.42) into (5.37) and (5.38) leads finally to

$$\bar{\mathbf{k}} = \frac{AE}{l}\begin{bmatrix} \mathbf{k}_0 & -\mathbf{k}_0 \\ -\mathbf{k}_0 & \mathbf{k}_0 \end{bmatrix} \tag{5.43}$$

where
$$\mathbf{k}_0 = \begin{bmatrix} l_{pq}^2 & l_{pq}m_{pq} & l_{pq}n_{pq} \\ m_{pq}l_{pq} & m_{pq}^2 & m_{pq}n_{pq} \\ n_{pq}l_{pq} & n_{pq}m_{pq} & n_{pq}^2 \end{bmatrix} \tag{5.44}$$

and
$$\bar{\mathbf{h}} = AE\{l_{pq} \quad m_{pq} \quad n_{pq} \quad -l_{pq} \quad -m_{pq} \quad -n_{pq}\} \tag{5.45}$$

Thus the matrix transformation given by Eq. (5.37) changes a 2×2 stiffness matrix $\mathbf{k}$ in a local coordinate system, measured along the length of the bar, into a 6×6 stiffness matrix $\bar{\mathbf{k}}$ in the datum system. Similarly, the transformation given by Eq. (5.38) changes a 2×1 matrix $\mathbf{h}$ into a 6×1 matrix $\bar{\mathbf{h}}$.

5.5 PIN-JOINTED BAR ELEMENTS

The stiffness matrix $\mathbf{k}$ and thermal force matrix $\mathbf{Q}$ for a pin-jointed bar element were derived in Sec. 5.2 using the unit-displacement theorem. An alternative method of deriving these matrices using Castigliano's theorem (part I) is illustrated in this section.

A pin-jointed bar (Fig. 5.2) is a one-dimensional element for which

$$e_{xx} = \frac{1}{l}[-1 \quad 1]\{u_1 \quad u_2\} \tag{5.19}$$

$$\sigma_{xx} = \frac{E}{l}[-1 \quad 1]\{u_1 \quad u_2\} \tag{5.46}$$

$$\boldsymbol{\varkappa} = E \tag{2.54}$$

$$\boldsymbol{\varkappa}_T = -E \tag{2.55}$$

$$\mathbf{e}_T = \alpha T \tag{5.47}$$

Therefore, from Eq. (5.28) it follows that

$$\begin{aligned} U_i &= \frac{1}{2}\int_v \left[E\frac{(u_2 - u_1)^2}{l^2} - 2\frac{u_2 - u_1}{l} E\alpha T + E\alpha^2 T^2 \right] dV \\ &= \frac{AE}{2l}[(u_2 - u_1) - \alpha T l]^2 \end{aligned} \tag{5.48}$$

Application of the Castigliano's theorem (part I) to Eq. (5.48) leads to equations for the element forces

$$S_1 = \frac{\partial U_i}{\partial u_1} = \frac{AE}{l}(u_1 - u_2) + AE\alpha T$$

$$S_2 = \frac{\partial U_i}{\partial u_2} = \frac{AE}{l}(-u_1 + u_2) - AE\alpha T$$

which can be combined into one matrix equation

$$\begin{bmatrix} S_1 \\ S_2 \end{bmatrix} = \frac{AE}{l}\begin{bmatrix} 1 & -1 \\ -1 & 1 \end{bmatrix}\begin{bmatrix} u_1 \\ u_2 \end{bmatrix} + AE\alpha T\begin{bmatrix} 1 \\ -1 \end{bmatrix} \tag{5.23}$$

This agrees with the previously derived force-displacement relationship in Eq. (5.23).

In deriving the force-displacement relationship (5.23) we have used a local coordinate system. The detailed calculation of the stiffness properties in an arbitrary datum system, using local system stiffnesses, was presented in Sec. 5.4.

5.6 BEAM ELEMENTS

The beam element will be assumed to be a straight bar of uniform cross section capable of resisting axial forces, bending moments about the two principal axes in the plane of its cross section, and twisting moments about its centroidal axis. The following forces are acting on the beam: axial forces S_1 and S_7; shearing forces S_2, S_3, S_8, and S_9; bending moments S_5, S_6, S_{11}, and S_{12}; and twisting moments (torques) S_4 and S_{10}. The location and positive direction of these forces are shown in Fig. 5.4. The corresponding displacements $u_1, \ldots, u_{12}$ will be taken, as before, to be positive in the positive directions of the forces. The position and attitude of the beam element in space will be specified by the coordinates of the pth end of the beam and by the direction cosines for the x axis (pq direction) and the y axis, both taken with respect to some convenient datum coordinate system, the latter being required to locate the directions of principal axes of the cross section.

The stiffness matrix for a beam element is of order 12×12, but if the local axes are chosen to coincide with the principal axes of the cross section, it is possible to construct the 12×12 stiffness matrix from 2×2 and 4×4 submatrices. It is obvious from the engineering bending and torsion theory of beams that the forces S_1 and S_7 depend only on their corresponding displacements; the same is true of the torques S_4 and S_{10}. However, for arbitrary choice of bending planes the bending moments and shearing forces in the xy plane would depend not only on their corresponding displacements but also on the displacements corresponding to the forces in the xz plane. Only if the xy

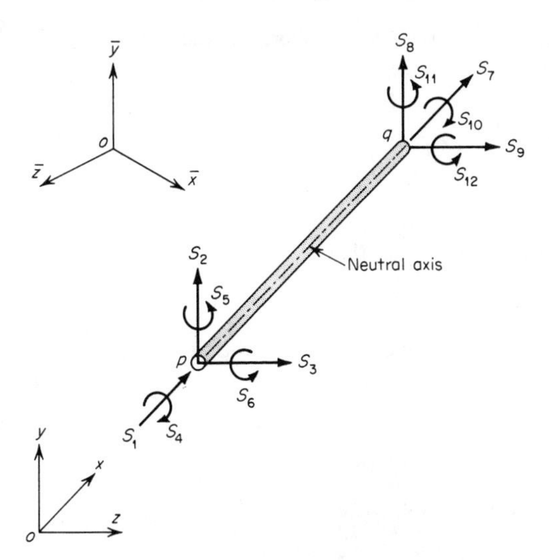

FIG. 5.4 Beam element.

and xz planes coincide with the principal axes of the cross section can the bending and shear in the two planes be considered independently of each other.

In order to demonstrate the third method for obtaining force-displacement relationships, the stiffness properties for a uniform beam element will be derived directly from the differential equations for beam displacements used in the engineering beam theory. The stiffness coefficients and thermal loads derived from these equations will be exact within the limits of the assumptions in the general engineering theory of beams subjected to loads and temperature gradients. Since the bending planes xy and xz will be assumed to coincide with the principal axes of the cross section and the ox axis will coincide with the centroidal axis of the beam, all forces can be separated into six groups, which can be considered independently of each other. These groups will now be considered, and the differential equations for each group will be derived. The temperature distribution through the beam cross section will be assumed not to vary along the length of the beam.

AXIAL FORCES (S_1 AND S_7)

The differential equation for the axial displacement u of the uniform beam shown in Fig. 5.5*a* is

$$S_1 = -\left(\frac{du}{dx} - \alpha T_m\right)EA \tag{5.49}$$

where $$T_m = \frac{1}{A}\int_A T\,dA \tag{5.50}$$

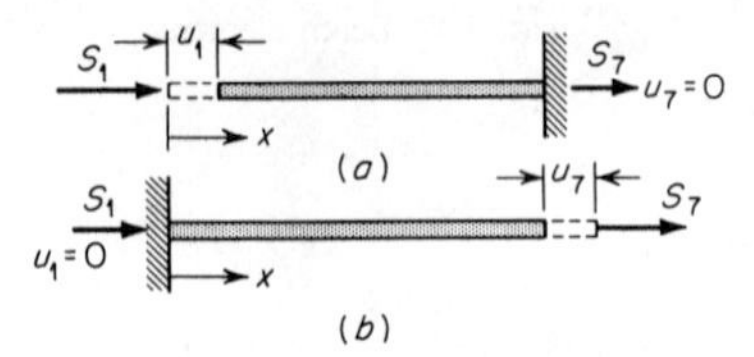

FIG. 5.5 Axial forces S_1 and S_7.

Equation (5.49) can be integrated directly, so that

$$S_1 x = -uEA + \alpha T_m EAx + C_1 \tag{5.51}$$

where C_1 is a constant of integration. We shall assume that the left end of the beam at $x = 0$ has displacement u_1 while the displacement is zero at $x = l$. Hence

$$C_1 = S_1 l - \alpha T_m EAl \tag{5.52}$$

Using Eqs. (5.51) and (5.52), for $x = 0$ we get

$$S_1 = \frac{EA}{l} u_1 + EA\alpha T_m \tag{5.53}$$

Also from the equation of equilibrium in the x direction it follows that

$$S_1 = -S_7 \tag{5.54}$$

Algebraic interpretation of the force-displacement relation $\mathbf{S} = \mathbf{ku} + \mathbf{Q}$ can be used to define individual stiffness coefficients k_{ij} and the thermal forces Q_i. For example, k_{ij} represents the element force S_i due to unit displacement u_j when all other displacements and the element temperature are equal to zero. The thermal force Q_i is equal to the element force S_i when all displacements are equal to zero and the element is subjected to a temperature change T. Hence

$$k_{11} = \left(\frac{S_1}{u_1}\right)_{T=0} = \frac{EA}{l} \tag{5.55}$$

and $$k_{71} = \left(\frac{S_7}{u_1}\right)_{T=0} = \frac{-EA}{l} \tag{5.56}$$

while all other coefficients in the first column of $\mathbf{k}$ are equal to zero. The thermal loads Q_1 and Q_7 are then found from

$$Q_1 = (S_1)_{u=0} = P_{Tx} \tag{5.57}$$

and $$Q_7 = (S_7)_{u=0} = -P_{Tx} \tag{5.58}$$

where $$P_{Tx} = EA\alpha T_m \tag{5.59}$$

Similarly, if $u_1 = 0$ and we allow u_7 to be nonzero (see Fig. 5.5*b*), it can be shown, either from symmetry or from the solution for u that

$$k_{77} = \frac{EA}{l} \tag{5.60}$$

TWISTING MOMENTS (S_4 AND S_{10})

The differential equation for the twist θ on the beam (see Fig. 5.6*a*) is

$$S_4 = -GJ\frac{d\theta}{dx} \tag{5.61}$$

where GJ is the torsional stiffness of the beam cross section. Integrating Eq. (5.61), we get

$$S_4 x = -GJ\theta + C_1 \tag{5.62}$$

and then by using the boundary condition $\theta = 0$ at $x = l$ we find that the constant of integration C_1 is given by

$$C_1 = S_4 l \tag{5.63}$$

Since $\theta = u_4$ at $x = 0$, it follows from (5.62) and (5.63) that

$$S_4 = \frac{GJ}{l}u_4 \tag{5.64}$$

Using the equilibrium condition for the twisting moments, we have

$$S_{10} = -S_4 \tag{5.65}$$

Hence $$k_{4,4} = \left(\frac{S_4}{u_4}\right)_{T=0} = \frac{GJ}{l} \tag{5.66}$$

and $$k_{10,4} = \left(\frac{S_{10}}{u_4}\right)_{T=0} = \frac{-GJ}{l} \tag{5.67}$$

while all other coefficients in the fourth column of $\mathbf{k}$ are equal to zero. Since the twisting moments S_4 and S_{10} are not affected by temperature, it follows that

$$Q_4 = Q_{10} = 0 \tag{5.68}$$

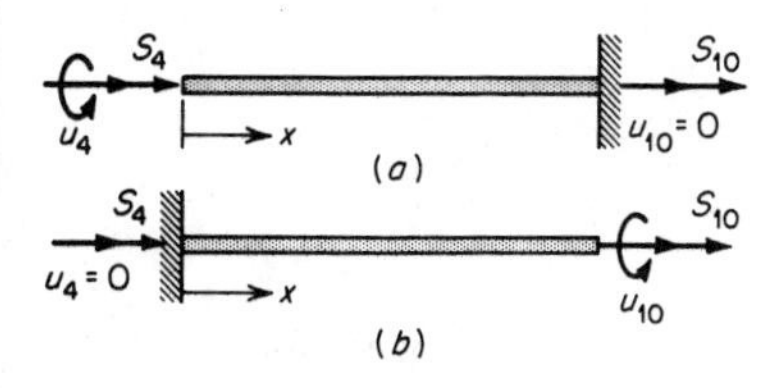

FIG. 5.6 Twisting moments S_4 and S_{10}.

Similarly, if $u_4 = 0$, as shown in Fig. 5.6*b*, it can be demonstrated that

$$k_{10,10} = \frac{GJ}{l} \tag{5.69}$$

SHEARING FORCES (S_2 AND S_8)

The lateral deflection v on the beam subjected to shearing forces and associated moments, as shown in Fig. 5.7*a*, is given by

$$v = v_b + v_s \tag{5.70}$$

where v_b is the lateral deflection due to bending strains and v_s is the additional deflection due to shearing strains, such that

$$\frac{dv_s}{dx} = \frac{-S_2}{GA_s} \tag{5.71}$$

with A_s representing the beam cross-sectional area effective in shear. The bending deflection for the beam shown in Fig. 5.7*a* is governed by the differential equation (see Boley and Weiner)

$$EI_z \frac{d^2v_b}{dx^2} = S_2x - S_6 - M_{T_z} \tag{5.72}$$

where $$M_{T_z} = \int_A \alpha ETy\, dA \tag{5.73}$$

From integration of Eqs. (5.71) and (5.72) it follows that

$$EI_zv = \frac{S_2x^3}{6} - \frac{S_6x^2}{2} - \frac{M_{T_z}x^2}{2} + \left(C_1 - \frac{S_2EI_z}{GA_s}\right)x + C_2 \tag{5.74}$$

where C_1 and C_2 are the constants of integration. Using the boundary conditions in Fig. 5.7*a*,

$$\frac{dv}{dx} = \frac{dv_s}{dx} = \frac{-S_2}{GA_s} \qquad \text{at } x = 0,\ x = l \tag{5.75}$$

$$v = 0 \qquad \text{at } x = l \tag{5.76}$$

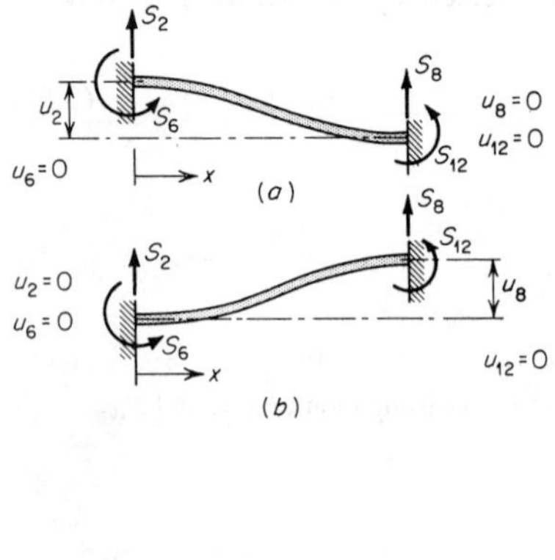

FIG. 5.7 Shear forces S_2 and S_8.

Eq. (5.74) becomes

$$EI_z v = \frac{S_2 x^3}{6} - \frac{S_6 x^2}{2} - \frac{M_{T_z} x^2}{2} - \frac{S_2 \Phi x l^2}{12} + (1 + \Phi)\frac{l^3 S_2}{12} \tag{5.77}$$

where $$S_6 = \frac{S_2 l}{2} - M_{T_z} \tag{5.78}$$

and $$\Phi = \frac{12EI_z}{GA_s l^2} \tag{5.79}$$

It should be noted here that the boundary condition for the built-in end in the engineering theory of bending when shear deformations v_s are included is taken as $dv_b/dx = 0$; that is, slope due to bending deformation is equal to zero.

The remaining forces acting on the beam can be determined from the equations of equilibrium; thus we have

$$S_8 = -S_2 \tag{5.80}$$

and $$S_{12} = -S_6 + S_2 l \tag{5.81}$$

Now at $x = 0$, $v = u_2$, and hence from Eq. (5.77)

$$u_2 = (1 + \Phi)\frac{l^3 S_2}{12EI_z} \tag{5.82}$$

Using Eqs. (5.78), and (5.80) to (5.82), we have

$$k_{2,2} = \left(\frac{S_2}{u_2}\right)_{T=0} = \frac{12EI_z}{(1 + \Phi)l^3} \tag{5.83}$$

$$k_{6,2} = \left(\frac{S_6}{u_2}\right)_{T=0} = \left(\frac{S_2 l}{2u_2}\right)_{T=0} = \frac{6EI_z}{(1 + \Phi)l^2} \tag{5.84}$$

$$k_{8,2} = \left(\frac{S_8}{u_2}\right)_{T=0} = \frac{-12EI_z}{(1 + \Phi)l^3} \tag{5.85}$$

$$k_{12,2} = \left(\frac{S_{12}}{u_2}\right)_{T=0} = \left(\frac{-S_6 + S_2 l}{u_2}\right)_{T=0} = \frac{6EI_z}{(1 + \Phi)l^2} \tag{5.86}$$

while the remaining coefficients in the second column are equal to zero. The thermal forces Q can be obtained from the condition $u = 0$, so that

$$Q_2 = (S_2)_{u=0} = 0 \tag{5.87}$$

$$Q_6 = (S_6)_{u=0} = -M_{T_z} \tag{5.88}$$

$$Q_8 = (S_8)_{u=0} = 0 \tag{5.89}$$

$$Q_{12} = (S_{12})_{u=0} = M_{T_z} \tag{5.90}$$

Similarly, if the left-hand end of the beam is built-in, as shown in Fig. 5.7*b*, then by use of the differential equations for the beam deflections or the condition of symmetry it can be demonstrated that

$$k_{8,8} = k_{2,2} = \frac{12EI_z}{(1+\Phi)l^3} \tag{5.91}$$

$$k_{12,8} = -k_{6,2} = \frac{-6EI_z}{(1+\Phi)l^2} \tag{5.92}$$

BENDING MOMENTS (S_6 AND S_{12})

In order to determine the stiffness coefficients associated with the rotations u_6 and u_{12}, the beam is subjected to bending moments and the associated shears, as shown in Fig. 5.8*a* and *b*. The deflections can be determined from Eq. (5.74), but the constants C_1 and C_2 in these equations must now be evaluated from a different set of boundary conditions. With the boundary conditions (Fig. 5.8*a*)

$$v = 0 \qquad \text{at } x = 0,\ x = l \tag{5.93}$$

and
$$\frac{dv}{dx} = \frac{dv_s}{dx} = -\frac{S_2}{GA_s} \qquad \text{at } x = l \tag{5.94}$$

Eq. (5.74) becomes

$$EI_z v = \frac{S_2}{6}(x^3 - l^2x) + \frac{M_{T_z}}{2}(lx - x^2) + \frac{S_6}{2}(lx - x^2) \tag{5.95}$$

and
$$S_2 = \frac{6S_6}{(4+\Phi)l} + \frac{6M_{T_z}}{(4+\Phi)l} \tag{5.96}$$

As before, the remaining forces acting on the beam can be determined from the equations of equilibrium, i.e., Eqs. (5.80) and (5.81).

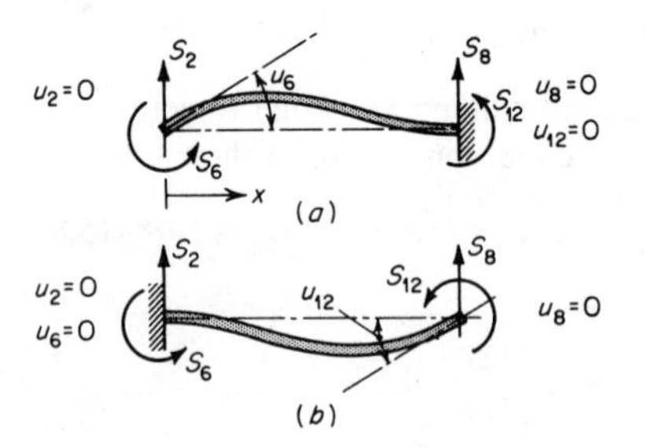

FIG. 5.8 Bending moments S_6 and S_{12}.

Now at $x = 0$

$$\frac{dv_b}{dx} = \frac{dv}{dx} - \frac{dv_s}{dx} = u_6$$

so that

$$u_6 = \frac{S_6(1+\Phi)l}{EI_z(4+\Phi)} + \frac{M_{T_z}(1+\Phi)l}{EI_z(4+\Phi)} \tag{5.97}$$

Hence, from Eqs. (5.80), (5.81), (5.96), and (5.97)

$$k_{6,6} = \left(\frac{S_6}{u_6}\right)_{T=0} = \frac{(4+\Phi)EI_z}{(1+\Phi)l} \tag{5.98}$$

$$k_{8,6} = \left(\frac{S_8}{u_6}\right)_{T=0} = \left(\frac{-S_2}{u_6}\right)_{T=0} = -\frac{6EI_z}{(1+\Phi)l^2} \tag{5.99}$$

$$k_{12,6} = \left(\frac{S_{12}}{u_6}\right)_{T=0} = \left(\frac{-S_6 + S_2 l}{u_6}\right)_{T=0} = \frac{(2-\Phi)EI_z}{(1+\Phi)l} \tag{5.100}$$

If the deflection of the left-hand end of the beam is equal to zero, as shown in Fig. 5.8*b*, it is evident from symmetry that

$$k_{12,12} = k_{6,6} = \frac{(4+\Phi)EI_z}{(1+\Phi)l} \tag{5.101}$$

SHEARING FORCES (S_3 AND S_9)

The stiffness coefficients associated with the displacements u_3 and u_9 can be derived directly from previous results. It should be observed, however, that with the sign convention adopted in Fig. 5.4 the directions of the positive bending moments in the *yx* and *zx* planes are different. This is illustrated clearly in Fig. 5.9, which shows that the positive direction of the bending moments S_5 and S_{11} is opposite to that of S_6 and S_{12}; it is therefore evident

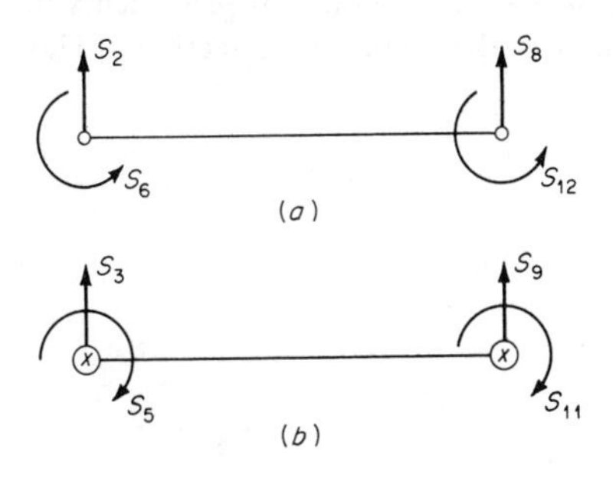

FIG. 5.9 Sign convention for shear forces and bending moments.

that

$$k_{3,3} = k_{2,2} \tag{5.102}$$

$$k_{5,3} = -k_{6,2} \tag{5.103}$$

$$k_{9,3} = k_{8,2} \tag{5.104}$$

$$k_{11,3} = -k_{12,2} \tag{5.105}$$

$$k_{9,9} = k_{8,8} \tag{5.106}$$

$$k_{11,9} = -k_{12,8} \tag{5.107}$$

$$Q_3 = 0 \tag{5.108}$$

$$Q_5 = M_{T_y} \tag{5.109}$$

$$Q_9 = 0 \tag{5.110}$$

$$Q_{11} = -M_{T_y} \tag{5.111}$$

where $$M_{T_y} = \int_A \alpha ETz\, dA \tag{5.112}$$

In Eqs. (5.102) to (5.107) it has been assumed that the cross section has the same properties about the z and y axes; however, when the complete 12×12 stiffness matrix is assembled, different values of I and A_s in the two bending planes must be allowed for.

BENDING MOMENTS (S_5 AND S_{11})

Here the same remarks apply as in the preceding section; thus we have

$$k_{5,5} = k_{6,6} \tag{5.113}$$

$$k_{9,5} = -k_{8,6} \tag{5.114}$$

$$k_{11,5} = k_{12,6} \tag{5.115}$$

The results obtained in these subsections can now be compiled into a matrix equation relating the element forces to their corresponding displacements in the presence of temperature gradients across the beam cross section. This relationship is given by

$$
\begin{bmatrix} S_1 \\ S_2 \\ S_3 \\ S_4 \\ S_5 \\ S_6 \\ S_7 \\ S_8 \\ S_9 \\ S_{10} \\ S_{11} \\ S_{12} \end{bmatrix}
=
\begin{bmatrix}
\frac{EA}{l} & & & & & & & & & & & \\
0 & \frac{12EI_z}{l^3(1+\Phi_y)} & & & & & & & \text{Symmetric} & & & \\
0 & 0 & \frac{12EI_y}{l^3(1+\Phi_z)} & & & & & & & & & \\
0 & 0 & 0 & \frac{GJ}{l} & & & & & & & & \\
0 & 0 & \frac{-6EI_y}{l^2(1+\Phi_z)} & 0 & \frac{(4+\Phi_z)EI_y}{l(1+\Phi_z)} & & & & & & & \\
0 & \frac{6EI_z}{l^2(1+\Phi_y)} & 0 & 0 & 0 & \frac{(4+\Phi_y)EI_z}{l(1+\Phi_y)} & & & & & & \\
\frac{-EA}{l} & 0 & 0 & 0 & 0 & 0 & \frac{AE}{l} & & & & & \\
0 & \frac{-12EI_z}{l^3(1+\Phi_y)} & 0 & 0 & 0 & \frac{-6EI_z}{l^2(1+\Phi_y)} & 0 & \frac{12EI_z}{l^3(1+\Phi_y)} & & & & \\
0 & 0 & \frac{-12EI_y}{l^3(1+\Phi_z)} & 0 & \frac{6EI_y}{l^2(1+\Phi_z)} & 0 & 0 & 0 & \frac{12EI_y}{l^3(1+\Phi_z)} & & & \\
0 & 0 & 0 & \frac{-GJ}{l} & 0 & 0 & 0 & 0 & 0 & \frac{GJ}{l} & & \\
0 & 0 & \frac{-6EI_y}{l^2(1+\Phi_z)} & 0 & \frac{(2-\Phi_z)EI_y}{l(1+\Phi_z)} & 0 & 0 & 0 & \frac{6EI_y}{l^2(1+\Phi_z)} & 0 & \frac{(4+\Phi_z)EI_y}{l(1+\Phi_z)} & \\
0 & \frac{6EI_z}{l^2(1+\Phi_y)} & 0 & 0 & 0 & \frac{(2-\Phi_y)EI_z}{l(1+\Phi_y)} & 0 & \frac{-6EI_z}{l^2(1+\Phi_y)} & 0 & 0 & 0 & \frac{(4+\Phi_y)EI_z}{l(1+\Phi_y)}
\end{bmatrix}
\begin{bmatrix} u_1 \\ u_2 \\ u_3 \\ u_4 \\ u_5 \\ u_6 \\ u_7 \\ u_8 \\ u_9 \\ u_{10} \\ u_{11} \\ u_{12} \end{bmatrix}
+
\begin{bmatrix} P_{T_x} \\ 0 \\ 0 \\ 0 \\ M_{T_y} \\ -M_{T_z} \\ -P_{T_x} \\ 0 \\ 0 \\ 0 \\ -M_{T_y} \\ M_{T_z} \end{bmatrix}
\tag{5.116}
$$

where $$\Phi_y = \frac{12EI_z}{GA_{s_y}l^2} = 24(1+\nu)\frac{A}{A_{s_y}}\left(\frac{r_z}{l}\right)^2 \tag{5.117}$$

and $$\Phi_z = \frac{12EI_y}{GA_{s_z}l^2} = 24(1+\nu)\frac{A}{A_{s_z}}\left(\frac{r_y}{l}\right)^2 \tag{5.118}$$

represent shear-deformation parameters. If r_z/l and r_y/l, the ratios of radius of gyration to element length, are small by comparison with unity, as is the case with a slender beam, both Φ_y and Φ_z can be taken as zero in Eq. (5.116). This leads then to a force-displacement relationship in which the effects of shear deformations are neglected.

For two-dimensional problems, the beam elements need only six forces and six displacements. We use the numbering system shown in Fig. 5.10, and it follows from the previous results that the stiffness matrix for these cases becomes

$$\mathbf{k} = \begin{bmatrix} \frac{EA}{l} & & & & & \\ 0 & \frac{12EI_z}{l^3(1+\Phi_y)} & & & \text{Symmetric} & \\ 0 & \frac{6EI_z}{l^2(1+\Phi_y)} & \frac{(4+\Phi_y)EI_z}{l(1+\Phi_y)} & & & \\ \frac{-EA}{l} & 0 & 0 & \frac{EA}{l} & & \\ 0 & \frac{-12EI_z}{l^3(1+\Phi_y)} & \frac{-6EI_z}{l^2(1+\Phi_y)} & 0 & \frac{12EI_z}{l^3(1+\Phi_y)} & \\ 0 & \frac{6EI_z}{l^2(1+\Phi_y)} & \frac{(2-\Phi_y)EI_z}{l(1+\Phi_y)} & 0 & \frac{-6EI_z}{l^2(1+\Phi_y)} & \frac{(4+\Phi_y)EI_z}{l(1+\Phi_y)} \end{bmatrix} \tag{5.119}$$

$$\mathbf{Q} = \begin{bmatrix} P_{T_x} \\ 0 \\ -M_{T_z} \\ -P_{T_x} \\ 0 \\ M_{T_z} \end{bmatrix} \tag{5.120}$$

If the shear deformations are neglected, that is, $\Phi_y = 0$, the stiffness matrix in

(5.119) simplifies to

$$\mathbf{k} = \frac{EI_z}{l^3}\begin{bmatrix} \dfrac{Al^2}{I_z} & & & & & \text{Symmetric} \\ 0 & 12 & & & & \\ 0 & 6l & 4l^2 & & & \\ -\dfrac{Al^2}{I_z} & 0 & 0 & \dfrac{Al^2}{I_z} & & \\ 0 & -12 & -6l & 0 & 12 & \\ 0 & 6l & 2l^2 & 0 & -6l & 4l^2 \end{bmatrix} \tag{5.121}$$

The matrix equation relating displacements in the local coordinate system to those in the datum system

$$\mathbf{u} = \boldsymbol{\lambda}\bar{\mathbf{u}} \tag{5.122}$$

can be derived, as before, by resolving element displacement vectors in one set of coordinates into displacements in another set. Thus it can easily be demonstrated that for a single beam element shown in Fig. 5.4, Eq. (5.122) is of the form

$$\begin{bmatrix} u_1 \\ u_2 \\ u_3 \\ u_4 \\ u_5 \\ u_6 \\ u_7 \\ u_8 \\ u_9 \\ u_{10} \\ u_{11} \\ u_{12} \end{bmatrix} = \left[\begin{array}{c|c|c|c} \boldsymbol{\lambda}_{ox} & & & \\ \boldsymbol{\lambda}_{oy} & \mathbf{0} & \mathbf{0} & \mathbf{0} \\ \boldsymbol{\lambda}_{oz} & & & \\ \hline & \boldsymbol{\lambda}_{ox} & & \\ \mathbf{0} & \boldsymbol{\lambda}_{oy} & \mathbf{0} & \mathbf{0} \\ & \boldsymbol{\lambda}_{oz} & & \\ \hline & & \boldsymbol{\lambda}_{ox} & \\ \mathbf{0} & \mathbf{0} & \boldsymbol{\lambda}_{oy} & \mathbf{0} \\ & & \boldsymbol{\lambda}_{oz} & \\ \hline & & & \boldsymbol{\lambda}_{ox} \\ \mathbf{0} & \mathbf{0} & \mathbf{0} & \boldsymbol{\lambda}_{oy} \\ & & & \boldsymbol{\lambda}_{oz} \end{array}\right] \begin{bmatrix} \bar{u}_1 \\ \bar{u}_2 \\ \bar{u}_3 \\ \bar{u}_4 \\ \bar{u}_5 \\ \bar{u}_6 \\ \bar{u}_7 \\ \bar{u}_8 \\ \bar{u}_9 \\ \bar{u}_{10} \\ \bar{u}_{11} \\ \bar{u}_{12} \end{bmatrix} \tag{5.123}$$

where $\boldsymbol{\lambda}_{ox} = [l_{ox} \quad m_{ox} \quad n_{ox}]$ (5.124)

$$\boldsymbol{\lambda}_{oy} = [l_{oy} \quad m_{oy} \quad n_{oy}] \tag{5.125}$$

$$\boldsymbol{\lambda}_{oz} = [l_{oz} \quad m_{oz} \quad n_{oz}] \tag{5.126}$$

represents matrices of direction cosines for the ox, oy, and oz directions, respectively, measured in the datum system $\bar{x}$, $\bar{y}$, and $\bar{z}$, and $\bar{u}_1, \ldots, \bar{u}_{12}$ represent element displacements in the datum system. Hence the transformation matrix $\boldsymbol{\lambda}$ is given by

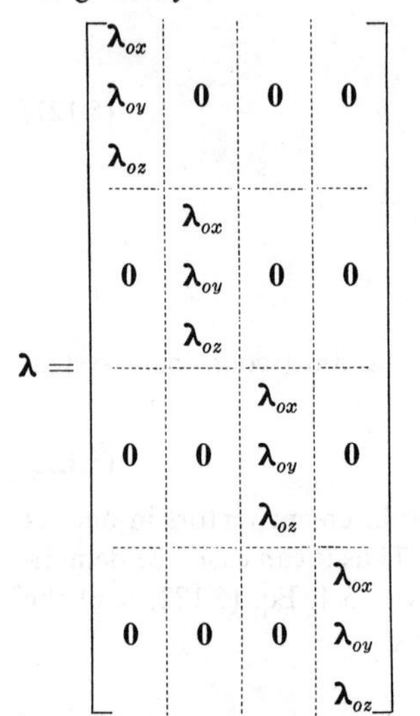

$$\boldsymbol{\lambda} = \begin{bmatrix} \boldsymbol{\lambda}_{ox} & & & \\ \boldsymbol{\lambda}_{oy} & \mathbf{0} & \mathbf{0} & \mathbf{0} \\ \boldsymbol{\lambda}_{oz} & & & \\ \hline & \boldsymbol{\lambda}_{ox} & & \\ \mathbf{0} & \boldsymbol{\lambda}_{oy} & \mathbf{0} & \mathbf{0} \\ & \boldsymbol{\lambda}_{oz} & & \\ \hline & & \boldsymbol{\lambda}_{ox} & \\ \mathbf{0} & \mathbf{0} & \boldsymbol{\lambda}_{oy} & \mathbf{0} \\ & & \boldsymbol{\lambda}_{oz} & \\ \hline & & & \boldsymbol{\lambda}_{ox} \\ \mathbf{0} & \mathbf{0} & \mathbf{0} & \boldsymbol{\lambda}_{oy} \\ & & & \boldsymbol{\lambda}_{oz} \end{bmatrix} \tag{5.127}$$

For two-dimensional problems the corresponding transformation matrix $\boldsymbol{\lambda}$ becomes

$$\boldsymbol{\lambda} = \begin{bmatrix} l_{ox} & m_{ox} & 0 & 0 & 0 & 0 \\ l_{oy} & m_{oy} & 0 & 0 & 0 & 0 \\ 0 & 0 & 1 & 0 & 0 & 0 \\ 0 & 0 & 0 & l_{ox} & m_{ox} & 0 \\ 0 & 0 & 0 & l_{oy} & m_{oy} & 0 \\ 0 & 0 & 0 & 0 & 0 & 1 \end{bmatrix} \tag{5.128}$$

where the numbering system corresponds to displacement numbers in Fig. 5.10.

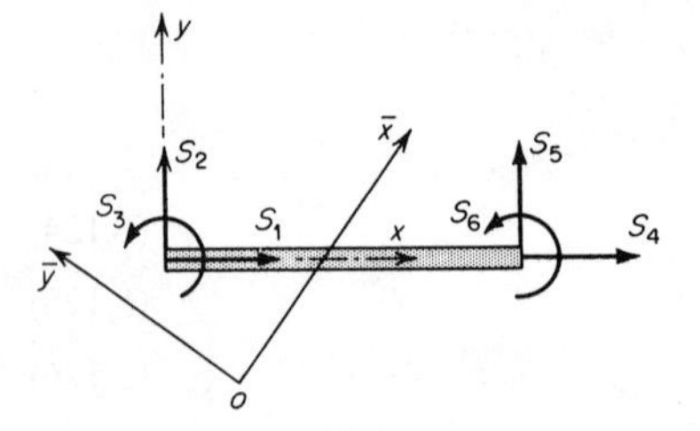

FIG. 5.10 Beam element for two-dimensional structures.

5.7 TRIANGULAR PLATE ELEMENTS (IN-PLANE FORCES)

In triangular plate elements the boundaries are attached continuously to the surrounding structure, and consequently no exact stiffness relationships can be derived, as explained in Chap. 4.

The assumed displacement variation will be taken as[329]

$$u_x = c_1 x + c_2 y + c_3 \qquad u_y = c_4 x + c_5 y + c_6 \tag{5.129}$$

where the six arbitrary coefficients $c_1, \ldots, c_6$ can be found from the displacements of the three vertices of the triangle (see Fig. 5.11). Using the boundary conditions

$$\begin{aligned} u_x &= u_1 \quad \text{and} \quad u_y = u_2 \quad \text{at } (x_1, y_1) \\ u_x &= u_3 \quad \text{and} \quad u_y = u_4 \quad \text{at } (x_2, y_2) \\ u_x &= u_5 \quad \text{and} \quad u_y = u_6 \quad \text{at } (x_3, y_3) \end{aligned} \tag{5.130}$$

in Eqs. (5.129) to evaluate the unknown coefficients we can show that

$$u_x = \frac{1}{2A_{123}} \{[y_{32}(x - x_2) - x_{32}(y - y_2)]u_1 + [-y_{31}(x - x_3) + x_{31}(y - y_3)]u_3 + [y_{21}(x - x_1) - x_{21}(y - y_1)]u_5\} \tag{5.131}$$

$$u_y = \frac{1}{2A_{123}} \{[y_{32}(x - x_2) - x_{32}(y - y_2)]u_2 + [-y_{31}(x - x_3) + x_{31}(y - y_3)]u_4 + [y_{21}(x - x_1) - x_{21}(y - y_1)]u_6\} \tag{5.132}$$

where

$$\begin{aligned} 2A_{123} &= x_{32}y_{21} - x_{21}y_{32} \\ &= 2(\text{area of the triangle 123}) \end{aligned} \tag{5.133}$$

and

$$x_{ij} = x_i - x_j \qquad y_{ij} = y_i - y_j \tag{5.134}$$

From Eqs. (5.131) and (5.132) it follows that the assumed displacements along any edge vary linearly, and they depend only on the displacements of the two vertices on the particular edge; this ensures the satisfaction of the compatibility of displacements on two adjacent triangular elements with a common boundary.

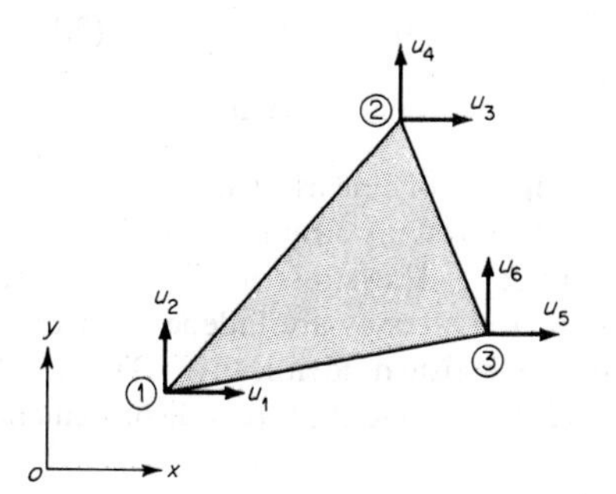

FIG. 5.11 Triangular plate element.

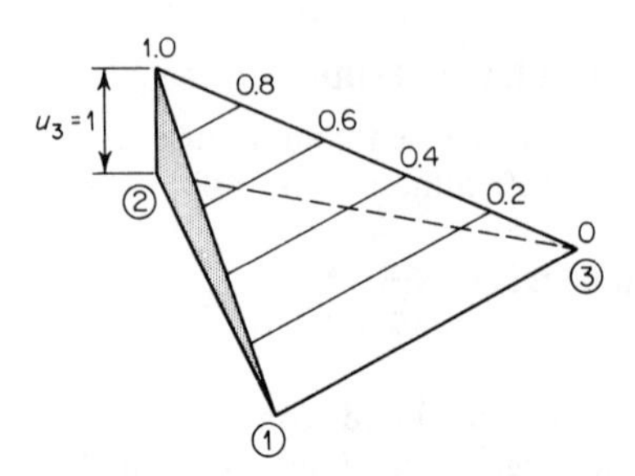

FIG. 5.12 Displacement distribution in a triangular plate due to $u_3 = 1$.

A typical displacement distribution due to $u_3 = 1$ with all other displacements kept zero is shown in Fig. 5.12.

Equations (5.131) and (5.132) can now be used to find the relationship between the total strains e_{xx}, e_{yy}, and e_{xy} and the six displacements $u_1, \ldots, u_6$. Differentiating these equations, we have

$$\mathbf{e} = \begin{bmatrix} e_{xx} \\ e_{yy} \\ e_{xy} \end{bmatrix} = \begin{bmatrix} \dfrac{\partial u_x}{\partial x} \\ \dfrac{\partial u_y}{\partial y} \\ \dfrac{\partial u_x}{\partial y} + \dfrac{\partial u_y}{\partial x} \end{bmatrix} = \frac{1}{2A_{123}} \begin{bmatrix} y_{32} & 0 & -y_{31} & 0 & y_{21} & 0 \\ 0 & -x_{32} & 0 & x_{31} & 0 & -x_{21} \\ -x_{32} & y_{32} & x_{31} & -y_{31} & -x_{21} & y_{21} \end{bmatrix} \begin{bmatrix} u_1 \\ u_2 \\ u_3 \\ u_4 \\ u_5 \\ u_6 \end{bmatrix} \tag{5.135}$$

$$\mathbf{e} = \underline{\mathbf{b}}\mathbf{u} \tag{5.136}$$

where

$$\mathbf{e} = \{e_{xx} \quad e_{yy} \quad e_{xy}\} \tag{5.137}$$

$$\mathbf{u} = \{u_1 \quad u_2 \quad \cdots \quad u_6\} \tag{5.138}$$

$$\underline{\mathbf{b}} = \frac{1}{2A_{123}} \begin{bmatrix} y_{32} & 0 & -y_{31} & 0 & y_{21} & 0 \\ 0 & -x_{32} & 0 & x_{31} & 0 & -x_{21} \\ -x_{32} & y_{32} & x_{31} & -y_{31} & -x_{21} & y_{21} \end{bmatrix} \tag{5.139}$$

Equation (5.139) indicates that the assumption of linearly varying displacements within the triangular element leads to constant strains, and hence, by Hooke's law, it also leads to constant stresses. The stress field satisfies the equations of strain compatibility, and since the stresses are independent of x and y, the stress-equilibrium equations are satisfied identically. The total strains $\mathbf{e}$ can now be substituted into Eq. (2.24) to give the stress-displacement relationship

$$
\begin{bmatrix} \sigma_{xx} \\ \sigma_{yy} \\ \sigma_{xy} \end{bmatrix} = \frac{E}{2A_{123}(1-\nu^2)} \begin{bmatrix} y_{32} & -\nu x_{32} & -y_{31} & \nu x_{31} & y_{21} & -\nu x_{21} \\ \nu y_{32} & -x_{32} & -\nu y_{31} & x_{31} & \nu y_{21} & -x_{21} \\ \frac{-(1-\nu)x_{32}}{2} & \frac{(1-\nu)y_{32}}{2} & \frac{(1-\nu)x_{31}}{2} & \frac{-(1-\nu)y_{31}}{2} & \frac{-(1-\nu)x_{21}}{2} & \frac{(1-\nu)y_{21}}{2} \end{bmatrix} \begin{bmatrix} u_1 \\ u_2 \\ u_3 \\ u_4 \\ u_5 \\ u_6 \end{bmatrix} - \frac{E\alpha T}{1-\nu} \begin{bmatrix} 1 \\ 1 \\ 0 \end{bmatrix} \qquad (5.140)
$$

The stiffness matrix **k** and thermal stiffness **h** can be found from Eqs. (5.16) and (5.17) using matrices $\boldsymbol{\varkappa}$ and $\boldsymbol{\varkappa}_T$ for the two-dimensional stress distribution. The evaluation of integrals for **k** and **h** presents no difficulty since none of the matrices involved is a function of the x and y coordinates; the matrix products $\underline{\mathbf{b}}^T\boldsymbol{\varkappa}\underline{\mathbf{b}}$ and $\underline{\mathbf{b}}^T\boldsymbol{\varkappa}_T$ can be taken outside the integration sign, and the resulting integrals are then equal to the element volume $A_{123}t$, where t is the plate thickness. For convenience of presentation the stiffness matrices **k** can be separated into two parts, so that

$$\mathbf{k} = \mathbf{k}_n + \mathbf{k}_s \tag{5.141}$$

where $\mathbf{k}_n$ represents stiffness due to normal stresses and $\mathbf{k}_s$ represents stiffness due to shearing stresses. The two component matrices, as derived from Eq. (5.16), are given by

$$\mathbf{k}_n = \frac{Et}{4A_{123}(1-\nu^2)}\begin{bmatrix} y_{32}^2 & & & & & \\ -\nu y_{32}x_{32} & x_{32}^2 & & \text{Symmetric} & & \\ -y_{32}y_{31} & \nu x_{32}y_{31} & y_{31}^2 & & & \\ \nu y_{32}x_{31} & -x_{32}x_{31} & -\nu y_{31}x_{31} & x_{31}^2 & & \\ y_{32}y_{21} & -\nu x_{32}y_{21} & -y_{31}y_{21} & \nu x_{31}y_{21} & y_{21}^2 & \\ -\nu y_{32}x_{21} & x_{32}x_{21} & \nu y_{31}x_{21} & -x_{31}x_{21} & -\nu y_{21}x_{21} & x_{21}^2 \end{bmatrix} \tag{5.141a}$$

$$\mathbf{k}_s = \frac{Et}{8A_{123}(1+\nu)}\begin{bmatrix} x_{32}^2 & & & & & \\ -x_{32}y_{32} & y_{32}^2 & & \text{Symmetric} & & \\ -x_{32}x_{31} & y_{32}x_{31} & x_{31}^2 & & & \\ x_{32}y_{31} & -y_{32}y_{31} & -x_{31}y_{31} & y_{31}^2 & & \\ x_{32}x_{21} & -y_{32}x_{21} & -x_{31}x_{21} & y_{31}x_{21} & x_{21}^2 & \\ -x_{32}y_{21} & y_{32}y_{21} & x_{31}y_{21} & -y_{31}y_{21} & -x_{21}y_{21} & y_{21}^2 \end{bmatrix} \tag{5.141b}$$

Similarly, from Eq. (5.17) the thermal stiffness **h** becomes

$$\mathbf{h} = \frac{Et}{2(1-\nu)}\begin{bmatrix} -y_{32} \\ x_{32} \\ y_{31} \\ -x_{31} \\ -y_{21} \\ x_{21} \end{bmatrix} \tag{5.142}$$

So far no restriction has been placed on the orientation of the local coordinate system; however, when calculating the transformation matrix $\boldsymbol{\lambda}$ for a triangular panel, it is preferable to select the oy direction parallel to, or coincident with,

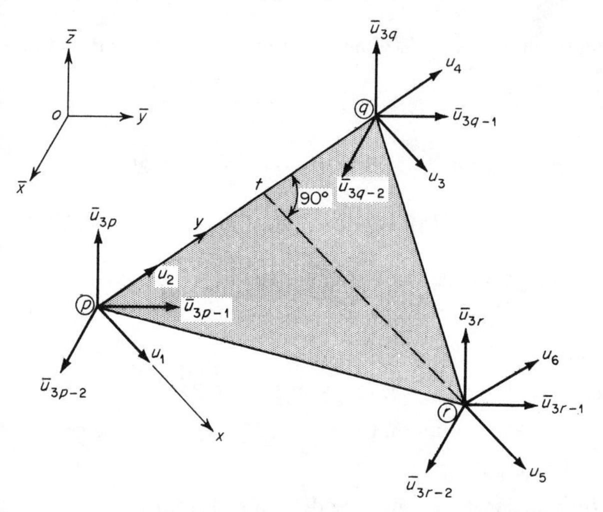

FIG. 5.13 Triangular plate displacements in local and datum coordinate systems.

the edge pq (edge 1,2), coinciding also with the direction of displacements u_2 and u_4 (see Fig. 5.13). The direction cosines of the edge pq (direction from p to q) can be specified by the matrix

$$\boldsymbol{\lambda}_{pq} = [l_{pq} \quad m_{pq} \quad n_{pq}] \tag{5.143}$$

whose elements are determined from the coordinates of p and q, that is,

$$l_{pq} = \frac{\bar{x}_q - \bar{x}_p}{d_{pq}} = \frac{\bar{x}_{qp}}{d_{pq}} \tag{5.144}$$

$$m_{pq} = \frac{\bar{y}_q - \bar{y}_p}{d_{pq}} = \frac{\bar{y}_{qp}}{d_{pq}} \tag{5.145}$$

$$n_{pq} = \frac{\bar{z}_q - \bar{z}_p}{d_{pq}} = \frac{\bar{z}_{qp}}{d_{pq}} \tag{5.146}$$

where $$d_{pq} = (\bar{x}_{qp}{}^2 + \bar{y}_{qp}{}^2 + \bar{z}_{qp}{}^2)^{\frac{1}{2}} \tag{5.147}$$

In addition, direction cosines for a direction perpendicular to pq in the plane of the triangle are required. To find these direction cosines a perpendicular to pq is drawn from the vertex r, and the point of intersection of this perpendicular with the line pq is denoted by t. The coordinates of the point t in the datum frame of reference can be expressed as

$$(\bar{x}_p + l_{pq}d_{pt}, \bar{y}_p + m_{pq}d_{pt}, \bar{z}_p + n_{pq}d_{pt})$$

where d_{pt} is the distance from p to t. Now the direction cosines of the direction tr are given by the matrix

$$\begin{aligned}\boldsymbol{\lambda}_{tr} &= [l_{tr} \quad m_{tr} \quad n_{tr}] \\ &= \frac{1}{d_{tr}}[(\bar{x}_r - \bar{x}_p - l_{pq}d_{pt}) \quad (\bar{y}_r - \bar{y}_p - m_{pq}d_{pt}) \quad (\bar{z}_r - \bar{z}_p - n_{pq}d_{pt})] \\ &= \frac{1}{d_{tr}}[(\bar{x}_{rp} - l_{pq}d_{pt}) \quad (\bar{y}_{rp} - m_{pq}d_{pt}) \quad (\bar{z}_{rp} - n_{pq}d_{pt})]\end{aligned} \tag{5.148}$$

The condition that pq is perpendicular to tr is expressed as

$$l_{pq}l_{tr} + m_{pq}m_{tr} + n_{pq}n_{tr} = 0 \tag{5.149}$$

Hence, from Eq. (5.148)

$$l_{pq}(\bar{x}_{rp} - l_{pq}d_{pt}) + m_{pq}(\bar{y}_{rp} - m_{pq}d_{pt}) + n_{pq}(\bar{z}_{rp} - n_{pq}d_{pt}) = 0 \tag{5.150}$$

Solving Eq. (5.150) for d_{pt}, and noting that $l_{pq}^2 + m_{pq}^2 + n_{pq}^2 = 1$, we get

$$d_{pt} = l_{pq}\bar{x}_{rp} + m_{pq}\bar{y}_{rp} + n_{pq}\bar{z}_{rp} \tag{5.151}$$

The length of the perpendicular d_{tr} can now be determined from

$$d_{tr} = (\bar{x}_{rp}^2 + \bar{y}_{rp}^2 + \bar{z}_{rp}^2 - d_{pt}^2)^{\frac{1}{2}} \tag{5.152}$$

and the direction cosines in Eq. (5.148) can be found from

$$l_{tr} = \frac{\bar{x}_{rp} - l_{pq}d_{pt}}{d_{tr}} \tag{5.153}$$

$$m_{tr} = \frac{\bar{y}_{rp} - m_{pq}d_{pt}}{d_{tr}} \tag{5.154}$$

$$n_{tr} = \frac{\bar{z}_{rp} - n_{pq}d_{pt}}{d_{tr}} \tag{5.155}$$

Finally the transformation matrix $\boldsymbol{\lambda}$ can be constructed using the direction cosines matrices $\boldsymbol{\lambda}_{pq}$ and $\boldsymbol{\lambda}_{tr}$. Resolving displacements in the two frames of reference, local and datum, leads to the following matrix relationship:

$$\begin{bmatrix} u_1 \\ u_2 \\ u_3 \\ u_4 \\ u_5 \\ u_6 \end{bmatrix} = \begin{bmatrix} \boldsymbol{\lambda}_{tr} & \mathbf{0} & \mathbf{0} \\ \boldsymbol{\lambda}_{pq} & \mathbf{0} & \mathbf{0} \\ \mathbf{0} & \boldsymbol{\lambda}_{tr} & \mathbf{0} \\ \mathbf{0} & \boldsymbol{\lambda}_{pq} & \mathbf{0} \\ \mathbf{0} & \mathbf{0} & \boldsymbol{\lambda}_{tr} \\ \mathbf{0} & \mathbf{0} & \boldsymbol{\lambda}_{pq} \end{bmatrix} \begin{bmatrix} \bar{u}_{3p-2} \\ \bar{u}_{3p-1} \\ \bar{u}_{3p} \\ \bar{u}_{3q-2} \\ \bar{u}_{3q-1} \\ \bar{u}_{3q} \\ \bar{u}_{3r-2} \\ \bar{u}_{3r-1} \\ \bar{u}_{3r} \end{bmatrix} \tag{5.156}$$

where $\bar{u}_{3p-2}, \ldots$ represent nodal displacements in the datum coordinate system (see Fig. 5.13). Hence

$$\boldsymbol{\lambda} = \begin{bmatrix} \boldsymbol{\lambda}_{tr} & \mathbf{0} & \mathbf{0} \\ \boldsymbol{\lambda}_{pq} & \mathbf{0} & \mathbf{0} \\ \mathbf{0} & \boldsymbol{\lambda}_{tr} & \mathbf{0} \\ \mathbf{0} & \boldsymbol{\lambda}_{pq} & \mathbf{0} \\ \mathbf{0} & \mathbf{0} & \boldsymbol{\lambda}_{tr} \\ \mathbf{0} & \mathbf{0} & \boldsymbol{\lambda}_{pq} \end{bmatrix} \tag{5.157}$$

where the submatrices $\boldsymbol{\lambda}_{tr}$ and $\boldsymbol{\lambda}_{pq}$ are given by Eqs. (5.148) and (5.143), respectively.

5.8 RECTANGULAR PLATE ELEMENTS (IN-PLANE FORCES)

LINEAR-EDGE-DISPLACEMENT ASSUMPTION

The origin of the local coordinate system will be taken at the lower left corner of the rectangle, as shown in Fig. 5.14, and to simplify subsequent analysis nondimensional coordinates

$$\xi = \frac{x}{a} \quad \text{and} \quad \eta = \frac{y}{b} \tag{5.158}$$

will be introduced, where a and b are the dimensions of the rectangular plate. The element displacements are represented by the displacements at the four corners. There are eight displacements $u_1, u_2, \ldots, u_8$, and their positive directions are the same as the positive directions of the x and y axes, as indicated in Fig. 5.14.

Simple displacement functions which satisfy the assumption of linearly varying boundary displacements may be taken as[10,277]

$$u_x = c_1\xi + c_2\xi\eta + c_3\eta + c_4 \tag{5.159}$$

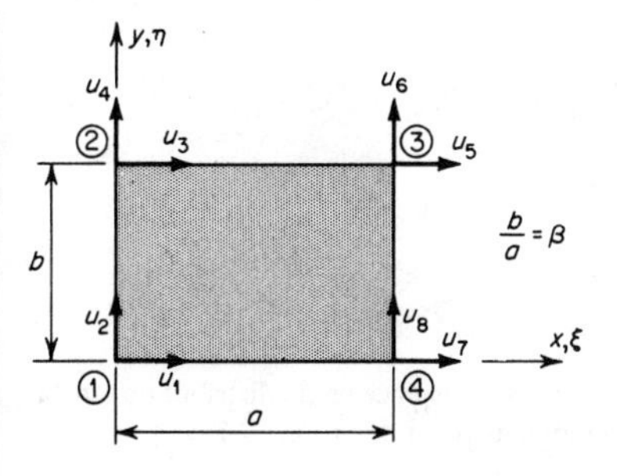

FIG. 5.14 Rectangular plate element.

and $\quad u_y = c_5\xi + c_6\xi\eta + c_7\eta + c_8 \quad (5.160)$

where the arbitrary constants $c_1, \ldots, c_8$ are determined from the known displacements in the x and y directions at the four corners of the rectangle. Thus, the assumed displacement distribution is represented by a second-degree surface, where for constant values of ξ (or η) the variation of displacement in the direction of η (or ξ) is linear.

The following boundary conditions are used to evaluate the unknown constants $c_1, \ldots, c_8$:

$$\begin{aligned} u_x &= u_1 \quad \text{and} \quad u_y = u_2 \quad \text{at } (0,0) \\ u_x &= u_3 \quad \text{and} \quad u_y = u_4 \quad \text{at } (0,1) \\ u_x &= u_5 \quad \text{and} \quad u_y = u_6 \quad \text{at } (1,1) \\ u_x &= u_7 \quad \text{and} \quad u_y = u_8 \quad \text{at } (1,0) \end{aligned} \qquad (5.161)$$

Substituting these boundary values into the equations for displacements, we determine the unknown constants $c_1, \ldots, c_8$; hence

$$u_x = (1-\xi)(1-\eta)u_1 + (1-\xi)\eta u_3 + \xi\eta u_5 + \xi(1-\eta)u_7 \qquad (5.162)$$

$$u_y = (1-\xi)(1-\eta)u_2 + (1-\xi)\eta u_4 + \xi\eta u_6 + \xi(1-\eta)u_8 \qquad (5.163)$$

Examining the form of Eqs. (5.162) and (5.163), we can see that the distribution of the u_x and u_y displacements along any edge is linear and that it depends only on the element displacements of the two corner points defining the particular edge. Thus, the assumed form of displacement distribution ensures that the compatibility of displacements on the boundaries of adjacent elements is satisfied. A typical displacement distribution due to $u_3 = 1$ while all other element displacements are kept zero is shown in Fig. 5.15.

The total strains corresponding to the assumed displacement functions can be obtained by differentiation of Eqs. (5.162) and (5.163). Noting that

$$e_{xx} = \frac{\partial u_x}{\partial x} = \frac{1}{a}\frac{\partial u_x}{\partial \xi} \qquad (5.164)$$

$$e_{yy} = \frac{\partial u_y}{\partial y} = \frac{1}{b}\frac{\partial u_y}{\partial \eta} \qquad (5.165)$$

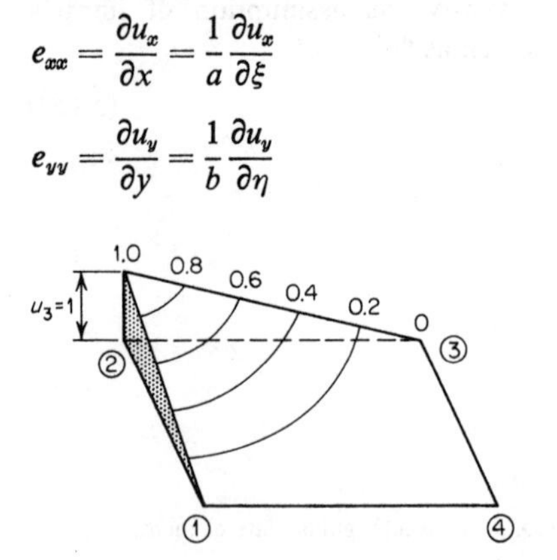

FIG. 5.15 Displacement distribution in a rectangular plate due to $u_3 = 1$.

and $$e_{xy} = \frac{\partial u_x}{\partial y} + \frac{\partial u_y}{\partial x} = \frac{1}{b}\frac{\partial u_x}{\partial \eta} + \frac{1}{a}\frac{\partial u_y}{\partial \xi} \tag{5.166}$$

we find that the total strain-displacement relationship for the rectangular plate becomes

$$\mathbf{e} = \begin{bmatrix} e_{xx} \\ e_{yy} \\ e_{xy} \end{bmatrix} = \begin{bmatrix} \frac{-(1-\eta)}{a} & 0 & \frac{-\eta}{a} & 0 & \frac{\eta}{a} & 0 & \frac{1-\eta}{a} & 0 \\ 0 & \frac{-(1-\xi)}{b} & 0 & \frac{1-\xi}{b} & 0 & \frac{\xi}{b} & 0 & \frac{-\xi}{b} \\ \frac{-(1-\xi)}{b} & \frac{-(1-\eta)}{a} & \frac{1-\xi}{b} & \frac{-\eta}{a} & \frac{\xi}{b} & \frac{\eta}{a} & \frac{-\xi}{b} & \frac{1-\eta}{a} \end{bmatrix} \begin{bmatrix} u_1 \\ u_2 \\ u_3 \\ u_4 \\ u_5 \\ u_6 \\ u_7 \\ u_8 \end{bmatrix} \tag{5.167}$$

or, in using matrix symbolism,

$$\mathbf{e} = \underline{\mathbf{b}}\mathbf{u} \tag{5.168}$$

where

$$\underline{\mathbf{b}} = \begin{bmatrix} \frac{-(1-\eta)}{a} & 0 & \frac{-\eta}{a} & 0 & \frac{\eta}{a} & 0 & \frac{1-\eta}{a} & 0 \\ 0 & \frac{-(1-\xi)}{b} & 0 & \frac{1-\xi}{b} & 0 & \frac{\xi}{b} & 0 & \frac{-\xi}{b} \\ \frac{-(1-\xi)}{b} & \frac{-(1-\eta)}{a} & \frac{1-\xi}{b} & \frac{-\eta}{a} & \frac{\xi}{b} & \frac{\eta}{a} & \frac{-\xi}{b} & \frac{1-\eta}{a} \end{bmatrix} \tag{5.169}$$

and $$\mathbf{u} = \{u_1 \quad u_2 \quad \cdots \quad u_8\} \tag{5.170}$$

Substituting Eq. (5.167) into (2.24) gives the following stress-displacement relationship:

$$
\begin{bmatrix} \sigma_{xx} \\ \sigma_{yy} \\ \sigma_{xy} \end{bmatrix} = \frac{E}{1-\nu^2}
\begin{bmatrix}
\frac{-(1-\eta)}{a} & \frac{-\nu(1-\xi)}{b} & \frac{-\eta}{a} & \frac{\nu(1-\xi)}{b} & \frac{\eta}{a} & \frac{\nu\xi}{b} & \frac{1-\eta}{a} & \frac{-\nu\xi}{b} \\
\frac{-\nu(1-\eta)}{a} & \frac{-(1-\xi)}{b} & \frac{-\nu\eta}{a} & \frac{1-\xi}{b} & \frac{\nu\eta}{a} & \frac{\xi}{b} & \frac{\nu(1-\eta)}{a} & \frac{-\xi}{b} \\
\frac{-(1-\nu)(1-\xi)}{2b} & \frac{-(1-\nu)(1-\eta)}{2a} & \frac{(1-\nu)(1-\xi)}{2b} & \frac{-(1-\nu)\eta}{2a} & \frac{(1-\nu)\xi}{2b} & \frac{(1-\nu)\eta}{2a} & \frac{-(1-\nu)\xi}{2b} & \frac{(1-\nu)(1-\eta)}{2a}
\end{bmatrix}
\begin{bmatrix} u_1 \\ u_2 \\ u_3 \\ u_4 \\ u_5 \\ u_6 \\ u_7 \\ u_8 \end{bmatrix}
- \frac{E\alpha T}{1-\nu} \begin{bmatrix} 1 \\ 1 \\ 0 \end{bmatrix} \quad (5.171)
$$

From Eq. (5.167) it is evident that for a given set of displacements **u** the e_{xx} strains are constant in the x direction (ξ direction) and that they vary linearly with y (η coordinate). Similarly, the e_{yy} strains are constant in the y direction, and they vary linearly with x (ξ coordinate). The shearing strains e_{xy}, on the other hand, vary linearly with both x and y (ξ and η). If the temperature of the element is assumed constant, it follows from Eq. (5.171) that all stress components in the panel vary linearly with x and y and that the stress distribution is such that, in general, it violates the stress-equilibrium equations within the rectangle.

Calculation of the stiffness matrix **k** and thermal stiffness **h** requires integration with respect to ξ and η since the matrix $\underline{\mathbf{b}}$, unlike the case of bar and triangular plate elements, is a function of the position variables. Substituting Eqs. (5.169) and (2.28) into (5.16), multiplying out the matrix product $\underline{\mathbf{b}}^T\boldsymbol{\varkappa}\underline{\mathbf{b}}$, and then integrating over the volume of the plate gives the stiffness matrix of Eq. (5.172),

$$
\mathbf{k} = \frac{Et}{12(1-\nu^2)}
\begin{array}{c|c|c|c|c|c|c|c|c|}
1 & 4\beta + 2(1-\nu)\beta^{-1} & & & & & & & \\
2 & \frac{3}{2}(1+\nu) & 4\beta^{-1} + 2(1-\nu)\beta & & & \text{Symmetric} & & & \\
3 & 2\beta - 2(1-\nu)\beta^{-1} & -\frac{3}{2}(1-3\nu) & 4\beta + 2(1-\nu)\beta^{-1} & & & & & \\
4 & \frac{3}{2}(1-3\nu) & -4\beta^{-1} + (1-\nu)\beta & -\frac{3}{2}(1+\nu) & 4\beta^{-1} + 2(1-\nu)\beta & & & & \\
5 & -2\beta - (1-\nu)\beta^{-1} & -\frac{3}{2}(1+\nu) & -4\beta + (1-\nu)\beta^{-1} & -\frac{3}{2}(1-3\nu) & 4\beta + 2(1-\nu)\beta^{-1} & & & \\
6 & -\frac{3}{2}(1+\nu) & -2\beta^{-1} - (1-\nu)\beta & \frac{3}{2}(1-3\nu) & 2\beta^{-1} - 2(1-\nu)\beta & \frac{3}{2}(1+\nu) & 4\beta^{-1} + 2(1-\nu)\beta & & \\
7 & -4\beta + (1-\nu)\beta^{-1} & \frac{3}{2}(1-3\nu) & -2\beta - (1-\nu)\beta^{-1} & \frac{3}{2}(1+\nu) & 2\beta - 2(1-\nu)\beta^{-1} & -\frac{3}{2}(1-3\nu) & 4\beta + 2(1-\nu)\beta^{-1} & \\
8 & -\frac{3}{2}(1-3\nu) & 2\beta^{-1} - 2(1-\nu)\beta & \frac{3}{2}(1+\nu) & -2\beta^{-1} - (1-\nu)\beta & \frac{3}{2}(1-3\nu) & -4\beta^{-1} + (1-\nu)\beta & -\frac{3}{2}(1+\nu) & 4\beta^{-1} + 2(1-\nu)\beta \\
\hline
 & 1 & 2 & 3 & 4 & 5 & 6 & 7 & 8
\end{array}
\tag{5.172}
$$

where to simplify the results the aspect ratio

$$\beta = \frac{b}{a} \tag{5.173}$$

has been introduced.

Similarly, substituting Eqs. (5.169) and (2.27) into (5.17), multiplying out the matrix product $\underline{\mathbf{b}}^T\boldsymbol{\varkappa}_T$, and then integrating over the whole volume of the plate, gives for the thermal stiffness $\mathbf{h}$

$$\mathbf{h} = \frac{Eta}{2(1-\nu)}\{\beta \quad 1 \quad \beta \quad -1 \quad -\beta \quad -1 \quad -\beta \quad 1\} \tag{5.174}$$

To determine the stiffness properties in the datum coordinate system the same procedure as in the case of other elements is adopted: the displacements in the local coordinate system are related to the displacement in the datum coordinate system, and from the relationship between the two sets of displacements the transformation matrix $\boldsymbol{\lambda}$ is formed. From Fig. 5.16 it is evident that the displacements $\mathbf{u} = \{u_1 \quad u_2 \quad \cdots \quad u_8\}$ can be expressed in terms of the datum coordinate system displacements $\bar{\mathbf{u}} = \{\bar{u}_{3p-2} \quad \bar{u}_{3p-1} \quad \bar{u}_{3p} \quad \cdots \quad \bar{u}_{3s}\}$ by the following equation:

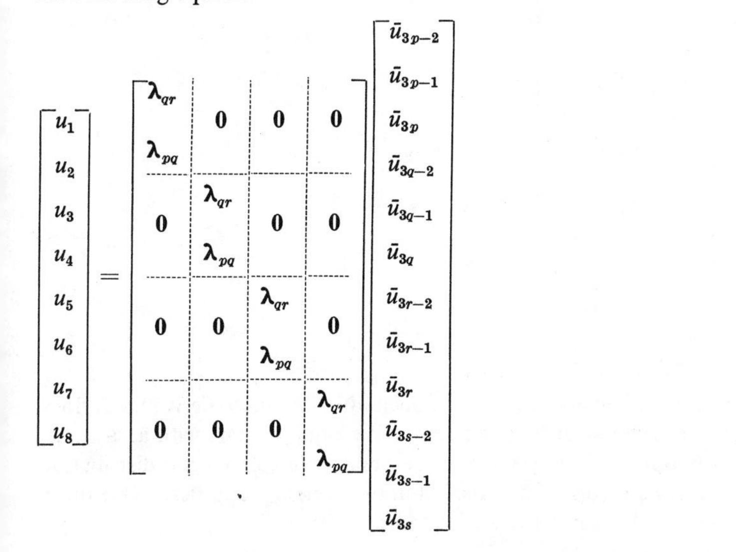

$$\begin{bmatrix} u_1 \\ u_2 \\ u_3 \\ u_4 \\ u_5 \\ u_6 \\ u_7 \\ u_8 \end{bmatrix} = \left[\begin{array}{c:c:c:c} \begin{matrix}\boldsymbol{\lambda}_{qr} \\ \boldsymbol{\lambda}_{pq}\end{matrix} & \mathbf{0} & \mathbf{0} & \mathbf{0} \\ \hdashline \mathbf{0} & \begin{matrix}\boldsymbol{\lambda}_{qr} \\ \boldsymbol{\lambda}_{pq}\end{matrix} & \mathbf{0} & \mathbf{0} \\ \hdashline \mathbf{0} & \mathbf{0} & \begin{matrix}\boldsymbol{\lambda}_{qr} \\ \boldsymbol{\lambda}_{pq}\end{matrix} & \mathbf{0} \\ \hdashline \mathbf{0} & \mathbf{0} & \mathbf{0} & \begin{matrix}\boldsymbol{\lambda}_{qr} \\ \boldsymbol{\lambda}_{pq}\end{matrix} \end{array}\right] \begin{bmatrix} \bar{u}_{3p-2} \\ \bar{u}_{3p-1} \\ \bar{u}_{3p} \\ \bar{u}_{3q-2} \\ \bar{u}_{3q-1} \\ \bar{u}_{3q} \\ \bar{u}_{3r-2} \\ \bar{u}_{3r-1} \\ \bar{u}_{3r} \\ \bar{u}_{3s-2} \\ \bar{u}_{3s-1} \\ \bar{u}_{3s} \end{bmatrix} \tag{5.175}$$

where $\boldsymbol{\lambda}_{pq}$ and $\boldsymbol{\lambda}_{qr}$ are the matrices of direction cosines for the pq and qr directions, respectively. In this case, the direction cosines can be found simply from the coordinates of the points p, q, and r. Thus, the transformation

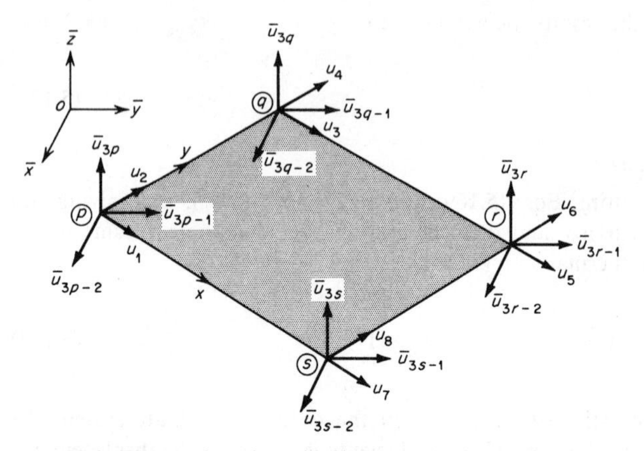

FIG. 5.16 Rectangular plate displacements in local and datum coordinate systems.

matrix $\boldsymbol{\lambda}$ for a rectangular panel is given by

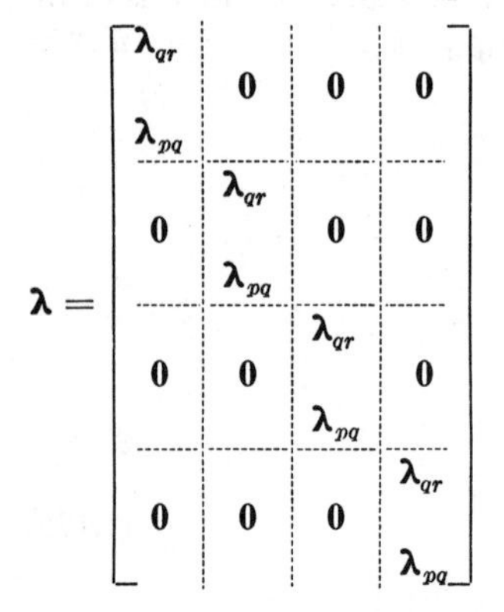

$$\boldsymbol{\lambda} = \begin{bmatrix} \begin{matrix}\boldsymbol{\lambda}_{qr}\\ \boldsymbol{\lambda}_{pq}\end{matrix} & \mathbf{0} & \mathbf{0} & \mathbf{0} \\ \mathbf{0} & \begin{matrix}\boldsymbol{\lambda}_{qr}\\ \boldsymbol{\lambda}_{pq}\end{matrix} & \mathbf{0} & \mathbf{0} \\ \mathbf{0} & \mathbf{0} & \begin{matrix}\boldsymbol{\lambda}_{qr}\\ \boldsymbol{\lambda}_{pq}\end{matrix} & \mathbf{0} \\ \mathbf{0} & \mathbf{0} & \mathbf{0} & \begin{matrix}\boldsymbol{\lambda}_{qr}\\ \boldsymbol{\lambda}_{pq}\end{matrix} \end{bmatrix} \tag{5.176}$$

LINEAR-STRESS ASSUMPTION

We have so far used an assumed displacement distribution to derive the stiffness properties of a rectangular plate element. We can also start with an assumed stress distribution and derive the corresponding displacement distribution, which can then be used for the calculation of stiffness properties. One of the simple stress distributions used is of the form[277,329]

$$\begin{aligned} \sigma_{xx} &= a_1 + a_2 y \\ \sigma_{yy} &= a_3 + a_4 x \\ \sigma_{xy} &= a_5 \end{aligned} \tag{5.177}$$

where $a_1, \ldots, a_5$ are constants and the orientation of the x and y axes is as shown in Fig. 5.14. This simple stress distribution, unlike the case of linearly varying edge displacements, satisfies identically the stress equilibrium within the rectangle; however, the resulting displacement distribution, which will be derived subsequently, violates the compatibility of boundary displacements on adjacent elements.

From Hooke's law for two-dimensional stress field and Eqs. (5.177) we have

$$\frac{\partial u_x}{\partial x} = \frac{1}{E}(a_1 + a_2 y - \nu a_3 - \nu a_4 x) + \alpha T \tag{5.178}$$

which, when integrated, becomes

$$u_x = \frac{1}{E}\left(a_1 x + a_2 xy - \nu a_3 x - \frac{\nu a_4 x^2}{2}\right) + \alpha T x + \frac{f(y)}{E} \tag{5.179}$$

where $f(y)$ is an arbitrary function of y. Similarly, starting with the strain ϵ_{yy}, we can show that

$$u_y = \frac{1}{E}\left(a_3 y + a_4 xy - \nu a_1 y - \frac{\nu a_2 y^2}{2}\right) + \alpha T y + \frac{g(x)}{E} \tag{5.180}$$

where $g(x)$ is a function of x only. Also, from the equation for the shearing strain we have

$$\frac{\partial u_x}{\partial y} + \frac{\partial u_y}{\partial x} = \frac{1}{G}\sigma_{xy} = 2(1 + \nu)\frac{a_5}{E} \tag{5.181}$$

Substituting Eqs. (5.179) and (5.180) into Eq. (5.181) and rearranging, we have

$$f'(y) + a_4 y = 2(1 + \nu)a_5 - [g'(x) + a_2 x] = a_6 \tag{5.182}$$

where a_6 represents a constant, which is the only possible condition that will satisfy Eq. (5.182). Primes in Eq. (5.182) represent derivatives with respect to the appropriate variables. Solving for $f(y)$ and $g(x)$, we have

$$f(y) = a_6 y - \frac{a_4 y^2}{2} + a_7 \tag{5.183}$$

$$g(x) = [2(1 + \nu)a_5 - a_6]x - \frac{a_2 x^2}{2} + a_8 \tag{5.184}$$

The constants of integration a_7 and a_8 simply represent rigid-body translations, while the previously introduced constants a_5 and a_6 define rigid-body rotation.

When we substitute Eqs. (5.183) and (5.184) into Eqs. (5.179) and (5.180) and rearrange constants $a_1, \ldots, a_8$ into new constants $c_1, \ldots, c_8$, it follows that

$$u_x = c_1 x + c_2 y - c_3(\nu x^2 + y^2) + 2c_4 xy + c_5 \tag{5.185}$$

$$u_y = c_6 x + c_7 y - c_4(x^2 + \nu y^2) + 2c_3 xy + c_8 \tag{5.186}$$

where

$$\begin{aligned}
&c_1 = \frac{1}{E}(a_1 - \nu a_3 + \alpha TE) \qquad c_2 = \frac{a_6}{E} \qquad c_3 = \frac{a_4}{2E} \\
&c_4 = \frac{a_2}{2E} \qquad c_5 = \frac{a_7}{E} \qquad c_6 = \frac{1}{E}[2(1 + \nu)a_5 - a_6] \\
&c_7 = \frac{1}{E}(a_3 - \nu a_1 + \alpha TE) \qquad c_8 = \frac{a_8}{E}
\end{aligned} \tag{5.187}$$

The unknown constants $c_1, \ldots, c_8$ can now be determined from the element displacements $u_1, \ldots, u_8$. Hence

$$\begin{aligned}
c_1 &= \frac{1}{a}(-u_1 + u_7) + \frac{\nu}{2b}(u_2 - u_4 + u_6 - u_8) \\
c_2 &= \frac{1}{b}(-u_1 + u_3) + \frac{1}{2a}(u_2 - u_4 + u_6 - u_8) \\
c_3 &= \frac{1}{2ab}(u_2 - u_4 + u_6 - u_8) \\
c_4 &= \frac{1}{2ab}(u_1 - u_3 + u_5 - u_7) \\
c_5 &= u_1 \\
c_6 &= \frac{1}{a}(-u_2 + u_8) + \frac{1}{2b}(u_1 - u_3 + u_5 - u_7) \\
c_7 &= \frac{1}{b}(-u_2 + u_4) + \frac{1}{2a}(u_1 - u_3 + u_5 - u_7) \\
c_8 &= u_2
\end{aligned} \tag{5.188}$$

An examination of the displacement functions given by Eqs. (5.185) and (5.186) reveals that the boundary displacements obtained from these functions are not compatible with the displacements on the adjacent elements. Nevertheless the use of such functions gives sufficient accuracy provided the size of elements is small in relation to the size of stress variations. The strain-displacement relationships can now be determined from Eqs. (5.185) and (5.186) and arranged into the matrix equation

$$\mathbf{e} = \bar{\mathbf{b}}\mathbf{u} \tag{5.189}$$

where

$$\bar{\mathbf{b}} = \begin{bmatrix} \frac{-(1-\eta)}{a} & \frac{\nu(1-2\xi)}{2b} & \frac{-\eta}{a} & \frac{-\nu(1-2\xi)}{2b} & \frac{\eta}{a} & \frac{\nu(1-2\xi)}{2b} & \frac{1-\eta}{a} & \frac{-\nu(1-2\xi)}{2b} \\ \frac{\nu(1-2\eta)}{2a} & \frac{-(1-\xi)}{b} & \frac{-\nu(1-2\eta)}{2a} & \frac{1-\xi}{b} & \frac{\nu(1-2\eta)}{2a} & \frac{\xi}{b} & \frac{-\nu(1-2\eta)}{2a} & \frac{-\xi}{b} \\ \frac{-1}{2b} & \frac{-1}{2a} & \frac{1}{2b} & \frac{-1}{2a} & \frac{1}{2b} & \frac{1}{2a} & \frac{-1}{2b} & \frac{1}{2a} \end{bmatrix} \tag{5.190}$$

Bars over the matrix $\bar{\mathbf{b}}$ are used here to indicate that the strain distribution satisfies equations of stress equilibrium. Substituting Eq. (5.189) into (2.24), we obtain the following stress-displacement equation:

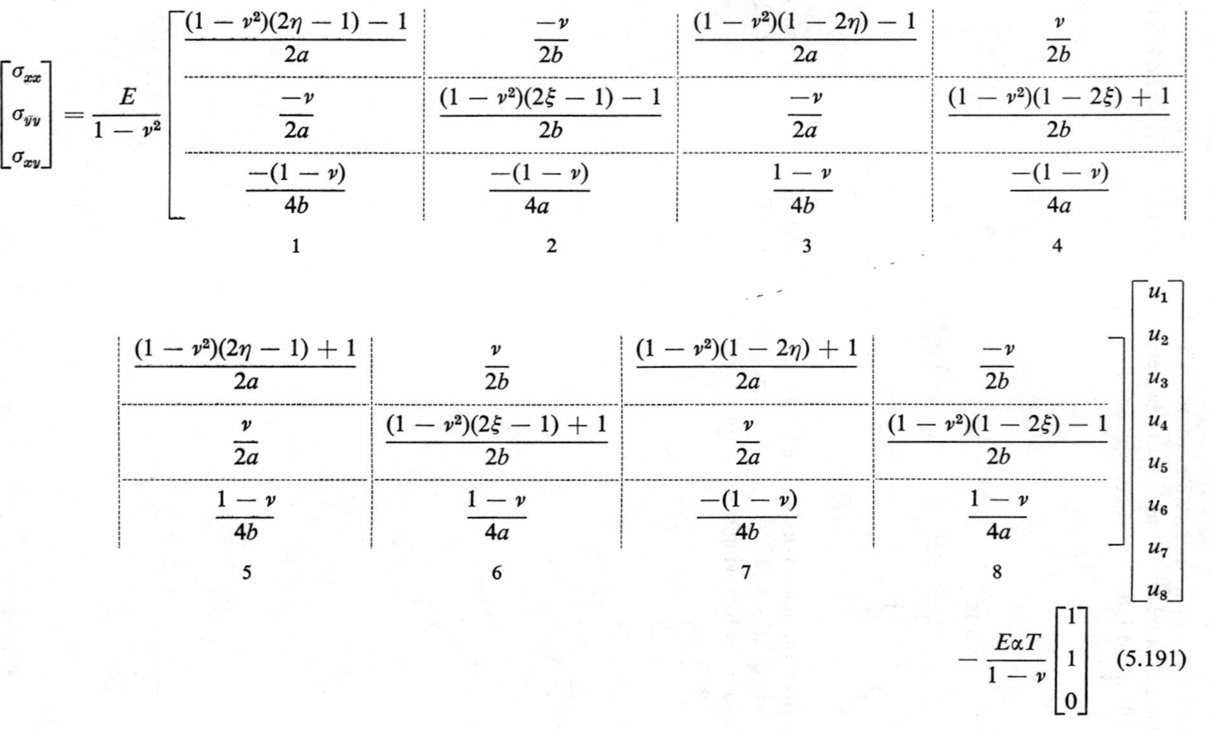

$$
\begin{bmatrix} \sigma_{xx} \\ \sigma_{yy} \\ \sigma_{xy} \end{bmatrix} = \frac{E}{1-\nu^2}
\left[
\underset{1}{\begin{matrix} \frac{(1-\nu^2)(2\eta-1)-1}{2a} \\ \frac{-\nu}{2a} \\ \frac{-(1-\nu)}{4b} \end{matrix}}
\;\Bigg|\;
\underset{2}{\begin{matrix} \frac{-\nu}{2b} \\ \frac{(1-\nu^2)(2\xi-1)-1}{2b} \\ \frac{-(1-\nu)}{4a} \end{matrix}}
\;\Bigg|\;
\underset{3}{\begin{matrix} \frac{(1-\nu^2)(1-2\eta)-1}{2a} \\ \frac{-\nu}{2a} \\ \frac{1-\nu}{4b} \end{matrix}}
\;\Bigg|\;
\underset{4}{\begin{matrix} \frac{\nu}{2b} \\ \frac{(1-\nu^2)(1-2\xi)+1}{2b} \\ \frac{-(1-\nu)}{4a} \end{matrix}}
\;\Bigg|\;
\underset{5}{\begin{matrix} \frac{(1-\nu^2)(2\eta-1)+1}{2a} \\ \frac{\nu}{2a} \\ \frac{1-\nu}{4b} \end{matrix}}
\;\Bigg|\;
\underset{6}{\begin{matrix} \frac{\nu}{2b} \\ \frac{(1-\nu^2)(2\xi-1)+1}{2b} \\ \frac{1-\nu}{4a} \end{matrix}}
\;\Bigg|\;
\underset{7}{\begin{matrix} \frac{(1-\nu^2)(1-2\eta)+1}{2a} \\ \frac{\nu}{2a} \\ \frac{-(1-\nu)}{4b} \end{matrix}}
\;\Bigg|\;
\underset{8}{\begin{matrix} \frac{-\nu}{2b} \\ \frac{(1-\nu^2)(1-2\xi)-1}{2b} \\ \frac{1-\nu}{4a} \end{matrix}}
\right]
\begin{bmatrix} u_1 \\ u_2 \\ u_3 \\ u_4 \\ u_5 \\ u_6 \\ u_7 \\ u_8 \end{bmatrix}
- \frac{E\alpha T}{1-\nu} \begin{bmatrix} 1 \\ 1 \\ 0 \end{bmatrix} \quad (5.191)
$$

The strain distribution matrix **б** due to unit displacements can now be used to determine the matrices **k** and **h** from the unit-displacement theorem. The stiffness matrix **k** is presented by

$$\mathbf{k} = \frac{Et}{12(1-\nu^2)} \times$$

	1	2	3	4	5	6	7	8
1	$(4-\nu^2)\beta + \frac{3}{2}(1-\nu)\beta^{-1}$							
2	$\frac{3}{2}(1+\nu)$	$(4-\nu^2)\beta^{-1} + \frac{3}{2}(1-\nu)\beta$			Symmetric			
3	$(2+\nu^2)\beta - \frac{3}{2}(1-\nu)\beta^{-1}$	$-\frac{3}{2}(1-3\nu)$	$(4-\nu^2)\beta + \frac{3}{2}(1-\nu)\beta^{-1}$					
4	$\frac{3}{2}(1-3\nu)$	$-(4-\nu^2)\beta^{-1} + \frac{3}{2}(1-\nu)\beta$	$-\frac{3}{2}(1+\nu)$	$(4-\nu^2)\beta^{-1} + \frac{3}{2}(1-\nu)\beta$				
5	$-(2+\nu^2)\beta - \frac{3}{2}(1-\nu)\beta^{-1}$	$-\frac{3}{2}(1+\nu)$	$-(4-\nu^2)\beta + \frac{3}{2}(1-\nu)\beta^{-1}$	$-\frac{3}{2}(1-3\nu)$	$(4-\nu^2)\beta + \frac{3}{2}(1-\nu)\beta^{-1}$			
6	$-\frac{3}{2}(1+\nu)$	$-(2+\nu^2)\beta^{-1} - \frac{3}{2}(1-\nu)\beta$	$\frac{3}{2}(1-3\nu)$	$(2+\nu^2)\beta^{-1} - \frac{3}{2}(1-\nu)\beta$	$\frac{3}{2}(1+\nu)$	$(4-\nu^2)\beta^{-1} + \frac{3}{2}(1-\nu)\beta$		
7	$-(4-\nu^2)\beta + \frac{3}{2}(1-\nu)\beta^{-1}$	$\frac{3}{2}(1-3\nu)$	$-(2+\nu^2)\beta - \frac{3}{2}(1-\nu)\beta^{-1}$	$\frac{3}{2}(1+\nu)$	$(2+\nu^2)\beta - \frac{3}{2}(1-\nu)\beta^{-1}$	$-\frac{3}{2}(1-3\nu)$	$(4-\nu^2)\beta + \frac{3}{2}(1-\nu)\beta^{-1}$	
8	$-\frac{3}{2}(1-3\nu)$	$(2+\nu^2)\beta^{-1} - \frac{3}{2}(1-\nu)\beta$	$\frac{3}{2}(1+\nu)$	$-(2+\nu^2)\beta^{-1} - \frac{3}{2}(1-\nu)\beta$	$\frac{3}{2}(1-3\nu)$	$-(4-\nu^2)\beta^{-1} + \frac{3}{2}(1-\nu)\beta$	$-\frac{3}{2}(1+\nu)$	$(4-\nu^2)\beta^{-1} + \frac{3}{2}(1-\nu)\beta$

(5.192)

and the thermal stiffness **h** is given by

$$\mathbf{h} = \frac{Eta}{2(1-\nu)}\{\beta \quad 1 \quad \beta \quad -1 \quad -\beta \quad -1 \quad -\beta \quad 1\} \tag{5.193}$$

It is interesting to note that the matrices **h** are identical for both the linear-displacement and linear-stress assumptions used for the rectangular plate elements. The stiffness matrices **k**, on the other hand, are different for the two assumptions.

5.9 QUADRILATERAL PLATE ELEMENTS (IN-PLANE FORCES)

The quadrilateral plate element is illustrated in Fig. 5.17, where the displacements u_1 to u_8 are referred to an arbitrary rectangular xy coordinate system. The stiffness properties of a general quadrilateral panel could be conveniently calculated by subdividing such panel into triangles and then combining the stiffness of individual triangular panels to form stiffness matrices of the quadrilateral panel. This method has been proposed by Turner et al.[329] There is, however, an alternative method, proposed by Taig,[315] which is somewhat analogous to the method used in the case of a rectangular panel. In this method a nonorthogonal coordinate system is used, which is indicated by the coordinates ξ and η in Fig. 5.17. The edges 1,2 and 3,4 on the quadrilateral are represented by $\xi = 0$ and $\xi = 1$, respectively, while the edges 1,4 and 2,3

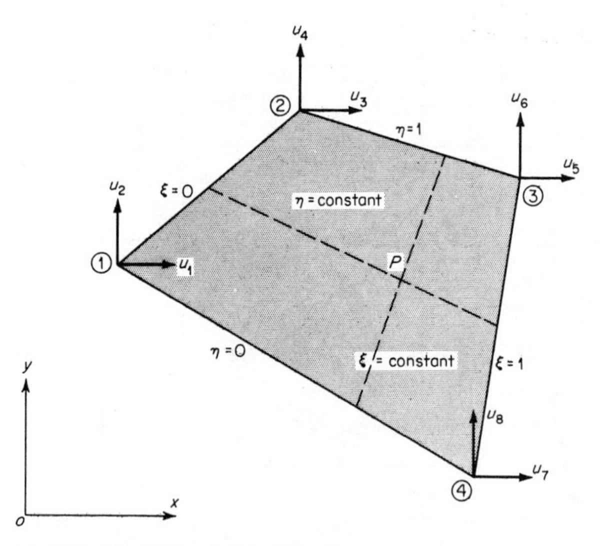

FIG. 5.17 Quadrilateral plate element.

correspond to $\eta = 0$ and $\eta = 1$. Thus, any arbitrary point P within the boundaries of the quadrilateral is defined by the intersection of two straight lines, $\xi =$ constant and $\eta =$ constant, which divide the two opposite sides of the panel in equal proportions. The coordinates ξ and η can be described as general quadrilateral coordinates. With this new coordinate system the rectangular coordinates (x,y) are related to the quadrilateral coordinates (ξ,η) by the following relationships:

$$x = x_1 + x_{41}\xi + x_{21}\eta + (x_{32} - x_{41})\xi\eta \tag{5.194}$$

$$y = y_1 + y_{41}\xi + y_{21}\eta + (y_{32} - y_{41})\xi\eta \tag{5.195}$$

The quadrilateral coordinate system is nondimensional, and when the general quadrilateral degenerates into rectangle, the quadrilateral coordinates (ξ,η) become identical with the nondimensional coordinates used on the rectangle in Sec. 5.8.

The assumed displacement functions for u_x and u_y will be taken to be of the same form as for the rectangular panel, as given by Eqs. (5.162) and (5.163), with the exception that the ξ and η coordinates refer now to the quadrilateral coordinate systems. This assumption will ensure, in view of the relations (5.194) and (5.195), that the boundary displacements will vary linearly in the rectangular coordinate system and also that the displacements on the adjacent elements will be compatible. Thus, for the subsequent development of the analysis, the displacements will be expressed in matrix notation as

$$\begin{bmatrix} u_x \\ u_y \end{bmatrix} = \begin{bmatrix} (1-\xi)(1-\eta) & 0 & (1-\xi)\eta & 0 & \xi\eta & 0 & \xi(1-\eta) & 0 \\ 0 & (1-\xi)(1-\eta) & 0 & (1-\xi)\eta & 0 & \xi\eta & 0 & \xi(1-\eta) \end{bmatrix} \begin{bmatrix} u_1 \\ u_2 \\ \cdot \\ \cdot \\ u_8 \end{bmatrix}$$

$$= \begin{bmatrix} f_1 & 0 & f_2 & 0 & f_3 & 0 & f_4 & 0 \\ 0 & f_1 & 0 & f_2 & 0 & f_3 & 0 & f_4 \end{bmatrix} \mathbf{u} \tag{5.196}$$

where $\quad \mathbf{u} = \{u_1 \quad u_2 \quad u_3 \quad u_4 \quad u_5 \quad u_6 \quad u_7 \quad u_8\}$ (5.197)

$$f_1 = (1-\xi)(1-\eta) \tag{5.198}$$

$$f_2 = (1-\xi)\eta \tag{5.199}$$

$$f_3 = \xi\eta \tag{5.200}$$

$$f_4 = \xi(1-\eta) \tag{5.201}$$

The three strain components e_{xx}, e_{yy}, and e_{xy} are obtained from the displacements u_x and u_y by partial differentiation with respect to the rectangular coordinates x and y; therefore, from Eq. (5.196) it follows that the strain matrix is given by

$$\mathbf{e} = \underline{\mathbf{b}}\mathbf{u} \tag{5.202}$$

where
$$\underline{\mathbf{b}} = \begin{bmatrix} \frac{\partial f_1}{\partial x} & 0 & \frac{\partial f_2}{\partial x} & 0 & \frac{\partial f_3}{\partial x} & 0 & \frac{\partial f_4}{\partial x} & 0 \\ 0 & \frac{\partial f_1}{\partial y} & 0 & \frac{\partial f_2}{\partial y} & 0 & \frac{\partial f_3}{\partial y} & 0 & \frac{\partial f_4}{\partial y} \\ \frac{\partial f_1}{\partial y} & \frac{\partial f_1}{\partial x} & \frac{\partial f_2}{\partial y} & \frac{\partial f_2}{\partial x} & \frac{\partial f_3}{\partial y} & \frac{\partial f_3}{\partial x} & \frac{\partial f_4}{\partial y} & \frac{\partial f_4}{\partial x} \end{bmatrix} \tag{5.203}$$

The partial derivatives of f_i are calculated from

$$\frac{\partial f_i}{\partial x} = \frac{\mathscr{J}\left(\dfrac{f_i, y}{\xi, \eta}\right)}{\mathscr{J}\left(\dfrac{x, y}{\xi, \eta}\right)} \tag{5.204}$$

$$\frac{\partial f_i}{\partial y} = \frac{\mathscr{J}\left(\dfrac{x, f_i}{\xi, \eta}\right)}{\mathscr{J}\left(\dfrac{x, y}{\xi, \eta}\right)} \tag{5.205}$$

$$\mathscr{J}\left(\frac{f_i, y}{\xi, \eta}\right) = \begin{vmatrix} \frac{\partial f_i}{\partial \xi} & \frac{\partial y}{\partial \xi} \\ \frac{\partial f_i}{\partial \eta} & \frac{\partial y}{\partial \eta} \end{vmatrix} = \frac{\partial f_i}{\partial \xi}\frac{\partial y}{\partial \eta} - \frac{\partial f_i}{\partial \eta}\frac{\partial y}{\partial \xi} \tag{5.206}$$

$$\mathscr{J}\left(\frac{x, f_i}{\xi, \eta}\right) = \begin{vmatrix} \frac{\partial x}{\partial \xi} & \frac{\partial f_i}{\partial \xi} \\ \frac{\partial x}{\partial \eta} & \frac{\partial f_i}{\partial \eta} \end{vmatrix} = \frac{\partial x}{\partial \xi}\frac{\partial f_i}{\partial \eta} - \frac{\partial x}{\partial \eta}\frac{\partial f_i}{\partial \xi} \tag{5.207}$$

$$\mathscr{J}\left(\frac{x, y}{\xi, \eta}\right) = \begin{vmatrix} \frac{\partial x}{\partial \xi} & \frac{\partial y}{\partial \xi} \\ \frac{\partial x}{\partial \eta} & \frac{\partial y}{\partial \eta} \end{vmatrix} = \frac{\partial x}{\partial \xi}\frac{\partial y}{\partial \eta} - \frac{\partial x}{\partial \eta}\frac{\partial y}{\partial \xi} \tag{5.208}$$

where the last equation represents the jacobian of the coordinate transformation (5.194) and (5.195). Evaluating expressions (5.206) to (5.208), we obtain

$$\mathscr{J}\left(\frac{f_1,y}{\xi,\eta}\right) = y_{42} - y_{32}\xi - y_{43}\eta \tag{5.209}$$

$$\mathscr{J}\left(\frac{f_2,y}{\xi,\eta}\right) = -y_{41} + y_{41}\xi + y_{43}\eta \tag{5.210}$$

$$\mathscr{J}\left(\frac{f_3,y}{\xi,\eta}\right) = -y_{41}\xi + y_{21}\eta \tag{5.211}$$

$$\mathscr{J}\left(\frac{f_4,y}{\xi,\eta}\right) = y_{21} + y_{32}\xi - y_{21}\eta \tag{5.212}$$

$$\mathscr{J}\left(\frac{x,f_1}{\xi,\eta}\right) = -x_{42} + x_{32}\xi + x_{43}\eta \tag{5.213}$$

$$\mathscr{J}\left(\frac{x,f_2}{\xi,\eta}\right) = x_{41} - x_{41}\xi - x_{43}\eta \tag{5.214}$$

$$\mathscr{J}\left(\frac{x,f_3}{\xi,\eta}\right) = x_{41}\xi - x_{21}\eta \tag{5.215}$$

$$\mathscr{J}\left(\frac{x,f_4}{\xi,\eta}\right) = -x_{21} - x_{32}\xi + x_{21}\eta \tag{5.216}$$

$$\mathscr{J}\left(\frac{x,y}{\xi,\eta}\right) = (x_{41}y_{21} - y_{41}x_{21}) + (x_{41}y_{32} - y_{41}x_{32})\xi + (x_{21}y_{43} - y_{21}x_{43})\eta \tag{5.217}$$

To evaluate the matrices **k** and **h** for the general quadrilateral panel the integration in Eqs. (5.16) and (5.17) must be carried out with respect to the quadrilateral coordinates ξ and η. Hence

$$\mathbf{k} = \int_v \underline{\mathbf{b}}^T \boldsymbol{\varkappa} \underline{\mathbf{b}}\, dV = t\int_0^1\int_0^1 \underline{\mathbf{b}}^T \boldsymbol{\varkappa} \underline{\mathbf{b}} \left| \mathscr{J}\left(\frac{x,y}{\xi,\eta}\right) \right| d\xi\, d\eta \tag{5.218}$$

and
$$\mathbf{h} = \int_v \underline{\mathbf{b}}^T \boldsymbol{\varkappa}_T\, dV = t\int_0^1\int_0^1 \underline{\mathbf{b}}^T \boldsymbol{\varkappa}_T \left| \mathscr{J}\left(\frac{x,y}{\xi,\eta}\right) \right| d\xi\, d\eta \tag{5.219}$$

When the expression for $\underline{\mathbf{b}}$ is substituted into (5.218), each coefficient in the matrix **k** is of the following form

$$I = \int_0^1\int_0^1 \frac{(A_1 + B_1\xi + C_1\eta)(A_2 + B_2\xi + C_2\eta)}{A_0 + B_0\xi + C_0\eta}\, d\xi\, d\eta \tag{5.220}$$

where $A_1, B_1, \ldots$ are constants expressed in terms of the rectangular coordinates of the four corners of the quadrilateral. Equation (5.220) can be integrated with respect to ξ; thus

$$I = \int_0^1 \left\{ \frac{B_1B_2}{2B_0} + \frac{A_1B_2 + B_1A_2 + (B_1C_2 + C_1B_2)\eta}{B_0} - \frac{B_1B_2(A_0 + C_0\eta)}{B_0^2} \right.$$

$$+ \frac{A_1A_2 + (A_1C_2 + C_1A_2)\eta + C_1C_2\eta^2}{B_0} \ln \frac{A_0 + B_0 + C_0\eta}{A_0 + C_0\eta}$$

$$- \frac{B_1B_2(A_0 + C_0\eta)}{B_0^2} \left[\frac{A_1B_2 + B_1A_2 + (B_1C_2 + C_1B_2)\eta}{B_1B_2} - \frac{A_0 + C_0\eta}{B_0} \right]$$

$$\left. \times \ln \frac{A_0 + B_0 + C_0\eta}{A_0 + C_0\eta} \right\} d\eta \qquad (5.220a)$$

Equation (5.220*a*) can be integrated further, but the results are too lengthy to be reproduced in full. A computer program for the calculation of integrals given by Eq. (5.220) has been compiled by Taig.[315] This program calculates directly all the necessary elements in the stiffness matrix **k**; however, the limiting case of a trapezoidal panel (including a parallelogram) for which the constant $C_0 = 0$ cannot be evaluated from the general program for quadrilateral panels. To circumvent this difficulty Taig used a special program for trapezoidal and parallelogram plate elements.

The calculation of the thermal stiffness **h** is considerably simpler. By using (5.203) and (2.27) in (5.219) it can be shown that

$$\mathbf{h} = \frac{Et}{2(1-\nu)} \{-y_{42} \quad x_{42} \quad y_{31} \quad -x_{31} \quad y_{42} \quad -x_{42} \quad -y_{31} \quad x_{31}\} \qquad (5.221)$$

The transformation matrix $\boldsymbol{\lambda}$ for a quadrilateral panel is of the same form as for the rectangle, that is,

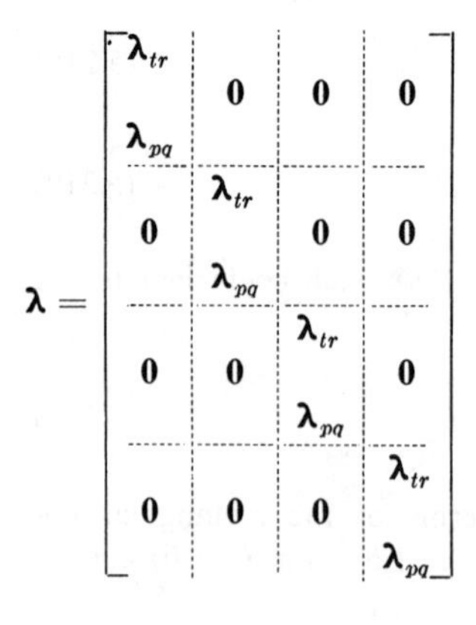

$$\boldsymbol{\lambda} = \begin{bmatrix} \begin{matrix}\boldsymbol{\lambda}_{tr} \\ \boldsymbol{\lambda}_{pq}\end{matrix} & \mathbf{0} & \mathbf{0} & \mathbf{0} \\ \mathbf{0} & \begin{matrix}\boldsymbol{\lambda}_{tr} \\ \boldsymbol{\lambda}_{pq}\end{matrix} & \mathbf{0} & \mathbf{0} \\ \mathbf{0} & \mathbf{0} & \begin{matrix}\boldsymbol{\lambda}_{tr} \\ \boldsymbol{\lambda}_{pq}\end{matrix} & \mathbf{0} \\ \mathbf{0} & \mathbf{0} & \mathbf{0} & \begin{matrix}\boldsymbol{\lambda}_{tr} \\ \boldsymbol{\lambda}_{pq}\end{matrix} \end{bmatrix} \qquad (5.222)$$

where the local y axis is assumed to be taken parallel to the pq edge, so that $\boldsymbol{\lambda}_{tr}$ represent the matrix direction cosines for a perpendicular from the point r to line pq, and $\boldsymbol{\lambda}_{pq}$ represents direction cosines for pq.

5.10 TETRAHEDRON ELEMENTS[109]

For a solid tetrahedron (see Fig. 5.18) the assumed displacement distribution will be taken to be linear in x, y, and z. Hence

$$u_x = c_1 x + c_2 y + c_3 z + c_4 \tag{5.223}$$

$$u_y = c_5 x + c_6 y + c_7 z + c_8 \tag{5.224}$$

$$u_z = c_9 x + c_{10} y + c_{11} z + c_{12} \tag{5.225}$$

where $c_1, \ldots, c_{12}$ are constants to be determined from the conditions

$$\begin{array}{llll} u_x = u_1 & u_y = u_2 & u_z = u_3 & \text{at } (x_1,y_1,z_1) \\ u_x = u_4 & u_y = u_5 & u_z = u_6 & \text{at } (x_2,y_2,z_2) \\ u_x = u_7 & u_y = u_8 & u_z = u_9 & \text{at } (x_3,y_3,z_3) \\ u_x = u_{10} & u_y = u_{11} & u_z = u_{12} & \text{at } (x_4,y_4,z_4) \end{array} \tag{5.226}$$

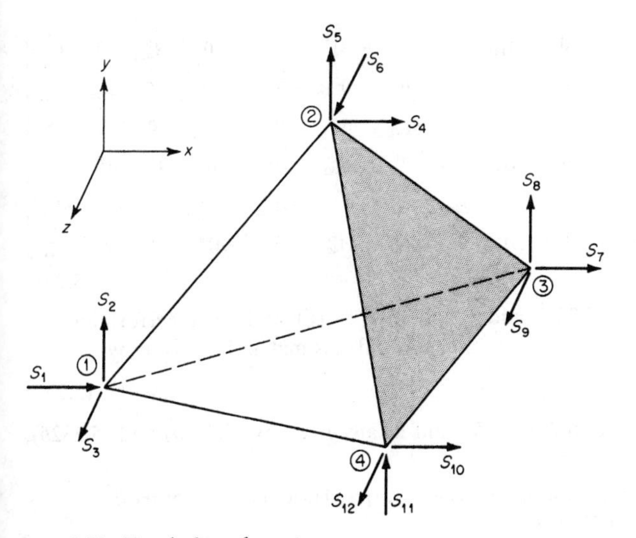

FIG. 5.18 Tetrahedron element.

From Eqs. (5.223) and (5.226) it can be shown that

$$\begin{aligned} u_x = &-\frac{1}{3V}(A^{yz}_{432}x + A^{zx}_{432}y + A^{xy}_{432}z - 3V_{4320})u_1 \\ &+\frac{1}{3V}(A^{yz}_{431}x + A^{zx}_{431}y + A^{xy}_{431}z - 3V_{4310})u_4 \\ &-\frac{1}{3V}(A^{yz}_{421}x + A^{zx}_{421}y + A^{xy}_{421}z - 3V_{4210})u_7 \\ &+\frac{1}{3V}(A^{yz}_{321}x + A^{zx}_{321}y + A^{xy}_{321}z - 3V_{3210})u_{10} \end{aligned} \tag{5.227}$$

where a typical term A^{ij}_{pqr} represents the area projection of the triangle pqr on the ij coordinate plane, V is the element volume, and V_{pqro} is the volume of the tetrahedron formed by the vertices pqr and the origin o.* Noting that Eqs. (5.223) to (5.225) are of identical form, we see that it follows that expressions for displacements u_y and u_z will be of the same form as (5.227), with u_1, u_4, u_7, and u_{10} replaced by the appropriate element displacements in the y and z directions.

On differentiation of the displacement relationships the following matrix equation

$$\mathbf{e} = \underline{\mathbf{b}}\mathbf{u} \tag{5.228}$$

is obtained, where in this case

$$\mathbf{e} = \{e_{xx} \quad e_{yy} \quad e_{zz} \quad e_{xy} \quad e_{yz} \quad e_{zx}\} \tag{5.229}$$

$$\mathbf{u} = \{u_1 \quad u_2 \quad \cdots \quad u_{12}\} \tag{5.230}$$

and

$$\underline{\mathbf{b}} = \frac{1}{3V}\begin{bmatrix} -A^{yz}_{432} & 0 & 0 & A^{yz}_{431} & 0 & 0 & -A^{yz}_{421} & 0 & 0 & A^{yz}_{321} & 0 & 0 \\ 0 & -A^{zx}_{432} & 0 & 0 & A^{zx}_{431} & 0 & 0 & -A^{zx}_{421} & 0 & 0 & A^{zx}_{321} & 0 \\ 0 & 0 & -A^{xy}_{432} & 0 & 0 & A^{xy}_{431} & 0 & 0 & -A^{xy}_{421} & 0 & 0 & A^{xy}_{321} \\ -A^{zx}_{432} & -A^{yz}_{432} & 0 & A^{zx}_{431} & A^{yz}_{431} & 0 & -A^{zx}_{421} & -A^{yz}_{421} & 0 & A^{zx}_{321} & A^{yz}_{321} & 0 \\ 0 & -A^{xy}_{432} & -A^{zx}_{432} & 0 & A^{xy}_{431} & A^{zx}_{431} & 0 & -A^{xy}_{421} & -A^{zx}_{421} & 0 & A^{xy}_{321} & A^{zx}_{321} \\ -A^{xy}_{432} & 0 & -A^{yz}_{432} & A^{xy}_{431} & 0 & A^{yz}_{431} & -A^{xy}_{421} & 0 & -A^{yz}_{421} & A^{xy}_{321} & 0 & A^{yz}_{321} \end{bmatrix} \tag{5.231}$$

Substituting Eqs. (5.231) and (2.18) into (5.16) and then performing the required integration demonstrate that the stiffness matrix $\mathbf{k}$ is given by

$$\mathbf{k} = \mathbf{k}_n + \mathbf{k}_s \tag{5.232}$$

where the component stiffnesses $\mathbf{k}_n$ and $\mathbf{k}_s$ are given by (5.232*a*) and (5.232*b*), respectively.†

*A^{ij}_{pqr}, V, and V_{pqro} are defined on the double-page spread following page 109; these quantities may be either positive or negative.

†Equation (5.232*a*) appears on the double-page spread following page 109. Equation (5.232*b*) appears on the double-page spread following that one.

Similarly, from Eqs. (5.231) and (2.27) it can be shown that the thermal stiffness **h** is given by

$$\mathbf{h} = \frac{E}{3(1-2\nu)}\{A^{yz}_{432} \quad A^{zx}_{432} \quad A^{xy}_{432} \quad -A^{yz}_{431} \quad -A^{zx}_{431} \quad -A^{xy}_{431} \quad A^{yz}_{421} \quad A^{zx}_{421} \quad A^{xy}_{421} \quad -A^{yz}_{321} \quad -A^{zx}_{321} \quad -A^{xy}_{321}\} \tag{5.233}$$

For tetrahedron elements there is no advantage in calculating **k** and **h** in the local coordinate system and then using $\boldsymbol{\lambda}$-matrix transformation to determine $\bar{\mathbf{k}}$ and $\bar{\mathbf{h}}$; the datum system of coordinates can be used here *ab initio* to determine $\bar{\mathbf{k}}$ and $\bar{\mathbf{h}}$ from Eqs. (5.232*a*), (5.232*b*), and (5.233).

5.11 TRIANGULAR PLATES IN BENDING

In the small-deflection theory of thin plates the transverse (normal) deflections u_z are uncoupled from the in-plane deflections u_x and u_y. Consequently, the stiffness matrices for the in-plane and transverse deflections are also uncoupled, and they can be calculated independently. The stiffness matrices for the in-plane displacements have been presented in Sec. 5.7. To calculate the stiffness matrix for the transverse deflections and rotations, as shown in Fig. 5.19, we may start with the assumed deflection u_z of the form

$$u_z = c_1 + c_2x + c_3y + c_4x^2 + c_5xy + c_6y^2 + c_7x^3 + c_8(xy^2 + x^2y) + c_9y^3 \tag{5.234}$$

where $c_1, \ldots, c_9$ are constants. These constants may be evaluated in terms of the displacements and slopes at the three corners of the triangular plate using

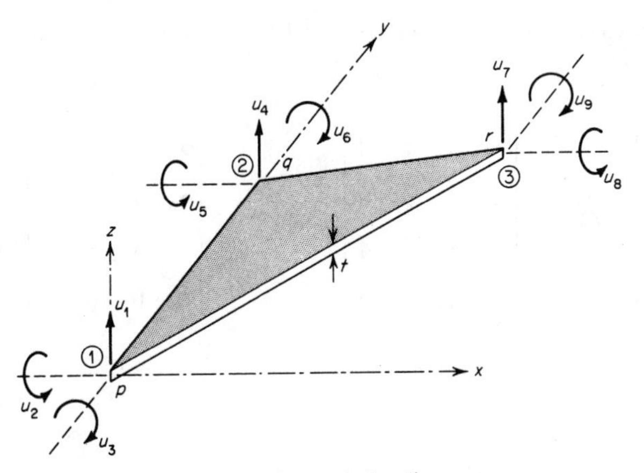

FIG. 5.19 Triangular plate element in bending.

$$\mathbf{k}_n = \frac{E}{9(1+\nu)(1-2\nu)V} \times$$

	1	2	3	4	5	6
1	$(1-\nu)A^{yz}_{432}A^{yz}_{432}$					
2	$\nu A^{zx}_{432}A^{yz}_{432}$	$(1-\nu)A^{zx}_{432}A^{zx}_{432}$				
3	$\nu A^{xy}_{432}A^{yz}_{432}$	$\nu A^{xy}_{432}A^{zx}_{432}$	$(1-\nu)A^{xy}_{432}A^{xy}_{432}$			
4	$-(1-\nu)A^{yz}_{431}A^{yz}_{432}$	$-\nu A^{yz}_{431}A^{zx}_{432}$	$-\nu A^{yz}_{431}A^{xy}_{432}$	$(1-\nu)A^{yz}_{431}A^{yz}_{431}$		
5	$-\nu A^{zx}_{431}A^{yz}_{432}$	$-(1-\nu)A^{zx}_{431}A^{zx}_{432}$	$-\nu A^{zx}_{431}A^{xy}_{432}$	$\nu A^{zx}_{431}A^{yz}_{431}$	$(1-\nu)A^{zx}_{431}A^{zx}_{431}$	
6	$-\nu A^{xy}_{431}A^{yz}_{432}$	$-\nu A^{xy}_{431}A^{zx}_{432}$	$-(1-\nu)A^{xy}_{431}A^{xy}_{432}$	$\nu A^{xy}_{431}A^{yz}_{431}$	$\nu A^{xy}_{431}A^{zx}_{431}$	$(1-\nu)A^{xy}_{431}A^{xy}_{431}$
7	$(1-\nu)A^{yz}_{421}A^{yz}_{432}$	$\nu A^{yz}_{421}A^{zx}_{432}$	$\nu A^{yz}_{421}A^{xy}_{432}$	$-(1-\nu)A^{yz}_{421}A^{yz}_{431}$	$-\nu A^{yz}_{421}A^{zx}_{431}$	$-\nu A^{yz}_{421}A^{xy}_{431}$
8	$\nu A^{zx}_{421}A^{yz}_{432}$	$(1-\nu)A^{zx}_{421}A^{zx}_{432}$	$\nu A^{zx}_{421}A^{xy}_{432}$	$-\nu A^{zx}_{421}A^{yz}_{431}$	$-(1-\nu)A^{zx}_{421}A^{zx}_{431}$	$-\nu A^{zx}_{421}A^{xy}_{431}$
9	$\nu A^{xy}_{421}A^{yz}_{432}$	$\nu A^{xy}_{421}A^{zx}_{432}$	$(1-\nu)A^{xy}_{421}A^{xy}_{432}$	$-\nu A^{xy}_{421}A^{yz}_{431}$	$-\nu A^{xy}_{421}A^{zx}_{431}$	$-(1-\nu)A^{xy}_{421}A^{xy}_{43}$
10	$-(1-\nu)A^{yz}_{321}A^{yz}_{432}$	$-\nu A^{yz}_{321}A^{zx}_{432}$	$-\nu A^{yz}_{321}A^{xy}_{432}$	$(1-\nu)A^{yz}_{321}A^{yz}_{431}$	$\nu A^{yz}_{321}A^{zx}_{431}$	$\nu A^{yz}_{321}A^{xy}_{431}$
11	$-\nu A^{zx}_{321}A^{yz}_{432}$	$-(1-\nu)A^{zx}_{321}A^{zx}_{432}$	$-\nu A^{zx}_{321}A^{xy}_{432}$	$\nu A^{zx}_{321}A^{yz}_{431}$	$(1-\nu)A^{zx}_{321}A^{zx}_{431}$	$\nu A^{zx}_{321}A^{xy}_{431}$
12	$-\nu A^{xy}_{321}A^{yz}_{432}$	$-\nu A^{xy}_{321}A^{zx}_{432}$	$-(1-\nu)A^{xy}_{321}A^{xy}_{432}$	$\nu A^{xy}_{321}A^{yz}_{431}$	$\nu A^{xy}_{321}A^{zx}_{431}$	$(1-\nu)A^{xy}_{321}A^{xy}_{431}$

where the projected areas A^{ij}_{pqr} and the volumes V and V_{pqro}, introduced in Eq. (5.227), are defined by:

$$2A^{ij}_{pqr} = \begin{vmatrix} i_p & j_p & 1 \\ i_q & j_q & 1 \\ i_r & j_r & 1 \end{vmatrix} \; ; \; 6V = \begin{vmatrix} x_1 & y_1 & z_1 & 1 \\ x_2 & y_2 & z_2 & 1 \\ x_3 & y_3 & z_3 & 1 \\ x_4 & y_4 & z_4 & 1 \end{vmatrix} \; ; \; 6V_{pqro} = \begin{vmatrix} x_p & y_p & z_p & 1 \\ x_q & y_q & z_q & 1 \\ x_r & y_r & z_r & 1 \\ 0 & 0 & 0 & 1 \end{vmatrix}$$

Symmetric

	7	8	9	10	11	12
7	$(1-\nu)A^{yz}_{421}A^{yz}_{421}$					
8	$\nu A^{zx}_{421}A^{yz}_{421}$	$(1-\nu)A^{zx}_{421}A^{zx}_{421}$				
9	$\nu A^{xy}_{421}A^{yz}_{421}$	$\nu A^{xy}_{421}A^{zx}_{421}$	$(1-\nu)A^{xy}_{421}A^{xy}_{421}$			
10	$-(1-\nu)A^{yz}_{321}A^{yz}_{421}$	$-\nu A^{yz}_{321}A^{zx}_{421}$	$-\nu A^{yz}_{321}A^{xy}_{421}$	$(1-\nu)A^{yz}_{321}A^{yz}_{321}$		
11	$-\nu A^{zx}_{321}A^{yz}_{421}$	$-(1-\nu)A^{zx}_{321}A^{zx}_{421}$	$-\nu A^{zx}_{321}A^{xy}_{421}$	$\nu A^{zx}_{321}A^{yz}_{321}$	$(1-\nu)A^{zx}_{321}A^{zx}_{321}$	
12	$-\nu A^{xy}_{321}A^{yz}_{421}$	$-\nu A^{xy}_{321}A^{zx}_{421}$	$-(1-\nu)A^{xy}_{321}A^{xy}_{421}$	$\nu A^{xy}_{321}A^{yz}_{321}$	$\nu A^{xy}_{321}A^{zx}_{321}$	$(1-\nu)A^{xy}_{321}A^{xy}_{321}$

(5.232*a*)

$$\mathbf{k}_s = \frac{E}{18(1+\nu)V} \times$$

	1	2	3	4	5	6
1	$(A^{zx}_{432})^2 + (A^{xy}_{432})^2$					
2	$A^{yz}_{432}A^{zx}_{432}$	$(A^{yz}_{432})^2 + (A^{xy}_{432})^2$				
3	$A^{yz}_{432}A^{xy}_{432}$	$A^{zx}_{432}A^{xy}_{432}$	$(A^{zx}_{432})^2 + (A^{yz}_{432})^2$			
4	$-A^{zx}_{431}A^{zx}_{432} - A^{xy}_{431}A^{xy}_{432}$	$-A^{zx}_{431}A^{yz}_{432}$	$-A^{xy}_{431}A^{yz}_{432}$	$(A^{zx}_{431})^2 + (A^{xy}_{431})^2$		
5	$-A^{yz}_{431}A^{zx}_{432}$	$-A^{yz}_{431}A^{yz}_{432} - A^{xy}_{431}A^{xy}_{432}$	$-A^{xy}_{431}A^{zx}_{432}$	$A^{yz}_{431}A^{zx}_{431}$	$(A^{yz}_{431})^2 + (A^{xy}_{431})^2$	
6	$-A^{yz}_{431}A^{xy}_{432}$	$-A^{zx}_{431}A^{xy}_{432}$	$-A^{zx}_{431}A^{zx}_{432} - A^{yz}_{431}A^{yz}_{432}$	$A^{yz}_{431}A^{xy}_{431}$	$A^{zx}_{431}A^{xy}_{431}$	$(A^{zx}_{431})^2 + (A^{yz}_{431})^2$
7	$A^{zx}_{421}A^{zx}_{432} + A^{xy}_{421}A^{xy}_{432}$	$A^{zx}_{421}A^{yz}_{432}$	$A^{xy}_{421}A^{yz}_{432}$	$-A^{zx}_{421}A^{zx}_{431} - A^{xy}_{421}A^{xy}_{431}$	$-A^{zx}_{421}A^{yz}_{431}$	$-A^{xy}_{421}A^{yz}_{431}$
8	$A^{yz}_{421}A^{zx}_{432}$	$A^{yz}_{421}A^{yz}_{432} + A^{xy}_{421}A^{xy}_{432}$	$A^{xy}_{421}A^{zx}_{432}$	$-A^{yz}_{421}A^{zx}_{431}$	$-A^{yz}_{421}A^{yz}_{431} - A^{xy}_{421}A^{xy}_{431}$	$-A^{xy}_{421}A^{zx}_{431}$
9	$A^{yz}_{421}A^{xy}_{432}$	$A^{zx}_{421}A^{xy}_{432}$	$A^{zx}_{421}A^{zx}_{432} + A^{yz}_{421}A^{yz}_{432}$	$-A^{yz}_{421}A^{xy}_{431}$	$-A^{zx}_{421}A^{xy}_{431}$	$-A^{zx}_{421}A^{zx}_{431} - A^{yz}_{421}A^{yz}_{431}$
10	$-A^{zx}_{321}A^{zx}_{432} - A^{xy}_{321}A^{xy}_{432}$	$-A^{zx}_{321}A^{yz}_{432}$	$-A^{xy}_{321}A^{yz}_{432}$	$A^{zx}_{321}A^{zx}_{431} + A^{xy}_{321}A^{xy}_{431}$	$A^{zx}_{321}A^{yz}_{431}$	$A^{xy}_{321}A^{yz}_{431}$
11	$-A^{yz}_{321}A^{zx}_{432}$	$-A^{yz}_{321}A^{yz}_{432} - A^{xy}_{321}A^{xy}_{432}$	$-A^{xy}_{321}A^{zx}_{432}$	$A^{yz}_{321}A^{zx}_{431}$	$A^{yz}_{321}A^{yz}_{431} + A^{xy}_{321}A^{xy}_{431}$	$A^{xy}_{321}A^{zx}_{431}$
12	$-A^{yz}_{321}A^{xy}_{432}$	$-A^{zx}_{321}A^{xy}_{432}$	$-A^{zx}_{321}A^{zx}_{432} - A^{yz}_{321}A^{yz}_{432}$	$A^{yz}_{321}A^{xy}_{431}$	$A^{zx}_{321}A^{xy}_{431}$	$A^{zx}_{321}A^{zx}_{431} + A^{yz}_{321}A^{yz}_{431}$

Symmetric

$(A_{421}^{zx})^2 + (A_{421}^{xy})^2$					
$A_{421}^{yz} A_{421}^{zx}$	$(A_{421}^{yz})^2 + (A_{421}^{xy})^2$				
$A_{421}^{yz} A_{421}^{xy}$	$A_{421}^{zx} A_{421}^{xy}$	$(A_{421}^{zx})^2 + (A_{421}^{yz})^2$			
$-A_{321}^{zx} A_{421}^{zx} - A_{321}^{xy} A_{421}^{xy}$	$-A_{321}^{zx} A_{421}^{yz}$	$-A_{321}^{xy} A_{421}^{yz}$	$(A_{321}^{zx})^2 + (A_{321}^{xy})^2$		
$-A_{321}^{yz} A_{421}^{zx}$	$-A_{321}^{yz} A_{421}^{yz} - A_{321}^{xy} A_{421}^{xy}$	$-A_{321}^{xy} A_{421}^{zx}$	$A_{321}^{yz} A_{321}^{zx}$	$(A_{321}^{yz})^2 + (A_{321}^{xy})^2$	
$-A_{321}^{yz} A_{421}^{xy}$	$-A_{321}^{zx} A_{421}^{xy}$	$-A_{321}^{zx} A_{421}^{zx} - A_{321}^{yz} A_{421}^{yz}$	$A_{321}^{yz} A_{321}^{xy}$	$A_{321}^{zx} A_{321}^{xy}$	$(A_{321}^{zx})^2 + (A_{321}^{yz})^2$
7	8	9	10	11	12

(5.232*b*)

$$\mathbf{u} = \mathbf{C}\mathbf{c} \tag{5.235}$$

where $\mathbf{u} = \{u_1 \quad u_2 \quad \cdots \quad u_9\}$ (5.236)

and $\mathbf{c} = \{c_1 \quad c_2 \quad \cdots \quad c_9\}$ (5.237)

while the matrix $\mathbf{C}$ is given by *

$$\mathbf{C} = \begin{array}{c} 1 \\ 2 \\ 3 \\ 4 \\ 5 \\ 6 \\ 7 \\ 8 \\ 9 \\ \\ \end{array} \begin{bmatrix} 1 & 0 & 0 & 0 & 0 & 0 & 0 & 0 & 0 \\ 0 & 0 & 1 & 0 & 0 & 0 & 0 & 0 & 0 \\ 0 & -1 & 0 & 0 & 0 & 0 & 0 & 0 & 0 \\ 1 & 0 & y_2 & 0 & 0 & y_2^2 & 0 & 0 & y_2^3 \\ 0 & 0 & 1 & 0 & 0 & 2y_2 & 0 & 0 & 3y_2^2 \\ 0 & -1 & 0 & 0 & -y_2 & 0 & 0 & -y_2^2 & 0 \\ 1 & x_3 & y_3 & x_3^2 & x_3y_3 & y_3^2 & x_3^3 & x_3y_3^2 + x_3^2y_3 & y_3^3 \\ 0 & 0 & 1 & 0 & x_3 & 2y_3 & 0 & 2x_3y_3 + x_3^2 & 3y_3^2 \\ 0 & -1 & 0 & -2x_3 & -y_3 & 0 & -3x_3^2 & -(y_3^2 + 2x_3y_3) & 0 \end{bmatrix}$$
$$\begin{array}{ccccccccc} 1 & 2 & 3 & 4 & 5 & 6 & 7 & 8 & 9 \end{array} \tag{5.238}$$

In deriving Eq. (5.235) the notation

$$(u_z)_{x_1,y_1} = u_1 \qquad \left(\frac{\partial u_z}{\partial y}\right)_{x_1,y_1} = u_2 \qquad \left(\frac{\partial u_z}{\partial x}\right)_{x_1,y_1} = -u_3 \tag{5.239}$$

and so on, has been used.

The deflection function represented by Eq. (5.234) was first introduced by Tocher,[322] in an attempt to retain symmetric form in x and y. Other functions have also been used to evaluate stiffness matrices of triangular plates in bending. For example, Gallagher suggested the function

$$u_z = c_1 + c_2x + c_3y + c_4x^2 + c_5xy + c_6y^2 + c_7x^3 + c_8xy^2 + c_9y^3 \tag{5.240}$$

which suffers from the lack of symmetry because of the presence of the xy^2 term. An excellent discussion of different forms of deflection functions employed for calculations of triangular plate stiffness is given by Clough and Tocher.[68] Further ideas on the subject are expounded by Bazeley et al.[35]

Returning to our assumed deflection function in Eq. (5.234), we can calculate

* The deflection function (5.234) can not be used when $2y_3 = -x_3 + y_2$ since for these cases $\mathbf{C}$ is singular. To avoid this difficulty deflection functions from Ref. 35 can be utilized.

the strains from the flat-plate theory, using

$$e_{xx} = -z\,\frac{\partial^2 u_z}{\partial x^2} \tag{5.241}$$

$$e_{yy} = -z\,\frac{\partial^2 u_z}{\partial y^2} \tag{5.242}$$

$$e_{xy} = -2z\,\frac{\partial^2 u_z}{\partial x\,\partial y} \tag{5.243}$$

Hence, using the above equations and (5.234), we have

$$\begin{bmatrix} e_{xx} \\ e_{yy} \\ e_{xy} \end{bmatrix} = -z\begin{bmatrix} 0 & 0 & 0 & 2 & 0 & 0 & 6x & 2y & 0 \\ 0 & 0 & 0 & 0 & 0 & 2 & 0 & 2x & 6y \\ 0 & 0 & 0 & 0 & 2 & 0 & 0 & 4(x+y) & 0 \end{bmatrix}\mathbf{c} \tag{5.244}$$

or symbolically

$$\mathbf{e} = \mathbf{D}\mathbf{c} \tag{5.244a}$$

where $\mathbf{D}$ stands for the 3×9 matrix in (5.244), including the premultiplying constant $-z$. Noting from Eq. (5.235) that

$$\mathbf{c} = \mathbf{C}^{-1}\mathbf{u} \tag{5.245}$$

we have $\quad \mathbf{e} = \mathbf{D}\mathbf{C}^{-1}\mathbf{u} = \underline{\mathbf{b}}\mathbf{u} \quad$ (5.246)

where $\quad \underline{\mathbf{b}} = \mathbf{D}\mathbf{C}^{-1} \quad$ (5.247)

Subsequent substitution of $\underline{\mathbf{b}}$ into (5.16) leads finally to the stiffness matrix

$$\mathbf{k} = (\mathbf{C}^{-1})^T \int_v \mathbf{D}^T \boldsymbol{\varkappa} \mathbf{D}\, dV\, \mathbf{C}^{-1} \tag{5.248}$$

The inverse of the matrix must be obtained numerically for each triangular element. The matrix product in the volume integral can be multiplied out to give

$$\int_v \mathbf{D}^T\boldsymbol{\varkappa}\mathbf{D}\, dV = \frac{Et^3}{12(1-\nu^2)}$$

$$\times \iint \left[\begin{array}{c|cccccc} \mathbf{0} & & & \mathbf{0} & & & \\ \hline & 4 & & & & & \\ & 0 & 2(1-\nu) & & \text{Symmetric} & & \\ & 4\nu & 0 & 4 & & & \\ \mathbf{0} & 12x & 0 & 12\nu x & 36x^2 & & \\ & 4(\nu x + y) & 4(1-\nu)(x+y) & 4(x+\nu y) & 12x(\nu x + y) & \begin{array}{c}(12-8\nu)(x+y)^2 \\ -8(1-\nu)xy\end{array} & \\ & 12\nu y & 0 & 12y & 36\nu xy & 12(x+\nu y)y & 36y^2 \end{array}\right] dx\,dy \tag{5.249}$$

The individual coefficients in (5.249) can be integrated using the area integrals listed below

$$I(x^0,y^0) = \iint dx\,dy = \tfrac{1}{2}x_3y_2 \qquad (5.250)$$

$$I(x^1,y^0) = \iint x\,dx\,dy = \tfrac{1}{6}x_3^2y_2 \qquad (5.251)$$

$$I(x^2,y^0) = \iint x^2\,dx\,dy = \tfrac{1}{12}x_3^3y_2 \qquad (5.252)$$

$$I(x^0,y^1) = \iint y\,dx\,dy = \tfrac{1}{6}x_3y_2(y_2 + y_3) \qquad (5.253)$$

$$I(x^0,y^2) = \iint y^2\,dx\,dy = \tfrac{1}{12}x_3y_2(y_2^2 + y_2y_3 + y_3^2) \qquad (5.254)$$

$$I(x^1,y^1) = \iint xy\,dx\,dy = \tfrac{1}{24}x_3^2y_2(y_2 + 2y_3) \qquad (5.255)$$

where the x and y axes and the location of the vertices of the triangle are as indicated in Fig. 5.19. It should also be noted that the oy direction has been selected to coincide with edge 1,2, with the origin placed at the vertex 1 only in order to simplify the presentation of the results.

To determine the thermal forces $\mathbf{Q}$ on the element we may assume that the temperature distribution is given by

$$T = T_m + \frac{z}{t}\,\Delta T \qquad (5.256)$$

where T_m is the mean temperature of the element and ΔT is the temperature difference between the upper ($z = t/2$) and lower ($z = -t/2$) surfaces. Both T_m and ΔT will be assumed constant over the element. This assumption is satisfactory provided that the size of elements permits adequate representation of variations in temperature over the whole field of the idealized elements.

Substituting Eqs. (5.247) and (5.256) into (5.17), we obtain

$$\mathbf{Q} = (\mathbf{C}^{-1})^T \int \frac{z}{t}\,\mathbf{D}^T \boldsymbol{\varkappa}_T\,dV\,\alpha\,\Delta T \qquad (5.257)$$

Here again the inverse of $\mathbf{C}$ must be found numerically, while the integral $\int_v \frac{z}{t}\,\mathbf{D}^T\boldsymbol{\varkappa}_T\,dV$ is determined from

$$\int_v \frac{z}{t}\,\mathbf{D}^T\boldsymbol{\varkappa}_T\,dV = \frac{-Et^2}{12(1-\nu)} \iint \begin{array}{c} 1 \\ 2 \\ 3 \\ 4 \\ 5 \\ 6 \\ 7 \\ 8 \\ 9 \end{array} \begin{bmatrix} 0 \\ 0 \\ 0 \\ 2 \\ 0 \\ 2 \\ 6x \\ 2(x+y) \\ 6y \end{bmatrix} dx\,dy = \frac{-Et^2A_{123}}{6(1-\nu)} \begin{array}{c} 1 \\ 2 \\ 3 \\ 4 \\ 5 \\ 6 \\ 7 \\ 8 \\ 9 \end{array} \begin{bmatrix} 0 \\ 0 \\ 0 \\ 1 \\ 0 \\ 1 \\ x_3 \\ \frac{1}{3}(x_3 + y_2 + y_3) \\ y_2 + y_3 \end{bmatrix} \qquad (5.258)$$

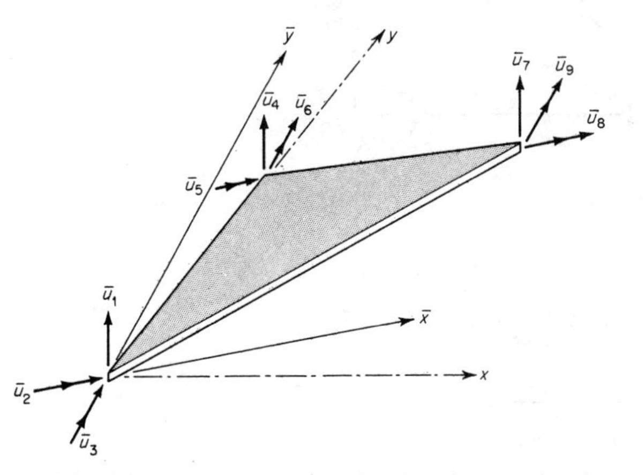

FIG. 5.20 Displacements on a triangular plate element referred to datum coordinate system.

The stiffness matrix **k** and thermal forces **Q** must now be transformed into the datum coordinate system corresponding to the displacements shown in Fig. 5.20. It can easily be demonstrated that the required transformation matrix $\boldsymbol{\lambda}$ is given by

$$\boldsymbol{\lambda} = \lfloor \boldsymbol{\Lambda} \quad \boldsymbol{\Lambda} \quad \boldsymbol{\Lambda} \rfloor \tag{5.259}$$

where
$$\boldsymbol{\Lambda} = \begin{bmatrix} 1 & 0 & 0 \\ 0 & l_{ox} & m_{ox} \\ 0 & l_{oy} & m_{oy} \end{bmatrix} \tag{5.260}$$

with $[l_{ox} \quad m_{ox}]$ and $[l_{oy} \quad m_{oy}]$ representing the direction cosines for oy and ox directions, respectively.

5.12 RECTANGULAR PLATES IN BENDING

DISPLACEMENT FUNCTION SATISFYING DEFLECTION COMPATIBILITY

One of the displacement functions used to calculate stiffness properties of rectangular plates in bending is of the form[365]

$$u_z = \mathbf{au} \tag{5.261}$$

where the positive directions of

$$\mathbf{u} = \{u_1 \quad u_2 \quad \cdots \quad u_{12}\} \tag{5.262}$$

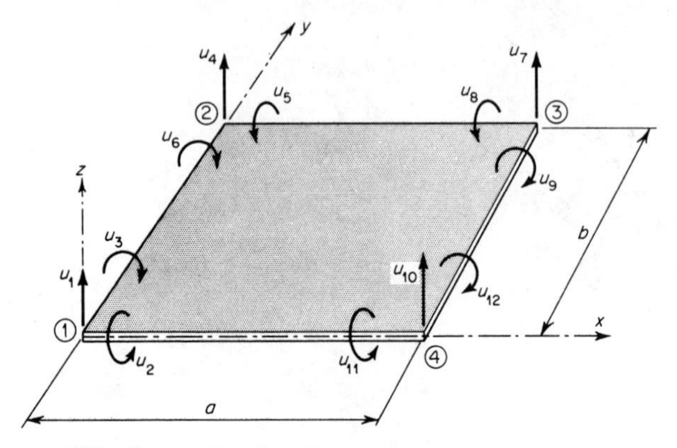

FIG. 5.21 Rectangular plate element.

are indicated in Fig. 5.21 and the matrix **a** is given by

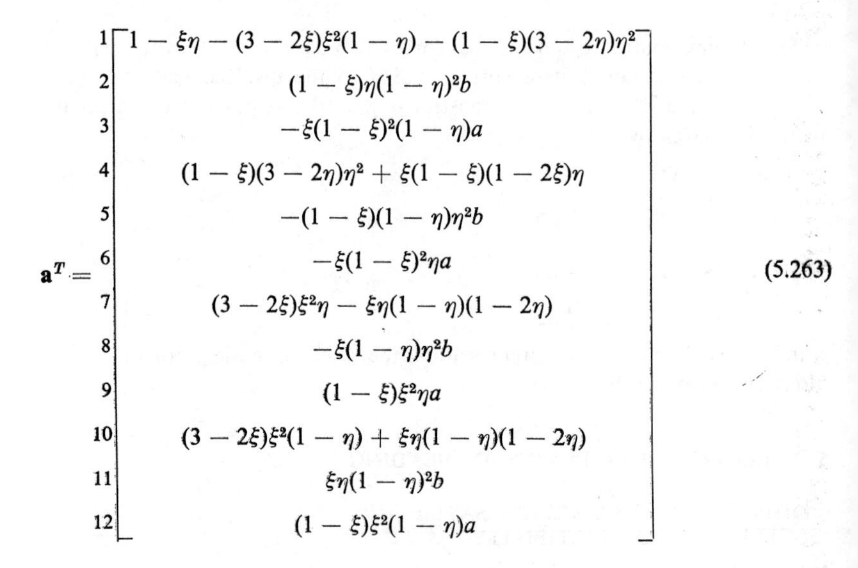

$$
\mathbf{a}^T = \begin{array}{r} 1 \\ 2 \\ 3 \\ 4 \\ 5 \\ 6 \\ 7 \\ 8 \\ 9 \\ 10 \\ 11 \\ 12 \end{array}
\begin{bmatrix}
1 - \xi\eta - (3 - 2\xi)\xi^2(1 - \eta) - (1 - \xi)(3 - 2\eta)\eta^2 \\
(1 - \xi)\eta(1 - \eta)^2 b \\
-\xi(1 - \xi)^2(1 - \eta)a \\
(1 - \xi)(3 - 2\eta)\eta^2 + \xi(1 - \xi)(1 - 2\xi)\eta \\
-(1 - \xi)(1 - \eta)\eta^2 b \\
-\xi(1 - \xi)^2\eta a \\
(3 - 2\xi)\xi^2\eta - \xi\eta(1 - \eta)(1 - 2\eta) \\
-\xi(1 - \eta)\eta^2 b \\
(1 - \xi)\xi^2\eta a \\
(3 - 2\xi)\xi^2(1 - \eta) + \xi\eta(1 - \eta)(1 - 2\eta) \\
\xi\eta(1 - \eta)^2 b \\
(1 - \xi)\xi^2(1 - \eta)a
\end{bmatrix}
\tag{5.263}
$$

The deflection function represented by (5.261) and (5.263) ensures that the boundary deflections on adjacent plate elements are compatible; however, rotations of the element edges on a common boundary are not compatible, and consequently discontinuities in slopes exist across the boundaries.

Using Eqs. (5.241) to (5.243) and (5.263), we have

$$
\mathbf{e} = \underline{\mathbf{b}}\mathbf{u} \tag{5.264}
$$

where

$$
\underline{\mathbf{b}}^T = \begin{array}{r} 1 \\ 2 \\ 3 \\ 4 \\ 5 \\ 6 \\ 7 \\ 8 \\ 9 \\ 10 \\ 11 \\ 12 \end{array}
\left[\begin{array}{c:c:c}
(1-2\xi)(1-\eta)\dfrac{6z}{a^2} & (1-\xi)(1-2\eta)\dfrac{6z}{b^2} & [1-6\xi(1-\xi) - 6\eta(1-\eta)]\dfrac{2z}{ab} \\
0 & (1-\xi)(2-3\eta)\dfrac{2z}{b} & (1-4\eta+3\eta^2)\dfrac{2z}{a} \\
-(2-3\xi)(1-\eta)\dfrac{2z}{a} & 0 & -(1-4\xi+3\xi^2)\dfrac{2z}{b} \\
(1-2\xi)\eta\dfrac{6z}{a^2} & -(1-\xi)(1-2\eta)\dfrac{6z}{b^2} & [-1+6\xi(1-\xi) + 6\eta(1-\eta)]\dfrac{2z}{ab} \\
0 & (1-\xi)(1-3\eta)\dfrac{2z}{b} & -\eta(2-3\eta)\dfrac{2z}{a} \\
-(2-3\xi)\eta\dfrac{2z}{a} & 0 & (1-4\xi+3\xi^2)\dfrac{2z}{b} \\
-(1-2\xi)\eta\dfrac{6z}{a^2} & -\xi(1-2\eta)\dfrac{6z}{b^2} & [1-6\xi(1-\xi) - 6\eta(1-\eta)]\dfrac{2z}{ab} \\
0 & \xi(1-3\eta)\dfrac{2z}{b} & \eta(2-3\eta)\dfrac{2z}{a} \\
-(1-3\xi)\eta\dfrac{2z}{a} & 0 & -\xi(2-3\xi)\dfrac{2z}{b} \\
-(1-2\xi)(1-\eta)\dfrac{6z}{a^2} & \xi(1-2\eta)\dfrac{6z}{b^2} & [-1+6\xi(1-\xi) + 6\eta(1-\eta)]\dfrac{2z}{ab} \\
0 & \xi(2-3\eta)\dfrac{2z}{b} & -(1-4\eta+3\eta^2)\dfrac{2z}{a} \\
-(1-3\xi)(1-\eta)\dfrac{2z}{a} & 0 & \xi(2-3\xi)\dfrac{2z}{b}
\end{array}\right] \tag{5.265}
$$

Now substituting $\underline{\mathbf{b}}$ from (5.265) into (5.16) and then performing the required operations, we obtain the stiffness matrix

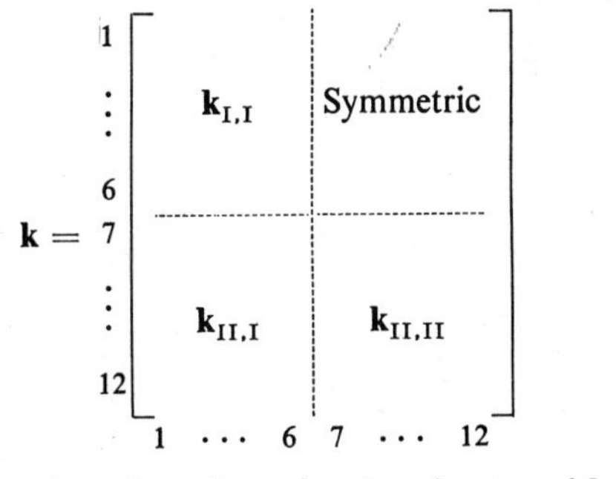

(5.266)

where the submatrices $\mathbf{k}_{\text{I,I}}$, $\mathbf{k}_{\text{II,I}}$, and $\mathbf{k}_{\text{II,II}}$ are presented separately in Tables 5.1 to 5.3.

TABLE 5.1 STIFFNESS MATRIX FOR RECTANGULAR PLATES IN BENDING: SUBMATRIX $\mathbf{k}_{I,I}$ BASED ON NONCOMPATIBLE DEFLECTIONS [ALL COEFFICIENTS TO BE MULTIPLIED BY $Et^3/12(1 - \nu^2)ab$]

1	$4(\beta^2 + \beta^{-2}) + \frac{1}{5}(14 - 4\nu)$			Symmetric		
2	$[2\beta^{-2} + \frac{1}{5}(1 + 4\nu)]b$	$[\frac{4}{3}\beta^{-2} + \frac{4}{15}(1 - \nu)]b^2$				
3	$-[2\beta^2 + \frac{1}{5}(1 + 4\nu)]a$	$-\nu ab$	$[\frac{4}{3}\beta^2 + \frac{4}{15}(1 - \nu)]a^2$			
4	$2(\beta^2 - 2\beta^{-2}) - \frac{1}{5}(14 - 4\nu)$	$-[2\beta^{-2} + \frac{1}{5}(1 - \nu)]b$	$[-\beta^2 + \frac{1}{5}(1 + 4\nu)]a$	$4(\beta^2 + \beta^{-2}) + \frac{1}{5}(14 - 4\nu)$		
5	$[2\beta^{-2} + \frac{1}{5}(1 - \nu)]b$	$[\frac{2}{3}\beta^{-2} - \frac{1}{15}(1 - \nu)]b^2$	0	$-[2\beta^{-2} + \frac{1}{5}(1 + 4\nu)]b$	$[\frac{4}{3}\beta^{-2} + \frac{4}{15}(1 - \nu)]b^2$	
6	$[-\beta^2 + \frac{1}{5}(1 + 4\nu)]a$	0	$[\frac{2}{3}\beta^2 - \frac{4}{15}(1 - \nu)]a^2$	$-[2\beta^2 + \frac{1}{5}(1 + 4\nu)]a$	νab	$[\frac{4}{3}\beta^2 + \frac{4}{15}(1 - \nu)]a^2$
	1	2	3	4	5	6

TABLE 5.2 STIFFNESS MATRIX FOR RECTANGULAR PLATES IN BENDING: SUBMATRIX $\mathbf{k}_{II,I}$ BASED ON NONCOMPATIBLE DEFLECTIONS [ALL COEFFICIENTS TO BE MULTIPLIED BY $Et^3/12(1 - \nu^2)ab$]

	1	2	3	4	5	6
7	$-2(\beta^2 + \beta^{-2}) + \frac{1}{5}(14 - 4\nu)$	$[-\beta^{-2} + \frac{1}{5}(1 - \nu)]b$	$[\beta^2 - \frac{1}{5}(1 - \nu)]a$	$-2(2\beta^2 - \beta^{-2}) - \frac{1}{5}(14 - 4\nu)$	$[-\beta^{-2} + \frac{1}{5}(1 + 4\nu)]b$	$[2\beta^2 + \frac{1}{5}(1 - \nu)]a$
8	$[\beta^{-2} - \frac{1}{5}(1 - \nu)]b$	$[\frac{1}{3}\beta^{-2} + \frac{1}{15}(1 - \nu)]b^2$	0	$[-\beta^{-2} + \frac{1}{5}(1 + 4\nu)]b$	$[\frac{2}{3}\beta^{-2} - \frac{4}{15}(1 - \nu)]b^2$	0
9	$[-\beta^2 + \frac{1}{5}(1 - \nu)]a$	0	$[\frac{1}{3}\beta^2 + \frac{1}{15}(1 - \nu)]a^2$	$-[2\beta^2 + \frac{1}{5}(1 - \nu)]a$	0	$[\frac{2}{3}\beta^2 - \frac{1}{15}(1 - \nu)]a^2$
10	$-2(2\beta^2 - \beta^{-2}) - \frac{1}{5}(14 - 4\nu)$	$[\beta^{-2} - \frac{1}{5}(1 + 4\nu)]b$	$[2\beta^2 + \frac{1}{5}(1 - \nu)]a$	$-2(\beta^2 + \beta^{-2}) + \frac{1}{5}(14 - 4\nu)$	$[\beta^{-2} - \frac{1}{5}(1 - \nu)]b$	$[\beta^2 - \frac{1}{5}(1 - \nu)]a$
11	$[\beta^{-2} - \frac{1}{5}(1 + 4\nu)]b$	$[\frac{2}{3}\beta^{-2} - \frac{4}{15}(1 - \nu)]b^2$	0	$[-\beta^{-2} + \frac{1}{5}(1 - \nu)]b$	$[\frac{1}{3}\beta^{-2} + \frac{1}{15}(1 - \nu)]b^2$	0
12	$-[2\beta^2 + \frac{1}{5}(1 - \nu)]a$	0	$[\frac{2}{3}\beta^2 - \frac{1}{15}(1 - \nu)]a^2$	$[-\beta^2 + \frac{1}{5}(1 - \nu)]a$	0	$[\frac{1}{3}\beta^2 + \frac{1}{15}(1 - \nu)]a^2$

TABLE 5.3 STIFFNESS MATRIX FOR RECTANGULAR PLATES IN BENDING: SUBMATRIX $\mathbf{k}_{II,II}$ BASED ON NONCOMPATIBLE DEFLECTIONS [ALL COEFFICIENTS TO BE MULTIPLIED BY $Et^3/12(1 - \nu^2)ab$]

	7	8	9	10	11	12
7	$4(\beta^2 + \beta^{-2}) + \frac{1}{5}(14 - 4\nu)$					
8	$-[2\beta^{-2} + \frac{1}{5}(1 + 4\nu)]b$	$[\frac{4}{3}\beta^{-2} + \frac{4}{15}(1 - \nu)]b^2$			Symmetric	
9	$[2\beta^2 + \frac{1}{5}(1 + 4\nu)]a$	$-\nu ab$	$[\frac{4}{3}\beta^2 + \frac{4}{15}(1 - \nu)]a^2$			
10	$2(\beta^2 - 2\beta^{-2}) - \frac{1}{5}(14 - 4\nu)$	$[2\beta^{-2} + \frac{1}{5}(1 - \nu)]b$	$[\beta^2 - \frac{1}{5}(1 + 4\nu)]a$	$4(\beta^2 + \beta^{-2}) + \frac{1}{5}(14 - 4\nu)$		
11	$-[2\beta^{-2} + \frac{1}{5}(1 - \nu)]b$	$[\frac{2}{3}\beta^{-2} - \frac{1}{15}(1 - \nu)]b^2$	0	$[2\beta^{-2} + \frac{1}{5}(1 + 4\nu)]b$	$[\frac{4}{3}\beta^{-2} + \frac{4}{15}(1 - \nu)]b^2$	
12	$[\beta^2 - \frac{1}{5}(1 + 4\nu)]a$	0	$[\frac{2}{3}\beta^2 - \frac{4}{15}(1 - \nu)]a^2$	$[2\beta^2 + \frac{1}{5}(1 + 4\nu)]a$	νab	$[\frac{4}{3}\beta^2 + \frac{4}{15}(1 - \nu)]a^2$

Assuming as before that the temperature distribution within the element is of the form $T = T_m + z\,\Delta T/t$, we can calculate the thermal forces $\mathbf{Q}$ from Eqs. (5.265) and (5.17). Hence it may be shown that

$$\mathbf{Q} = \frac{Et^2\alpha\,\Delta T}{24(1-\nu)}\{\overset{1}{0}\quad \overset{2}{-a}\quad \overset{3}{b}\quad \overset{4}{0}\quad \overset{5}{a}\quad \overset{6}{b}\quad \overset{7}{0}\quad \overset{8}{a}\quad \overset{9}{-b}\quad \overset{10}{0}\quad \overset{11}{-a}\quad \overset{12}{-b}\} \tag{5.267}$$

The transformation matrix $\boldsymbol{\lambda}$ for the rectangular plate elements in bending is given by

$$\boldsymbol{\lambda} = \lceil \boldsymbol{\Lambda} \quad \boldsymbol{\Lambda} \quad \boldsymbol{\Lambda} \quad \boldsymbol{\Lambda} \rfloor \tag{5.268}$$

where the matrix $\boldsymbol{\Lambda}$ is the same as for the triangular plate [see Eq. (5.260)].

DISPLACEMENT FUNCTION SATISFYING DEFLECTION AND SLOPE COMPATIBILITY

A deflection function that ensures both the deflection and slope compatibility on adjacent elements was introduced by Bogner et al.,[55] who used the deflection u_z expressed by means of the hermitian interpolation formula for two dimensions.* Using the sign convention previously established in Fig. 5.21, we have the matrix $\mathbf{a}$ for such compatible deflections and slopes given by

$$\mathbf{a}^T = \begin{matrix} 1 \\ 2 \\ 3 \\ 4 \\ 5 \\ 6 \\ 7 \\ 8 \\ 9 \\ 10 \\ 11 \\ 12 \end{matrix} \begin{bmatrix} (1+2\xi)(1-\xi)^2(1+2\eta)(1-\eta)^2 \\ (1+2\xi)(1-\xi)^2\eta(1-\eta)^2 b \\ -\xi(1-\xi)^2(1+2\eta)(1-\eta)^2 a \\ (1+2\xi)(1-\xi)^2(3-2\eta)\eta^2 \\ -(1+2\xi)(1-\xi)^2(1-\eta)\eta^2 b \\ -\xi(1-\xi)^2(3-2\eta)\eta^2 a \\ (3-2\xi)\xi^2(3-2\eta)\eta^2 \\ -(3-2\xi)\xi^2(1-\eta)\eta^2 b \\ (1-\xi)\xi^2(3-2\eta)\eta^2 a \\ (3-2\xi)\xi^2(1+2\eta)(1-\eta)^2 \\ (3-2\xi)\xi^2\eta(1-\eta)^2 b \\ (1-\xi)\xi^2(1+2\eta)(1-\eta)^2 a \end{bmatrix} \tag{5.269}$$

* This deflection function, unlike the first function used in this section, does not contain terms capable of representing constant values of the twist of the middle surface of the plate $\partial^2 u_z/\partial x\,\partial y$. Another function satisfying the deflection and slope compatibility and using "generalized" displacements $\partial^2 u_z/\partial x\,\partial y$ at the four corners, in addition to the 12 displacements in the present derivation, was introduced by Bogner et al. in Ref. 54. This function can represent constant value of the twist over the whole field of the rectangular plate, and it appears to offer distinct advantages over other deflection patterns.

Equation (5.269) can then be used to determine the matrix $\underline{\mathbf{b}}$ expressing the total strains due to unit displacements. It can be easily demonstrated that this matrix is given by

$$
\underline{\mathbf{b}}^T = \begin{array}{r} 1\\2\\3\\4\\5\\6\\7\\8\\9\\10\\11\\12 \end{array}
\begin{bmatrix}
(1-2\xi)(1+2\eta)(1-\eta)^2\dfrac{6z}{a^2} & (1+2\xi)(1-\xi)^2(1-2\eta)\dfrac{6z}{b^2} & -\xi(1-\xi)\eta(1-\eta)\dfrac{72z}{ab} \\
(1-2\xi)\eta(1-\eta)^2\dfrac{6bz}{a^2} & (1+2\xi)(1-\xi)^2(2-3\eta)\dfrac{2z}{b} & \xi(1-\xi)(1-\eta)(1-3\eta)\dfrac{12z}{a} \\
-(2-3\xi)(1+2\eta)(1-\eta)^2\dfrac{2z}{a} & -\xi(1-\xi)^2(1-2\eta)\dfrac{6az}{b^2} & -(1-\xi)(1-3\xi)\eta(1-\eta)\dfrac{12z}{b} \\
(1-2\xi)(3-2\eta)\eta^2\dfrac{6z}{a^2} & -(1+2\xi)(1-\xi)^2(1-2\eta)\dfrac{6z}{b^2} & \xi(1-\xi)\eta(1-\eta)\dfrac{72z}{ab} \\
-(1-2\xi)(1-\eta)\eta^2\dfrac{6bz}{a^2} & (1+2\xi)(1-\xi)^2(1-3\eta)\dfrac{2z}{b} & -\xi(1-\xi)\eta(2-3\eta)\dfrac{12z}{a} \\
-(2-3\xi)(3-2\eta)\eta^2\dfrac{2z}{a} & \xi(1-\xi)^2(1-2\eta)\dfrac{6az}{b^2} & (1-\xi)(1-3\xi)\eta(1-\eta)\dfrac{12z}{b} \\
-(1-2\xi)(3-2\eta)\eta^2\dfrac{6z}{a^2} & -(3-2\xi)\xi^2(1-2\eta)\dfrac{6z}{b^2} & -\xi(1-\xi)\eta(1-\eta)\dfrac{72z}{ab} \\
(1-2\xi)(1-\eta)\eta^2\dfrac{6bz}{a^2} & (3-2\xi)\xi^2(1-3\eta)\dfrac{2z}{b} & \xi(1-\xi)\eta(2-3\eta)\dfrac{12z}{a} \\
-(1-3\xi)(3-2\eta)\eta^2\dfrac{2z}{a} & -(1-\xi)\xi^2(1-2\eta)\dfrac{6az}{b^2} & -\xi(2-3\xi)\eta(1-\eta)\dfrac{12z}{b} \\
-(1-2\xi)(1+2\eta)(1-\eta)^2\dfrac{6z}{a^2} & (3-2\xi)\xi^2(1-2\eta)\dfrac{6z}{b^2} & \xi(1-\xi)\eta(1-\eta)\dfrac{72z}{ab} \\
-(1-2\xi)(1+2\eta)(1-\eta)^2\dfrac{6bz}{a^2} & (3-2\xi)\xi^2(2-3\eta)\dfrac{2z}{b} & -\xi(1-\xi)(1-\eta)(1-3\eta)\dfrac{12z}{a} \\
-(1-3\xi)(1+2\eta)(1-\eta)^2\dfrac{2z}{a} & (1-\xi)\xi^2(1-2\eta)\dfrac{6az}{b^2} & \xi(1-\xi)\eta(1-\eta)\dfrac{12z}{b}
\end{bmatrix}
\tag{5.270}
$$

The matrix $\underline{\mathbf{b}}$ is then substituted into Eq. (5.16) to evaluate the stiffness matrix based on the compatible deflection function. For convenience, the results are presented again as submatrices of (5.266) in Tables 5.4 to 5.6.

The thermal-force matrix $\mathbf{Q}$ based on the compatible deflection distribution leads to the identical result obtained in Eq. (5.267) for noncompatible distribution.

5.13 METHOD FOR IMPROVING STIFFNESS MATRICES

In deriving stiffness properties for elements requiring an assumed deflection form we have used displacement functions with the number of free constants equal to the number of displacements at the node points. These constants were evaluated in terms of the element displacements, and the strains were then determined for unit values of the displacements. This allowed the use

TABLE 5.4 STIFFNESS MATRIX FOR RECTANGULAR PLATES IN BENDING: SUBMATRIX $\mathbf{k}_{I,I}$ BASED ON COMPATIBLE DEFLECTIONS [ALL COEFFICIENTS TO BE MULTIPLIED BY $Et^3/12(1-\nu^2)ab$]

	1	2	3	4	5	6
1	$\frac{156}{35}(\beta^2+\beta^{-2})+\frac{72}{25}$			Symmetric		
2	$[\frac{22}{35}\beta^2+\frac{78}{35}\beta^{-2}+\frac{6}{25}(1+5\nu)]b$	$(\frac{4}{35}\beta^2+\frac{52}{35}\beta^{-2}+\frac{8}{25})b^2$				
3	$-[\frac{78}{35}\beta^2+\frac{22}{35}\beta^{-2}+\frac{6}{25}(1+5\nu)]a$	$-[\frac{11}{35}(\beta^2+\beta^{-2})+\frac{1}{50}(1+60\nu)]ab$	$(\frac{52}{35}\beta^2+\frac{4}{35}\beta^{-2}+\frac{8}{25})a^2$			
4	$\frac{54}{35}\beta^2-\frac{156}{35}\beta^{-2}-\frac{72}{25}$	$(\frac{13}{35}\beta^2-\frac{78}{35}\beta^{-2}-\frac{6}{25})b$	$[-\frac{27}{35}\beta^2+\frac{22}{35}\beta^{-2}+\frac{6}{25}(1+5\nu)]a$	$\frac{156}{35}(\beta^2+\beta^{-2})+\frac{72}{25}$		
5	$(-\frac{13}{35}\beta^2+\frac{78}{35}\beta^{-2}+\frac{6}{25})b$	$(-\frac{3}{35}\beta^2+\frac{26}{35}\beta^{-2}-\frac{2}{25})b^2$	$[\frac{13}{70}\beta^2-\frac{11}{35}\beta^{-2}-\frac{1}{50}(1+5\nu)]ab$	$-[\frac{22}{35}\beta^2+\frac{78}{35}\beta^{-2}+\frac{6}{25}(1+5\nu)]b$	$(\frac{4}{35}\beta^2+\frac{52}{35}\beta^{-2}+\frac{8}{25})b^2$	
6	$[-\frac{27}{35}\beta^2+\frac{22}{35}\beta^{-2}+\frac{6}{25}(1+5\nu)]a$	$[-\frac{13}{70}\beta^2+\frac{11}{35}\beta^{-2}+\frac{1}{50}(1+5\nu)]ab$	$(\frac{18}{35}\beta^2-\frac{4}{35}\beta^{-2}-\frac{8}{25})a^2$	$-[\frac{78}{35}\beta^2+\frac{22}{35}\beta^{-2}+\frac{6}{25}(1+5\nu)]a$	$[\frac{11}{35}(\beta^2+\beta^{-2})+\frac{1}{50}(1+60\nu)]ab$	$(\frac{52}{35}\beta^2+\frac{4}{35}\beta^{-2}+\frac{8}{25})a^2$

TABLE 5.5 STIFFNESS MATRIX FOR RECTANGULAR PLATES IN BENDING: SUBMATRIX $\mathbf{k}_{II,I}$ BASED ON COMPATIBLE DEFLECTIONS [ALL COEFFICIENTS TO BE MULTIPLIED BY $Et^3/12(1 - \nu^2)ab$]

	1	2	3	4	5	6
7	$-\frac{54}{35}(\beta^2 + \beta^{-2}) + \frac{72}{25}$	$(-\frac{13}{35}\beta^2 - \frac{27}{35}\beta^{-2} + \frac{6}{25})b$	$(\frac{27}{35}\beta^2 + \frac{13}{35}\beta^{-2} - \frac{6}{25})a$	$-\frac{156}{35}\beta^2 + \frac{54}{35}\beta^{-2} - \frac{72}{25}$	$[\frac{22}{35}\beta^2 - \frac{27}{35}\beta^{-2} + \frac{6}{25}(1 + 5\nu)]b$	$(\frac{78}{35}\beta^2 - \frac{13}{35}\beta^{-2} + \frac{6}{25})a$
8	$(\frac{13}{35}\beta^2 + \frac{27}{35}\beta^{-2} - \frac{6}{25})b$	$(\frac{3}{35}\beta^2 + \frac{9}{35}\beta^{-2} + \frac{2}{25})b^2$	$[-\frac{13}{70}(\beta^2 + \beta^{-2}) + \frac{1}{50}]ab$	$[\frac{22}{35}\beta^2 - \frac{27}{35}\beta^{-2} + \frac{6}{25}(1 + 5\nu)]b$	$(-\frac{4}{35}\beta^2 + \frac{18}{35}\beta^{-2} - \frac{8}{25})b^2$	$[-\frac{11}{35}\beta^2 + \frac{13}{70}\beta^{-2} - \frac{1}{50}(1 + 5\nu)]ab$
9	$(-\frac{27}{35}\beta^2 - \frac{13}{35}\beta^{-2} + \frac{6}{25})a$	$[-\frac{13}{70}(\beta^2 + \beta^{-2}) + \frac{1}{50}]ab$	$(\frac{9}{35}\beta^2 + \frac{3}{35}\beta^{-2} + \frac{2}{25})a^2$	$(-\frac{78}{35}\beta^2 + \frac{13}{35}\beta^{-2} - \frac{6}{25})a$	$[\frac{11}{35}\beta^2 - \frac{13}{70}\beta^{-2} + \frac{1}{50}(1 + 5\nu)]ab$	$(\frac{26}{35}\beta^2 - \frac{3}{35}\beta^{-2} - \frac{2}{25})a^2$
10	$-\frac{156}{35}\beta^2 + \frac{54}{35}\beta^{-2} - \frac{72}{25}$	$[-\frac{22}{35}\beta^2 + \frac{27}{35}\beta^{-2} - \frac{6}{25}(1 + 5\nu)]b$	$(\frac{78}{35}\beta^2 - \frac{13}{35}\beta^{-2} + \frac{6}{25})a$	$-\frac{54}{35}(\beta^2 + \beta^{-2}) + \frac{72}{25}$	$(\frac{13}{35}\beta^2 + \frac{27}{35}\beta^{-2} - \frac{6}{25})b$	$(\frac{27}{35}\beta^2 + \frac{13}{35}\beta^{-2} - \frac{6}{25})a$
11	$[-\frac{22}{35}\beta^2 + \frac{27}{35}\beta^{-2} - \frac{6}{25}(1 + 5\nu)]b$	$(-\frac{4}{35}\beta^2 + \frac{18}{35}\beta^{-2} - \frac{8}{25})b^2$	$[\frac{11}{35}\beta^2 - \frac{13}{70}\beta^{-2} + \frac{1}{50}(1 + 5\nu)]ab$	$(-\frac{13}{35}\beta^2 - \frac{27}{35}\beta^{-2} + \frac{6}{25})b$	$(\frac{3}{35}\beta^2 + \frac{9}{35}\beta^{-2} + \frac{2}{25})b^2$	$[\frac{13}{70}(\beta^2 + \beta^{-2}) - \frac{1}{50}]ab$
12	$(-\frac{78}{35}\beta^2 + \frac{13}{35}\beta^{-2} - \frac{6}{25})a$	$[-\frac{11}{35}\beta^2 + \frac{13}{70}\beta^{-2} - \frac{1}{50}(1 + 5\nu)]ab$	$(\frac{26}{35}\beta^2 - \frac{3}{35}\beta^{-2} - \frac{2}{25})a^2$	$(-\frac{27}{35}\beta^2 - \frac{13}{35}\beta^{-2} + \frac{6}{25})a$	$[\frac{13}{70}(\beta^2 + \beta^{-2}) - \frac{1}{50})ab$	$(\frac{9}{35}\beta^2 + \frac{3}{35}\beta^{-2} + \frac{2}{25})a^2$

TABLE 5.6 STIFFNESS MATRIX FOR RECTANGULAR PLATES IN BENDING: SUBMATRIX $\mathbf{k}_{II,II}$ BASED ON COMPATIBLE DEFLECTIONS [ALL COEFFICIENTS TO BE MULTIPLIED BY $Et^3/12(1-\nu^2)ab$]

	7	8	9	10	11	12
7	$\frac{156}{35}(\beta^2+\beta^{-2})+\frac{72}{25}$			Symmetric		
8	$-[\frac{22}{35}\beta^2+\frac{78}{35}\beta^{-2}+\frac{6}{25}(1+5\nu)]b$	$(\frac{4}{35}\beta^2+\frac{52}{35}\beta^{-2}+\frac{8}{25})b^2$				
9	$[\frac{78}{35}\beta^2+\frac{22}{35}\beta^{-2}+\frac{6}{25}(1+5\nu)]a$	$-[\frac{11}{35}(\beta^2+\beta^{-2})+\frac{1}{50}(1+60\nu)]ab$	$(\frac{52}{35}\beta^2+\frac{4}{35}\beta^{-2}+\frac{8}{25})a^2$			
10	$\frac{54}{35}\beta^2-\frac{156}{35}\beta^{-2}-\frac{72}{25}$	$(-\frac{13}{35}\beta^2+\frac{78}{35}\beta^{-2}+\frac{6}{25})b$	$[\frac{27}{35}\beta^2-\frac{22}{35}\beta^{-2}-\frac{6}{25}(1+5\nu)]a$	$\frac{156}{35}(\beta^2+\beta^{-2})+\frac{72}{25}$		
11	$(\frac{13}{35}\beta^2-\frac{78}{35}\beta^{-2}-\frac{6}{25})b$	$(-\frac{3}{35}\beta^2+\frac{26}{35}\beta^{-2}-\frac{2}{25})b^2$	$[\frac{13}{70}\beta^2-\frac{11}{35}\beta^{-2}-\frac{1}{50}(1+5\nu)]ab$	$[\frac{22}{35}\beta^2+\frac{78}{35}\beta^{-2}+\frac{6}{25}(1+5\nu)]b$	$(\frac{4}{35}\beta^2+\frac{52}{35}\beta^{-2}+\frac{8}{25})b^2$	
12	$[\frac{27}{35}\beta^2-\frac{22}{35}\beta^{-2}-\frac{6}{25}(1+5\nu)]a$	$[-\frac{13}{70}\beta^2+\frac{11}{35}\beta^{-2}+\frac{1}{50}(1+5\nu)]ab$	$(\frac{18}{35}\beta^2-\frac{4}{35}\beta^{-2}-\frac{8}{25})a^2$	$[\frac{78}{35}\beta^2+\frac{22}{35}\beta^{-2}+\frac{6}{25}(1+5\nu)]a$	$[\frac{11}{35}(\beta^2+\beta^{-2})+\frac{1}{50}(1+60\nu)]ab$	$(\frac{52}{35}\beta^2+\frac{4}{35}\beta^{-2}+\frac{8}{25})a^2$

of the unit-displacement theorem to determine stiffness and thermal-force matrices. Since the assumed displacements were compatible within the element (not necessarily satisfying compatibility of displacements and slopes on the adjacent elements), the calculated thermal-force matrices were exact [see Eq. (5.17)]. However, not all assumed displacement distributions did satisfy the stress-equilibrium equations.

If the assumed displacement function contains more free constants than the number of displacements at the node points, these additional constants may be determined from the condition of minimum potential energy. Thus by taking more terms in the displacement function we can obtain an improvement in the satisfaction of equilibrium conditions.[253] The analysis which follows may be applied to any type of element.

We assume first that the displacement function for $\{u_x \quad u_y \quad u_z\}$ is given by

$$\begin{bmatrix} u_x \\ u_y \\ u_z \end{bmatrix} = \mathbf{G}\mathbf{c} \tag{5.271}$$

where $\mathbf{G}$ is a rectangular matrix whose elements are the assumed displacement functions and

$$\mathbf{c} = \{c_1 \quad c_2 \quad \cdots \quad c_m\} \tag{5.272}$$

By using the strain-displacement relations we can obtain from (5.271)

$$\mathbf{e} = \mathbf{H}\mathbf{c} \tag{5.273}$$

where $\mathbf{H}$ is derived from $\mathbf{G}$ by partial differentiation in accordance with relations (2.2).

From Eq. (5.271) we can express the n displacements $\mathbf{u}$ at the node points in terms of the m undetermined constants c:

$$\mathbf{u} = \mathbf{C}\mathbf{c} \tag{5.274}$$

which will be partitioned as

$$\mathbf{u} = [\mathbf{C}_a \quad \mathbf{C}_b]\begin{bmatrix} \mathbf{c}_a \\ \mathbf{c}_b \end{bmatrix} \tag{5.275}$$

Assuming that $|\mathbf{C}_a| \neq 0$, we have

$$\mathbf{c}_a = \mathbf{C}_a^{-1}\mathbf{u} - \mathbf{C}_a^{-1}\mathbf{C}_b\mathbf{c}_b \tag{5.276}$$

which can be combined with the identity $\mathbf{c}_b = \mathbf{c}_b$ to yield

$$\mathbf{c} = \begin{bmatrix} \mathbf{c}_a \\ \mathbf{c}_b \end{bmatrix} = \begin{bmatrix} \mathbf{C}_a^{-1} & -\mathbf{C}_a^{-1}\mathbf{C}_b \\ \mathbf{0} & \mathbf{I} \end{bmatrix}\begin{bmatrix} \mathbf{u} \\ \mathbf{c}_b \end{bmatrix} = \mathbf{W}\hat{\mathbf{u}} \tag{5.277}$$

where $$\mathbf{W} = \begin{bmatrix} \mathbf{C}_a^{-1} & -\mathbf{C}_a^{-1}\mathbf{C}_b \\ \mathbf{0} & \mathbf{I} \end{bmatrix} \tag{5.278}$$

and $$\hat{\mathbf{u}} = \begin{bmatrix} \mathbf{u} \\ \mathbf{c}_b \end{bmatrix} \tag{5.279}$$

From Eq. (5.28) with $T = 0$ and Eqs. (5.273) and (5.277) we can write

$$\begin{aligned} U_i &= \frac{1}{2}\int_v \mathbf{e}^T \boldsymbol{\varkappa} \mathbf{e}\, dV \\ &= \tfrac{1}{2}\mathbf{c}^T \int_v \mathbf{H}^T \boldsymbol{\varkappa} \mathbf{H}\, dV\, \mathbf{c} \\ &= \tfrac{1}{2}\hat{\mathbf{u}}^T \mathbf{W}^T \int_v \mathbf{H}^T \boldsymbol{\varkappa} \mathbf{H}\, dV\, \mathbf{W}\hat{\mathbf{u}} \\ &= \tfrac{1}{2}\hat{\mathbf{u}}^T \hat{\mathbf{K}} \hat{\mathbf{u}} \end{aligned} \tag{5.280}$$

where $$\hat{\mathbf{K}} = \mathbf{W}^T \int_v \mathbf{H}^T \boldsymbol{\varkappa} \mathbf{H}\, dV\, \mathbf{W} \tag{5.281}$$

The total potential energy U can be written as

$$\begin{aligned} U &= U_i - \mathbf{u}^T \mathbf{S} \\ &= \tfrac{1}{2}\hat{\mathbf{u}}^T \hat{\mathbf{K}} \hat{\mathbf{u}} - \mathbf{u}^T \mathbf{S} \\ &= \tfrac{1}{2}[\mathbf{u}^T \quad \mathbf{c}_b^T] \begin{bmatrix} \mathbf{K}_{aa} & \mathbf{K}_{ab} \\ \mathbf{K}_{ba} & \mathbf{K}_{bb} \end{bmatrix} \begin{bmatrix} \mathbf{u} \\ \mathbf{c}_b \end{bmatrix} - [\mathbf{u}^T \quad \mathbf{c}_b^T] \begin{bmatrix} \mathbf{S} \\ \mathbf{0} \end{bmatrix} \end{aligned} \tag{5.282}$$

where $\mathbf{S}$ is a column matrix of the element forces corresponding with the displacements $\mathbf{u}$. Now the condition of minimum potential energy requires that

$$\frac{\partial U}{\partial \hat{\mathbf{u}}} = \mathbf{0} \tag{5.283}$$

leading to

$$\begin{bmatrix} \mathbf{K}_{aa} & \mathbf{K}_{ab} \\ \mathbf{K}_{ba} & \mathbf{K}_{bb} \end{bmatrix} \begin{bmatrix} \mathbf{u} \\ \mathbf{c}_b \end{bmatrix} - \begin{bmatrix} \mathbf{S} \\ \mathbf{0} \end{bmatrix} = \begin{bmatrix} \mathbf{0} \\ \mathbf{0} \end{bmatrix} \tag{5.284}$$

The matrix $\mathbf{c}_b$ can be expressed in terms of $\mathbf{u}$ by solving the second row of the $m - n$ equations in (5.284), so that

$$\mathbf{c}_b = -\mathbf{K}_{bb}^{-1}\mathbf{K}_{ba}\mathbf{u} \tag{5.285}$$

which, when substituted into the first row of the n equations in (5.284), results in

$$(\mathbf{K}_{aa} - \mathbf{K}_{ab}\mathbf{K}_{bb}^{-1}\mathbf{K}_{ba})\mathbf{u} = \mathbf{S} \tag{5.286}$$

Hence, by definition, the element stiffness matrix is given by

$$\mathbf{K} = (\mathbf{K}_{aa} - \mathbf{K}_{ab}\mathbf{K}_{bb}^{-1}\mathbf{K}_{ba}) \tag{5.287}$$

Some simplification will result if we select the displacement functions in two sets, n functions satisfying the interelement compatibility on boundary deflections and slopes and $m - n$ functions vanishing on the boundaries. For such functions $\mathbf{C}_b = \mathbf{0}$, and the $m - n$ functions can be regarded as perturbations on the first set to ensure satisfaction of the stress-equilibrium conditions. For example, the compatible displacement function for the rectangular plate element in bending could be improved with perturbation functions in the form of products of normal modes of vibration of a clamped-clamped beam.

PROBLEMS

5.1 Derive the stiffness and thermal-force matrices for a pin-jointed bar element with the cross-sectional area varying linearly from A_1 to A_2.

5.2 Derive the stiffness matrix $\mathbf{k}$ for a constant-shear-flow panel with midpoint attachments, as shown in Fig. 4.15.

5.3 Explain why the row sums of coefficients of the stiffness matrices for triangular or rectangular plates with only in-plane forces are equal to zero, that is,

$$\sum_j k_{ij} = 0 \qquad \text{for any row}$$

5.4 Using the unit-displacement theorem, derive the stiffness matrix $\mathbf{k}$ for a beam element with the second moment of area varying linearly from I_1 to I_2. The assumed deflection shape is to be taken as a cubic curve.

5.5 The exact stress distribution in a pin-jointed bar element is given by

$$\sigma = \frac{E(u_2 - u_1)}{l}$$

where u_2 and u_1 are the displacements of the two ends (see Fig. 5.2), E is the Young's modulus, and l is the length. Using a compatible displacement distribution

$$u_x = c_1 + c_2 x + c_3 x^2$$

and the exact stresses, demonstrate that the unit-displacement theorem gives the correct expression for the stiffness matrix. To determine the constants c_1, c_2, and c_3 we must use $u_x = u_1$ at $x = 0$ and $u_x = u_2$ at $x = l$, while the third condition may be selected arbitrarily as $u_x = 2u_1$ at $x = l/2$.

5.6 Verify Eq. (5.221) for the thermal-stiffness matrix $\mathbf{h}$ of a quadrilateral plate element with in-plane forces only.

5.7 Verify Eq. (5.267) for the thermal-force matrix $\mathbf{Q}$ for a rectangular plate element in bending.

5.8 Derive an improved stiffness matrix for a rectangular plate element with in-plane forces only using the compatible displacement distribution given by Eqs. (5.159) and (5.160) and a corrective distribution of the form

$$u_x = c_x \sin \pi\xi \sin \pi\eta \qquad \text{and} \qquad u_y = c_y \sin \pi\xi \sin \pi\eta$$

CHAPTER 6 THE MATRIX DISPLACEMENT METHOD

In Chap. 5 we determined stiffness properties of individual structural elements that can be used in the idealization of continuous structures. To determine the displacements of the idealized structure under some specified external loading and temperature distribution, we must obtain stiffness properties of the assembled structure made up from the idealized elements. In this chapter we shall discuss how the stiffness matrices for individual elements can be combined to form the matrix equation relating the applied mechanical forces and equivalent thermal forces to the corresponding displacements on the assembled structure. Since the displacements appear as the unknowns, this formulation is, therefore, described as the matrix displacement method.

6.1 MATRIX FORMULATION OF THE DISPLACEMENT ANALYSIS

The fundamental assumption used in the analysis is that the structure can be satisfactorily represented by an assembly of discrete elements having simplified elastic properties and that these elements are interconnected so as to represent the actual continuous structure. The boundary displacements are compatible at least at the node points, where the elements are joined, and the stresses within each element are equilibrated by a set of element forces $\bar{\mathbf{S}}^{(i)}$ in the directions of element displacements $\bar{\mathbf{u}}^{(i)}$. The element forces are related to the corresponding displacements by the matrix equation

$$\bar{\mathbf{S}}^{(i)} = \bar{\mathbf{k}}^{(i)}\bar{\mathbf{u}}^{(i)} + \bar{\mathbf{Q}}^{(i)} \tag{6.1}$$

where the superscript (i) denotes the ith element and all matrices refer to a datum coordinate system. Equation (6.1) can be determined for each element separately, and for the complete structure all these equations can be combined into a single matrix equation of the form

$$\bar{\mathbf{S}} = \bar{\mathbf{k}}\bar{\mathbf{u}} + \bar{\mathbf{Q}} \tag{6.2}$$

where

$$\bar{\mathbf{S}} = \{\bar{\mathbf{S}}^{(1)} \quad \bar{\mathbf{S}}^{(2)} \quad \cdots \quad \bar{\mathbf{S}}^{(i)} \quad \cdots\} \tag{6.3}$$

$$\bar{\mathbf{k}} = [\bar{\mathbf{k}}^{(1)} \quad \bar{\mathbf{k}}^{(2)} \quad \cdots \quad \bar{\mathbf{k}}^{(i)} \quad \cdots] \tag{6.4}$$

$$\bar{\mathbf{u}} = \{\bar{\mathbf{u}}^{(1)} \quad \bar{\mathbf{u}}^{(2)} \quad \cdots \quad \bar{\mathbf{u}}^{(i)} \quad \cdots\} \tag{6.5}$$

$$\bar{\mathbf{Q}} = \{\bar{\mathbf{Q}}^{(1)} \quad \bar{\mathbf{Q}}^{(2)} \quad \cdots \quad \bar{\mathbf{Q}}^{(i)} \quad \cdots\} \tag{6.6}$$

It should be noted here that in Chap. 5 all matrices pertaining to individual elements were used without superscripts indicating the element number.

For subsequent analysis it is convenient to introduce a matrix of displacements on the assembled structure

$$\mathbf{U} = \{U_1 \quad U_2 \quad \cdots \quad U_j \quad \cdots \quad U_m\} \tag{6.7}$$

where U_j represents a typical nodal displacement, referred to a datum coordinate system. It is evident from the form of Eq. (6.5) that the element displacements $\bar{\mathbf{u}}$ can be expressed in terms of the structure displacements $\mathbf{U}$ by the equation

$$\bar{\mathbf{u}} = \mathbf{A}\mathbf{U} \tag{6.8}$$

where $\mathbf{A}$ is a rectangular matrix in which every row consists of zeros except for a single term of unity, the position of which identifies that element of $\bar{\mathbf{u}}$ which corresponds to the particular element of $\mathbf{U}$. The external loading corresponding to the displacements $\mathbf{U}$ will be denoted by the matrix $\mathbf{P}$, such that

$$\mathbf{P} = \{P_1 \quad P_2 \quad \cdots \quad P_j \quad \cdots \quad P_m\} \tag{6.9}$$

where P_j represents an external force in the direction of the displacement U_j.

To relate the external forces $\mathbf{P}$ to the corresponding displacements $\mathbf{U}$ virtual displacements $\delta\mathbf{U}$ are introduced. The virtual displacements on individual elements can then be obtained from Eq. (6.8), so that

$$\delta\bar{\mathbf{u}} = \mathbf{A}\,\delta\mathbf{U} \tag{6.10}$$

The virtual work for the displacements $\delta\mathbf{U}$ is given by

$$\delta W = \delta\mathbf{U}^T\,\mathbf{P} \tag{6.11}$$

while the virtual strain energy can be determined from

$$\delta U_i = \sum_i\,[\delta\bar{\mathbf{u}}^{(i)}]^T\bar{\mathbf{S}}^{(i)} = \delta\bar{\mathbf{u}}^T\,\bar{\mathbf{S}} \tag{6.12}$$

From the principle of virtual work $\delta W = \delta U_i$ it follows that

$$\delta\mathbf{U}^T\,\mathbf{P} = \delta\bar{\mathbf{u}}^T\,\bar{\mathbf{S}} \tag{6.13}$$

Substituting Eq. (6.10) into (6.13), we have

$$\delta \mathbf{U}^T(\mathbf{P} - \mathbf{A}^T\bar{\mathbf{S}}) = \mathbf{0} \tag{6.14}$$

and since $\delta\mathbf{U}$ represents arbitrary virtual displacements,

$$\mathbf{P} = \mathbf{A}^T\bar{\mathbf{S}} \tag{6.15}$$

Equation (6.15) is in fact the equation of equilibrium relating the external forces $\mathbf{P}$ to the internal forces $\bar{\mathbf{S}}$ (element forces). Finally, substitution of Eqs. (6.2) and (6.8) into (6.15) results in the matrix equation

$$\mathbf{P} = \mathbf{A}^T\bar{\mathbf{k}}\mathbf{A}\mathbf{U} + \mathbf{A}^T\bar{\mathbf{Q}}$$

or $$\mathbf{K}\mathbf{U} = \mathbf{P} - \mathbf{Q} \tag{6.16}$$

where $$\mathbf{K} = \mathbf{A}^T\bar{\mathbf{k}}\mathbf{A} \tag{6.17}$$

and $$\mathbf{Q} = \mathbf{A}^T\bar{\mathbf{Q}} \tag{6.18}$$

The matrix $\mathbf{K}$ is the stiffness matrix for the complete structure regarded as a free body, and Eq. (6.16) represents equations of equilibrium for element forces acting at all joints. This implies that the load matrix $\mathbf{P}$ must constitute a set of forces in static equilibrium, and it includes the reaction forces. However, all these forces are not independent. From the consideration of overall equilibrium of the structure it is clear that there must be six dependent equations, corresponding to the six rigid-body degrees of freedom, relating the forces $\mathbf{P}$. This dependence will render the matrix $\mathbf{K}$ singular and thus must be eliminated from Eq. (6.16). This is accomplished by assuming that six displacements at certain selected points on the structure are equal to zero and eliminating the corresponding rows and columns from the complete stiffness matrix $\mathbf{K}$. Only then is it possible to obtain the solution for displacements from

$$\mathbf{U}_r = \mathbf{K}_r^{-1}(\mathbf{P}_r - \mathbf{Q}_r) \tag{6.19}$$

where the subscript r is used to indicate that all matrices have been reduced in size to exclude forces and displacements at the selected points. The six displacements must be chosen in such a way as to ensure that all rigid-body degrees of freedom are completely restrained (three translations and three rotations). Naturally, additional displacements may also be eliminated from Eq. (6.16). For example, in the analysis of symmetrical structures under symmetrical loading the out-of-plane displacements at the plane of symmetry must be equal to zero, and they must also be eliminated.

The calculation of the stiffness matrix $\mathbf{K}$ in Eq. (6.17), as presented in the general theory, involves the matrix multiplication $\mathbf{A}^T\bar{\mathbf{k}}\mathbf{A}$. In practice, however, this multiplication is never carried out, since this operation is equivalent to the placing of elements from $\bar{\mathbf{k}}^{(i)}$ in their correct positions in the larger framework of the matrix $\mathbf{K}$ and then summing all the overlapping terms. This is a simpler operation than the matrix product $\mathbf{A}^T\bar{\mathbf{k}}\mathbf{A}$, and it can be programmed directly for a digital computer without actual setting up the transformation matrix $\mathbf{A}$. Similarly, the multiplication $\mathbf{A}^T\bar{\mathbf{Q}}$ is equivalent to placing elements from $\bar{\mathbf{Q}}$ in their correct positions in the larger framework of the matrix $\mathbf{Q}$ and then summing all the overlapping terms.

In order to prove the above statements, $\mathbf{A}$ must be considered in its partitioned form, that is,

$$\mathbf{A} = \{\mathbf{A}^{(1)} \quad \mathbf{A}^{(2)} \quad \cdots \quad \mathbf{A}^{(r)} \quad \cdots \quad \mathbf{A}^{(N)}\} \tag{6.20}$$

where N denotes number of structural elements. From Eqs. (6.4) and (6.20) it follows that the matrix product $\mathbf{A}^T\bar{\mathbf{k}}\mathbf{A}$ is given by

$$\mathbf{A}^T\bar{\mathbf{k}}\mathbf{A} = [(\mathbf{A}^{(1)})^T \quad (\mathbf{A}^{(2)})^T \quad \cdots \quad (\mathbf{A}^{(r)})^T \quad \cdots \quad (\mathbf{A}^{(N)})^T]$$

$$\times \begin{bmatrix} \bar{\mathbf{k}}^{(1)} & & & & & & \\ & \bar{\mathbf{k}}^{(2)} & & & \mathbf{0} & & \\ & & \cdot & & & & \\ & & & \bar{\mathbf{k}}^{(r)} & & & \\ & & \mathbf{0} & & \cdot & & \\ & & & & & \cdot & \\ & & & & & & \bar{\mathbf{k}}^{(N)} \end{bmatrix} \begin{bmatrix} \mathbf{A}^{(1)} \\ \mathbf{A}^{(2)} \\ \cdot \\ \cdot \\ \mathbf{A}^{(r)} \\ \cdot \\ \cdot \\ \mathbf{A}^{(N)} \end{bmatrix}$$

$$= \sum_{r=1}^{N} (\mathbf{A}^{(r)})^T\bar{\mathbf{k}}^{(r)}\mathbf{A}^{(r)} \tag{6.21}$$

It has therefore been shown that the matrix multiplication $\mathbf{A}^T\bar{\mathbf{k}}\mathbf{A}$ can be reduced to summation of the products $(\mathbf{A}^{(r)})^T\bar{\mathbf{k}}^{(r)}\mathbf{A}^{(r)}$ taken over all elements. For a pin-jointed bar element $\bar{\mathbf{k}}^{(r)}$ is a 6×6 matrix relating the element forces to their corresponding displacements in the directions g, h, i, j, k, l, which without loss of generality may be assumed to be taken in the same order as in the matrix $\mathbf{U}$. Thus the matrix $\bar{\mathbf{k}}^{(r)}$ can be represented as

$$\bar{\mathbf{k}}^{(r)} = \begin{bmatrix} k_{gg} & k_{gh} & \cdots & k_{gl} \\ k_{hg} & k_{hh} & \cdots & k_{hl} \\ \cdots & \cdots & \cdots & \cdots \\ k_{lg} & k_{lh} & \cdots & k_{ll} \end{bmatrix} \tag{6.22}$$

and the corresponding transformation matrix $\mathbf{A}^{(r)}$ becomes

$$\mathbf{A}^{(r)} = \begin{array}{c} \\ 1 \\ 2 \\ 3 \\ 4 \\ 5 \\ 6 \end{array} \begin{array}{c} \begin{array}{ccc} 1\,2\,3\cdots & g\,h\,i\,j\,k\,l & \cdots n \end{array} \\ \left[\begin{array}{c|c|c} & & \\ \mathbf{0} & \mathbf{I} & \mathbf{0} \\ & & \end{array} \right] \end{array} \tag{6.23}$$

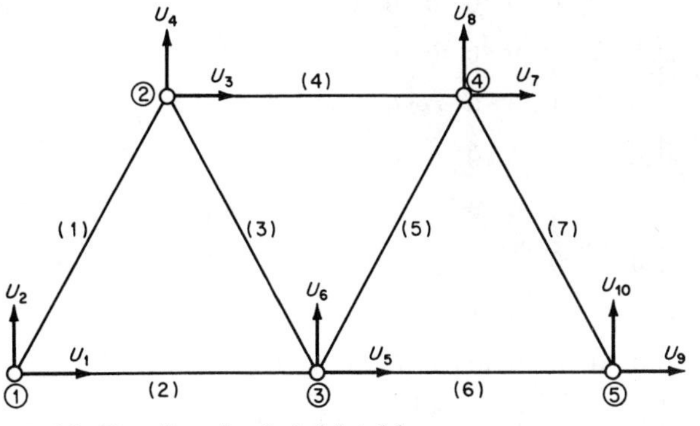

FIG. 6.1 Two-dimensional pin-jointed frame.

From the multiplication $(\mathbf{A}^{(r)})^T\bar{\mathbf{k}}^{(r)}\mathbf{A}^{(r)}$ it follows that

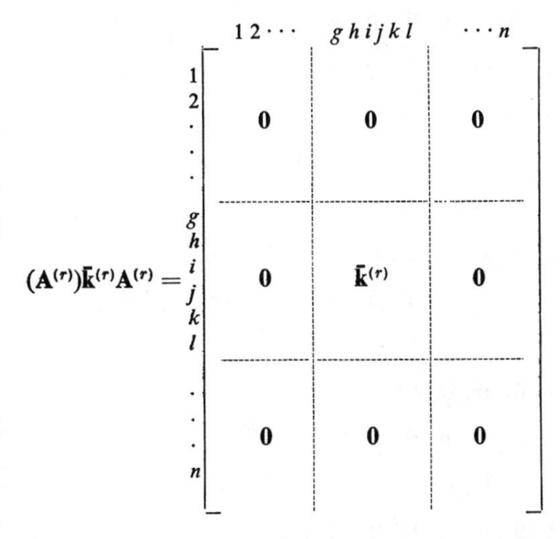

(6.24)

from which it is evident that the alternative method of calculation of the stiffness matrix $\mathbf{K}$ is justified. The proof that $\mathbf{A}^T\overline{\mathbf{Q}}$ can be obtained by the summation procedure is similar and need not be discussed in detail.

To emphasize the advantages of the summation of stiffness as compared with the matrix multiplication in Eq. (6.17) consider a two-dimensional truss formed by seven pin-jointed bar elements, as shown in Fig. 6.1. The joints are numbered consecutively from 1 to 5, and the horizontal and vertical displacements of joints are numbered from 1 to 10. For this case the relationship $\bar{\mathbf{u}} = \mathbf{AU}$ is presented by Eq. (6.25):

$$
\begin{bmatrix} \bar{\mathbf{u}}^{(1)} \\ \bar{\mathbf{u}}^{(2)} \\ \bar{\mathbf{u}}^{(3)} \\ \bar{\mathbf{u}}^{(4)} \\ \bar{\mathbf{u}}^{(5)} \\ \bar{\mathbf{u}}^{(6)} \\ \bar{\mathbf{u}}^{(7)} \end{bmatrix}
=
\begin{bmatrix} U_1 \\ U_2 \\ U_3 \\ U_4 \\ U_1 \\ U_2 \\ U_5 \\ U_6 \\ U_3 \\ U_4 \\ U_5 \\ U_6 \\ U_3 \\ U_4 \\ U_7 \\ U_8 \\ U_5 \\ U_6 \\ U_7 \\ U_8 \\ U_5 \\ U_6 \\ U_9 \\ U_{10} \\ U_7 \\ U_8 \\ U_9 \\ U_{10} \end{bmatrix}
=
\begin{array}{r}1\\2\\3\\4\\5\\6\\7\\8\\9\\10\\11\\12\\13\\14\\15\\16\\17\\18\\19\\20\\21\\22\\23\\24\\25\\26\\27\\28\\ \end{array}
\begin{bmatrix}
1 & 0 & 0 & 0 & 0 & 0 & 0 & 0 & 0 & 0 \\
0 & 1 & 0 & 0 & 0 & 0 & 0 & 0 & 0 & 0 \\
0 & 0 & 1 & 0 & 0 & 0 & 0 & 0 & 0 & 0 \\
0 & 0 & 0 & 1 & 0 & 0 & 0 & 0 & 0 & 0 \\
1 & 0 & 0 & 0 & 0 & 0 & 0 & 0 & 0 & 0 \\
0 & 1 & 0 & 0 & 0 & 0 & 0 & 0 & 0 & 0 \\
0 & 0 & 0 & 0 & 1 & 0 & 0 & 0 & 0 & 0 \\
0 & 0 & 0 & 0 & 0 & 1 & 0 & 0 & 0 & 0 \\
0 & 0 & 1 & 0 & 0 & 0 & 0 & 0 & 0 & 0 \\
0 & 0 & 0 & 1 & 0 & 0 & 0 & 0 & 0 & 0 \\
0 & 0 & 0 & 0 & 1 & 0 & 0 & 0 & 0 & 0 \\
0 & 0 & 0 & 0 & 0 & 1 & 0 & 0 & 0 & 0 \\
0 & 0 & 1 & 0 & 0 & 0 & 0 & 0 & 0 & 0 \\
0 & 0 & 0 & 1 & 0 & 0 & 0 & 0 & 0 & 0 \\
0 & 0 & 0 & 0 & 0 & 0 & 1 & 0 & 0 & 0 \\
0 & 0 & 0 & 0 & 0 & 0 & 0 & 1 & 0 & 0 \\
0 & 0 & 0 & 0 & 1 & 0 & 0 & 0 & 0 & 0 \\
0 & 0 & 0 & 0 & 0 & 1 & 0 & 0 & 0 & 0 \\
0 & 0 & 0 & 0 & 0 & 0 & 1 & 0 & 0 & 0 \\
0 & 0 & 0 & 0 & 0 & 0 & 0 & 1 & 0 & 0 \\
0 & 0 & 0 & 0 & 1 & 0 & 0 & 0 & 0 & 0 \\
0 & 0 & 0 & 0 & 0 & 1 & 0 & 0 & 0 & 0 \\
0 & 0 & 0 & 0 & 0 & 0 & 0 & 0 & 1 & 0 \\
0 & 0 & 0 & 0 & 0 & 0 & 0 & 0 & 0 & 1 \\
0 & 0 & 0 & 0 & 0 & 0 & 1 & 0 & 0 & 0 \\
0 & 0 & 0 & 0 & 0 & 0 & 0 & 1 & 0 & 0 \\
0 & 0 & 0 & 0 & 0 & 0 & 0 & 0 & 1 & 0 \\
0 & 0 & 0 & 0 & 0 & 0 & 0 & 0 & 0 & 1
\end{bmatrix}
\begin{bmatrix} U_1 \\ U_2 \\ U_3 \\ U_4 \\ U_5 \\ U_6 \\ U_7 \\ U_8 \\ U_9 \\ U_{10} \end{bmatrix}
\qquad (6.25)
$$

Column numbers: 1 2 3 4 5 6 7 8 9 10

It is clear that in view of the size of the matrix **A** (order 28 × 10), the multiplication $\mathbf{A}^T\bar{\mathbf{k}}\mathbf{A}$ is time-consuming and that the summation of stiffnesses is by far the easiest and fastest method of compiling the stiffness matrix **K** for the complete structure.

Although the numbering system for node points can be selected arbitrarily, it is generally preferable to number nodes first in one bay or segment of the structure and then to proceed with the numbering of nodes in the next bay, and so on, since this particular numbering scheme leads to a stiffness matrix of the band type, for which special time-saving programs may be employed.

The displacements at each node point (in general six, i.e., three translations and three rotations) are also numbered consecutively. Thus, for a typical node p the corresponding displacements in the matrix **U** are arranged in the following manner:

U_{6p-5} = displacement in the x direction
U_{6p-4} = displacement in the y direction
U_{6p-3} = displacement in the z direction
U_{6p-2} = rotation about the x axis
U_{6p-1} = rotation about the y axis
U_{6p} = rotation about the z axis

Naturally, if only translational displacements are considered, only three displacements for each node point need be specified. For these cases

U_{3p-2} = displacement in the x direction
U_{3p-1} = displacement in the y direction
U_{3p} = displacement in the z direction

The basic steps in any matrix displacement method are outlined in the flow diagram of analysis in Fig. 6.2. The following steps leading to the calculation of displacements on the complete structure can be identified:

1. Compilation of basic data on the idealized structure: position of node points and element orientation (topology of the structure), element material and geometrical characteristics (stiffness characteristics of the structure)
2. Determination of local coordinates (x,y,z)
3. Calculation of element stiffnesses in local coordinates $\mathbf{k}^{(i)}$
4. Determination of direction cosines for each element $\boldsymbol{\lambda}^{(i)}$
5. Calculation of element stiffnesses in datum coordinates

$$\bar{\mathbf{k}}^{(i)} = (\boldsymbol{\lambda}^{(i)})^T\mathbf{k}^{(i)}\boldsymbol{\lambda}^{(i)}$$

6. Addition of element stiffnesses to form the stiffness matrix for the complete structure **K**
7. Elimination of rigid-body degrees of freedom to establish the reduced stiffness matrix $\mathbf{K}_r$
8. Inversion of the stiffness matrix $\mathbf{K}_r$

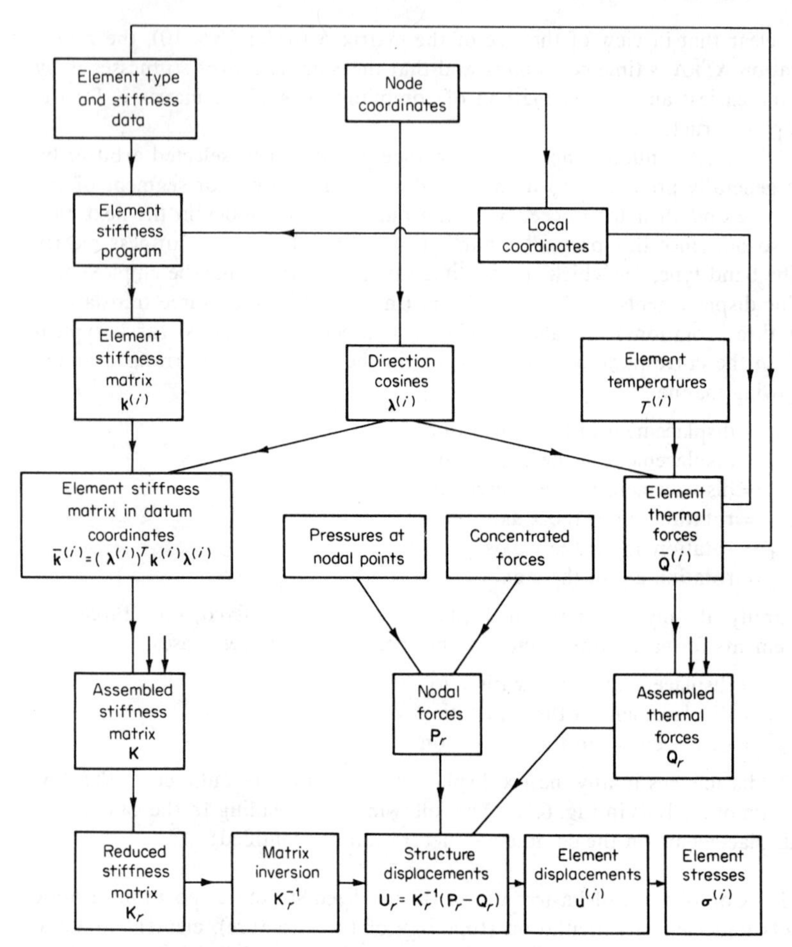

FIG. 6.2 Flow diagram for the matrix displacement analysis.

9. Conversion of pressure distribution into concentrated forces
10. Calculation of element thermal forces $\mathbf{Q}^{(i)}$ and $\bar{\mathbf{Q}}^{(i)}$
11. Addition of element thermal forces to form $\mathbf{Q}_r$
12. Calculation of the resultant loading $\mathbf{P}_r - \mathbf{Q}_r$
13. Calculation of structure displacement $\mathbf{U}_r = \mathbf{K}_r^{-1}(\mathbf{P}_r - \mathbf{Q}_r)$ and hence element displacement, forces, and stresses

The elimination of rows and columns to remove the rigid-body degrees of freedom is cumbersome inasmuch as it reduces the size of matrices. It has been found preferable to modify the stiffness matrix $\mathbf{K}$ by replacing the affected rows and columns with zeros, with the exception of the diagonal terms, which are replaced with ones.[277] Similarly the affected rows in $\mathbf{P}$ and $\mathbf{Q}$ are replaced with zeros.

Hence Eq. (6.16) becomes

$$\begin{bmatrix} \mathbf{P}_r \\ \mathbf{0} \end{bmatrix} - \begin{bmatrix} \mathbf{Q}_r \\ \mathbf{0} \end{bmatrix} = \begin{bmatrix} \mathbf{K}_r & \mathbf{0} \\ \mathbf{0} & \mathbf{I} \end{bmatrix} \begin{bmatrix} \mathbf{U}_r \\ \mathbf{0} \end{bmatrix}$$

or $$\mathbf{P}_m = \mathbf{K}_m \mathbf{U}_m \tag{6.26}$$

where $$\mathbf{P}_m = \begin{bmatrix} \mathbf{P}_r \\ \mathbf{0} \end{bmatrix} - \begin{bmatrix} \mathbf{Q}_r \\ \mathbf{0} \end{bmatrix} \tag{6.27}$$

$$\mathbf{K}_m = \begin{bmatrix} \mathbf{K}_r & \mathbf{0} \\ \mathbf{0} & \mathbf{I} \end{bmatrix} \tag{6.28}$$

$$\mathbf{U}_m = \begin{bmatrix} \mathbf{U}_r \\ \mathbf{0} \end{bmatrix} \tag{6.29}$$

Hence $$\mathbf{U}_m = \mathbf{K}_m^{-1} \mathbf{P}_m \tag{6.30}$$

Thus the retention of a unit matrix in $\mathbf{K}_m$ ensures the correct solution for $\mathbf{U}_r$ without rearranging the basic matrices appearing in the analysis.

6.2 ELIMINATION OF THE RIGID-BODY DEGREES OF FREEDOM: CHOICE OF REACTIONS

The displacements $\mathbf{U}$ of all node points in an unconstrained structure may be partitioned into two submatrices such that

$$\mathbf{U} = \begin{bmatrix} \mathbf{w} \\ \mathbf{U}_r \end{bmatrix} \tag{6.31}$$

where $\mathbf{w}$ represents displacements restraining the rigid-body degrees of freedom and $\mathbf{U}_r$ represents all remaining displacements (degrees of freedom). Thus the matrix $\mathbf{w}$ is of order 1×1 for one-dimensional structures, 3×1 for two-dimensional structures, and 6×1 for three-dimensional structures. It can comprise both deflections and rotations or deflections only. The external forces $\mathbf{P}$ can likewise be partitioned so that

$$\mathbf{P} = \begin{bmatrix} \mathbf{P}_w \\ \mathbf{P}_u \end{bmatrix} \tag{6.32}$$

where $\mathbf{P}_w$ and $\mathbf{P}_u$ are the forces in the directions of $\mathbf{w}$ and $\mathbf{U}_r$, respectively. The forces $\mathbf{P}_w$ may be considered to be reaction forces due to $\mathbf{P}_u$.

Consider now a kinematic relationship, valid only for small displacements,

$$\mathbf{U}_r = \mathbf{r} + \mathbf{T}\mathbf{w} \tag{6.33}$$

where $\mathbf{r}$ represents the displacements $\mathbf{U}_r$ relative to the rigid frame of reference, established by $\mathbf{w} = \mathbf{0}$, and $\mathbf{T}$ is a transformation matrix. If we apply arbitrary virtual displacements $\delta\mathbf{w}$, then from Eq. (6.33)

$$\delta\mathbf{U}_r = \mathbf{T}\,\delta\mathbf{w} \tag{6.34}$$

and the virtual work

$$\delta W = \delta\mathbf{w}^T\,\mathbf{P}_w + \delta\mathbf{U}_r^{\,T}\,\mathbf{P}_u \tag{6.35}$$

From the principle of virtual work it is clear that if the virtual displacements are only those representing rigid-body degrees of freedom, then

$$\delta W = 0 \tag{6.36}$$

and hence from Eqs. (6.34) to (6.36) it follows that

$$\delta\mathbf{w}^T\,(\mathbf{P}_w + \mathbf{T}^T\mathbf{P}_u) = 0 \tag{6.37}$$

Since the virtual displacements $\delta\mathbf{w}$ have been chosen arbitrarily, $\delta\mathbf{w} \neq \mathbf{0}$, and we must have that

$$\mathbf{P}_w + \mathbf{T}^T\mathbf{P}_u = \mathbf{0} \tag{6.38}$$

This last equation is in fact the equation of equilibrium for the rigid-body degrees of freedom, i.e., overall static equilibrium, and it can be used to determine the reactions $\mathbf{P}_w$ for a given set of externally applied forces $\mathbf{P}_u$. However, the more usual form of the overall equilibrium equations is

$$\mathbf{P}_w + \mathbf{V}\mathbf{P}_u = \mathbf{0} \tag{6.39}$$

where $$\mathbf{V} = \mathbf{T}^T \tag{6.40}$$

is determined from statics. It should be noted that the above analysis is also applicable to a single structural element. In such cases, however, we should use the symbols $\mathbf{S}_w$ and $\mathbf{S}_u$ instead of $\mathbf{P}_w$ and $\mathbf{P}_u$, and $\mathbf{v}$ instead of $\mathbf{r}$.

In the matrix displacement method of analysis the rigid-body degrees of freedom are eliminated by making $\mathbf{w} = \mathbf{0}$. This establishes a rigid frame of reference with respect to which all displacements $\mathbf{U}_r$ are measured. Thus when $\mathbf{w} = \mathbf{0}$, it follows from Eq. (6.33) that

$$\mathbf{U}_r = \mathbf{r} \tag{6.41}$$

The transformation matrix $\mathbf{T}$ depends on the choice of the displacements $\mathbf{w}$. For certain choices of $\mathbf{w}$, however, the matrix $\mathbf{T}$ does not exist, and the stiffness matrix $\mathbf{K}_r$ corresponding to the displacements $\mathbf{U}_r$ is singular. The next two sections discuss methods of selecting the correct rigid frame of reference using either equilibrium equations or kinematics relations. The problem of finding a suitable rigid frame of reference occurs in aeronautical and aerospace analyses since the structures have no real externally applied constraints. In civil engineering the frame of reference is the foundation with zero displacements. The transformation matrix $\mathbf{T}$ will also be utilized in the vibration analysis of unconstrained structures (Chap. 12).

6.3 DERIVATION OF THE TRANSFORMATION MATRIX V FROM EQUILIBRIUM EQUATIONS

ROTATIONAL AND TRANSLATIONAL DEGREES OF FREEDOM

Consider a three-dimensional structure (Fig. 6.3) in which each nodal point has six degrees of freedom, three translations, and three rotations. Resolving forces in the directions of the datum axes and then taking moments about the same axes gives the following equilibrium equations:

FORCES

$$\begin{aligned} P_1 + P_7 + P_{13} + \cdots &= 0 \\ P_2 + P_8 + P_{14} + \cdots &= 0 \\ P_3 + P_9 + P_{15} + \cdots &= 0 \end{aligned} \tag{6.42}$$

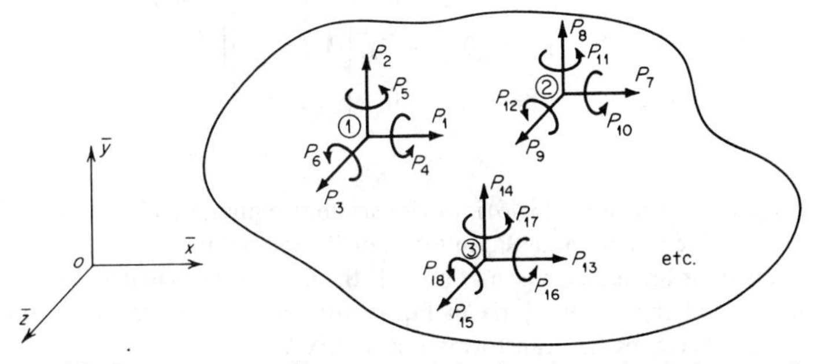

FIG. 6.3 Forces corresponding to translational and rotational degrees of freedom.

MOMENTS

$$
\begin{aligned}
&-P_2\bar{z}_1 + P_3\bar{y}_1 + P_4 - P_8\bar{z}_2 + P_9\bar{y}_2 + P_{10} - P_{14}\bar{z}_3 + P_{15}\bar{y}_3 + P_{16} - \cdots = 0\\
&P_1\bar{z}_1 - P_3\bar{x}_1 + P_5 + P_7\bar{z}_2 - P_9\bar{x}_2 + P_{11} + P_{13}\bar{z}_3 - P_{15}\bar{x}_3 + P_{17} + \cdots = 0\\
&-P_1\bar{y}_1 + P_2\bar{x}_1 + P_6 - P_7\bar{y}_2 + P_8\bar{x}_2 + P_{12} - P_{13}\bar{y}_3 + P_{14}\bar{x}_3 + P_{18} - \cdots = 0
\end{aligned} \tag{6.43}
$$

where the subscripts with the $\bar{x}$, $\bar{y}$, and $\bar{z}$ coordinates refer here to the nodal numbers.

If we take

$$\mathbf{P}_w = \{P_1 \quad P_2 \quad P_3 \quad P_4 \quad P_5 \quad P_6\} \tag{6.44}$$

$$\mathbf{P}_u = \{P_7 \quad P_8 \quad P_9 \quad \cdots\} \tag{6.45}$$

the equations of equilibrium can then be written in matrix form as

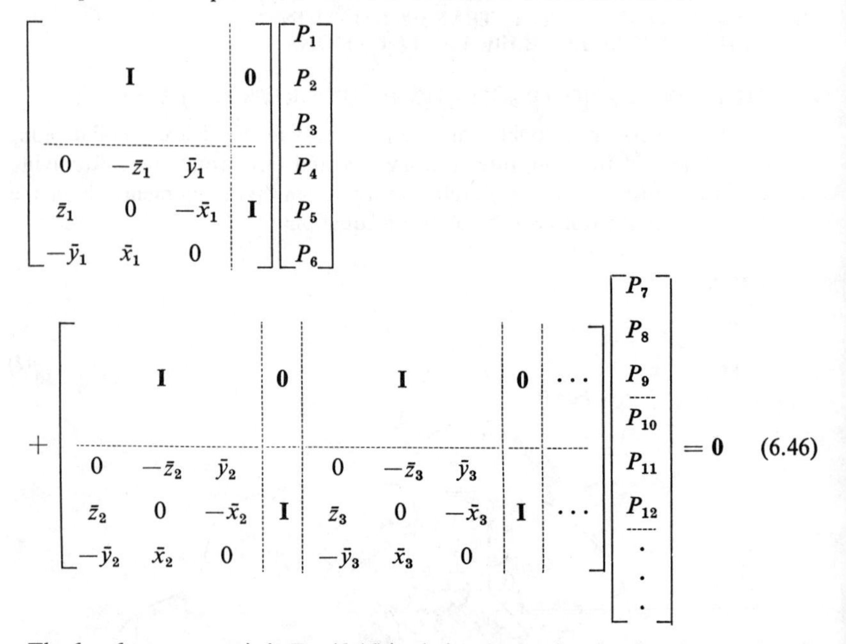

$$
\left[\begin{array}{ccc|c}
 & \mathbf{I} & & \mathbf{0}\\
\hline
0 & -\bar{z}_1 & \bar{y}_1 & \\
\bar{z}_1 & 0 & -\bar{x}_1 & \mathbf{I}\\
-\bar{y}_1 & \bar{x}_1 & 0 &
\end{array}\right]
\begin{bmatrix} P_1\\ P_2\\ P_3\\ \hline P_4\\ P_5\\ P_6\end{bmatrix}
+
\left[\begin{array}{ccc|c|ccc|c|c}
 & \mathbf{I} & & \mathbf{0} & & \mathbf{I} & & \mathbf{0} & \cdots\\
\hline
0 & -\bar{z}_2 & \bar{y}_2 & & 0 & -\bar{z}_3 & \bar{y}_3 & & \\
\bar{z}_2 & 0 & -\bar{x}_2 & \mathbf{I} & \bar{z}_3 & 0 & -\bar{x}_3 & \mathbf{I} & \cdots\\
-\bar{y}_2 & \bar{x}_2 & 0 & & -\bar{y}_3 & \bar{x}_3 & 0 & &
\end{array}\right]
\begin{bmatrix} P_7\\ P_8\\ P_9\\ \hline P_{10}\\ P_{11}\\ P_{12}\\ \hline \cdot\\ \cdot\\ \cdot\end{bmatrix}
= \mathbf{0} \tag{6.46}
$$

The 6×6 square matrix in Eq. (6.46) is obviously nonsingular,* and consequently it is always possible to determine $\mathbf{P}_w$ in terms of $\mathbf{P}_u$ and obtain $\mathbf{V}$. If the origin for the datum coordinates is chosen at node 1, the 6×6 square matrix becomes a unit matrix, and the $6 \times n$ matrix in Eq. (6.46), where n represents number of forces in $\mathbf{P}_u$, becomes the transformation matrix $\mathbf{V}$.

* See matrix inversion by partitioning in Appendix A.

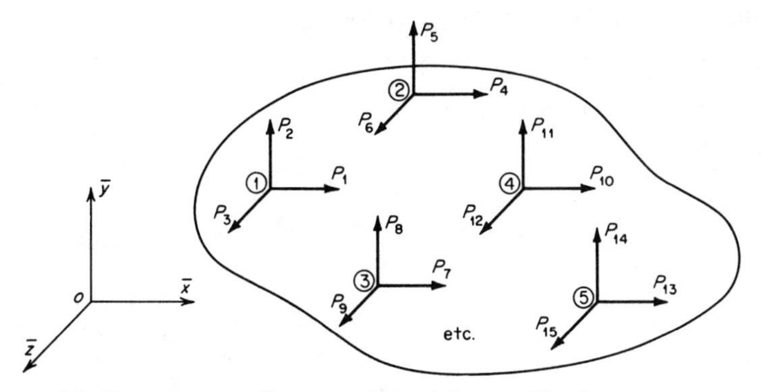

FIG. 6.4 Forces corresponding to translational degrees of freedom.

TRANSLATIONAL DEGREES OF FREEDOM ONLY

Consider a three-dimensional structure (Fig. 6.4) in which each nodal point has only three translational degrees of freedom, i.e., three deflections. Here the following equations of overall equilibrium are obtained:

FORCES

$$\begin{aligned} P_1 + P_4 + P_7 + P_{10} + P_{13} + \cdots &= 0 \\ P_2 + P_5 + P_8 + P_{11} + P_{14} + \cdots &= 0 \\ P_3 + P_6 + P_9 + P_{12} + P_{15} + \cdots &= 0 \end{aligned} \tag{6.47}$$

MOMENTS

$$\begin{aligned} -P_2\bar{z}_1 + P_3\bar{y}_1 - P_5\bar{z}_2 + P_6\bar{y}_2 - P_8\bar{z}_3 + P_9\bar{y}_3 \\ - P_{11}\bar{z}_4 + P_{12}\bar{y}_4 - P_{14}\bar{z}_5 + P_{15}\bar{y}_5 - \cdots = 0 \\ P_1\bar{z}_1 - P_3\bar{x}_1 + P_4\bar{z}_2 - P_6\bar{x}_2 + P_7\bar{z}_3 - P_9\bar{x}_3 \\ + P_{10}\bar{z}_4 - P_{12}\bar{x}_4 + P_{13}\bar{z}_5 - P_{15}\bar{x}_5 + \cdots = 0 \\ -P_1\bar{y}_1 + P_2\bar{x}_1 - P_4\bar{y}_2 + P_5\bar{x}_2 - P_7\bar{y}_3 + P_8\bar{x}_3 \\ - P_{10}\bar{y}_4 + P_{11}\bar{x}_4 - P_{13}\bar{y}_5 + P_{14}\bar{x}_5 - \cdots = 0 \end{aligned} \tag{6.48}$$

where, as before, the subscripts with the $\bar{x}$, $\bar{y}$, and $\bar{z}$ coordinates refer to the nodal numbers associated with the corresponding forces. Taking

$$\mathbf{P}_w = \{P_1 \quad P_2 \quad P_3 \quad P_4 \quad P_8 \quad P_{12}\} \tag{6.49}$$

and $$\mathbf{P}_u = \{P_5 \quad P_6 \quad P_7 \quad P_9 \quad P_{10} \quad P_{11} \quad P_{13} \quad P_{14} \quad P_{15} \quad \cdots\} \tag{6.50}$$

we can write Eqs. (6.47) and (6.48) in matrix form as

$$
\left[\begin{array}{ccc|ccc}
1 & 0 & 0 & 1 & 0 & 0 \\
0 & 1 & 0 & 0 & 1 & 0 \\
0 & 0 & 1 & 0 & 0 & 1 \\
\hline
0 & -\bar{z}_1 & \bar{y}_1 & 0 & -\bar{z}_3 & \bar{y}_4 \\
\bar{z}_1 & 0 & -\bar{x}_1 & \bar{z}_2 & 0 & -\bar{x}_4 \\
-\bar{y}_1 & \bar{x}_1 & 0 & -\bar{y}_2 & \bar{x}_3 & 0
\end{array}\right]
\left[\begin{array}{c}
P_1 \\ P_2 \\ P_3 \\ \hline P_4 \\ P_8 \\ P_{12}
\end{array}\right]
$$

$$
+\left[\begin{array}{ccc|ccc|ccc|c}
0 & 0 & 1 & 0 & 1 & 0 & 1 & 0 & 0 & \cdots \\
1 & 0 & 0 & 0 & 0 & 1 & 0 & 1 & 0 & \cdots \\
0 & 1 & 0 & 1 & 0 & 0 & 0 & 0 & 1 & \cdots \\
\hline
-\bar{z}_2 & \bar{y}_2 & 0 & \bar{y}_3 & 0 & -\bar{z}_4 & 0 & -\bar{z}_5 & \bar{y}_5 & \cdots \\
0 & -\bar{x}_2 & \bar{z}_3 & -\bar{x}_3 & \bar{z}_4 & 0 & \bar{z}_5 & 0 & -\bar{x}_5 & \cdots \\
\bar{x}_2 & 0 & -\bar{y}_3 & 0 & -\bar{y}_4 & \bar{x}_4 & -\bar{y}_5 & \bar{x}_5 & 0 & \cdots
\end{array}\right]
\left[\begin{array}{c}
P_5 \\ P_6 \\ P_7 \\ \hline P_9 \\ P_{10} \\ P_{11} \\ \hline P_{13} \\ P_{14} \\ P_{15} \\ \hline \cdot \\ \cdot \\ \cdot
\end{array}\right] = \mathbf{0} \tag{6.51}
$$

The solution for $\mathbf{P}_w$ will exist provided the 6×6 square matrix in the first term of Eq. (6.51) is nonsingular. When we introduce

$$
\mathbf{A}_{21} = \begin{bmatrix} 0 & -\bar{z}_1 & \bar{y}_1 \\ \bar{z}_1 & 0 & -\bar{x}_1 \\ -\bar{y}_1 & \bar{x}_1 & 0 \end{bmatrix} \tag{6.52}
$$

$$
\mathbf{A}_{22} = \begin{bmatrix} 0 & -\bar{z}_3 & \bar{y}_4 \\ \bar{z}_2 & 0 & -\bar{x}_4 \\ -\bar{y}_2 & \bar{x}_3 & 0 \end{bmatrix} \tag{6.53}
$$

and
$$
\mathbf{Z} = \mathbf{A}_{22} - \mathbf{A}_{21} = \begin{bmatrix} 0 & -\bar{z}_{31} & \bar{y}_{41} \\ \bar{z}_{21} & 0 & -\bar{x}_{41} \\ -\bar{y}_{21} & \bar{x}_{31} & 0 \end{bmatrix} \tag{6.54}
$$

where
$$
\bar{z}_{21} = \bar{z}_2 - \bar{z}_1 \tag{6.55}
$$

and so forth, it can be shown that the inverse of the 6×6 matrix is given by

$$\begin{bmatrix} \mathbf{I} & \mathbf{I} \\ \mathbf{A}_{21} & \mathbf{A}_{22} \end{bmatrix}^{-1} = \begin{bmatrix} \mathbf{I} + \mathbf{Z}^{-1}\mathbf{A}_{21} & -\mathbf{Z}^{-1} \\ -\mathbf{Z}^{-1}\mathbf{A}_{21} & \mathbf{Z}^{-1} \end{bmatrix} \tag{6.56}$$

provided $\qquad |\mathbf{Z}| \neq 0 \qquad (6.57)$

which finally reduces to

$$\bar{z}_{21}\bar{x}_{31}\bar{y}_{41} - \bar{y}_{21}\bar{z}_{31}\bar{x}_{41} \neq 0 \tag{6.58}$$

$|\mathbf{Z}| = 0$ represents the condition for the lines of action of the constraining displacement vectors, corresponding to P_4, P_8, and P_{12}, to pass through a straight line in space drawn from node 1. This implies that the three forces P_4, P_8, and P_{12} would then have a zero moment about this line, and consequently it would not be possible to restrain rotation about this line using the selected force reactions.

Premultiplication of Eq. (6.51) by (6.56) leads directly to the solution for $\mathbf{P}_w$ and $\mathbf{V}$ [see Eq. (6.39)]. The form of $\mathbf{V}$ naturally depends on the choice of $\mathbf{P}_w$. However, it is generally advantageous to select the first three forces in $\mathbf{P}_w$ as three orthogonal forces at a node, while the three remaining forces may be selected in many different ways. In civil engineering structures this latter choice is governed by the degrees of freedom, which are restrained by the rigid foundation to which the structure is attached. If forces other than P_4, P_8, and P_{12} are selected for $\mathbf{P}_w$, we shall obtain a different transformation matrix $\mathbf{V}$, and the criterion $|\mathbf{Z}| \neq 0$ for the correct choice of reactive forces will naturally have different coefficients from those given by Eq. (6.54). The derivation of the determinant $|\mathbf{Z}|$ in such cases is analogous to the method presented for $\mathbf{P}_w = \{P_1 \quad P_2 \quad P_3 \quad P_4 \quad P_8 \quad P_{12}\}$.

6.4 DERIVATION OF THE TRANSFORMATION MATRIX T FROM KINEMATICS

Combining the kinematic relationship (6.33) with the identity $\mathbf{w} = \mathbf{w}$, we have

$$\begin{bmatrix} \mathbf{w} \\ \mathbf{U}_r \end{bmatrix} = \begin{bmatrix} \mathbf{0} \\ \mathbf{r} \end{bmatrix} + \begin{bmatrix} \mathbf{I} \\ \mathbf{T} \end{bmatrix} \mathbf{w} \tag{6.59}$$

If the displacement $\mathbf{w}$ consist of three deflections w_x, w_y, and w_z and three rotations θ_x, θ_y, and θ_z, the transformation matrix $\mathbf{T}$ can be determined directly from kinematics. However, if only translational displacements are considered, the matrix $\mathbf{T}$ cannot be evaluated directly, and several auxiliary matrices must be used for its determination. To illustrate this we shall consider the alternative form of Eq. (6.59) as

$$\begin{bmatrix} \mathbf{w} \\ \hdashline \mathbf{U}_r \end{bmatrix} = \begin{bmatrix} \mathbf{0} \\ \hdashline \mathbf{r} \end{bmatrix} + \left[\begin{array}{c:c} \mathbf{T}_{ww} & \mathbf{T}_{w\theta} \\ \hdashline \mathbf{T}_{uw} & \mathbf{T}_{u\theta} \end{array}\right] \begin{bmatrix} w_x \\ w_y \\ w_z \\ \theta_x \\ \theta_y \\ \theta_z \end{bmatrix} \tag{6.60}$$

where it will be assumed that $w_x = w_1$, $w_y = w_2$, $w_z = w_3$ refer to three orthogonal displacements of a node and θ_x, θ_y, θ_z represent respective notations about the w_x, w_y, and w_z displacement vectors. If the rigid frame of reference is established by the translational displacements

$$\mathbf{w} = \{w_1 \quad \cdots \quad w_6\} \tag{6.61}$$

the displacements $\{w_4 \quad w_5 \quad w_6\}$ should be expressible in terms of $\{\theta_x \quad \theta_y \quad \theta_z\}$. From Eq. (6.60) it follows that

$$\begin{aligned} \mathbf{w} &= \mathbf{T}_{ww} \begin{bmatrix} w_1 \\ w_2 \\ w_3 \end{bmatrix} + \mathbf{T}_{w\theta} \begin{bmatrix} \theta_x \\ \theta_y \\ \theta_z \end{bmatrix} \\ &= \begin{bmatrix} \mathbf{A} \\ \mathbf{B} \end{bmatrix} \begin{bmatrix} w_1 \\ w_2 \\ w_3 \end{bmatrix} + \begin{bmatrix} \mathbf{C} \\ \mathbf{D} \end{bmatrix} \begin{bmatrix} \theta_x \\ \theta_y \\ \theta_z \end{bmatrix} \end{aligned} \tag{6.62}$$

$$\text{where} \qquad \mathbf{T}_{ww} = \begin{bmatrix} \mathbf{A} \\ \mathbf{B} \end{bmatrix} \qquad \text{and} \qquad \mathbf{T}_{w\theta} = \begin{bmatrix} \mathbf{C} \\ \mathbf{D} \end{bmatrix} \tag{6.63}$$

Separating Eq. (6.62) into two equations, we have

$$\mathbf{C} \begin{bmatrix} \theta_x \\ \theta_y \\ \theta_z \end{bmatrix} = (\mathbf{I} - \mathbf{A}) \begin{bmatrix} w_1 \\ w_2 \\ w_3 \end{bmatrix} \tag{6.64}$$

$$\mathbf{D} \begin{bmatrix} \theta_x \\ \theta_y \\ \theta_z \end{bmatrix} = \begin{bmatrix} w_4 \\ w_5 \\ w_6 \end{bmatrix} - \mathbf{B} \begin{bmatrix} w_1 \\ w_2 \\ w_3 \end{bmatrix} \tag{6.65}$$

As Eq. (6.64) must be satisfied for all values of $\{\theta_x \quad \theta_y \quad \theta_z\}$ and $\{w_1 \quad w_2 \quad w_3\}$, it follows immediately that

$$\mathbf{C} = \mathbf{0} \qquad \text{and} \qquad \mathbf{A} = \mathbf{I} \tag{6.66}$$

Also from Eq. (6.65) we have

$$\begin{bmatrix} \theta_x \\ \theta_y \\ \theta_z \end{bmatrix} = \mathbf{D}^{-1} \begin{bmatrix} w_4 \\ w_5 \\ w_6 \end{bmatrix} - \mathbf{D}^{-1}\mathbf{B} \begin{bmatrix} w_1 \\ w_2 \\ w_3 \end{bmatrix} \tag{6.67}$$

with the proviso that $|\mathbf{D}| \neq 0$.

The column matrix $\{w_x \quad w_y \quad w_z \quad \theta_x \quad \theta_y \quad \theta_z\}$ in Eq. (6.60) can now be related to the six displacements $\{w_1 \quad w_2 \quad w_3 \quad w_4 \quad w_5 \quad w_6\}$; thus using Eq. (6.67), we derive

$$\begin{bmatrix} w_x \\ w_y \\ w_z \\ \hline \theta_x \\ \theta_y \\ \theta_z \end{bmatrix} = \left[\begin{array}{c|c} \mathbf{I} & \mathbf{0} \\ \hline -\mathbf{D}^{-1}\mathbf{B} & \mathbf{D}^{-1} \end{array}\right] \begin{bmatrix} w_1 \\ w_2 \\ w_3 \\ \hline w_4 \\ w_5 \\ w_6 \end{bmatrix} \tag{6.68}$$

Substituting Eq. (6.68) into (6.60), we have

$$\begin{bmatrix} \mathbf{w} \\ \hline \mathbf{U}_r \end{bmatrix} = \begin{bmatrix} \mathbf{0} \\ \hline \mathbf{r} \end{bmatrix} + \left[\begin{array}{c|c} \mathbf{T}_{ww} - \mathbf{T}_{w\theta}\mathbf{D}^{-1}\mathbf{B} & \mathbf{T}_{w\theta}\mathbf{D}^{-1} \\ \hline \mathbf{T}_{uw} - \mathbf{T}_{u\theta}\mathbf{D}^{-1}\mathbf{B} & \mathbf{T}_{u\theta}\mathbf{D}^{-1} \end{array}\right] \begin{bmatrix} w_1 \\ w_2 \\ w_3 \\ \hline w_4 \\ w_5 \\ w_6 \end{bmatrix}$$

$$= \begin{bmatrix} \mathbf{0} \\ \hline \mathbf{r} \end{bmatrix} + \left[\begin{array}{c|c} \multicolumn{2}{c}{\mathbf{I}} \\ \hline \mathbf{T}_{uw} - \mathbf{T}_{u\theta}\mathbf{D}^{-1}\mathbf{B} & \mathbf{T}_{u\theta}\mathbf{D}^{-1} \end{array}\right] \begin{bmatrix} w_1 \\ w_2 \\ w_3 \\ \hline w_4 \\ w_5 \\ w_6 \end{bmatrix} \tag{6.69}$$

where $\quad [\mathbf{T}_{ww} - \mathbf{T}_{w\theta}\mathbf{D}^{-1}\mathbf{B} \mid \mathbf{T}_{w\theta}\mathbf{D}^{-1}] = \mathbf{I}$ (6.70)

may be easily verified by substituting Eqs. (6.63) for $\mathbf{T}_{ww}$ and $\mathbf{T}_{w\theta}$ with $\mathbf{A} = \mathbf{I}$ and $\mathbf{C} = \mathbf{0}$.

Comparing now Eqs. (6.59) and (6.69), we see clearly that

$$\mathbf{T} = [\mathbf{T}_{uw} - \mathbf{T}_{u\theta}\mathbf{D}^{-1}\mathbf{B} \mid \mathbf{T}_{u\theta}\mathbf{D}^{-1}] \tag{6.71}$$

From Eq. (6.60)

$$\mathbf{U}_r = \mathbf{r} + \mathbf{T}_{uw}\begin{bmatrix} w_1 \\ w_2 \\ w_3 \end{bmatrix} + \mathbf{T}_{u\theta}\begin{bmatrix} \theta_x \\ \theta_y \\ \theta_z \end{bmatrix} \tag{6.72}$$

If we assume that the relative displacements $\mathbf{r} = \mathbf{0}$, that is, the displacements $\mathbf{U}_r$, are derived only from the rigid-body translations and rotations,

$$[\mathbf{U}_r]_{\mathbf{r}=\mathbf{0}} = \mathbf{T}_{uw}\begin{bmatrix} w_1 \\ w_2 \\ w_3 \end{bmatrix} + \mathbf{T}_{u\theta}\begin{bmatrix} \theta_x \\ \theta_y \\ \theta_z \end{bmatrix} \tag{6.73}$$

where the first and second term of the right-hand side of this equation represent displacements due to rigid-body translations and rotations, respectively. Equation (6.73) can be used to derive the submatrices $\mathbf{T}_{uw}$ and $\mathbf{T}_{u\theta}$. Rewriting Eq. (6.65) as

$$\begin{bmatrix} w_4 \\ w_5 \\ w_6 \end{bmatrix} = \mathbf{B}\begin{bmatrix} w_1 \\ w_2 \\ w_3 \end{bmatrix} + \mathbf{D}\begin{bmatrix} \theta_x \\ \theta_y \\ \theta_z \end{bmatrix} \tag{6.65a}$$

makes it evident that we can interpret this equation kinematically to obtain the submatrices **B** and **D**. For example, if the displacements w_4, w_5, and w_6 are taken in the $\bar{x}$, $\bar{y}$, and $\bar{z}$ directions, respectively, then

$$\mathbf{B} = \mathbf{I} \tag{6.74}$$

and
$$\mathbf{D} = \begin{bmatrix} 0 & \bar{z}_{21} & -\bar{y}_{21} \\ -\bar{z}_{31} & 0 & \bar{x}_{31} \\ \bar{y}_{41} & -\bar{x}_{41} & 0 \end{bmatrix} \tag{6.75}$$

Since the matrix **D** must be nonsingular,

$$|\mathbf{D}| \neq 0 \tag{6.76}$$

Noting also that $\mathbf{D} = \mathbf{Z}^T$, it follows that the criterion for selecting the displacements $\mathbf{w}$ to establish the rigid frame of reference based on kinematics is the same as the criterion based on equilibrium consideration.

If instead of the displacement w_4, w_5, and w_6 we use rotations θ_x, θ_y, and θ_z, then from Eq. (6.65*a*)

$$\mathbf{B} = \mathbf{0} \tag{6.77}$$

$$\mathbf{D} = \mathbf{I} \tag{6.78}$$

and $$\mathbf{T} = [\mathbf{T}_{uw} \quad \mathbf{T}_{u\theta}] \tag{6.79}$$

6.5 CONDENSATION OF STIFFNESS MATRICES

The stiffness matrix **K** for the complete structure relates *all* forces to their corresponding displacements. In many applications, however, only a limited number of external forces are applied to the structure. If only the displacements in the directions of these forces are required, it is generally preferable to evaluate a new stiffness matrix relating the forces and the corresponding displacements which are of interest in the analysis (see Fig. 6.5).

Suppose that the force-displacement equations for the whole structure are partitioned so that

$$\begin{bmatrix} \mathbf{P}_x \\ \mathbf{P}_y \end{bmatrix} = \begin{bmatrix} \mathbf{K}_{xx} & \mathbf{K}_{xy} \\ \mathbf{K}_{yx} & \mathbf{K}_{yy} \end{bmatrix} \begin{bmatrix} \mathbf{U}_x \\ \mathbf{U}_y \end{bmatrix} + \begin{bmatrix} \mathbf{Q}_x \\ \mathbf{Q}_y \end{bmatrix} \tag{6.80}$$

and $$\mathbf{P}_y = \mathbf{0} \tag{6.81}$$

while $$\mathbf{P}_x \neq \mathbf{0} \tag{6.82}$$

From Eqs. (6.80) and (6.81) it follows that

$$\mathbf{K}_{yx}\mathbf{U}_x + \mathbf{K}_{yy}\mathbf{U}_y + \mathbf{Q}_y = \mathbf{0} \tag{6.83}$$

and hence $$\mathbf{U}_y = -\mathbf{K}_{yy}^{-1}\mathbf{K}_{yx}\mathbf{U}_x - \mathbf{K}_{yy}^{-1}\mathbf{Q}_y \tag{6.84}$$

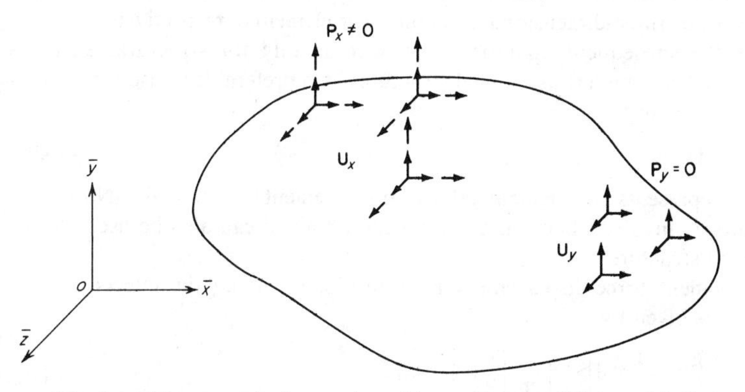

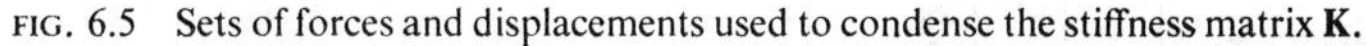

FIG. 6.5 Sets of forces and displacements used to condense the stiffness matrix **K**.

provided

$$|\mathbf{K}_{yy}| \neq 0 \tag{6.85}$$

Substituting (6.84) into (6.80), we have that

$$\begin{aligned}\mathbf{P}_x &= (\mathbf{K}_{xx} - \mathbf{K}_{xy}\mathbf{K}_{yy}^{-1}\mathbf{K}_{yx})\mathbf{U}_x + (\mathbf{Q}_x - \mathbf{K}_{xy}\mathbf{K}_{yy}^{-1}\mathbf{Q}_y) \\ &= \mathbf{K}_c\mathbf{U}_x + \mathbf{Q}_c\end{aligned} \tag{6.86}$$

where $\mathbf{K}_c = \mathbf{K}_{xx} - \mathbf{K}_{xy}\mathbf{K}_{yy}^{-1}\mathbf{K}_{yx}$ (6.87)

represents the condensed stiffness matrix and

$$\mathbf{Q}_c = \mathbf{Q}_x - \mathbf{K}_{xy}\mathbf{K}_{yy}^{-1}\mathbf{Q}_y \tag{6.88}$$

represents the condensed thermal-loading matrix.

Eliminating all rigid-body degrees of freedom from $\mathbf{U}_x$ means that Eq. (6.86) is reduced to (by eliminating appropriate rows and columns)

$$\mathbf{P}_{x_r} = \mathbf{K}_{c_r}\mathbf{U}_{x_r} + \mathbf{Q}_{c_r} \tag{6.89}$$

and hence $\mathbf{U}_{x_r} = \mathbf{K}_{c_r}^{-1}\mathbf{P}_{x_r} - \mathbf{K}_{c_r}^{-1}\mathbf{Q}_{c_r}$ (6.90)

The elimination of the rigid-body degrees of freedom from $\mathbf{U}_x$ can also be carried out before the matrix condensation.

6.6 DERIVATION OF STIFFNESS MATRICES FROM FLEXIBILITY

In some applications it is necessary to determine the stiffness matrix from the corresponding flexibility matrix. For example, experimental flexibility coefficients obtained by a direct measurement of displacements can be used to obtain experimental stiffness coefficients. The stiffness matrices can be derived from the corresponding flexibility matrices using a suitable transformation matrix, which raises the order of the flexibility matrix by one, three, or six for one-, two-, or three-dimensional structures (or elements), respectively.

Since the subsequent analysis is intended mainly for structural elements rather than for complete assembled structures, it is preferable to use $\mathbf{v}$ instead of $\mathbf{r}$ in Eq. (6.33) so that

$$\mathbf{u}_v = \mathbf{v} + \mathbf{T}\mathbf{w} \tag{6.91}$$

where $\mathbf{v}$ represents the element relative displacements for $\mathbf{w} = \mathbf{0}$. No loss of generality is involved here, and the results obtained can also be used for the assembled structure.

The element force-displacement relationship, including the effects of temperature, is given by

$$\begin{bmatrix}\mathbf{S}_w \\ \mathbf{S}_u\end{bmatrix} = \begin{bmatrix}\mathbf{k}_{ww} & \mathbf{k}_{wu} \\ \mathbf{k}_{uw} & \mathbf{k}_{uu}\end{bmatrix}\begin{bmatrix}\mathbf{w} \\ \mathbf{u}_v\end{bmatrix} + \begin{bmatrix}\mathbf{Q}_w \\ \mathbf{Q}_u\end{bmatrix} \tag{6.92}$$

or symbolically

$$\mathbf{S} = \mathbf{k}\mathbf{u} + \mathbf{Q} \tag{6.93}$$

The element forces $\mathbf{S} = \{\mathbf{S}_w \quad \mathbf{S}_u\}$ must be in equilibrium, and they must satisfy the equation [see Eq. (6.38)]

$$\mathbf{S}_w + \mathbf{T}^T\mathbf{S}_u = \mathbf{0} \tag{6.94}$$

Substituting $\mathbf{S}_w$ and $\mathbf{S}_u$ from Eq. (6.92) into (6.94) and noting that

$$\mathbf{k}_{wu} = \mathbf{k}_{uw}{}^T \tag{6.95}$$

we have

$$\mathbf{k}_{ww}\mathbf{w} + \mathbf{k}_{uw}{}^T\mathbf{u}_v + \mathbf{Q}_w + \mathbf{T}^T\mathbf{k}_{uw}\mathbf{w} + \mathbf{T}^T\mathbf{k}_{uu}\mathbf{u}_v + \mathbf{T}^T\mathbf{Q}_u = \mathbf{0} \tag{6.96}$$

which upon substitution of Eq. (6.91) for $\mathbf{u}_v$ leads to

$$(\mathbf{T}^T\mathbf{k}_{uu} + \mathbf{k}_{uw}{}^T)\mathbf{v} + (\mathbf{T}^T\mathbf{k}_{uu}\mathbf{T} + \mathbf{k}_{uw}{}^T\mathbf{T} + \mathbf{T}^T\mathbf{k}_{uw} + \mathbf{k}_{ww})\mathbf{w} + \mathbf{T}^T\mathbf{Q}_u + \mathbf{Q}_w = \mathbf{0} \tag{6.97}$$

The above equation must be true for any values of $\mathbf{v}$, $\mathbf{w}$, and temperature of the element. Hence

$$\mathbf{k}_{uw}{}^T = -\mathbf{T}^T\mathbf{k}_{uu} \tag{6.98}$$

$$\begin{aligned}\mathbf{k}_{ww} &= -\mathbf{T}^T\mathbf{k}_{uu}\mathbf{T} - \mathbf{k}_{uw}{}^T\mathbf{T} - \mathbf{T}^T\mathbf{k}_{uw} \\ &= \mathbf{T}^T\mathbf{k}_{uu}\mathbf{T}\end{aligned} \tag{6.99}$$

and $$\mathbf{Q}_w = -\mathbf{T}^T\mathbf{Q}_u \tag{6.100}$$

Substituting Eqs. (6.98) to (6.100) into (6.92), we have

$$\begin{bmatrix}\mathbf{S}_w \\ \mathbf{S}_u\end{bmatrix} = \begin{bmatrix}\mathbf{T}^T\mathbf{k}_{uu}\mathbf{T} & -\mathbf{T}^T\mathbf{k}_{uu} \\ -\mathbf{k}_{uu}\mathbf{T} & \mathbf{k}_{uu}\end{bmatrix}\begin{bmatrix}\mathbf{w} \\ \mathbf{u}_v\end{bmatrix} + \begin{bmatrix}-\mathbf{T}^T\mathbf{Q}_u \\ \mathbf{Q}_u\end{bmatrix} \tag{6.101}$$

When $\mathbf{w} = \mathbf{0}$, then $\mathbf{u}_v = \mathbf{v}$, and

$$\mathbf{S}_u = \mathbf{k}_{uu}\mathbf{v} + \mathbf{Q}_u \tag{6.102}$$

$$\mathbf{v} = \mathbf{k}_{uu}{}^{-1}\mathbf{S}_u - \mathbf{k}_{uu}{}^{-1}\mathbf{Q}_u = \mathbf{f}_{uu}\mathbf{S}_u + \mathbf{v}_T \tag{6.103}$$

where $$\mathbf{f}_{uu} = \mathbf{k}_{uu}{}^{-1} \tag{6.104}$$

and $$\mathbf{v}_T = -\mathbf{k}_{uu}{}^{-1}\mathbf{Q}_u = -\mathbf{f}_{uu}\mathbf{Q}_u \tag{6.105}$$

Using Eqs. (6.104) and (6.105), we can write the force-displacement relationship as

$$\begin{aligned}\begin{bmatrix}\mathbf{S}_w \\ \mathbf{S}_u\end{bmatrix} &= \begin{bmatrix}\mathbf{T}^T\mathbf{f}_{uu}{}^{-1}\mathbf{T} & -\mathbf{T}^T\mathbf{f}_{uu}{}^{-1} \\ -\mathbf{f}_{uu}{}^{-1}\mathbf{T} & \mathbf{f}_{uu}{}^{-1}\end{bmatrix}\begin{bmatrix}\mathbf{w} \\ \mathbf{u}_v\end{bmatrix} + \begin{bmatrix}\mathbf{T}^T\mathbf{f}_{uu}{}^{-1}\mathbf{v}_T \\ -\mathbf{f}_{uu}{}^{-1}\mathbf{v}_T\end{bmatrix} \\ &= \begin{bmatrix}\mathbf{T}^T \\ -\mathbf{I}\end{bmatrix}\mathbf{f}_{uu}{}^{-1}[\mathbf{T} \quad -\mathbf{I}]\begin{bmatrix}\mathbf{w} \\ \mathbf{u}_v\end{bmatrix} + \begin{bmatrix}\mathbf{T}^T \\ -\mathbf{I}\end{bmatrix}\mathbf{f}_{uu}{}^{-1}\mathbf{v}_T\end{aligned} \tag{6.106}$$

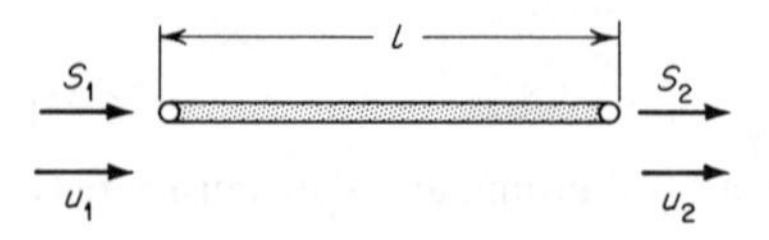

FIG. 6.6 Pin-jointed bar element.

Introducing now

$$\mathbf{N} = [\mathbf{T} \quad -\mathbf{I}] \tag{6.107}$$

it is clear that the stiffness matrix **k** is expressible by

$$\mathbf{k} = \mathbf{N}^T\mathbf{f}_{uu}{}^{-1}\mathbf{N} \tag{6.108}$$

while the thermal loading matrix **Q** becomes

$$\mathbf{Q} = \mathbf{N}^T\mathbf{f}_{uu}{}^{-1}\mathbf{v}_T \tag{6.109}$$

As an example of the application of Eqs. (6.108) and (6.109) consider a pin-jointed bar element as shown in Fig. 6.6. For this case we have

$$u_1 = w \qquad \text{and} \qquad u_2 = v + u_1 \tag{6.110}$$

Hence $\quad \mathbf{T} = 1 \qquad (6.111)$

and $\quad \mathbf{N} = [1 \quad -1] \qquad (6.112)$

From the flexibility analysis of bar elements (see Chap. 7)

$$\mathbf{f} = \frac{l}{AE} \tag{6.113}$$

and $\quad \mathbf{v}_T = \alpha T l \qquad (6.114)$

Thus

$$\mathbf{k} = \mathbf{N}^T\mathbf{f}_{uu}{}^{-1}\mathbf{N}$$

$$= \begin{bmatrix} 1 \\ -1 \end{bmatrix} \frac{AE}{l} [1 \quad -1] = \frac{AE}{l} \begin{bmatrix} 1 & -1 \\ -1 & 1 \end{bmatrix} \tag{6.115}$$

and

$$\mathbf{Q} = \mathbf{N}^T\mathbf{f}_{uu}{}^{-1}\mathbf{v}_T = \begin{bmatrix} 1 \\ -1 \end{bmatrix} \frac{AE}{l} \alpha T l = AE\alpha T \begin{bmatrix} 1 \\ -1 \end{bmatrix} \tag{6.116}$$

which agrees with the previously derived results in Sec. 5.2.

6.7 STIFFNESS MATRIX FOR CONSTANT-SHEAR-FLOW PANELS

One of the idealizations used in the matrix force method is a constant-shear-flow panel with edge members carrying linearly varying axial forces. This concept was actually inherited from the early methods of analysis of aircraft stressed-skin construction used since World War II. A typical body-force diagram for such an idealization is shown in Fig. 6.7. Although this idealization technique has been used quite extensively for the force methods, no papers have

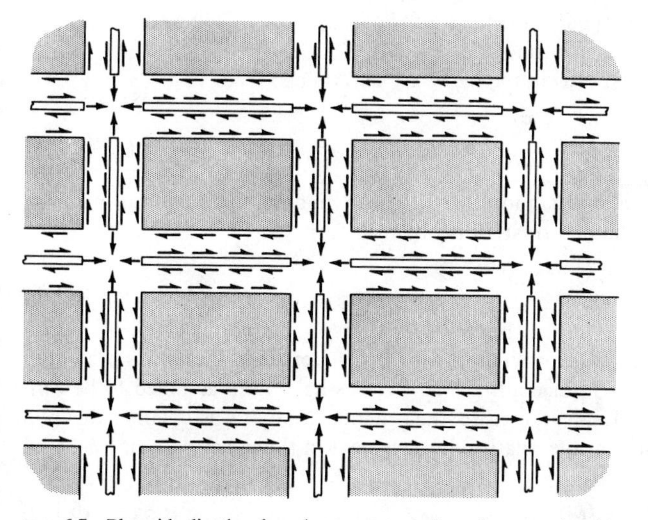

FIG. 6.7 Plate idealization based on constant-shear-flow panels with edge members carrying linearly varying axial forces.

been published on its application to the displacement methods. The reasons for this have perhaps been the restrictive assumption of constant shearing stress without any normal stresses and the difficulty of including the Poisson's ratio effects, all of which can be avoided in other panel idealizations, as discussed in Sec. 5.8.

To demonstrate further the derivation of stiffness from flexibility, we shall consider the constant-shear-flow panel element. The panel will be assumed to be of rectangular shape and of constant thickness. The forces acting on the panel are shown in Fig. 6.8. The shear flow F_1 will be used as the element

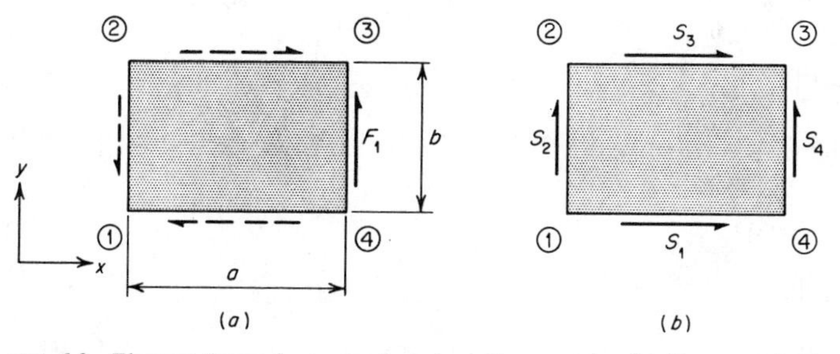

FIG. 6.8 Element forces in a constant-shear-flow panel. (*a*) Force method; (*b*) displacement method.

force in the force method, while the shear flows $S_1, \ldots, S_4$ will be designated as forces for the displacement method.

The stresses and strains in the panel are determined from

$$\sigma_{xy} = \frac{F_1}{t} \quad \text{and} \quad \epsilon_{xy} = \frac{F_1}{Gt} \tag{6.117}$$

where t is panel thickness and G is the shear modulus. The complementary strain energy of total deformation is obtained from

$$U_d^* = \frac{1}{2}\iiint \frac{F_1^2}{Gt^2}\, dx\, dy\, dz = \frac{ab}{2Gt} F_1^2 \tag{6.118}$$

where a and b denote the panel dimensions. No temperature term is present in the expression for U_d^* since all normal stresses have been assumed to be zero within the panel. Using, therefore, the generalization of Castigliano's theorem (part II), we have that the relative displacement in the panel is given by

$$v = \frac{\partial U_d^*}{\partial F_1} = \frac{ab}{Gt} F_1 = fF_1 \tag{6.119}$$

where $$f = \frac{ab}{Gt} \tag{6.120}$$

is the flexibility matrix of the constant-shear-flow panel.

The equations of equilibrium for the shear flows $S_1, \ldots, S_4$ are

$$\begin{bmatrix} 1 & 0 & 1 \\ 0 & 1 & 0 \\ 0 & 0 & 1 \end{bmatrix} \begin{bmatrix} S_1 \\ S_2 \\ S_3 \end{bmatrix} + \begin{bmatrix} 0 \\ 1 \\ -1 \end{bmatrix} S_4 = \mathbf{0} \tag{6.121}$$

which after premultiplying by the inverse of the 3×3 matrix with $\{S_1 \quad S_2 \quad S_3\}$ become

$$\begin{bmatrix} S_1 \\ S_2 \\ S_3 \end{bmatrix} + \begin{bmatrix} 1 \\ 1 \\ -1 \end{bmatrix} S_4 = \mathbf{0} \tag{6.121a}$$

Comparing now (6.121*a*) with (6.38), we note that the transformation matrix $\mathbf{T}$ is given by

$$\mathbf{T} = [1 \quad 1 \quad -1] \tag{6.122}$$

Therefore, using

$$\mathbf{N} = [\mathbf{T} \quad -\mathbf{I}] \tag{6.107}$$

we have from (6.108) that

$$\mathbf{k} = \mathbf{N}^T\mathbf{f}^{-1}\mathbf{N} = \begin{bmatrix} 1 \\ 1 \\ -1 \\ -1 \end{bmatrix} \frac{Gt}{ab} [1 \quad 1 \quad -1 \quad -1] = \frac{Gt}{ab} \begin{bmatrix} 1 & 1 & -1 & -1 \\ 1 & 1 & -1 & -1 \\ -1 & -1 & 1 & 1 \\ -1 & -1 & 1 & 1 \end{bmatrix} \tag{6.123}$$

Equation (6.123) represents the stiffness matrix for a constant-shear-flow panel. The thermal forces $\mathbf{Q}$ are absent for this element because $\mathbf{v}_T = \mathbf{0}$.

6.8 STIFFNESS MATRIX FOR LINEARLY VARYING AXIAL-FORCE MEMBERS

The positive directions of the element forces for an axial-force element are shown in Fig. 6.9. The axial force F in the element is given by

$$F = -\left(1 - \frac{x}{l}\right)F_1 + \frac{x}{l} F_2 \tag{6.124}$$

The temperature variation will also be assumed to vary linearly, so that

$$T = \left(1 - \frac{x}{l}\right)T_1 + \frac{x}{l} T_2 \tag{6.125}$$

where T_1 and T_2 denote the temperatures at $x = 0$ and l, respectively. The energy U_d^* is then evaluated from

$$U_d^* = \frac{1}{2}\iint \frac{F^2}{EA^2}\, dx\, dA + \iint \frac{F}{A}\, \alpha T\, dx\, dA \tag{6.126}$$

Subsequent differentiation of (6.126) with respect to F_1 and F_2 leads to

$$\begin{bmatrix} v_1 \\ v_2 \end{bmatrix} = \frac{l}{6EA}\begin{bmatrix} 2 & -1 \\ -1 & 2 \end{bmatrix}\begin{bmatrix} F_1 \\ F_2 \end{bmatrix} + \frac{\alpha l}{6}\begin{bmatrix} -(2T_1 + T_2) \\ T_1 + 2T_2 \end{bmatrix}$$
$$= \mathbf{fF} + \mathbf{v}_T \tag{6.127}$$

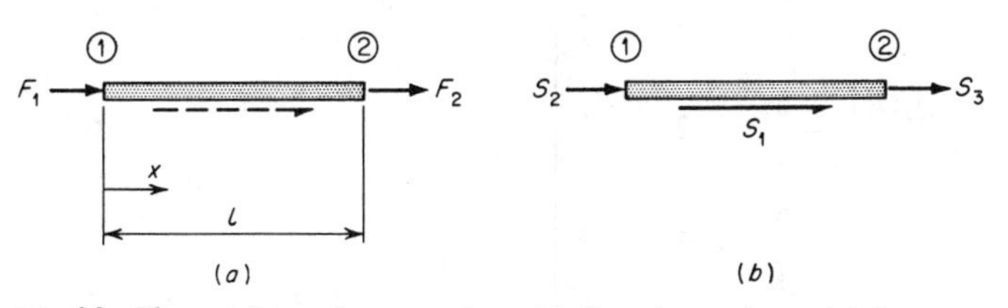

FIG. 6.9 Element forces in a member with linearly varying axial force. (*a*) Force method; (*b*) displacement method.

where $$\mathbf{f} = \frac{l}{6EA}\begin{bmatrix} 2 & -1 \\ -1 & 2 \end{bmatrix} \tag{6.128}$$

is the element flexibility and

$$\mathbf{v}_T = \frac{\alpha l}{6}\begin{bmatrix} -(2T_1 + T_2) \\ T_1 + 2T_2 \end{bmatrix} \tag{6.129}$$

represents the relative thermal expansion.*

The equation of equilibrium for the shear flow S_1 and concentrated forces S_2 and S_3 in Fig. 6.9 is given by

$$S_1 l + S_2 + S_3 = 0 \tag{6.130}$$

or in matrix form

$$S_1 + \begin{bmatrix} \frac{1}{l} & \frac{1}{l} \end{bmatrix}\begin{bmatrix} S_2 \\ S_3 \end{bmatrix} = 0 \tag{6.130a}$$

from which we obtain

$$\mathbf{T} = \begin{bmatrix} \frac{1}{l} \\ \frac{1}{l} \end{bmatrix} \tag{6.131}$$

and $$\mathbf{N} = [\mathbf{T} \quad -\mathbf{I}] = \begin{bmatrix} \frac{1}{l} & -1 & 0 \\ \frac{1}{l} & 0 & -1 \end{bmatrix} \tag{6.132}$$

Hence, using (6.108) and (6.109), we get

$$\mathbf{k} = \mathbf{N}^T\mathbf{f}^{-1}\mathbf{N} = \begin{bmatrix} \frac{1}{l} & \frac{1}{l} \\ -1 & 0 \\ 0 & -1 \end{bmatrix} \frac{2AE}{l}\begin{bmatrix} 2 & 1 \\ 1 & 2 \end{bmatrix}\begin{bmatrix} \frac{1}{l} & -1 & 0 \\ \frac{1}{l} & 0 & -1 \end{bmatrix}$$

$$= \frac{2AE}{l}\begin{bmatrix} \frac{6}{l^2} & -\frac{3}{l} & -\frac{3}{l} \\ -\frac{3}{l} & 2 & 1 \\ -\frac{3}{l} & 1 & 2 \end{bmatrix} \tag{6.133}$$

* The matrices $\mathbf{f}$ and $\mathbf{v}_T$ can also be determined using the force F_1 in the opposite direction to that shown in Fig. 6.9.

and $$\mathbf{Q} = \mathbf{N}^T\mathbf{f}^{-1}\mathbf{v}_T$$

$$= \begin{bmatrix} \frac{1}{l} & \frac{1}{l} \\ -1 & 0 \\ 0 & -1 \end{bmatrix} \frac{2AE}{l} \begin{bmatrix} 2 & 1 \\ 1 & 2 \end{bmatrix} \frac{\alpha l}{6} \begin{bmatrix} -(2T_1 + T_2) \\ T_1 + 2T_2 \end{bmatrix}$$

$$= \alpha EA \begin{bmatrix} \frac{-T_1 + T_2}{l} \\ T_1 \\ -T_2 \end{bmatrix} \tag{6.134}$$

6.9 ANALYSIS OF A PIN-JOINTED TRUSS BY THE DISPLACEMENT METHOD

As the first example of the displacement analysis, we shall consider the pin-jointed truss shown in Fig. 6.10. The structure is loaded by a vertical force of 1,000 lb at node 1, and all members are kept at the temperature at which the structure was initially assembled, with the exception of member 3, whose temperature is T. The numbering system for the nodes, members, and displacements is indicated in Fig. 6.10, while all other pertinent data are given in Table 6.1. In Table 6.1, location of the p-q direction is identified by the node numbers, and the direction cosines have been obtained from the node coordinates in the datum system $\bar{x}o\bar{y}$. The Young's modulus of the material will be assumed to be 10×10^6 lb/in.^{2}

The stiffness matrix for a pin-jointed bar, in a local coordinate system, is given by

$$\mathbf{k}^{(i)} = \frac{A^{(i)}E^{(i)}}{l^{(i)}} \begin{bmatrix} 1 & -1 \\ -1 & 1 \end{bmatrix}$$

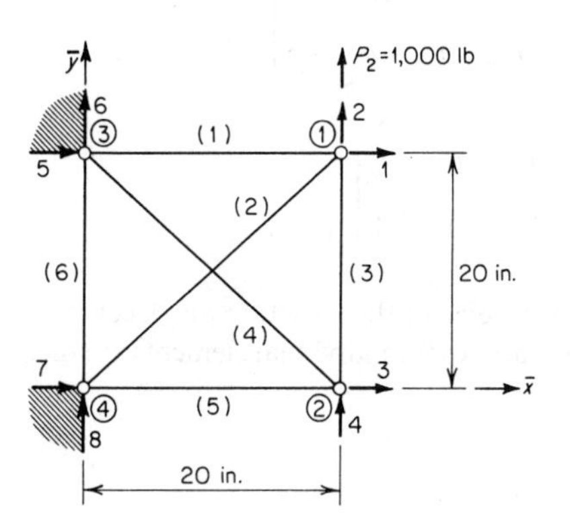

FIG. 6.10 Element and displacement numbering system.

TABLE 6.1

Member No. i	Cross-sectional area, in.^{2}	Length, in.	Location p-q	Direction cosines l_{pq}	m_{pq}
1	1.0	20	3,1	1.0	0
2	$\sqrt{2}/2$	$20\sqrt{2}$	4,1	$\sqrt{2}/2$	$\sqrt{2}/2$
3	1.0	20	1,2	0	-1.0
4	$\sqrt{2}/2$	$20\sqrt{2}$	3,2	$\sqrt{2}/2$	$-\sqrt{2}/2$
5	1.0	20	4,2	1.0	0
6	1.0	20	3,4	0	-1.0

To obtain stiffnesses in the datum system, we require the transformation matrices $\boldsymbol{\lambda}^{(i)}$. For bar elements

$$\boldsymbol{\lambda}^{(i)} = \begin{bmatrix} l_{pq} & m_{pq} & 0 & 0 \\ 0 & 0 & l_{pq} & m_{pq} \end{bmatrix}$$

Hence, using the information in Table 6.1, we obtain matrices

$$\boldsymbol{\lambda}^{(1)} = \overset{\begin{matrix} 5 & 6 & 1 & 2 \end{matrix}}{\begin{bmatrix} 1 & 0 & 0 & 0 \\ 0 & 0 & 1 & 0 \end{bmatrix}} \qquad \boldsymbol{\lambda}^{(2)} = \overset{\begin{matrix} 7 & 8 & 1 & 2 \end{matrix}}{\begin{bmatrix} c & c & 0 & 0 \\ 0 & 0 & c & c \end{bmatrix}}$$

$$\boldsymbol{\lambda}^{(3)} = \overset{\begin{matrix} 1 & 2 & 3 & 4 \end{matrix}}{\begin{bmatrix} 0 & -1 & 0 & 0 \\ 0 & 0 & 0 & -1 \end{bmatrix}} \qquad \boldsymbol{\lambda}^{(4)} = \overset{\begin{matrix} 5 & 6 & 3 & 4 \end{matrix}}{\begin{bmatrix} c & -c & 0 & 0 \\ 0 & 0 & c & -c \end{bmatrix}}$$

$$\boldsymbol{\lambda}^{(5)} = \overset{\begin{matrix} 7 & 8 & 3 & 4 \end{matrix}}{\begin{bmatrix} 1 & 0 & 0 & 0 \\ 0 & 0 & 1 & 0 \end{bmatrix}} \qquad \boldsymbol{\lambda}^{(6)} = \overset{\begin{matrix} 5 & 6 & 7 & 8 \end{matrix}}{\begin{bmatrix} 0 & -1 & 0 & 0 \\ 0 & 0 & 0 & -1 \end{bmatrix}}$$

where $c = \sqrt{2}/2$ and the column numbers refer to the structure displacements, as indicated in Fig. 6.10. The stiffness matrices for individual elements are then computed from

$$\bar{\mathbf{k}}^{(i)} = (\boldsymbol{\lambda}^{(i)})^T \mathbf{k}^{(i)} \boldsymbol{\lambda}^{(i)}$$

Hence

$$\bar{\mathbf{k}}^{(1)} = \begin{matrix} & 5 & 6 & 1 & 2 \\ 5 & 1 & 0 & -1 & 0 \\ 6 & 0 & 0 & 0 & 0 \\ 1 & -1 & 0 & 1 & 0 \\ 2 & 0 & 0 & 0 & 0 \end{matrix} \; 0.5 \times 10^6 \text{ lb/in.}$$

$$\bar{\mathbf{k}}^{(2)} = \begin{matrix} & 7 & 8 & 1 & 2 \\ 7 & 1 & 1 & -1 & -1 \\ 8 & 1 & 1 & -1 & -1 \\ 1 & -1 & -1 & 1 & 1 \\ 2 & -1 & -1 & 1 & 1 \end{matrix} \; 0.125 \times 10^6 \text{ lb/in.}$$

$$\bar{\mathbf{k}}^{(3)} = \begin{matrix} & 1 & 2 & 3 & 4 \\ 1 & 0 & 0 & 0 & 0 \\ 2 & 0 & 1 & 0 & -1 \\ 3 & 0 & 0 & 0 & 0 \\ 4 & 0 & -1 & 0 & 1 \end{matrix} \; 0.5 \times 10^6 \text{ lb/in.}$$

$$\bar{\mathbf{k}}^{(4)} = \begin{matrix} & 5 & 6 & 3 & 4 \\ 5 & 1 & -1 & -1 & 1 \\ 6 & -1 & 1 & 1 & -1 \\ 3 & -1 & 1 & 1 & -1 \\ 4 & 1 & -1 & -1 & 1 \end{matrix} \; 0.125 \times 10^6 \text{ lb/in.}$$

$$\bar{\mathbf{k}}^{(5)} = \begin{matrix} & 7 & 8 & 3 & 4 \\ 7 & 1 & 0 & -1 & 0 \\ 8 & 0 & 0 & 0 & 0 \\ 3 & -1 & 0 & 1 & 0 \\ 4 & 0 & 0 & 0 & 0 \end{matrix} \; 0.5 \times 10^6 \text{ lb/in.}$$

$$\bar{\mathbf{k}}^{(6)} = \begin{matrix} & 5 & 6 & 7 & 8 \\ 5 & 0 & 0 & 0 & 0 \\ 6 & 0 & 1 & 0 & -1 \\ 7 & 0 & 0 & 0 & 0 \\ 8 & 0 & -1 & 0 & 1 \end{matrix} \; 0.5 \times 10^6 \text{ lb/in.}$$

In the present example we have only one nonzero thermal-force matrix

$$\mathbf{Q}^{(3)} = \begin{bmatrix} 1 \\ -1 \end{bmatrix} 10^7 \alpha T$$

which in the datum system becomes

$$\bar{\mathbf{Q}}^{(3)} = (\boldsymbol{\lambda}^{(3)})^T\mathbf{Q}^{(3)} = \begin{matrix}1\\2\\3\\4\end{matrix}\begin{bmatrix}0\\-1\\0\\1\end{bmatrix}10^7\alpha T \qquad \text{lb}$$

The element stiffness matrices are then combined into the stiffness matrix of the assembled structure. This leads to

$$\mathbf{K} = \begin{matrix}1\\2\\3\\4\\5\\6\\7\\8\\ \\\end{matrix}\begin{bmatrix}5 & & & & & & & \\ 1 & 5 & & & \text{Symmetric} & & & \\ 0 & 0 & 5 & & & & & \\ 0 & -4 & -1 & 5 & & & & \\ -4 & 0 & -1 & 1 & 5 & & & \\ 0 & 0 & 1 & -1 & -1 & 5 & & \\ -1 & -1 & -4 & 0 & 0 & 0 & 5 & \\ -1 & -1 & 0 & 0 & 0 & -4 & 1 & 5 \\ 1 & 2 & 3 & 4 & 5 & 6 & 7 & 8\end{bmatrix} 0.125 \times 10^6 \text{ lb/in.}$$

Similarly, the thermal-force matrices $\bar{\mathbf{Q}}^{(i)}$ are combined to form

$$\mathbf{Q} = \{\overset{1}{0} \quad \overset{2}{-1} \quad \overset{3}{0} \quad \overset{4}{1} \quad \overset{5}{0} \quad \overset{6}{0} \quad \overset{7}{0} \quad \overset{8}{0}\}10^7\alpha T \qquad \text{lb}$$

Noting that displacements 5 to 8 are all equal to zero, we may write the equilibrium equation

$$\mathbf{P}_r = \mathbf{K}_r\mathbf{U}_r + \mathbf{Q}_r$$

as

$$\begin{matrix}1\\2\\3\\4\end{matrix}\begin{bmatrix}0\\10^3\\0\\0\end{bmatrix} = \begin{matrix}1\\2\\3\\4\end{matrix}\overset{\begin{matrix}1 & 2 & 3 & 4\end{matrix}}{\begin{bmatrix}5 & 1 & 0 & 0\\1 & 5 & 0 & -4\\0 & 0 & 5 & -1\\0 & -4 & -1 & 5\end{bmatrix}} 0.125 \times 10^6 \begin{bmatrix}U_1\\U_2\\U_3\\U_4\end{bmatrix} + \begin{matrix}1\\2\\3\\4\end{matrix}\begin{bmatrix}0\\-1\\0\\1\end{bmatrix}10^7\alpha T$$

Hence $\quad \mathbf{U}_r = \mathbf{K}_r^{-1}(\mathbf{P}_r - \mathbf{Q}_r) \qquad (6.19)$

becomes

$$\begin{bmatrix} U_1 \\ U_2 \\ U_3 \\ U_4 \end{bmatrix} = \tfrac{2}{11} \times 10^{-6} \begin{bmatrix} 10 & -6 & -1 & -5 \\ -6 & 30 & 5 & 25 \\ -1 & 5 & 10 & 6 \\ -5 & 25 & 6 & 30 \end{bmatrix} \left\{ \begin{bmatrix} 0 \\ 10^3 \\ 0 \\ 0 \end{bmatrix} + \begin{bmatrix} 0 \\ 1 \\ 0 \\ -1 \end{bmatrix} 10^7 \alpha T \right\}$$

$$= \tfrac{2}{11} \times 10^{-3} \left\{ \begin{bmatrix} -6 \\ 30 \\ 5 \\ 25 \end{bmatrix} + \begin{bmatrix} -1 \\ 5 \\ -1 \\ -5 \end{bmatrix} 10^4 \alpha T \right\} \quad \text{in.}$$

The displacements of the nodes (joints) can now be used to compute forces and stresses in individual members in the structure.

It should be understood that this example and all other examples in the text have been introduced to illustrate particular points in the general theory, and in order to simplify the presentation they have had to be restricted to relatively trivial problems. For complex structures with a large number of degrees of freedom the required matrix operations would have to be performed on a computer.

6.10 ANALYSIS OF A CANTILEVER BEAM BY THE DISPLACEMENT METHOD

As the second example we shall analyze the cantilever beam shown in Fig. 6.11. The beam is idealized into two beam elements. The applied loading consists of a transverse load P_3, concentrated moments P_4 and P_6, and a temperature distribution through the cross section but constant along the length of the beam.

By selecting the datum axis $o\bar{x}$ to coincide with the beam axis we can use the stiffness matrices directly in the local system to obtain the stiffness of the assembled structure.

Since no axial forces are applied, we can eliminate the corresponding displacements and use condensed stiffness matrices for the transverse deflections and rotations. Formal matrix condensation as explained in Sec. 6.5, however,

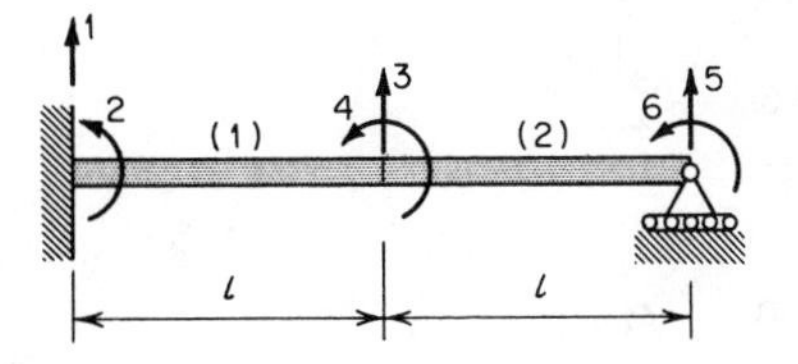

FIG. 6.11 Element and displacement numbering system.

is not required, because the coupling coefficients for axial and transverse displacements in the complete matrix are all equal to zero [see Eq. (5.116)]. Hence the stiffness matrices for the two elements are given by

$$\mathbf{k}^{(1)} = \frac{EI}{l^3}\begin{array}{c} \\ 1 \\ 2 \\ 3 \\ 4 \end{array}\begin{array}{c} \begin{array}{cccc} 1 & 2 & 3 & 4 \end{array} \\ \begin{bmatrix} 12 & 6l & -12 & 6l \\ 6l & 4l^2 & -6l & 2l^2 \\ -12 & -6l & 12 & -6l \\ 6l & 2l^2 & -6l & 4l^2 \end{bmatrix} \end{array}$$

$$\mathbf{k}^{(2)} = \frac{EI}{l^3}\begin{array}{c} \\ 3 \\ 4 \\ 5 \\ 6 \end{array}\begin{array}{c} \begin{array}{cccc} 3 & 4 & 5 & 6 \end{array} \\ \begin{bmatrix} 12 & 6l & -12 & 6l \\ 6l & 4l^2 & -6l & 2l^2 \\ -12 & -6l & 12 & -6l \\ 6l & 2l^2 & -6l & 4l^2 \end{bmatrix} \end{array}$$

where, as before, the row and column numbers refer to the force and displacement numbering system.

The thermal-force matrices are determined from

$$\mathbf{Q}^{(1)} = \begin{array}{c} 1 \\ 2 \\ 3 \\ 4 \end{array}\begin{bmatrix} 0 \\ -M_{T_z} \\ 0 \\ M_{T_z} \end{bmatrix} \quad \text{and} \quad \mathbf{Q}^{(2)} = \begin{array}{c} 3 \\ 4 \\ 5 \\ 6 \end{array}\begin{bmatrix} 0 \\ -M_{T_z} \\ 0 \\ M_{T_z} \end{bmatrix}$$

where $\quad M_{T_z} = \int_A \alpha E T y \, dA$

Assembling **K** and **Q** for the complete beam, we have

$$\mathbf{K} = \frac{EI}{l^3}\begin{array}{c} 1 \\ 2 \\ 3 \\ 4 \\ 5 \\ 6 \\ \\ \end{array}\begin{array}{c} \begin{bmatrix} 12 & & & & & \\ 6l & 4l^2 & & \text{Symmetric} & & \\ -12 & -6l & 24 & & & \\ 6l & 2l^2 & 0 & 8l^2 & & \\ 0 & 0 & -12 & -6l & 12 & \\ 0 & 0 & 6l & 2l^2 & -6l & 4l^2 \end{bmatrix} \\ \begin{array}{cccccc} 1 & 2 & 3 & 4 & 5 & 6 \end{array} \end{array}$$

and $\quad \mathbf{Q} = \{\overset{1}{0} \quad \overset{2}{-M_{T_z}} \quad \overset{3}{0} \quad \overset{4}{0} \quad \overset{5}{0} \quad \overset{6}{M_{T_z}}\}$

Hence $\mathbf{P}_r = \mathbf{K}_r \mathbf{U}_r + \mathbf{Q}_r$

is therefore written as

$$\begin{bmatrix} P_3 \\ P_4 \\ P_6 \end{bmatrix} = \frac{EI}{l^3} \begin{matrix} & 3 & 4 & 6 \\ 3 & 24 & 0 & 6l \\ 4 & 0 & 8l^2 & 2l^2 \\ 6 & 6l & 2l^2 & 4l^2 \end{matrix} \begin{bmatrix} U_3 \\ U_4 \\ U_6 \end{bmatrix} + \begin{bmatrix} 0 \\ 0 \\ M_{T_z} \end{bmatrix}$$

Inversion of the 3×3 matrix in the above equation leads to

$$\begin{bmatrix} U_3 \\ U_4 \\ U_6 \end{bmatrix} = \frac{l}{96EI} \begin{matrix} 3 \\ 4 \\ 6 \end{matrix} \begin{bmatrix} 7l^2 & 3l & -12l \\ 3l & 15 & -12 \\ -12l & -12 & 48 \end{bmatrix} \left\{ \begin{bmatrix} P_3 \\ P_4 \\ P_6 \end{bmatrix} - \begin{bmatrix} 0 \\ 0 \\ M_{T_z} \end{bmatrix} \right\}$$

6.11 EQUIVALENT CONCENTRATED FORCES

The actual applied loads are usually distributed, e.g., pressure loading, on structural elements, and therefore a technique is required for determining equivalent concentrated forces at the location and direction of the element forces. These equivalent forces can be resolved in the direction of the datum coordinate system and then summed up at each node to obtain the total applied-loading matrix **P**. Furthermore, concentrated forces may be applied at points other than the nodes of an element, and such cases also require the determination of equivalent forces at the nodes. The equivalent concentrated forces can be obtained directly from an energy approach which is consistent with the determination of stiffness matrices for structural elements.

The distributed loading will be represented by the matrix of surface forces $\mathbf{\Phi}$, and the equivalent concentrated forces in the directions of the element displacements **U** or forces **S** will be represented by $\mathbf{P}_{\text{equivalent}}$. Equating the virtual work δW of the applied distributed forces to the virtual work of the equivalent concentrated forces, we have

$$\delta W = \int_S \delta \mathbf{u}^T \mathbf{\Phi} \, dS = \delta \mathbf{U}^T \mathbf{P}_{\text{equivalent}} \tag{6.135}$$

where $\delta \mathbf{u}$ denotes the distribution of virtual displacements and $\delta \mathbf{U}$ represents virtual displacements in the directions of the forces **S**. From the relationship

$$\mathbf{u} = \mathbf{a}\mathbf{U} \tag{4.1}$$

it follows immediately that

$$\delta\mathbf{u} = \mathbf{a}\,\delta\mathbf{U} \tag{6.136}$$

Substitution of Eq. (6.136) into (6.135) leads to

$$\delta\mathbf{U}^T\left(\int_S \mathbf{a}^T\boldsymbol{\Phi}\,dS - \mathbf{P}_{\text{equivalent}}\right) = 0 \tag{6.137}$$

Since the virtual displacements $\delta\mathbf{U}$ are arbitrary, we must have that

$$\mathbf{P}_{\text{equivalent}} = \int_S \mathbf{a}^T\boldsymbol{\Phi}\,dS \tag{6.138}$$

Equation (6.138) can be used to convert into concentrated forces any distributed loading $\boldsymbol{\Phi}$ acting on a structural element. The special case of concentrated forces acting at intermediate points can also be included if we note that a concentrated load may be represented by an infinite pressure acting over zero area with the proviso that the product $\boldsymbol{\Phi}\,dS$ is equal to the value of the concentrated load.

As an example we shall consider a beam element subjected to a distributed loading p_y lb/in., as shown in Fig. 6.12. Using the results of Sec. 5.6, we can show that for this element the matrix $\mathbf{a}$ relating the transverse deflections $\mathbf{u} = u_y$ to the displacements $\mathbf{U} = \{U_1 \quad \cdots \quad U_4\}$ is given by

$$\mathbf{a} = \frac{1}{1+\Phi_s}\Big[[1 - 3\xi^2 + 2\xi^3 + (1-\xi)\Phi_s] \quad [\xi - 2\xi^2 + \xi^3 + \tfrac{1}{2}(\xi - \xi^2)\Phi_s]l \quad [3\xi^2 - 2\xi^3 + \xi\Phi_s] \quad [-\xi^2 + \xi^3 - \tfrac{1}{2}(\xi - \xi^2)\Phi_s]l\Big] \tag{6.139}$$

where $\xi = x/l$ and Φ_s is the shear-deformation parameter used in Sec. 5.6. In the present example $\boldsymbol{\Phi}\,dS = p_y l\,d\xi$, so that Eq. (6.138) becomes

$$\mathbf{P}_{\text{equivalent}} = \int_0^1 \mathbf{a}^T p_y l\,d\xi \tag{6.140}$$

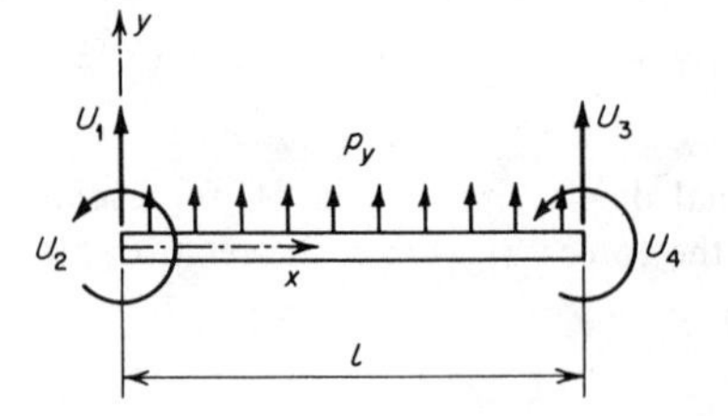

FIG. 6.12 Beam element with transverse loading p_y lb/in.

Substituting Eq. (6.139) into (6.140) and then integrating lead to

$$\mathbf{P}_{\text{equivalent}} = p_y l \begin{bmatrix} \frac{1}{2} \\ \frac{l}{12} \\ \frac{1}{2} \\ -\frac{l}{12} \end{bmatrix} \tag{6.141}$$

Equation (6.141) indicates that the equivalent concentrated forces consist not only of transverse loads $p_y l/2$ at the ends of the beam but also of end moments of magnitude $p_y l^2/12$.

PROBLEMS

6.1 Considering only the transverse deflections and rotations in a beam (Fig. 6.13), we can eliminate the rigid-body degrees of freedom by assuming that $\mathbf{w} = \{U_1 \quad U_2\} = \mathbf{0}$. Using this assumption, calculate the transformation matrix $\mathbf{T}$ relating the total displacements to the rigid-body displacements in the equation

$$\mathbf{U}_r = \mathbf{r} + \mathbf{T}\mathbf{w}$$

6.2 Solve Prob. 6.1 with $\mathbf{w} = \{U_1 \quad U_5\} = \mathbf{0}$.

6.3 The flexibility matrix for the curved-beam element shown in Fig. 6.14 is given by

$$\mathbf{f} = \frac{R}{4EI} \begin{bmatrix} \pi R^2 & 2R^2 & 4R \\ 2R^2 & (3\pi - 8)R^2 & 2(\pi - 2)R \\ 4R & 2(\pi - 2)R & 2\pi \end{bmatrix}$$

(rows and columns numbered 1, 2, 3)

where EI is the flexural stiffness and R is the radius of the curved (circular arc) element, while the row and column numbers in $\mathbf{f}$ refer to the element relative displacements v_1, v_2, v_3 and the element forces F_1, F_2, F_3, respectively. Using the flexibility matrix $\mathbf{f}$, calculate the stiffness matrix $\mathbf{k}$ for this curved-beam element.

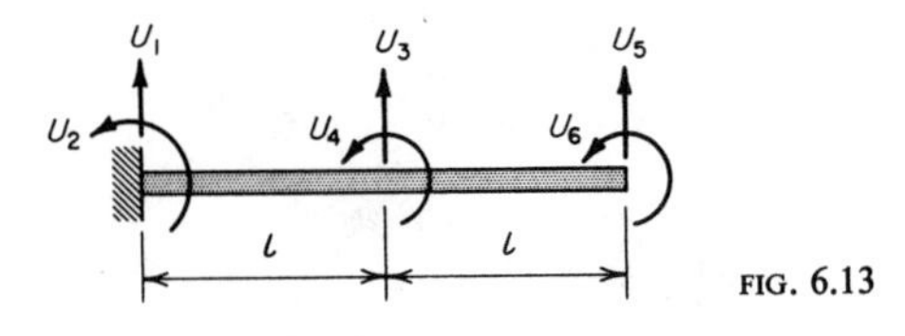

FIG. 6.13

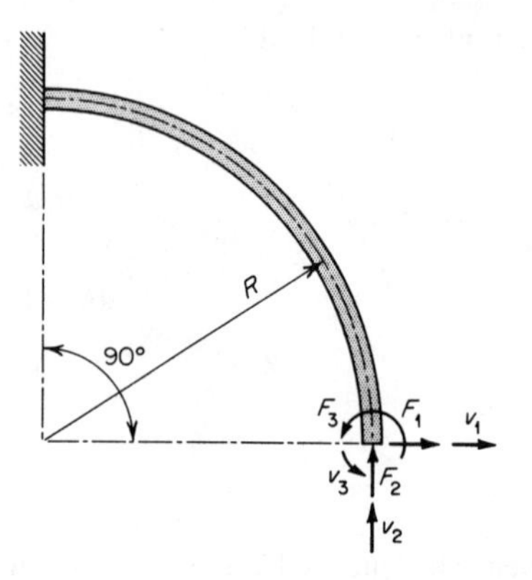

FIG. 6.14

6.4 Calculate the condensed stiffness matrix $\mathbf{K}_c$ for deflections 3 and 5 in the cantilever beam shown in Fig. 6.13.

6.5 Solve Prob. 6.4 for the deflections 5 and 6. Comment on the result.

6.6 Using Eqs. (6.133) and (6.134), verify that when the element force S_1 is eliminated and $T_1 = T_2 = T$, the condensed stiffness and thermal-force matrices are reduced to those for a pin-jointed bar element.

6.7 Using the matrix displacement method, determine rotations at points B, C, and D due to the applied moment M on a uniform beam with multiple supports shown in Fig. 6.15.

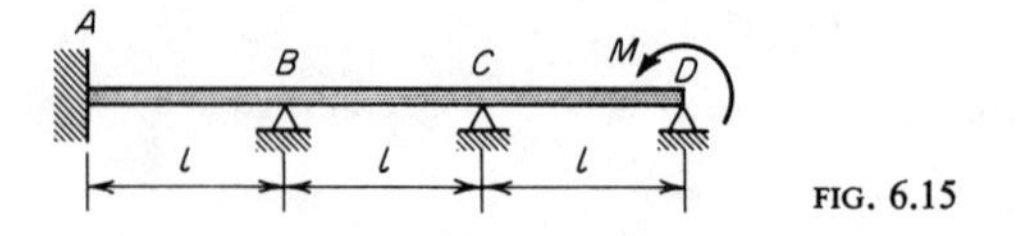

FIG. 6.15

6.8 A simple two-dimensional pin-jointed truss, shown in Fig. 6.16, is loaded by a vertical force of 1,000 lb. The cross-sectional areas of all bars are 0.5 in.^{2}, and Young's modulus is 10×10^6 lb/in.^{2} Determine all joint displacements and forces in the members by the displacement method. Check the results using equilibrium at joint 1.

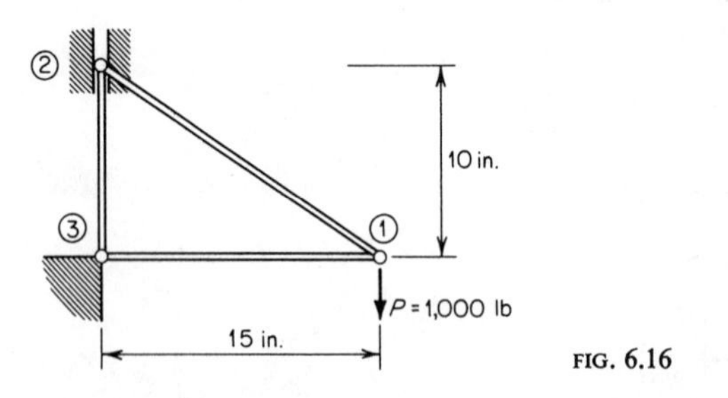

FIG. 6.16

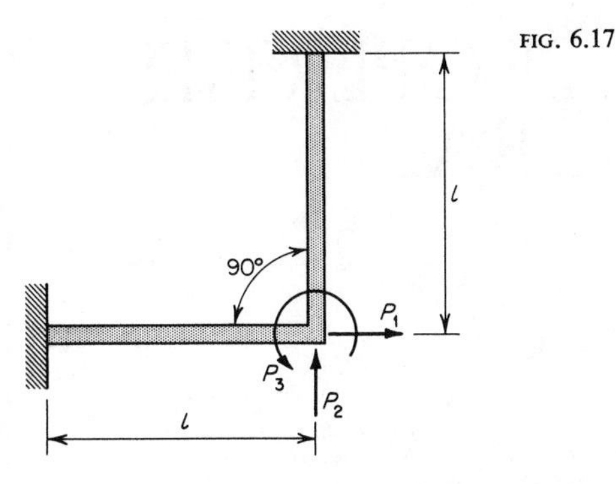

FIG. 6.17

6.9 A uniform bracket of constant cross-sectional area A and moment of inertia I is built-in at both ends, as shown in Fig. 6.17. The applied loading consists of forces P_1 and P_2 and a moment P_3. Using the matrix displacement method of analysis, determine displacements in the directions of P_1, P_2, and P_3.

6.10 Determine the equivalent concentrated-force matrix $\mathbf{P}_{\text{equivalent}}$ for a beam element subjected to transverse loading varying linearly from p_1 at $x = 0$ to p_2 at $x = l$.

6.11 Determine an equivalent concentrated-force matrix for a rectangular plate element subjected to constant pressure.

CHAPTER 7
FLEXIBILITY PROPERTIES OF STRUCTURAL ELEMENTS

In the displacement method of structural analysis we require the stiffness property of individual elements, while in the force method we require the inverse of the stiffness property of the elements. This reciprocal property is described as the *element flexibility*, and it may be obtained directly from the stiffness relationships derived in Chap. 5. In this chapter flexibility properties are derived for the following typical structural elements: pin-jointed bars, beams, triangular and rectangular plates with in-plane forces, solid tetrahedra, constant-shear-flow panels, linearly varying axial-load members, and rectangular plates in bending.

7.1 METHODS OF DETERMINING ELEMENT DISPLACEMENT-FORCE RELATIONSHIPS

The fundamental consideration in the matrix force method of analysis is the determination of the flexibility properties of structural elements. A number of alternative methods are available for obtaining displacement-force relationships describing the flexibility properties of the elements. The following methods may be used:

1. Inversion of the force-displacement equations
2. Unit-load theorem
3. Castigliano's theorem (part II)
4. Solution of differential equations for element displacements

In order to emphasize the dual character of the matrix force and displacement methods, the flexibility properties of structural elements will be derived first from the corresponding stiffness equations.

7.2 INVERSION OF THE FORCE-DISPLACEMENT EQUATIONS: FLEXIBILITY PROPERTIES OF PIN-JOINTED BARS AND BEAM ELEMENTS

We start with the equation relating forces to displacements

$$\mathbf{S}^{(i)} = \mathbf{k}^{(i)}\mathbf{u}^{(i)} + \mathbf{h}^{(i)}\alpha T^{(i)} \tag{7.1}$$

where the superscript i denotes the ith element. The matrix $\mathbf{S}^{(i)}$ constitutes a complete set of forces equal to the number of degrees of freedom assumed on the element. These forces must be in equilibrium with themselves, which implies that Eqs. (7.1) are linearly dependent, with the number of dependent relationships being equal to the number of equations of overall equilibrium relevant to a particular structural element. Thus, for a one-dimensional element there is only one dependent relationship; for a two-dimensional element, three relationships; and for a three-dimensional element, six relationships. This linear dependence in Eq. (7.1) makes the stiffness matrix $\mathbf{k}^{(i)}$ singular, and consequently the solution for the displacements can be determined only if this dependence is eliminated. The elimination procedure to be followed is the same as in the case of the stiffness matrix $\mathbf{K}$ for the complete structure. A number of displacements (equal to the number of rigid-body degrees of freedom) are selected for that purpose, and the corresponding rows and columns in $\mathbf{k}^{(i)}$ are eliminated, and thus all matrices in Eq. (7.1) are reduced in size. To check whether all rigid-body degrees of freedom have been eliminated, general expressions derived for the complete assembled structure (see Chap. 6) may be used.

After the elimination of rows and columns Eq. (7.1) becomes

$$\mathbf{S}_r^{(i)} = \mathbf{k}_r^{(i)}\mathbf{u}_r^{(i)} + \mathbf{h}_r^{(i)}\alpha T^{(i)} \tag{7.2}$$

where the subscript r has been added to indicate that the order of the original matrices, as given in Eq. (7.1), is reduced. Solving Eq. (7.2) for displacements $\mathbf{u}_r^{(i)}$, we get

$$\mathbf{u}_r^{(i)} = \mathbf{v}^{(i)} = (\mathbf{k}_r^{(i)})^{-1}\mathbf{S}_r^{(i)} - (\mathbf{k}_r^{(i)})^{-1}\mathbf{h}_r^{(i)}\alpha T^{(i)} \tag{7.3}$$

where $\mathbf{v}^{(i)}$ can be interpreted as the matrix of displacements relative to the rigid frame of reference established by the zero displacements eliminated from the matrix $\mathbf{u}^{(i)}$. Equation (7.3) can now be rewritten as

$$\mathbf{v}^{(i)} = \mathbf{f}^{(i)}\mathbf{F}^{(i)} + \mathbf{v}_T^{(i)} \tag{7.4}$$

where

$$\mathbf{f}^{(i)} = (\mathbf{k}_r^{(i)})^{-1} \tag{7.5}$$

$$\mathbf{F}^{(i)} = \mathbf{S}_r^{(i)} \tag{7.6}$$

$$\mathbf{v}_T^{(i)} = -(\mathbf{k}_r^{(i)})^{-1}\mathbf{h}_r^{(i)}\alpha T^{(i)} \tag{7.7}$$

The matrix $\mathbf{f}^{(i)}$ represents the element flexibility referred to the selected reference frame, and the matrix $\mathbf{v}_T^{(i)}$ represents relative thermal expansions in the element when the forces $\mathbf{F}^{(i)}$ are equal to zero. It should be noted that the flexibility matrix $\mathbf{f}^{(i)}$ and the relative thermal expansions $\mathbf{v}_T^{(i)}$ of an unassembled element can take many different forms, depending on the choice of the rigid frame of reference.

Equation (7.4) can be determined for each structural element separately, and for the complete structure all these equations can be combined into a single matrix equation of the form

$$\mathbf{v} = \mathbf{f}\mathbf{F} + \mathbf{v}_T \tag{7.8}$$

where

$$\mathbf{v} = \{\mathbf{v}^{(i)} \quad \mathbf{v}^{(2)} \quad \cdots \quad \mathbf{v}^{(i)} \quad \cdots\} \tag{7.9}$$

$$\mathbf{f} = [\mathbf{f}^{(1)} \quad \mathbf{f}^{(2)} \quad \cdots \quad \mathbf{f}^{(i)} \quad \cdots] \tag{7.10}$$

$$\mathbf{F} = \{\mathbf{F}^{(1)} \quad \mathbf{F}^{(2)} \quad \cdots \quad \mathbf{F}^{(i)} \quad \cdots\} \tag{7.11}$$

$$\mathbf{v}_T = \{\mathbf{v}_T^{(1)} \quad \mathbf{v}_T^{(2)} \quad \cdots \quad \mathbf{v}_T^{(i)} \quad \cdots\} \tag{7.12}$$

To demonstrate the method consider the pin-jointed bar element shown in Fig. 7.1, for which the stiffness relationship is

$$\begin{bmatrix} S_1 \\ S_2 \end{bmatrix} = \frac{AE}{l}\begin{bmatrix} 1 & -1 \\ -1 & 1 \end{bmatrix}\begin{bmatrix} u_1 \\ u_2 \end{bmatrix} + AE\alpha T\begin{bmatrix} 1 \\ -1 \end{bmatrix} \tag{7.13}$$

where for simplicity the superscripts (i) have been dropped. To eliminate the rigid-body degree of freedom (only the rigid-body translation along the element axis is present) we can set either $u_1 = 0$ or $u_2 = 0$; thus the element forces $\mathbf{F}$ used in the flexibility relationship will be as shown in Fig. 7.1. When we select $u_1 = 0$ to establish the rigid frame of reference, it follows from Eqs. (7.4) to (7.7) and (7.13) that

$$\mathbf{v} = \frac{l}{AE}F + \alpha Tl \tag{7.14}$$

$$\mathbf{f} = \frac{l}{AE} \tag{7.15}$$

$$\mathbf{v}_T = \alpha Tl \tag{7.16}$$

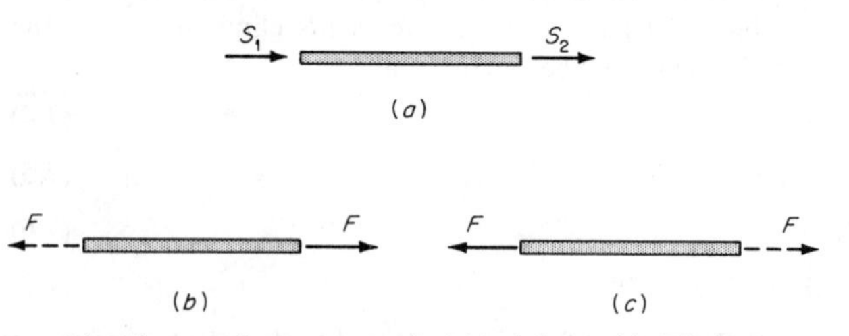

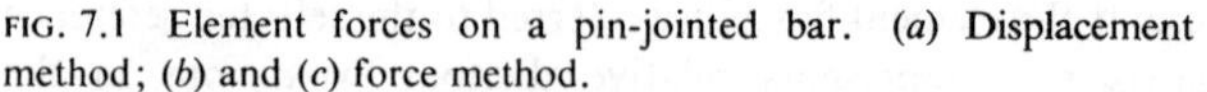

FIG. 7.1 Element forces on a pin-jointed bar. (*a*) Displacement method; (*b*) and (*c*) force method.

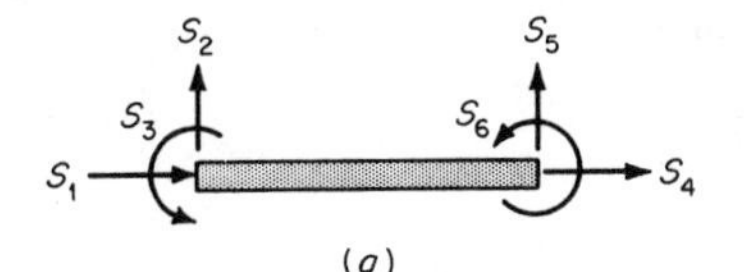

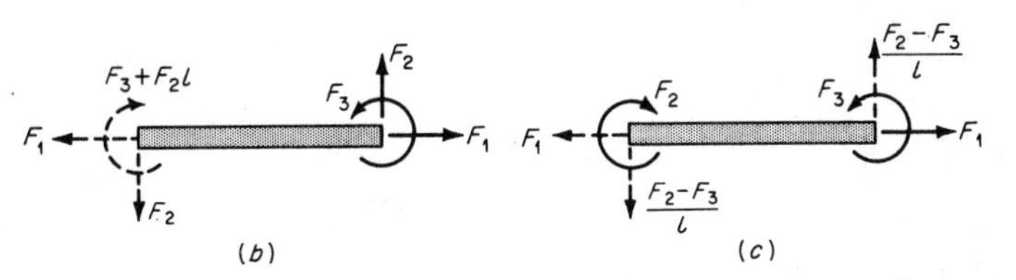

FIG. 7.2 Element forces on a beam element. (*a*) Displacement method; (*b*) and (*c*) force method.

With $u_1 = 0$, $\mathbf{v}$ in Eq. (7.14) represents relative displacement of joint 2 with respect to joint 1; thus, a positive value of $\mathbf{v}$ represents extension of the member while a negative value represents contraction. Similarly, a positive value of $\mathbf{v}_T$ (for positive temperature T) represents relative thermal expansion of the bar element.

For subsequent analysis it will be necessary to use a relationship between the element forces $\mathbf{S}$ and $\mathbf{F}$. It is clear that for the bar element we have

$$\begin{bmatrix} S_1 \\ S_2 \end{bmatrix} = \begin{bmatrix} -1 \\ 1 \end{bmatrix} F \tag{7.17}$$

or in matrix notation

$$\mathbf{S}^{(i)} = \mathbf{B}^{(i)}\mathbf{F}^{(i)} \tag{7.18}$$

where superscripts (i) have been added to denote the ith element and $\mathbf{B}^{(i)}$ is the transformation matrix. Equation (7.18) can be written for each element, and then all such equations can be assembled into a single matrix equation

$$\mathbf{S} = \mathbf{BF} \tag{7.19}$$

where $$\mathbf{B} = \lfloor \mathbf{B}^{(1)} \quad \mathbf{B}^{(2)} \quad \cdots \quad \mathbf{B}^{(i)} \quad \cdots \rfloor \tag{7.20}$$

For a beam element there are many possible ways of selecting the frame of reference: two commonly used ones are shown in Fig. 7.2. The stiffness matrix $\mathbf{k}$ and thermal-load matrix $\mathbf{Q}$ for the beam element in two dimensions are given by Eqs. (5.119) and (5.120). Selecting the independent forces F_1, F_2, and F_3, as in

Fig. 7.2*b*, it can be shown that

$$\mathbf{f} = \begin{bmatrix} \dfrac{l}{AE} & 0 & 0 \\ 0 & \dfrac{(4+\Phi_y)l^3}{12EI_z} & \dfrac{l^2}{2EI_z} \\ 0 & \dfrac{l^2}{2EI_z} & \dfrac{l}{EI_z} \end{bmatrix} \tag{7.21}$$

$$\mathbf{v}_T = \begin{bmatrix} \dfrac{\alpha l}{A} \displaystyle\int_A T\, dA \\ \dfrac{-\alpha l^2}{2I_z} \displaystyle\int_A Ty\, dA \\ \dfrac{-\alpha l}{I_z} \displaystyle\int_A Ty\, dA \end{bmatrix} \tag{7.22}$$

and

$$\begin{bmatrix} S_1 \\ S_2 \\ S_3 \\ S_4 \\ S_5 \\ S_6 \end{bmatrix} = \begin{bmatrix} -1 & 0 & 0 \\ 0 & -1 & 0 \\ 0 & -l & -1 \\ 1 & 0 & 0 \\ 0 & 1 & 0 \\ 0 & 0 & 1 \end{bmatrix} \begin{bmatrix} F_1 \\ F_2 \\ F_3 \end{bmatrix} \tag{7.23}$$

Similarly, selecting F_1, F_2, and F_3, as shown in Fig. 7.2*c*, and noting that $S_3 = -F_2$, we obtain

$$\mathbf{f} = \begin{bmatrix} \dfrac{l}{AE} & 0 & 0 \\ 0 & \dfrac{(4+\Phi_y)l}{12EI_z} & \dfrac{(2-\Phi_y)l}{12EI_z} \\ 0 & \dfrac{(2-\Phi_y)l}{12EI_z} & \dfrac{(4+\Phi_y)l}{12EI_z} \end{bmatrix} \tag{7.24}$$

$$\mathbf{v}_T = \begin{bmatrix} \dfrac{\alpha l}{A} \displaystyle\int_A T\, dA \\ \dfrac{-\alpha l}{2I_z} \displaystyle\int_A Ty\, dA \\ \dfrac{-\alpha l}{2I_z} \displaystyle\int_A Ty\, dA \end{bmatrix} \tag{7.25}$$

and
$$\begin{bmatrix} S_1 \\ S_2 \\ S_3 \\ S_4 \\ S_5 \\ S_6 \end{bmatrix} = \begin{bmatrix} -1 & 0 & 0 \\ 0 & \dfrac{-1}{l} & \dfrac{1}{l} \\ 0 & -1 & 0 \\ 1 & 0 & 0 \\ 0 & \dfrac{1}{l} & \dfrac{-1}{l} \\ 0 & 0 & 1 \end{bmatrix} \begin{bmatrix} F_1 \\ F_2 \\ F_3 \end{bmatrix} \tag{7.26}$$

Naturally, other combinations of the independent forces on the beam element are possible, but they will not be discussed here.

7.3 DETERMINATION OF ELEMENT FLEXIBILITY PROPERTIES BY THE UNIT-LOAD THEOREM

Consider an elastic element, shown in Fig. 7.3, subjected to a set of n forces

$$\mathbf{F} = \{F_1 \quad F_2 \quad \cdots \quad F_i \quad F_j \quad \cdots \quad F_n\} \tag{7.27}$$

and some specified temperature distribution

$$T = T(x,y,z) \tag{7.28}$$

The forces $\mathbf{F}$ are reacted by a set of statically determinate reactions. The displacements in the direction of these reactions will be used to establish the rigid frame of reference. Let the displacements, relative to this frame of reference, corresponding to the forces $\mathbf{F}$ be denoted by

$$\mathbf{v} = \{v_1 \quad v_2 \quad \cdots \quad v_i \quad v_j \quad \cdots \quad v_n\} \tag{7.29}$$

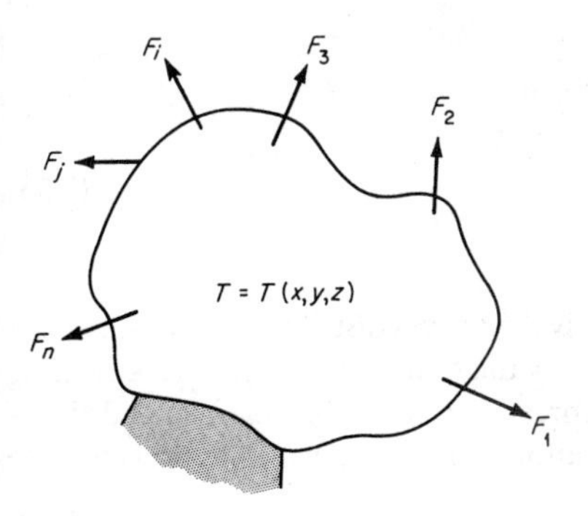

FIG. 7.3 Elastic element subjected to forces $\mathbf{F} = \{F_1 \quad F_2 \quad \cdots \quad F_n\}$.

To determine a typical deflection v_i we may use the unit-load theorem, so that [see Eq. (3.88)]

$$v_i = \int_v \mathbf{e}^T \bar{\boldsymbol{\sigma}}_i \, dV = \int_v \bar{\boldsymbol{\sigma}}_i^T \mathbf{e} \, dV \tag{7.30}$$

where $\bar{\boldsymbol{\sigma}}_i$ represents the matrix of statically equivalent stresses due to a unit load in the direction of F_i and $\mathbf{e}$ is the exact strain matrix due to all applied forces $\mathbf{F}$ and the temperature distribution T. The unit loads can be applied in turn at all points where the forces are impressed, and hence

$$\mathbf{v} = \int_v \bar{\boldsymbol{\sigma}}^T \mathbf{e} \, dV \tag{7.31}$$

where $\quad \bar{\boldsymbol{\sigma}} = [\bar{\boldsymbol{\sigma}}_1 \quad \bar{\boldsymbol{\sigma}}_2 \quad \cdots \quad \bar{\boldsymbol{\sigma}}_i \quad \bar{\boldsymbol{\sigma}}_j \quad \cdots \quad \bar{\boldsymbol{\sigma}}_n] \qquad (7.32)$

For a linear system

$$\boldsymbol{\sigma} = \mathbf{cF} \tag{7.33}$$

where $\mathbf{c}$ represents exact stress distribution due to unit forces $\mathbf{F}$. From Eqs. (2.19) and (7.33) it follows therefore that

$$\begin{aligned} \mathbf{e} &= \boldsymbol{\phi}\boldsymbol{\sigma} + \mathbf{e}_T \\ &= \boldsymbol{\phi}\mathbf{cF} + \mathbf{e}_T \end{aligned} \tag{7.34}$$

which when substituted into Eq. (7.31) leads to

$$\mathbf{v} = \int_v \bar{\boldsymbol{\sigma}}^T \boldsymbol{\phi} \mathbf{c} \, dV \, \mathbf{F} + \int_v \bar{\boldsymbol{\sigma}}^T \mathbf{e}_T \, dV \tag{7.35}$$

or $\quad \mathbf{v} = \mathbf{fF} + \mathbf{v}_T \qquad (7.36)$

where $\quad \mathbf{f} = \int_v \bar{\boldsymbol{\sigma}}^T \boldsymbol{\phi} \mathbf{c} \, dV \qquad (7.37)$

represents the elemental flexibility matrix and

$$\mathbf{v}_T = \int_v \bar{\boldsymbol{\sigma}}^T \mathbf{e}_T \, dV \tag{7.38}$$

is the matrix of relative thermal expansions.

The determination of the statically equivalent stress distributions $\bar{\boldsymbol{\sigma}}$ presents no special difficulty. On the other hand, the evaluation of the matrix $\mathbf{c}$ representing the exact stress distribution due to the forces $\mathbf{F}$ may not be possible, and therefore we can only use approximate relationships. The usual approach is to select the matrix **c** so that it will satisfy at least the equations of equilibrium. Denoting this approximate matrix by $\bar{\mathbf{c}}$ and noting that $\bar{\boldsymbol{\sigma}} = \bar{\mathbf{c}}$, we have

$$\mathbf{f} = \int_v \bar{\mathbf{c}}^T \boldsymbol{\phi} \bar{\mathbf{c}} \, dV \tag{7.39}$$

and $$\mathbf{v}_T = \int_v \bar{\mathbf{c}}^T \mathbf{e}_T \, dV \tag{7.40}$$

Application of the above equations to typical structural elements will be illustrated in later sections.

7.4 APPLICATION OF CASTIGLIANO'S THEOREM (PART II) TO DERIVE FLEXIBILITY PROPERTIES

Applying Castigliano's theorem (part II) [Eq. (3.82)] to a structural element in Fig. 7.3, we have

$$v_i = \left(\frac{\partial U_d^*}{\partial F_i}\right)_{T=\text{const}} \tag{7.41}$$

By varying the subscript i from 1 to n we obtain the complete set of relative displacement-force equations. Symbolically this may be represented by the differentiation with respect to the force matrix **F**, that is,

$$\mathbf{v} = \left(\frac{\partial U_d^*}{\partial \mathbf{F}}\right)_{T=\text{const}} \tag{7.42}$$

In performing the differentiation it is important, however, that the complementary energy of total deformation U_d^* be expressed only in terms of the element forces **F**.

From the results of Chap. 3 it is clear that for linearly elastic structures

$$U_d^* = \frac{1}{2}\int_v \boldsymbol{\sigma}^T \boldsymbol{\epsilon} \, dV + \int_v \boldsymbol{\sigma}^T \mathbf{e}_T \, dV \tag{7.43}$$

Noting that the elastic strain $\boldsymbol{\epsilon}$ is given by

$$\boldsymbol{\epsilon} = \boldsymbol{\varkappa}^{-1}\boldsymbol{\sigma} = \boldsymbol{\phi}\boldsymbol{\sigma} \tag{7.44}$$

it follows that

$$U_d^* = \frac{1}{2}\int_v \boldsymbol{\sigma}^T \boldsymbol{\phi} \boldsymbol{\sigma} \, dV + \int_v \boldsymbol{\sigma}^T \mathbf{e}_T \, dV \tag{7.45}$$

Assuming now that the stresses $\boldsymbol{\sigma}$ are related to the forces $\mathbf{F}$ by the linear equation

$$\boldsymbol{\sigma} = \mathbf{cF} \tag{7.33}$$

we find the complementary energy of total deformation to be

$$U_d^* = \frac{1}{2}\int_v \mathbf{F}^T\mathbf{c}^T\boldsymbol{\phi}\mathbf{cF}\,dV + \int_v \mathbf{F}^T\mathbf{c}^T\mathbf{e}_T\,dV \tag{7.46}$$

Hence*
$$\mathbf{v} = \left(\frac{\partial U_d^*}{\partial \mathbf{F}}\right)_{T=\text{const}} = \int_v \mathbf{c}^T\boldsymbol{\phi}\mathbf{c}\,dV\,\mathbf{F} + \int_v \mathbf{c}^T\mathbf{e}_T\,dV \tag{7.47}$$

which agrees with the result obtained by the unit-load theorem for $\bar{\boldsymbol{\sigma}} = \mathbf{c}$ in Eq. (7.35).

7.5 SOLUTION OF DIFFERENTIAL EQUATIONS FOR ELEMENT DISPLACEMENTS TO DERIVE FLEXIBILITY PROPERTIES

The displacement-force relationship for an element can be written fully as

$$\begin{bmatrix} v_1 \\ v_2 \\ \cdot \\ v_i \\ \cdot \\ v_n \end{bmatrix} = \begin{bmatrix} f_{11} & f_{12} & \cdots & f_{1i} & \cdots & f_{1n} \\ f_{21} & f_{22} & \cdots & f_{2i} & \cdots & f_{2n} \\ \cdots & \cdots & \cdots & \cdots & \cdots & \cdots \\ f_{i1} & f_{i2} & \cdots & f_{ii} & \cdots & f_{in} \\ \cdots & \cdots & \cdots & \cdots & \cdots & \cdots \\ f_{n1} & f_{n2} & \cdots & f_{ni} & \cdots & f_{nn} \end{bmatrix} \begin{bmatrix} F_1 \\ F_2 \\ \cdot \\ F_i \\ \cdot \\ F_n \end{bmatrix} + \begin{bmatrix} v_{T_1} \\ v_{T_2} \\ \cdot \\ v_{T_i} \\ \cdot \\ v_{T_n} \end{bmatrix} \tag{7.48}$$

A typical flexibility coefficient f_{ij} is interpreted as the displacement in the ith direction relative to the selected rigid frame of reference due to a unit force in the jth direction while all other forces (except the reactions) are equal to zero. A typical coefficient v_{T_i} is simply the thermal deflection in the ith direction due to the specified temperature distribution in the element while all element forces are equal to zero. Relying on these two definitions, we can use the solutions of differential equations for displacements due to forces $\mathbf{F}$ and the temperature T to determine the flexibility coefficients and thermal displacements.

7.6 PIN-JOINTED BAR ELEMENTS

The flexibility properties of bar elements have been derived in Sec. 7.2 and are given by Eqs. (7.15) and (7.16). The transformation matrix $\mathbf{B}$ for element forces in the force and displacement methods is given by Eq. (7.17).

* See Sec. A.21 for differentiation of matrices.

7.7 BEAM ELEMENTS

For a beam element used for three-dimensional problems, the deflections and rotations will be in the principal planes. Assuming that *oy* and *oz* refer to the principal axes of the beam cross section and selecting a set of independent element forces, as shown in Fig. 7.4, we obtain the following flexibility matrix, using the results of Eq. (7.21),

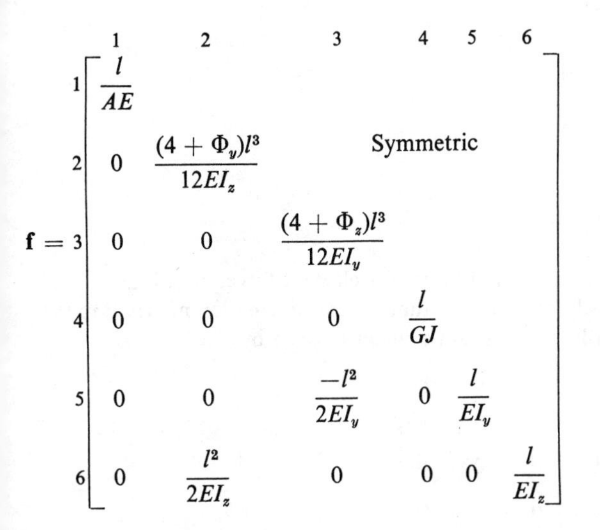

$$\mathbf{f} = \begin{bmatrix} \frac{l}{AE} & & & & & \\ 0 & \frac{(4+\Phi_y)l^3}{12EI_z} & & \text{Symmetric} & & \\ 0 & 0 & \frac{(4+\Phi_z)l^3}{12EI_y} & & & \\ 0 & 0 & 0 & \frac{l}{GJ} & & \\ 0 & 0 & \frac{-l^2}{2EI_y} & 0 & \frac{l}{EI_y} & \\ 0 & \frac{l^2}{2EI_z} & 0 & 0 & 0 & \frac{l}{EI_z} \end{bmatrix} \tag{7.49}$$

where G is the shear modulus and J is the Saint-Venant torsion constant of the cross section. The relative thermal expansions (and rotations) in the direction

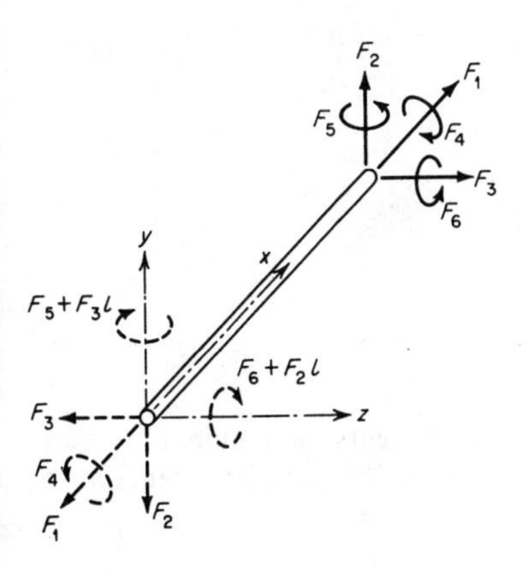

FIG. 7.4 Independent forces on a beam element (force method).

of forces $F_1, \ldots, F_6$ can be obtained from the results of Eq. (7.22). Hence

$$
\mathbf{v}_T = \begin{matrix} 1 \\ 2 \\ 3 \\ 4 \\ 5 \\ 6 \end{matrix}
\begin{bmatrix}
\dfrac{\alpha l}{A} \displaystyle\int_A T\,dA \\
-\dfrac{\alpha l^2}{2I_z} \displaystyle\int_A Ty\,dA \\
-\dfrac{\alpha l^2}{2I_y} \displaystyle\int_A Tz\,dA \\
0 \\
\dfrac{\alpha l}{I_y} \displaystyle\int_A Tz\,dA \\
-\dfrac{\alpha l}{I_z} \displaystyle\int_A Ty\,dA
\end{bmatrix}
\tag{7.50}
$$

Using the equations of overall equilibrium for element forces **S** in Fig. 5.4 and **F** in Fig. 7.4, we can easily demonstrate that the transformation matrix **B** in the matrix equations $\mathbf{S} = \mathbf{BF}$ for the beam element is given by

$$
\mathbf{B} = \begin{matrix} \\ 1 \\ 2 \\ 3 \\ 4 \\ 5 \\ 6 \\ 7 \\ 8 \\ 9 \\ 10 \\ 11 \\ 12 \end{matrix}
\begin{matrix} \begin{matrix} 1 & 2 & 3 & 4 & 5 & 6 \end{matrix} \\
\begin{bmatrix}
-1 & 0 & 0 & 0 & 0 & 0 \\
0 & -1 & 0 & 0 & 0 & 0 \\
0 & 0 & -1 & 0 & 0 & 0 \\
0 & 0 & 0 & -1 & 0 & 0 \\
0 & 0 & -l & 0 & -1 & 0 \\
0 & -l & 0 & 0 & 0 & -1 \\
1 & 0 & 0 & 0 & 0 & 0 \\
0 & 1 & 0 & 0 & 0 & 0 \\
0 & 0 & 1 & 0 & 0 & 0 \\
0 & 0 & 0 & 1 & 0 & 0 \\
0 & 0 & 0 & 0 & 1 & 0 \\
0 & 0 & 0 & 0 & 0 & 1
\end{bmatrix} \end{matrix}
\tag{7.51}
$$

For two-dimensional problems involving beam elements the required matrices are given by Eqs. (7.21) to (7.23), which were derived for the force system shown in Fig. 7.2*b*.

7.8 TRIANGULAR PLATE ELEMENTS (IN-PLANE FORCES)[271]

The flexibility properties for triangular plate elements with no bending stiffness will be determined using the unit-load theorem. As in the displacement method, we shall assume that the stresses within the triangular plate are constant. The first step is to obtain the element forces, as shown in Fig. 7.5, which are statically equivalent to the assumed constant-stress field. The procedure for doing this will be illustrated for the normal stress σ_{xx}. In Fig. 7.6*a* the constant-stress field σ_{xx} acting on the triangle may be replaced by statically equivalent forces acting at the midpoints of the triangle. These forces can then be transferred to the adjacent node points, and hence the S forces become

$$\begin{aligned} S_1 &= \tfrac{1}{2}(y_{31} - y_{21})t\sigma_{xx} = \tfrac{1}{2}y_{32}t\sigma_{xx} \qquad & S_2 &= 0 \\ S_3 &= \tfrac{1}{2}(y_{23} - y_{21})t\sigma_{xx} = \tfrac{1}{2}y_{31}t\sigma_{xx} \qquad & S_4 &= 0 \\ S_5 &= \tfrac{1}{2}(y_{23} + y_{31})t\sigma_{xx} = \tfrac{1}{2}y_{21}t\sigma_{xx} \qquad & S_6 &= 0 \end{aligned} \tag{7.52}$$

where, as before, t is the plate thickness (assumed to be constant) and

$$\begin{aligned} x_{ij} &= x_i - x_j \\ y_{ij} &= y_i - y_j \end{aligned} \qquad i, j = 1, 2, 3 \tag{7.53}$$

Repeating the same procedure for the stresses σ_{yy} and σ_{xy} in Fig. 7.6*b* and *c* and then collecting the results into a matrix equation, we have

$$\begin{bmatrix} S_1 \\ S_2 \\ S_3 \\ S_4 \\ S_5 \\ S_6 \end{bmatrix} = \frac{t}{2} \begin{bmatrix} y_{32} & 0 & -x_{32} \\ 0 & -x_{32} & y_{32} \\ -y_{31} & 0 & x_{31} \\ 0 & x_{31} & -y_{31} \\ y_{21} & 0 & -x_{21} \\ 0 & -x_{21} & y_{21} \end{bmatrix} \begin{bmatrix} \sigma_{xx} \\ \sigma_{yy} \\ \sigma_{xy} \end{bmatrix} \tag{7.54}$$

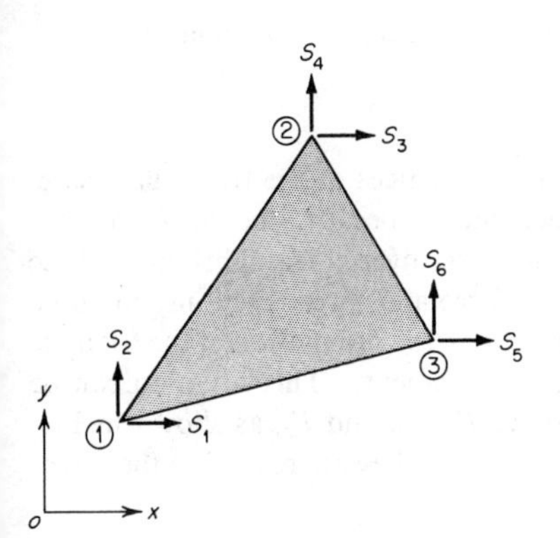

FIG. 7.5 Element forces on a triangular plate element (displacement method).

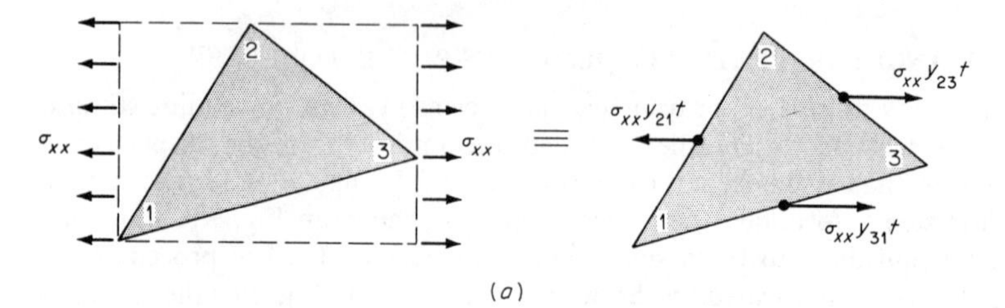

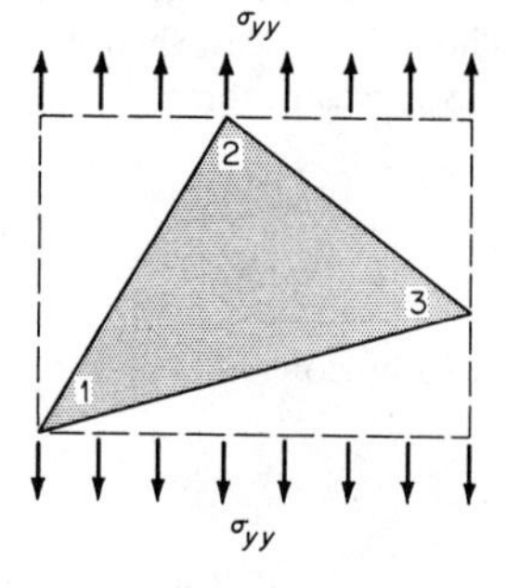

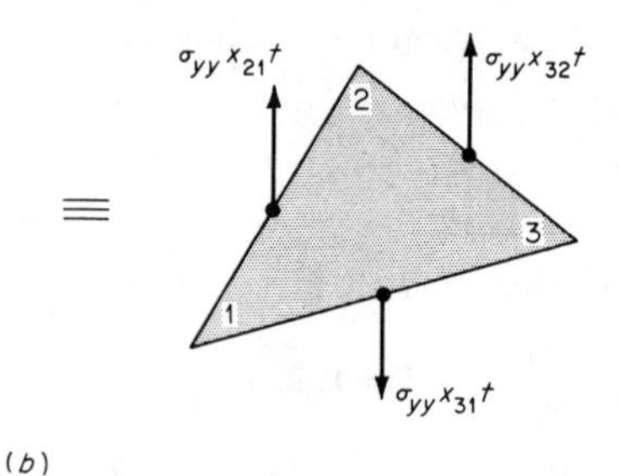

(b)

(c)

FIG. 7.6 Statically equivalent forces on triangular plate elements. (*a*) Normal stress σ_{xx}; (*b*) normal stress σ_{yy}; (*c*) shearing stress σ_{xy}.

Out of the six forces $S_1, \ldots, S_6$ acting at the vertices of the triangular plate, only three force systems are linearly independent, because the six forces obviously must be related by three equations of overall equilibrium. Three independent sets of four forces, one applied force and three reactions, could be selected, but their selection would be dependent on the orientation of the triangle, and, in general, it would be different for each element. This situation can be avoided by selecting three sets of edge forces F_1, F_2, and F_3, as shown in Fig. 7.7; these forces are independent of each other, and they are related to the forces

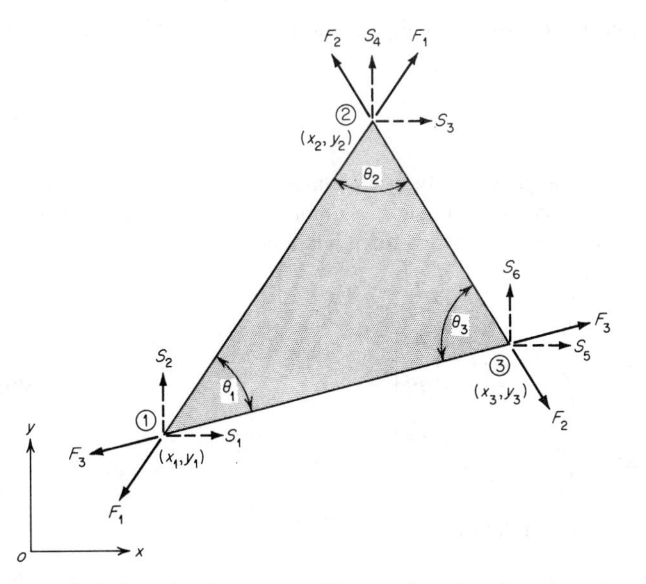

FIG. 7.7 Independent-force system **F** on a triangular plate element (force method).

$S_1, \ldots, S_6$ by the matrix equation

$$\begin{bmatrix} S_1 \\ S_2 \\ S_3 \\ S_4 \\ S_5 \\ S_6 \end{bmatrix} = \begin{bmatrix} -l_{12} & 0 & l_{31} \\ -m_{12} & 0 & m_{31} \\ l_{12} & -l_{23} & 0 \\ m_{12} & -m_{23} & 0 \\ 0 & l_{23} & -l_{31} \\ 0 & m_{23} & -m_{31} \end{bmatrix} \begin{bmatrix} F_1 \\ F_2 \\ F_3 \end{bmatrix} \tag{7.55}$$

where l_{ij} and m_{ij} denote direction cosines for the direction along the edge ij. Equation (7.55) is used to define the matrix **B** in the equation $\mathbf{S} = \mathbf{BF}$.

If only F_1 forces are applied, the stresses due to this force system can be determined from Eqs. (7.54) and (7.55). Hence,

$$\sigma_{xx} = \frac{2l_{12}{}^2F_1}{th_3} \qquad \sigma_{yy} = \frac{2m_{12}{}^2F_1}{th_3} \qquad \sigma_{xy} = \frac{2l_{12}m_{12}F_1}{th_3} \tag{7.56}$$

where h_3 denotes the triangle height measured from vertix 3.

It can be demonstrated easily that Eqs. (7.56) represent, in fact, a constant stress

$$\sigma_{12} = \frac{2F_1}{th_3} \tag{7.57}$$

in the direction of the edge 1,2. By cyclic changes of indices in Eqs. (7.56) stresses due to F_2 and F_3 can be obtained. Combining the stress equations into a single matrix equation leads to stress-force relationship

$$\begin{bmatrix} \sigma_{xx} \\ \sigma_{yy} \\ \sigma_{xy} \end{bmatrix} = \frac{2}{t} \begin{bmatrix} \dfrac{l_{12}^2}{h_3} & \dfrac{l_{23}^2}{h_1} & \dfrac{l_{31}^2}{h_2} \\ \dfrac{m_{12}^2}{h_3} & \dfrac{m_{23}^2}{h_1} & \dfrac{m_{31}^2}{h_2} \\ \dfrac{l_{12}m_{12}}{h_3} & \dfrac{l_{23}m_{23}}{h_1} & \dfrac{l_{31}m_{31}}{h_2} \end{bmatrix} \begin{bmatrix} F_1 \\ F_2 \\ F_3 \end{bmatrix} \tag{7.58}$$

or symbolically in matrix notation,

$$\boldsymbol{\sigma} = \bar{\mathbf{c}}\mathbf{F} \tag{7.59}$$

where

$$\bar{\mathbf{c}} = \frac{2}{t} \begin{bmatrix} \dfrac{l_{12}^2}{h_3} & \dfrac{l_{23}^2}{h_1} & \dfrac{l_{31}^2}{h_2} \\ \dfrac{m_{12}^2}{h_3} & \dfrac{m_{23}^2}{h_1} & \dfrac{m_{31}^2}{h_2} \\ \dfrac{l_{12}m_{12}}{h_3} & \dfrac{l_{23}m_{23}}{h_1} & \dfrac{l_{31}m_{31}}{h_2} \end{bmatrix} \tag{7.60}$$

The matrix $\bar{\mathbf{c}}$ represents statically equivalent stress system due to unit forces $\mathbf{F}$. Using Eqs. (7.39), (7.40), and (7.60) together with two-dimensional expressions for $\boldsymbol{\phi}$ and $\mathbf{e}_T$, we can demonstrate that

$$\mathbf{f} = \int_v \bar{\mathbf{c}}^T \boldsymbol{\phi} \bar{\mathbf{c}}\, dV$$

$$= \frac{2}{Et} \begin{bmatrix} \dfrac{\sin\theta_3}{\sin\theta_1 \sin\theta_2} & \cos\theta_2 \cot\theta_2 - \nu \sin\theta_2 & \cos\theta_1 \cot\theta_1 - \nu \sin\theta_1 \\ \cos\theta_2 \cot\theta_2 - \nu \sin\theta_2 & \dfrac{\sin\theta_1}{\sin\theta_2 \sin\theta_3} & \cos\theta_3 \cot\theta_3 - \nu \sin\theta_3 \\ \cos\theta_1 \cot\theta_1 - \nu \sin\theta_1 & \cos\theta_3 \cot\theta_3 - \nu \sin\theta_3 & \dfrac{\sin\theta_2}{\sin\theta_3 \sin\theta_1} \end{bmatrix} \tag{7.61}$$

and $$\mathbf{v}_T = \int_v \bar{\mathbf{c}}^T \mathbf{e}_T \, dV$$

$$= \alpha T \begin{bmatrix} s_{12} \\ s_{23} \\ s_{31} \end{bmatrix} \tag{7.62}$$

where θ_1, θ_2, and θ_3 are the triangle angles shown in Fig. 7.7 and s_{12}, s_{23}, and s_{31} represent the lengths of the three sides of the triangle. It is interesting to note that the relative thermal displacements $\mathbf{v}_T$ simply represent elongations of the three sides of the triangle due to the temperature change T, as could be expected from physical reasoning.

The structural idealization based on the concept of constant stress in triangular plate elements can be represented by a two-dimensional pin-jointed framework made up by the sides of triangular elements such that when adjacent sides of two triangular plate elements meet, the corresponding framework elements are represented by two parallel pin-jointed bars. However, in contrast to a real pin-jointed framework, where the flexibility of each unassembled bar element is independent of other elements, the flexibilities of bar elements in the idealized framework are coupled in sets of three bars representing the three sides of a triangular plate element.

Other force systems can also be used to derive the flexibility properties of triangular plate elements, but it appears that the edge-force systems described here are advantageous from a computational point of view.

7.9 RECTANGULAR PLATE ELEMENTS (IN-PLANE FORCES)

Figure 7.8 shows a rectangular plate with the element forces $S_1, \ldots, S_8$. These forces are linearly dependent since they must be related by the three equations for the overall equilibrium. Thus it is possible, using the forces $S_1, \ldots, S_8$, to select five sets of independent force systems. The five independent force systems $F_1, \ldots, F_5$ selected for the determination of

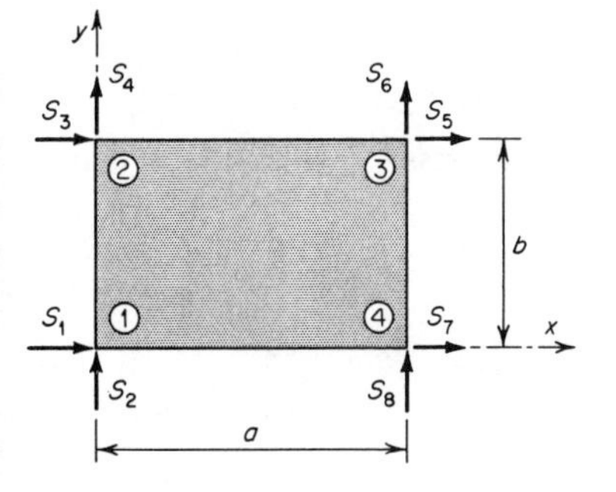

FIG. 7.8 Element forces on a rectangular plate element (displacement method).

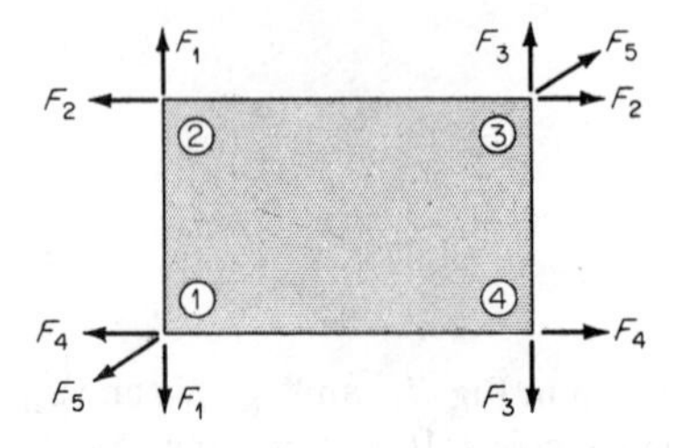

FIG. 7.9 Independent-force systems on a rectangular plate element (force method).

flexibility properties of rectangular plate elements are shown in Fig. 7.9. Four force systems are acting along the four sides of the rectangle, and one system is acting along the diagonal running from the upper right to the lower left corner. These force systems are independent of each other, and they are related to the forces $S_1, \ldots, S_8$ by the matrix equation

$$\begin{bmatrix} S_1 \\ S_2 \\ S_3 \\ S_4 \\ S_5 \\ S_6 \\ S_7 \\ S_8 \end{bmatrix} = \begin{bmatrix} 0 & 0 & 0 & -1 & \dfrac{-1}{(1+\beta^2)^{\frac{1}{2}}} \\ -1 & 0 & 0 & 0 & \dfrac{-\beta}{(1+\beta^2)^{\frac{1}{2}}} \\ 0 & -1 & 0 & 0 & 0 \\ 1 & 0 & 0 & 0 & 0 \\ 0 & 1 & 0 & 0 & \dfrac{1}{(1+\beta^2)^{\frac{1}{2}}} \\ 0 & 0 & 1 & 0 & \dfrac{\beta}{(1+\beta^2)^{\frac{1}{2}}} \\ 0 & 0 & 0 & 1 & 0 \\ 0 & 0 & -1 & 0 & 0 \end{bmatrix} \begin{bmatrix} F_1 \\ F_2 \\ F_3 \\ F_4 \\ F_5 \end{bmatrix} \tag{7.63}$$

where $$\beta = \frac{b}{a} \tag{7.64}$$

The direct stresses σ_{xx} and σ_{yy} and the shearing stress σ_{xy} within the rectangle will be assumed to be given by

$$\begin{aligned} \sigma_{xx} &= c_1 + c_2\eta \\ \sigma_{yy} &= c_3 + c_4\xi \\ \sigma_{xy} &= c_5 \end{aligned} \tag{7.65}$$

where $c_1, \ldots, c_5$ are constants and

$$\xi = \frac{x}{a} \qquad \eta = \frac{y}{b} \tag{7.66}$$

Applying now in turn the force systems F_1 to F_5, we can derive statically equivalent stress distributions based on Eqs. (7.65) for each force system acting separately. The resulting stress-force equations can be combined into a single matrix equation

$$\begin{bmatrix} \sigma_{xx} \\ \sigma_{yy} \\ \sigma_{xy} \end{bmatrix} = \frac{2}{bt} \begin{bmatrix} 0 & -1 + 3\eta & 0 & 2 - 3\eta & \dfrac{1}{2(1 + \beta^2)^{\frac{1}{2}}} \\ \beta(2 - 3\xi) & 0 & \beta(-1 + 3\xi) & 0 & \dfrac{\beta^2}{2(1 + \beta^2)^{\frac{1}{2}}} \\ 0 & 0 & 0 & 0 & \dfrac{\beta}{2(1 + \beta^2)^{\frac{1}{2}}} \end{bmatrix} \begin{bmatrix} F_1 \\ F_2 \\ F_3 \\ F_4 \\ F_5 \end{bmatrix} \tag{7.67}$$

which is represented symbolically as

$$\boldsymbol{\sigma} = \bar{\mathbf{c}}\mathbf{F} \tag{7.68}$$

where

$$\bar{\mathbf{c}} = \frac{2}{bt} \begin{bmatrix} 0 & -1 + 3\eta & 0 & 2 - 3\eta & \dfrac{1}{2(1 + \beta^2)^{\frac{1}{2}}} \\ \beta(2 - 3\xi) & 0 & \beta(-1 + 3\xi) & 0 & \dfrac{\beta^2}{2(1 + \beta^2)^{\frac{1}{2}}} \\ 0 & 0 & 0 & 0 & \dfrac{\beta}{2(1 + \beta^2)^{\frac{1}{2}}} \end{bmatrix} \tag{7.69}$$

Using now Eqs. (7.39), (7.40), and (7.69), we obtain

$$\mathbf{f} = \frac{1}{Et} \begin{bmatrix} 4\beta & & & & \\ -\nu & \dfrac{4}{\beta} & & \text{Symmetric} & \\ -2\beta & -\nu & 4\beta & & \\ -\nu & \dfrac{-2}{\beta} & -\nu & \dfrac{4}{\beta} & \\ \dfrac{\beta^2 - \nu}{(1 + \beta^2)^{\frac{1}{2}}} & \dfrac{1 - \nu\beta^2}{\beta(1 + \beta^2)^{\frac{1}{2}}} & \dfrac{\beta^2 - \nu}{(1 + \beta^2)^{\frac{1}{2}}} & \dfrac{1 - \nu\beta^2}{\beta(1 + \beta^2)^{\frac{1}{2}}} & \dfrac{1 + \beta^2}{\beta} \end{bmatrix} \tag{7.70}$$

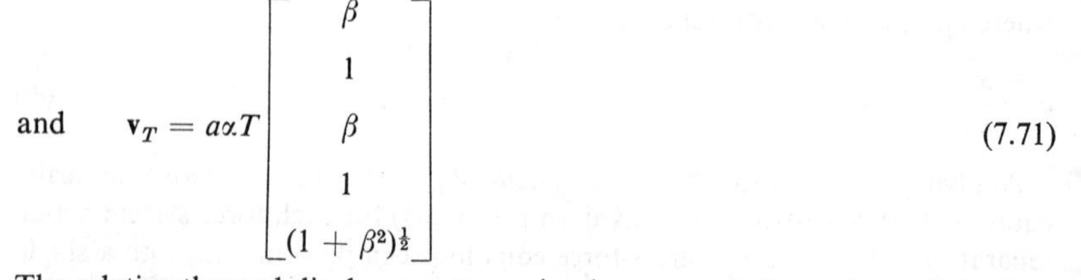

and $$\mathbf{v}_T = a\alpha T \begin{bmatrix} \beta \\ 1 \\ \beta \\ 1 \\ (1+\beta^2)^{\frac{1}{2}} \end{bmatrix} \tag{7.71}$$

The relative thermal displacements $\mathbf{v}_T$ simply represent elongations of the four edges and the diagonal due to the temperature change T.

The structural idealization of structures made up from rectangular plates may be thought of as being represented by a pin-jointed framework with bars representing sides of the rectangular plates and one of their diagonals such that when adjacent sides of two rectangular plate elements meet, the corresponding framework elements are represented by two parallel pin-jointed bars. The flexibilities of such idealized bar elements, however, are coupled in sets of five bars corresponding to the five independent force systems $F_1, \ldots, F_5$ on each rectangular plate element.

7.10 TETRAHEDRON ELEMENTS[273]

The assumption of constant stresses within the element can also be used on the solid tetrahedron to establish its flexibility properties required in the matrix force method of analysis. The flexibility properties of the tetrahedron element can be determined most conveniently for a set of edge-force systems acting along the six edges of the tetrahedron.

The six independent force systems $F_1, \ldots, F_6$ will be assumed to be located along the edges identified according to Table 7.1. The numbering i for the force systems also will be used to specify directions of the edges on which these systems are acting. Sequences other than those given in Table 7.1 may be selected, but once a sequence has been chosen, it must be adhered to throughout the analysis.

TABLE 7.1 LOCATION OF INDEPENDENT FORCE SYSTEMS ON TETRAHEDRON ELEMENTS

Force system or direction i	Location (nodes)
1	1,2
2	2,3
3	3,4
4	4,1
5	1,3
6	2,4

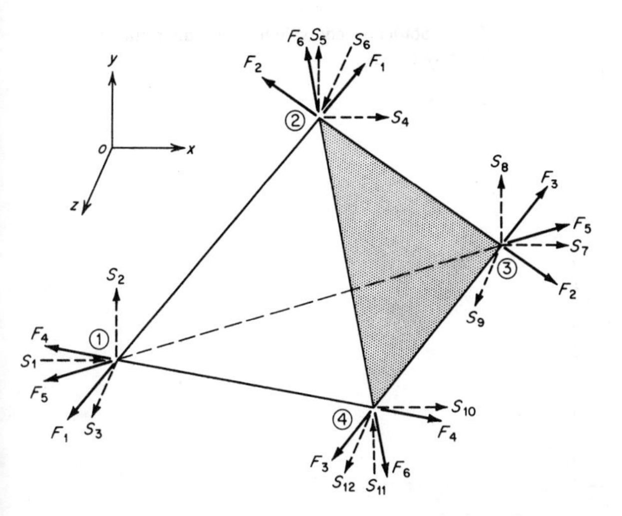

FIG. 7.10 Independent-force system **F** on a tetrahedron element.

The forces F and S are shown in Fig. 7.10. The relationship $\mathbf{S} = \mathbf{BF}$ is given by

$$\begin{bmatrix} S_1 \\ S_2 \\ S_3 \\ S_4 \\ S_5 \\ S_6 \\ S_7 \\ S_8 \\ S_9 \\ S_{10} \\ S_{11} \\ S_{12} \end{bmatrix} = \begin{bmatrix} -l_1 & 0 & 0 & l_4 & -l_5 & 0 \\ -m_1 & 0 & 0 & m_4 & -m_5 & 0 \\ -n_1 & 0 & 0 & n_4 & -n_5 & 0 \\ l_1 & -l_2 & 0 & 0 & 0 & -l_6 \\ m_1 & -m_2 & 0 & 0 & 0 & -m_6 \\ n_1 & -n_2 & 0 & 0 & 0 & -n_6 \\ 0 & l_2 & -l_3 & 0 & l_5 & 0 \\ 0 & m_2 & -m_3 & 0 & m_5 & 0 \\ 0 & n_2 & -n_3 & 0 & n_5 & 0 \\ 0 & 0 & l_3 & -l_4 & 0 & l_6 \\ 0 & 0 & m_3 & -m_4 & 0 & m_6 \\ 0 & 0 & n_3 & -n_4 & 0 & n_6 \end{bmatrix} \begin{bmatrix} F_1 \\ F_2 \\ F_3 \\ F_4 \\ F_5 \\ F_6 \end{bmatrix} \tag{7.72}$$

where l_i, m_i, and n_i denote here the direction cosines for the directions i, as specified in Table 7.1.

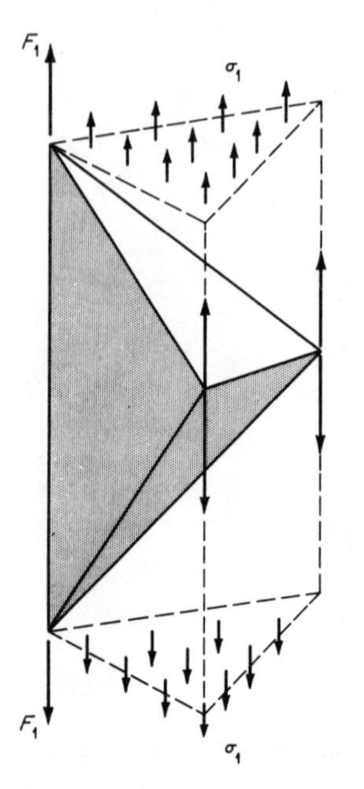

FIG. 7.11 Statically equivalent stress distribution due to the force system F_1.

If the force system F_1 is applied alone, an equivalent stress system, in equilibrium with the applied forces, may be taken as a constant tensile stress in the direction 1. Figure 7.11 makes it evident that this stress is given by

$$\sigma_1 = \frac{3F_1}{A_1} \tag{7.73}$$

in which A_1 represents area of the projection of the tetrahedron onto a plane normal to direction 1. The stress σ_1 can then be resolved into stress components in the x, y, z coordinate system. By cyclic changes of subscripts, contributions from other force systems can be obtained, and the results can be expressed by the matrix equation

$$\boldsymbol{\sigma} = \bar{\mathbf{c}}\mathbf{F} \tag{7.74}$$

where $$\boldsymbol{\sigma} = \{\sigma_{xx} \quad \sigma_{yy} \quad \sigma_{zz} \quad \sigma_{xy} \quad \sigma_{yz} \quad \sigma_{zx}\} \tag{7.75}$$

$$\mathbf{F} = \{F_1 \quad \cdots \quad F_6\} \tag{7.76}$$

and

$$\bar{\mathbf{c}} = 3\begin{bmatrix} \frac{l_1^2}{A_1} & \frac{l_2^2}{A_2} & \frac{l_3^2}{A_3} & \frac{l_4^2}{A_4} & \frac{l_5^2}{A_5} & \frac{l_6^2}{A_6} \\ \frac{m_1^2}{A_1} & \frac{m_2^2}{A_2} & \frac{m_3^2}{A_3} & \frac{m_4^2}{A_4} & \frac{m_5^2}{A_5} & \frac{m_6^2}{A_6} \\ \frac{n_1^2}{A_1} & \frac{n_2^2}{A_2} & \frac{n_3^2}{A_3} & \frac{n_4^2}{A_4} & \frac{n_5^2}{A_5} & \frac{n_6^2}{A_6} \\ \frac{l_1m_1}{A_1} & \frac{l_2m_2}{A_2} & \frac{l_3m_3}{A_3} & \frac{l_4m_4}{A_4} & \frac{l_5m_5}{A_5} & \frac{l_6m_6}{A_6} \\ \frac{m_1n_1}{A_1} & \frac{m_2n_2}{A_2} & \frac{m_3n_3}{A_3} & \frac{m_4n_4}{A_4} & \frac{m_5n_5}{A_5} & \frac{m_6n_6}{A_6} \\ \frac{n_1l_1}{A_1} & \frac{n_2l_2}{A_2} & \frac{n_3l_3}{A_3} & \frac{n_4l_4}{A_4} & \frac{n_5l_5}{A_5} & \frac{n_6l_6}{A_6} \end{bmatrix} \tag{7.77}$$

Using Eqs. (7.39) and (7.77) together with the three-dimensional expression for $\boldsymbol{\phi}$ we can demonstrate that

$$\mathbf{f} = [f_{ij}] \qquad i, j = 1, 2, \ldots, 6 \tag{7.78}$$

with typical coefficient

$$f_{ij} = \frac{s_i s_j}{EV}[(1 + \nu)\cos^2\theta_{i,j} - \nu] \tag{7.79}$$

where s_i represents the length of the edge corresponding to the ith force system and $\theta_{i,j}$ is the angle between the i and j directions.

Similarly, from Eqs. (7.40) and (7.77) it can be shown that

$$\mathbf{v}_T = \alpha T\{s_1 \quad s_2 \quad s_3 \quad s_4 \quad s_5 \quad s_6\} \tag{7.80}$$

In deriving Eqs. (7.78) and (7.80), the following equation for the volume of the element was used:

$$V = \frac{s_i A_i}{3} \qquad i = 1, 2, \ldots, 6$$

A single equation valid for all the f_{ij} coefficients in the flexibility matrix $\mathbf{f}$ greatly simplifies the necessary computer programming. Furthermore, the thermal-displacement matrix $\mathbf{v}_T$ is also easy to evaluate, since it simply represents elongations of the six sides of the tetrahedron caused by the temperature change T.

The structural idealization based on the concept of constant stress distribution within solid tetrahedron elements can be represented by a three-dimensional pin-jointed framework made up of the sides of tetrahedron elements such that when adjacent sides of two tetrahedron elements meet, the corresponding framework elements are represented by two parallel pin-jointed bars. The flexibilities of bar elements in the idealized framework structure are coupled in sets of six bars representing the six edges of a tetrahedron element.

7.11 CONSTANT-SHEAR-FLOW PANELS

The determination of flexibility properties for a constant-shear-flow panel has been discussed in Sec. 6.7. The flexibility matrix is given by Eq. (6.120). No thermal deformation matrix $\mathbf{v}_T$ exists for this element since no direct stresses were allowed in this idealization.

7.12 LINEARLY VARYING AXIAL-FORCE MEMBERS

The flexibility properties for this element have been presented in Sec. 6.8. The flexibility and thermal deformation matrices are given by Eqs. (6.128) and (6.129), respectively.

7.13 RECTANGULAR PLATES IN BENDING

The bending and in-plane deformations for small deflections in flat plates are uncoupled, and therefore the bending and in-plane flexibilities can be considered separately. For rectangular plates in bending we considered (Sec. 5.12) twelve element forces $S_1, \ldots, S_{12}$, shown in Fig. 7.12. These twelve forces are

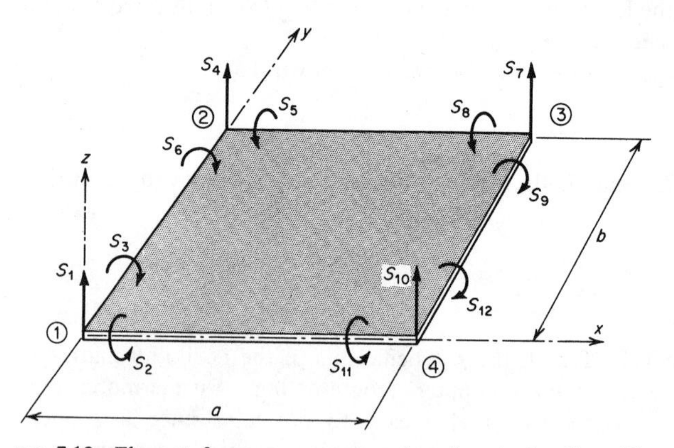

FIG. 7.12 Element forces on a rectangular plate in bending (displacement method).

related by three equations of equilibrium, and consequently only nine of these forces are linearly independent. Many choices are possible for selecting the independent forces to be used for the flexibility matrix. The independent force systems we shall select are shown in Fig. 7.13. These systems consist of four sets of symmetric moments F_1, F_3, F_5, and F_7 applied at the four sides of the rectangle, four sets of antisymmetric moments F_2, F_4, F_6, and F_8 also applied at the sides, and one set of four forces normal to the plate and applied at the corners. It should be noted from Fig. 7.13 that the antisymmetric moments require corner forces for equilibrium.

It can be easily seen from Figs. 7.12 and 7.13 that the equation $\mathbf{S} = \mathbf{BF}$ is simply

$$\begin{bmatrix} S_1 \\ S_2 \\ S_3 \\ S_4 \\ S_5 \\ S_6 \\ S_7 \\ S_8 \\ S_9 \\ S_{10} \\ S_{11} \\ S_{12} \end{bmatrix} = \begin{array}{c} \\ 1 \\ 2 \\ 3 \\ 4 \\ 5 \\ 6 \\ 7 \\ 8 \\ 9 \\ 10 \\ 11 \\ 12 \end{array} \begin{array}{c} \begin{array}{ccccccccc} 1 & 2 & 3 & 4 & 5 & 6 & 7 & 8 & 9 \end{array} \\ \begin{bmatrix} 0 & \frac{2}{b} & 0 & 0 & 0 & 0 & 0 & \frac{-2}{a} & -1 \\ 1 & 1 & 0 & 0 & 0 & 0 & 0 & 0 & 0 \\ 0 & 0 & 0 & 0 & 0 & 0 & -1 & 1 & 0 \\ 0 & \frac{-2}{b} & 0 & \frac{2}{a} & 0 & 0 & 0 & 0 & 1 \\ -1 & 1 & 0 & 0 & 0 & 0 & 0 & 0 & 0 \\ 0 & 0 & -1 & -1 & 0 & 0 & 0 & 0 & 0 \\ 0 & 0 & 0 & \frac{-2}{a} & 0 & \frac{2}{b} & 0 & 0 & -1 \\ 0 & 0 & 0 & 0 & -1 & -1 & 0 & 0 & 0 \\ 0 & 0 & 1 & -1 & 0 & 0 & 0 & 0 & 0 \\ 0 & 0 & 0 & 0 & 0 & \frac{-2}{b} & 0 & \frac{2}{a} & 1 \\ 0 & 0 & 0 & 0 & 1 & -1 & 0 & 0 & 0 \\ 0 & 0 & 0 & 0 & 0 & 0 & 1 & 1 & 0 \end{bmatrix} \end{array} \begin{bmatrix} F_1 \\ F_2 \\ F_3 \\ F_4 \\ F_5 \\ F_6 \\ F_7 \\ F_8 \\ F_9 \end{bmatrix} \tag{7.81}$$

Assuming that the normal stresses vary linearly and the shearing stresses are constant, we can obtain a relationship between the statically equivalent stresses $\boldsymbol{\sigma} = \{\sigma_{xx} \quad \sigma_{yy} \quad \sigma_{xy}\}$ and the element forces $\mathbf{F} = \{F_1 \quad \cdots \quad F_9\}$. This relationship is of the form

$$\boldsymbol{\sigma} = \bar{\mathbf{c}}\mathbf{F} \tag{7.82}$$

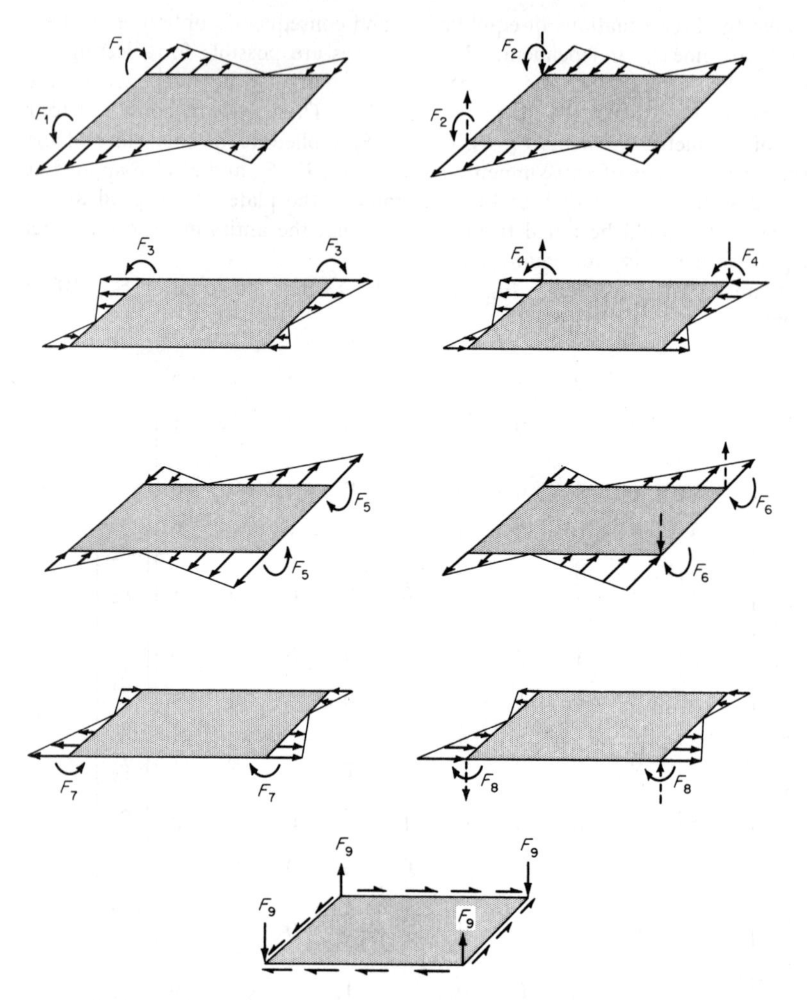

FIG. 7.13 Statically equivalent forces on rectangular plates in bending (stresses shown are for the surface at $z = t/2$).

where $\bar{\mathbf{c}}$ can be shown to be given by

$$\bar{\mathbf{c}}^T = \frac{12z}{t^3} \begin{bmatrix} 0 & \frac{2}{a}(2-3\xi) & 0 \\ 0 & \frac{2}{a}(2-3\xi)(1-2\eta) & 0 \\ \frac{2}{b}(-1+3\eta) & 0 & 0 \\ \frac{2}{b}(-1+3\eta)(1-2\xi) & 0 & 0 \\ 0 & \frac{2}{a}(-1+3\xi) & 0 \\ 0 & \frac{2}{a}(-1+3\xi)(-1+2\eta) & 0 \\ \frac{2}{b}(2-3\eta) & 0 & 0 \\ \frac{2}{b}(2-3\eta)(-1+2\xi) & 0 & 0 \\ 0 & 0 & 1 \end{bmatrix} \begin{matrix} 1 \\ 2 \\ 3 \\ 4 \\ 5 \\ 6 \\ 7 \\ 8 \\ 9 \end{matrix} \tag{7.83}$$

Using Eqs. (7.39) and (7.83), we obtain the following flexibility matrix

$$\mathbf{f} = \frac{12}{Et^3} \begin{bmatrix} \frac{4b}{a} & & & & & & & & \\ 0 & \frac{4b}{3a} & & & & \text{Symmetric} & & & \\ -\nu & \nu & \frac{4a}{b} & & & & & & \\ -\nu & \nu & 0 & \frac{4a}{3b} & & & & & \\ \frac{-2b}{a} & 0 & -\nu & \nu & \frac{4b}{a} & & & & \\ 0 & \frac{2b}{3a} & -\nu & \nu & 0 & \frac{4b}{3a} & & & \\ -\nu & -\nu & \frac{-2a}{b} & 0 & -\nu & \nu & \frac{4a}{b} & & \\ \nu & \nu & 0 & \frac{2a}{3b} & -\nu & \nu & 0 & \frac{4a}{3b} & \\ 0 & 0 & 0 & 0 & 0 & 0 & 0 & 0 & 2(1+\nu)ab \end{bmatrix} \tag{7.84}$$

(rows and columns numbered 1 to 9)

Assuming that the temperature throughout the element is given by

$$T = T_m + \Delta T \, \frac{z}{t} \tag{7.85}$$

where T_m is a constant mean temperature and ΔT is a constant temperature difference between upper and lower surfaces, we can determine the thermal deformation matrix $\mathbf{v}_T$ from Eq. (7.40). Hence

$$\mathbf{v}_T = \frac{\alpha\,\Delta T}{t}\overset{\begin{matrix}1 & 2 & 3 & 4 & 5 & 6 & 7 & 8 & 9\end{matrix}}{\{\begin{matrix}b & 0 & a & 0 & b & 0 & a & 0 & 0\end{matrix}\}} \tag{7.86}$$

PROBLEMS

7.1 Derive the flexibility matrix for a pin-jointed bar element with the cross-sectional area varying linearly from A_1 to A_2.

7.2 Using Castigliano's theorem (part II), demonstrate that if the shear and axial deformations are neglected, the flexibility matrix for the curved-beam element in Fig. 6.14 is given by

$$\mathbf{f} = \frac{R}{4EI}\begin{matrix} & 1 & 2 & 3 \\ 1 & \\ 2 & \\ 3 & \end{matrix}\begin{bmatrix} \pi R^2 & 2R^2 & 4R \\ 2R^2 & (3\pi - 8)R^2 & 2(\pi - 2)R \\ 4R & 2(\pi - 2)R & 2\pi \end{bmatrix}$$

7.3 Determine the thermal deformations matrix $\mathbf{v}_T$ for the curved-beam element in Fig. 6.14.

7.4 Solve Prob. 7.2 when the effects of shear and axial deformations are included.

7.5 Discuss the derivation of a flexibility matrix for the triangular plate element in bending using concepts similar to those for the rectangular plate in Sec. 7.13.

7.6 Explain why the stiffness coefficients k_{ij} depend on the number of displacements considered in the stiffness matrix, whereas the flexibility coefficients f_{ij} (influence coefficients) are independent of the number of displacements.

CHAPTER 8 THE MATRIX FORCE METHOD

The force method of analysis is based on the equations of equilibrium expressed in terms of the element forces **F** introduced in Chap. 7. For some structures these equations are sufficient to determine all the forces **F** and hence the element stresses and displacements. Such structures are said to be *statically determinate*. For general applications, however, the number of element forces **F** exceeds the number of available equations of equilibrium, and the structure is then said to be *statically indeterminate* (or redundant). For such cases the equations of equilibrium are insufficient to obtain solutions for the element forces, and therefore additional equations are required. These additional equations are supplied by the compatibility conditions on displacements. In this chapter the general formulation of the force method of analysis is presented, leading to the equilibrium and compatibility matrix equations. The derivation of the latter equations is obtained from the unit-load theorem generalized for the discrete-element structural idealization.

8.1 MATRIX FORMULATION OF THE UNIT-LOAD THEOREM FOR EXTERNAL-FORCE SYSTEMS

The formulation of the unit-load theorem in Chap. 3 was based on a continuous elastic system subjected to a set of external forces and known temperature distribution. Although the general theorem for continuous systems could be modified directly so as to be applicable to discrete-element systems, it is preferable to derive the equivalent matrix form of the unit-load theorem *ab initio*. To derive this theorem, a virtual load δP in the direction of the displacement r

is applied in a discrete-element structure subjected to a system of forces and temperature changes which induce internal-element forces given by the matrix $\mathbf{F}$. The complementary virtual work is then

$$\delta W^* = r\,\delta P \tag{8.1}$$

If we now consider a typical structural element, it can easily be observed that because of the virtual force δP the virtual complementary energy of total deformation on the ith element is given by

$$[\delta U_d^*]^{(i)} = [\delta \mathbf{F}^{(i)}]^T \mathbf{v}^{(i)} \tag{8.2}$$

Hence for the complete structure

$$\delta U_d^* = \delta \mathbf{F}^T\,\mathbf{v} \tag{8.3}$$

where $$\delta \mathbf{F} = \{\delta \mathbf{F}^{(1)} \quad \delta \mathbf{F}^{(2)} \quad \cdots \quad \delta \mathbf{F}^{(i)} \quad \cdots\} \tag{8.4}$$

which need only satisfy equations of equilibrium and be in balance with the virtual force δP.

For linear elasticity Eq. (8.4) can be expressed as

$$\delta \mathbf{F} = \{\bar{\mathbf{F}}^{(1)} \quad \bar{\mathbf{F}}^{(2)} \quad \cdots \quad \bar{\mathbf{F}}^{(i)} \quad \cdots\}\,\delta P = \bar{\mathbf{b}}\,\delta P \tag{8.5}$$

where $\bar{\mathbf{b}}$ represents element forces due to $\delta P = 1$. From the principle of virtual forces $\delta W^* = \delta U_d^*$, it follows that

$$r\,\delta P = \bar{\mathbf{b}}^T \mathbf{v}\,\delta P \tag{8.6}$$

and hence $$r = \bar{\mathbf{b}}^T \mathbf{v} \tag{8.7}$$

which, using Eq. (7.8), becomes

$$r = \bar{\mathbf{b}}^T(\mathbf{f}\mathbf{F} + \mathbf{v}_T) \tag{8.8}$$

Equation (8.8) represents the matrix form of the unit-load theorem for a single displacement r. If, however, n displacements are required as a result of some specified loading and temperature condition, all these displacements can still be obtained from Eq. (8.8) provided $\bar{\mathbf{b}}$ consists of n columns, representing statically equivalent element forces due to unit loads applied in turn in the specified n directions. For such cases, the displacements are determined directly from

$$\mathbf{r} = \bar{\mathbf{b}}^T \mathbf{v} = \bar{\mathbf{b}}^T(\mathbf{f}\mathbf{F} + \mathbf{v}_T) \tag{8.9}$$

where $\mathbf{r}$ is a column matrix of the required displacements due to some specified loading conditions for which $\mathbf{F}$ and $\mathbf{v}_T$ are known and $\bar{\mathbf{b}}$ represents statically equivalent element forces due to unit external forces applied in the directions of $\mathbf{r}$.

It is important to realize that $\bar{\mathbf{b}}$ need satisfy only the internal and external equations of equilibrium, a fact which can greatly reduce the amount of computing in determining the displacements. The advantages of using a statically

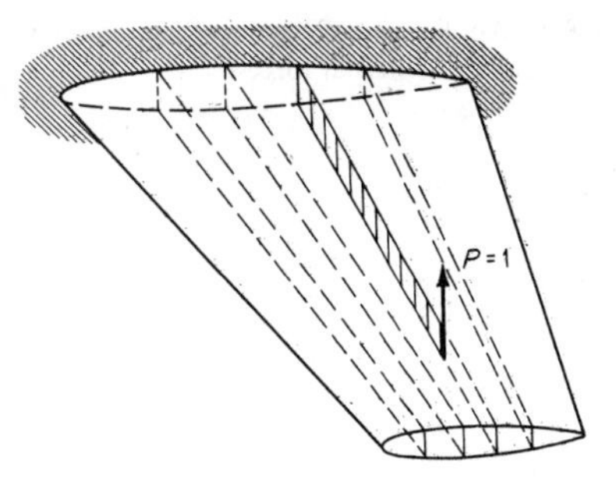

FIG. 8.1 Application of the unit-load theorem to a multispar wing structure.

equivalent force system for $\bar{\mathbf{b}}$ will be apparent from the example to be discussed next. Suppose it is required to determine vertical deflection of a point on the wing structure shown in Fig. 8.1 subjected to a system of external forces and temperature distribution for which the exact distribution of element forces and relative thermal expansions is known. The simplest statically equivalent force system $\bar{\mathbf{b}}$ due to a unit load $P = 1$ applied in the direction of the required displacement can be evaluated for a single cantilever shown by full lines in Fig. 8.1. Clearly the products $\bar{\mathbf{b}}^T\mathbf{fF}$ and $\bar{\mathbf{b}}^T\mathbf{v}_T$ in Eq. (8.9) have only to be taken over the structural elements in the cantilever, since $\bar{\mathbf{b}}$ is elsewhere zero, and this naturally results in a considerable saving in computing effort.

The unit-load theorem is also applicable to a whole system of external forces, but the physical interpretation of the results in such cases may be possible only for specific distributions of the externally applied unit force system. One such system, which has practical applications, is a set of three parallel self-equilibrating forces acting in one plane. To illustrate this, consider the wing structure shown in Fig. 8.2 subjected to a specified loading and temperature for which the element forces $\mathbf{F}$ and relative thermal expansions $\mathbf{v}_T$ are known. A system of three self-equilibrating forces P_A, P_B, and P_C is then applied at the wing tip, and it will be assumed that

$$P_C = 1 \tag{8.10}$$

Since the system is a self-equilibrating one, we must have

$$P_A + P_B + P_C = 0 \tag{8.11}$$

and $$P_C a + P_B c = 0 \tag{8.12}$$

Solving for P_A and P_B gives

$$P_A = -\frac{b}{c} \tag{8.13}$$

$$P_B = -\frac{a}{c} \tag{8.14}$$

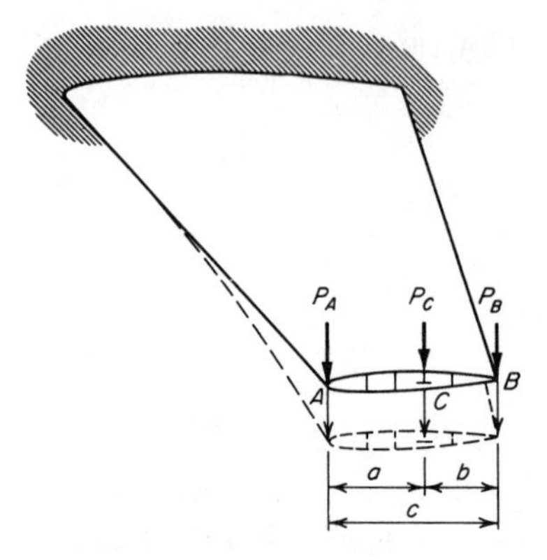

FIG. 8.2 Application of the unit-load theorem to a system of external forces.

Furthermore, it will be assumed that the actual deflections at the points A, B, and C are r_A, r_B, and r_C, respectively. Thus, if $\bar{\mathbf{b}}_A$ is the statically equivalent element force distribution due to P_A, it is clear that the application of Eq. (8.9) leads to

$$\bar{\mathbf{b}}_A{}^T(\mathbf{fF} + \mathbf{v}_T) = -\frac{b}{c}r_A \tag{8.15}$$

Similarly, for the P_B and P_C forces applied separately

$$\bar{\mathbf{b}}_B{}^T(\mathbf{fF} + \mathbf{v}_T) = -\frac{a}{c}r_B \tag{8.16}$$

and $$\bar{\mathbf{b}}_C{}^T(\mathbf{fF} + \mathbf{v}_T) = r_C \tag{8.17}$$

Adding Eqs. (8.15) to (8.17), we have

$$(\bar{\mathbf{b}}_A{}^T + \bar{\mathbf{b}}_B{}^T + \bar{\mathbf{b}}_C{}^T)(\mathbf{fF} + \mathbf{v}_T) = \bar{\mathbf{b}}_{ABC}^T(\mathbf{fF} + \mathbf{v}_T)$$
$$= r_C - \frac{a}{c}r_B - \frac{b}{c}r_A \tag{8.18}$$

where $\bar{\mathbf{b}}_{ABC}$ represents a statically equivalent distribution of element forces due to P_A, P_B, and P_C applied simultaneously. By rearranging Eq. (8.18) it can be shown that

$$\bar{\mathbf{b}}_{ABC}^T(\mathbf{fF} + \mathbf{v}_T) = r_C - r_A - (r_B - r_A)\frac{a}{c} \tag{8.19}$$

The right side of Eq. (8.19) represents deflection of the point C relative to a straight line joining the displaced positions of points A and B (see Fig. 8.3).

A somewhat similar result could be obtained if the unit load were reacted by three parallel noncoplanar forces. Here the deflection obtained from the

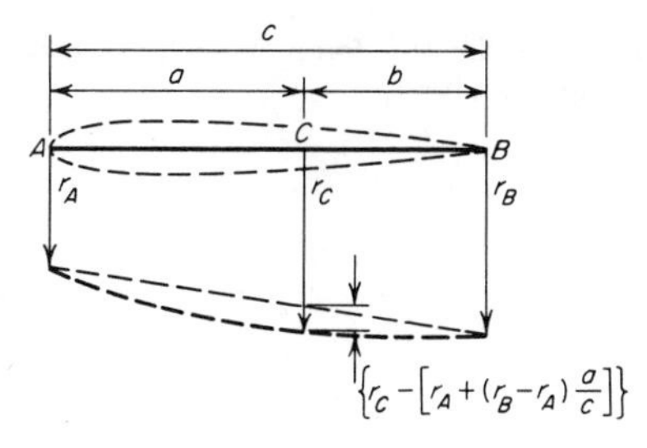

FIG. 8.3 Deflection of point C relative to a straight line joining deflected positions of A and B.

unit-load theorem based on such a force system would be equal to the deflection of the point at which the unit load is applied relative to a plane passing through the deflected positions of the reaction points.

8.2 MATRIX FORMULATION OF THE UNIT-LOAD THEOREM FOR INTERNAL-FORCE SYSTEMS: SELF-EQUILIBRATING FORCE SYSTEMS

So far, the unit-load theorem has been applied to external forces. This theorem, however, can be generalized so as to be applicable to internal forces. In such cases the resulting displacements, obtained from the theorem, represent internal relative displacements which in order to satisfy continuity of deformations (compatibility conditions) must be equal to zero. To illustrate this, consider a two-dimensional redundant pin-jointed truss subjected to external loads P_1 and P_2 and some specified temperature distribution T (see Fig. 8.4). The relative displacements $\mathbf{v}$ on individual elements due to P_1 and P_2 and the temperature T applied to the structure are given by

$$\mathbf{v} = \mathbf{fF} + \mathbf{v}_T \tag{8.20}$$

where $\mathbf{f}$ is the diagonal matrix of element flexibilities, $\mathbf{F}$ is the column matrix of element forces, and $\mathbf{v}_T$ is the column matrix of relative thermal displacements calculated for the temperature distribution T. If we now introduce a fictitious cut in the diagonal member 1,4 near joint 1, then in order to maintain equilibrium with the external loading and thermal forces which may arise from the

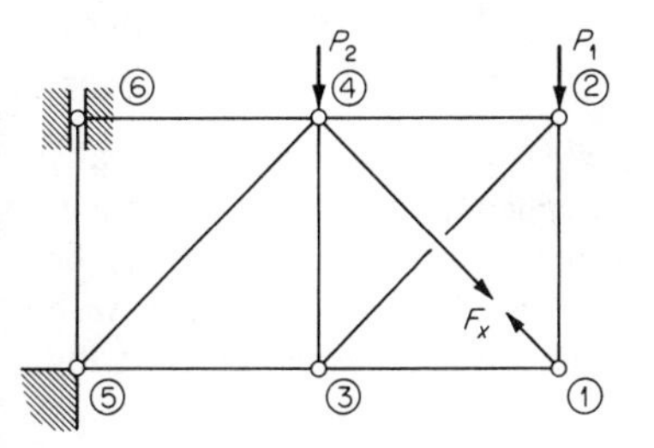

FIG. 8.4 Redundant pin-jointed truss under external loads P_1 and P_2.

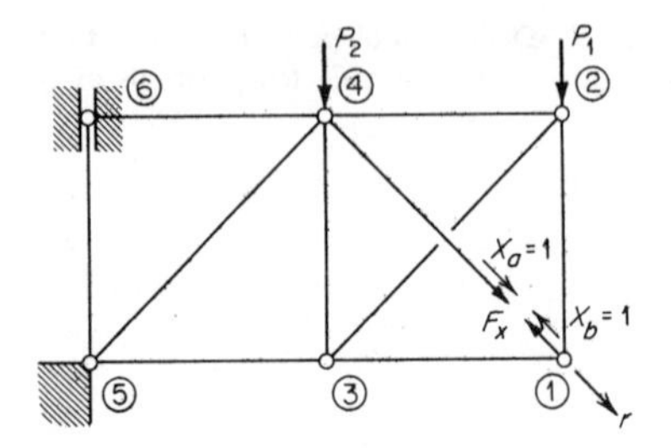

FIG. 8.5 Redundant forces X applied across the cut in member 1,4.

temperature distribution the force F_x, which existed in the member before the cut was introduced, must be supplied by some external means. It should be observed that by introducing the cut in the diagonal member 1,4 this particular structure is reduced to a statically determinate one.

If a unit load X_a is applied at the cut in the direction 4,1, as shown in Fig. 8.5, the unit-load theorem gives a deflection r_a which is

$$r_a = \bar{\mathbf{b}}_a{}^T(\mathbf{fF} + \mathbf{v}_T) \tag{8.21}$$

where $\bar{\mathbf{b}}_a$ represents the matrix of statically equivalent element forces due to $X_a = 1$ (note that $\mathbf{F}$ includes force F_x in the cut member 1,4). If instead of X_a a unit force X_b is applied in the direction 1,4, then from the unit-load theorem it follows that

$$r_b = \bar{\mathbf{b}}_b{}^T(\mathbf{fF} + \mathbf{v}_T) \tag{8.22}$$

where $\bar{\mathbf{b}}_b$ represents the matrix of statically equivalent element forces due to $X_b = 1$. Since the direction of X_b is opposite to that of X_a,

$$r_a = -r_b = r \tag{8.23}$$

where r represents the actual displacement of node 1 in the direction 4,1.

From Eq. (8.23) it follows that

$$r_a + r_b = 0 \tag{8.24}$$

and this may be interpreted as representing the relative displacement across the cut. Substitution of Eqs. (8.21) and (8.22) into (8.24) leads to

$$(\bar{\mathbf{b}}_a + \bar{\mathbf{b}}_b)^T(\mathbf{fF} + \mathbf{v}_T) = 0 \tag{8.25}$$

or $$\bar{\mathbf{b}}_c{}^T\mathbf{fF} + \bar{\mathbf{b}}_c{}^T\mathbf{v}_T = 0 \tag{8.26}$$

where $$\bar{\mathbf{b}}_c = \bar{\mathbf{b}}_a + \bar{\mathbf{b}}_b \tag{8.27}$$

The matrix $\bar{\mathbf{b}}_c$ can be interpreted as *internal-force* system representing statically equivalent element forces due to a unit force applied across the fictitious cut. This system requires no external reaction forces, and it may therefore be described as a self-equilibrating force system. It need not be associated with any real physical cuts, since the cuts are introduced only as a means of identifying the location of the unit force.

Any self-equilibrating force system may be thought of as being introduced as the result of self-straining action, which incidentally is possible only in a redundant structure. For example, if in the two-dimensional truss in Fig. 8.4 all elements (pin-jointed bars) are assembled to form the truss structure with the exception of the diagonal member 1,4, which is found to be too short, then by stretching this member to the required length (by pre-tension), attaching it to the structure, and releasing the pre-tension, a self-equilibrating force system will have been induced. Naturally, the amount of lack of fit can be adjusted to make the final value of the element force equal to a unit value, and the internal-force distribution will then be exactly represented by the matrix $\mathfrak{b}_c$.

The concept of self-equilibrating force systems can be applied to any redundant structure. In a statically determinate structure, self-equilibrating systems are not possible, since any cuts reduce the structure to a mechanism. Another example of a self-equilibrating system is shown in Fig. 8.6, where a unit internal force is introduced in the flange of a box-beam structure.

If the structure is redundant, we can introduce a number of cuts equal to the degree of redundancy, while the structure is reduced to a statically determinate system. For each cut Eq. (8.26) may be formulated, and thus for n cuts the following set of equations is obtained

$$
\begin{aligned}
&\mathfrak{b}_{X_1}{}^T(\mathbf{fF} + \mathbf{v}_T) = 0 \\
&\mathfrak{b}_{X_2}{}^T(\mathbf{fF} + \mathbf{v}_T) = 0 \\
&\cdots\cdots\cdots\cdots \\
&\mathfrak{b}_{X_n}{}^T(\mathbf{fF} + \mathbf{v}_T) = 0
\end{aligned}
\tag{8.28}
$$

where the second subscript with $\mathfrak{b}$ identifies the unit internal load on the structure. The above equations can be combined into a single matrix equation

$$\mathfrak{b}_X{}^T\mathbf{fF} + \mathbf{b}_X{}^T\mathbf{v}_T = \mathbf{0} \tag{8.29}$$

or $$\mathbf{b}_X{}^T\mathbf{v} = \mathbf{0} \tag{8.29a}$$

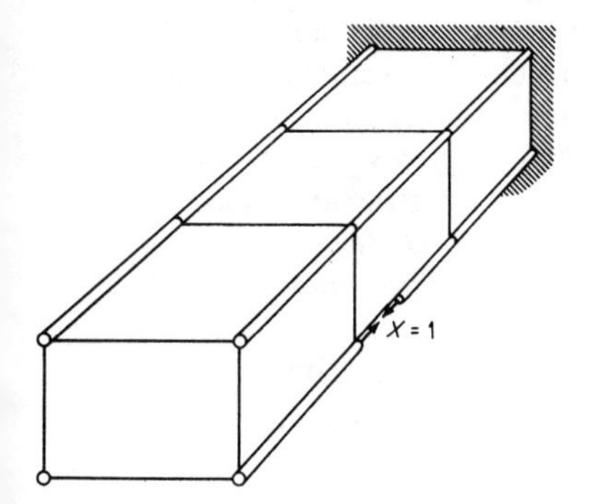

FIG. 8.6 Unit self-equilibrating force system in a box-beam structure.

where $\quad \mathbf{b}_X = [\bar{\mathbf{b}}_{X_1} \quad \bar{\mathbf{b}}_{X_2} \quad \cdots \quad \bar{\mathbf{b}}_{X_n}]$ (8.30)

If the cut structure is subjected to a set of externally applied forces and temperature distribution T, the element forces can be calculated directly from equations of statics (equations of equilibrium). The element forces are naturally proportional to the applied loading, and hence

$$\mathbf{F}_{\text{cut structure}} = \mathbf{b}_0\mathbf{P} \tag{8.31}$$

where $\quad \mathbf{P} = \{P_1 \quad P_2 \quad \cdots\}$ (8.32)

and $\mathbf{b}_0$ is a rectangular matrix in which columns represent element forces due to $P_1 = 1$, $P_2 = 1$, . . . , respectively. In order to preserve continuity of deformations forces X_1, X_2, . . . must be applied across the cuts, and the magnitudes of these forces will depend on the amount of "gap" across the cut caused by the external loading and temperature. The actual element forces in the uncut structure must therefore be given by

$$\mathbf{F} = \mathbf{b}_0\mathbf{P} + \mathbf{b}_X\mathbf{X} = [\mathbf{b}_0 \quad \mathbf{b}_X]\begin{bmatrix}\mathbf{P} \\ \mathbf{X}\end{bmatrix} \tag{8.33}$$

Substituting now Eq. (8.33) into (8.29), we have

$$\mathbf{b}_X{}^T\mathbf{f}\mathbf{b}_0\mathbf{P} + \mathbf{b}_X{}^T\mathbf{f}\mathbf{b}_X\mathbf{X} + \mathbf{b}_X{}^T\mathbf{v}_T = \mathbf{0} \tag{8.34}$$

Equation (8.34) may be interpreted as the compatibility equation for a statically indeterminate structure. The unknown forces $\mathbf{X}$ across the cuts may be determined directly from Eq. (8.34) by premultiplying this equation by $(\mathbf{b}_X{}^T\mathbf{f}\mathbf{b}_X)^{-1}$. This leads to

$$\mathbf{X} = -(\mathbf{b}_X{}^T\mathbf{f}\mathbf{b}_X)^{-1}(\mathbf{b}_X{}^T\mathbf{f}\mathbf{b}_0\mathbf{P} + \mathbf{b}_X{}^T\mathbf{v}_T) \tag{8.35}$$

8.3 MATRIX FORMULATION OF THE FORCE ANALYSIS: JORDANIAN ELIMINATION TECHNIQUE

The first step in the force method of analysis is the formulation of equilibrium equations. The components of the element forces in the directions of all the degrees of freedom at a node joint of the idealized structure are algebraically summed up and then equated to the corresponding components of the externally applied forces. This is carried out for all joints, including those used to establish either a rigid frame of reference or structure supports. The structure reactions, either statically determinate or redundant, are entered into the equations as internal forces, i.e., as reaction complements. Since this is a relatively new concept in structural analysis,[278] the reaction complements are illustrated in Fig. 8.7 for a pinned joint attached to a rigid foundation.

The equilibrium equations can be combined into a single matrix equation

$$\mathbf{n}_F\mathbf{F} + \mathbf{n}_R\mathbf{R} = \mathbf{P} \tag{8.36}$$

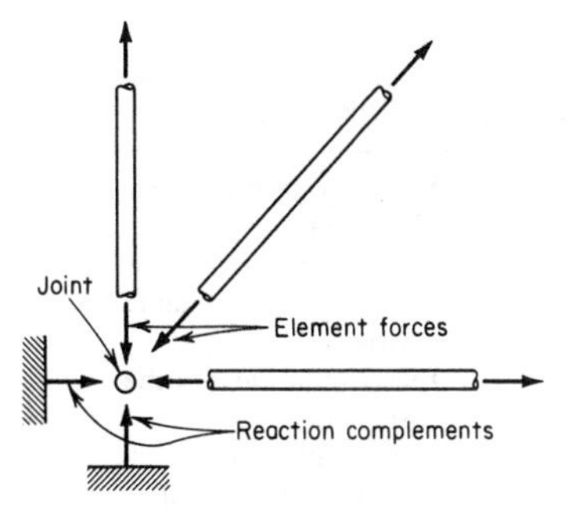

FIG. 8.7 Reaction complements.

where $\mathbf{n}_F$ and $\mathbf{n}_R$ are rectangular matrices whose coefficients are the direction cosines used in resolving the element forces **F** and reaction complements **R**, and **P** is a column matrix of external forces applied in the directions of the joint degrees of freedom. Equation (8.36) may be written alternatively as

$$\bar{\mathbf{n}}\bar{\mathbf{F}} = \mathbf{P} \tag{8.37}$$

where $$\bar{\mathbf{F}} = \begin{bmatrix} \mathbf{F} \\ \mathbf{R} \end{bmatrix} \tag{8.38}$$

and $$\bar{\mathbf{n}} = [\mathbf{n}_F \quad \mathbf{n}_R] \tag{8.39}$$

If the structure is statically determinate, the element forces **F** and reaction complements **R** can be found directly from Eq. (8.37). For this special case

$$\bar{\mathbf{F}} = \bar{\mathbf{n}}^{-1}\mathbf{P} \tag{8.40}$$

For statically indeterminate structures the matrix $\bar{\mathbf{n}}$ is singular with the number of columns greater than the number of rows, and the equations of equilibrium are not sufficient to determine the forces $\bar{\mathbf{F}}$. The required additional equations are supplied by the compatibility conditions, but the detailed discussion of these equations will be deferred in order to demonstrate first the formulation of Eq. (8.36) from the equilibrium equations used in the displacement method (Chap. 6).

Mathematically the so-called statical indeterminacy of an idealized structure may be defined in terms of the rank of the matrix $\bar{\mathbf{n}}$ in Eq. (8.37). This is illustrated in Table 8.1 for the various possible cases arising when the matrix $\bar{\mathbf{n}}$ is formulated.

From Table 8.1 it follows that the statical indeterminacy (or degree of redundancy) of a structure is $n - m$, with the restriction that the rank of the matrix $\bar{\mathbf{n}}$ is equal to m. If this condition is not satisfied, we must look for errors either in assembling the structural members or in the location of assigned reactions. In general, if $m > n$ the structure is a mechanism; however, if

TABLE 8.1 STATICAL INDETERMINACY

Case	Order $m \times n$ of matrix $\bar{\mathbf{n}}$	Rank r of matrix $\bar{\mathbf{n}}$	Type of structure
1	$m = n$	$r = m$	Statically determinate structure
2	$m < n$	$r = m$	Statically indeterminate structure (redundant structure)
3	$m \leq n$	$r < m$	Either some parts of the structure form a mechanism, or rigid-body degrees of freedom are not constrained properly by reactions
4	$m > n$	$r \leq n$	

external loads are applied only in certain directions, it is permissible to delete certain rows from the equation of equilibrium (8.37) to make $m \leq n$ in $\bar{\mathbf{n}}$. Examples of all these cases will be given in Sec. 8.4.

We now turn to the formulation of the matrix $\bar{\mathbf{n}}$. To obtain this matrix we consider first the relationship between the element forces in the displacement and force methods expressed by

$$\mathbf{S}^{(i)} = \mathbf{B}^{(i)}\mathbf{F}^{(i)} \tag{8.41}$$

In this equation both sets of element forces $\mathbf{S}^{(i)}$ and $\mathbf{F}^{(i)}$ are based on a local coordinate system; however, in order to form the equations of equilibrium we must use the element forces $\bar{\mathbf{S}}^{(i)}$ (see Chap. 6) given by

$$\bar{\mathbf{S}}^{(i)} = (\boldsymbol{\lambda}^{(i)})^T\mathbf{S}^{(i)} \tag{8.42}$$

Hence, using Eqs. (8.41) and (8.42), we have

$$\bar{\mathbf{S}}^{(i)} = \bar{\mathbf{B}}^{(i)}\mathbf{F}^{(i)} \tag{8.43}$$

where $$\bar{\mathbf{B}}^{(i)} = (\boldsymbol{\lambda}^{(i)})^T\mathbf{B}^{(i)} \tag{8.44}$$

Equation (8.44) can be written collectively for all structural elements as

$$\bar{\mathbf{S}} = \bar{\mathbf{B}}\mathbf{F} \tag{8.45}$$

where $$\bar{\mathbf{S}} = \{\bar{\mathbf{S}}^{(1)} \quad \bar{\mathbf{S}}^{(2)} \quad \cdots \quad \bar{\mathbf{S}}^{(i)} \quad \cdots\} \tag{8.46}$$

$$\bar{\mathbf{B}} = \lfloor\bar{\mathbf{B}}^{(1)} \quad \bar{\mathbf{B}}^{(2)} \quad \cdots \quad \bar{\mathbf{B}}^{(i)} \quad \cdots\rfloor \tag{8.47}$$

$$\mathbf{F} = \{\mathbf{F}^{(1)} \quad \mathbf{F}^{(2)} \quad \cdots \quad \mathbf{F}^{(i)} \quad \cdots\} \tag{8.48}$$

Using the results of Chap. 6, we can write the equations of equilibrium as

$$\mathbf{A}^T\bar{\mathbf{S}} + \mathbf{n}_R\mathbf{R} = \mathbf{P} \tag{8.49}$$

where the term $\mathbf{n}_R\mathbf{R}$ is included to account for the reaction complements $\mathbf{R}$ resolved in the directions of the externally applied loads $\mathbf{P}$. It should be noted that the forces $\mathbf{R}$ represent both the statically determinate and redundant reactions. Substituting Eq. (8.45) into (8.49), we obtain

$$\mathbf{n}_F\mathbf{F} + \mathbf{n}_R\mathbf{R} = \mathbf{P} \tag{8.36}$$

where $$\mathbf{n}_F = \mathbf{A}^T\bar{\mathbf{B}} \tag{8.50}$$

The element forces $\mathbf{F}$ will now be partitioned symbolically into statically determinate forces $\mathbf{F}_0$ and redundant forces $\mathbf{F}_1$, that is,

$$\mathbf{F} = \begin{bmatrix} \mathbf{F}_0 \\ \mathbf{F}_1 \end{bmatrix} \tag{8.51}$$

Similarly, the reaction complements $\mathbf{R}$ will be represented by

$$\mathbf{R} = \begin{bmatrix} \mathbf{R}_0 \\ \mathbf{R}_1 \end{bmatrix} \tag{8.52}$$

where $\mathbf{R}_0$ and $\mathbf{R}_1$ represent statically determinate and redundant reaction complements, respectively. Using Eqs. (8.36), (8.51), and (8.52), we obtain

$$\begin{bmatrix} \mathbf{n}_1 & \mathbf{n}_2 & \mathbf{n}_3 & \mathbf{n}_4 \end{bmatrix} \begin{bmatrix} \mathbf{F}_0 \\ \mathbf{F}_1 \\ \mathbf{R}_0 \\ \mathbf{R}_1 \end{bmatrix} = \mathbf{P} \tag{8.53}$$

where $$\begin{bmatrix} \mathbf{n}_1 & \mathbf{n}_2 \end{bmatrix} = \mathbf{n}_F \tag{8.54}$$

and $$\begin{bmatrix} \mathbf{n}_3 & \mathbf{n}_4 \end{bmatrix} = \mathbf{n}_R \tag{8.55}$$

Collecting the statically determinate and redundant forces into separate matrices, so that

$$\bar{\mathbf{F}}_0 = \begin{bmatrix} \mathbf{F}_0 \\ \mathbf{R}_0 \end{bmatrix} \tag{8.56}$$

$$\bar{\mathbf{F}}_1 = \begin{bmatrix} \mathbf{F}_1 \\ \mathbf{R}_1 \end{bmatrix} = \mathbf{X} \tag{8.57}$$

we can transform the equilibrium equation (8.53) into

$$[\mathbf{n}_0 \quad \mathbf{n}_X]\begin{bmatrix}\bar{\mathbf{F}}_0 \\ \mathbf{X}\end{bmatrix} = \mathbf{P} \tag{8.58}$$

where $\mathbf{n}_0 = [\mathbf{n}_1 \quad \mathbf{n}_3]$ (8.59)

and $\mathbf{n}_X = [\mathbf{n}_2 \quad \mathbf{n}_4]$ (8.60)

It should be noted that the rectangular matrix $[\mathbf{n}_0 \quad \mathbf{n}_X]$ is obtained from $\bar{\mathbf{n}}$ by interchanging columns in order to follow the sequence of forces in $\{\bar{\mathbf{F}}_0 \quad \mathbf{X}\}$.

Solving now Eq. (8.58) for $\bar{\mathbf{F}}_0$, we have

$$\bar{\mathbf{F}}_0 = [\mathbf{n}_0^{-1} \quad -\mathbf{n}_0^{-1}\mathbf{n}_X]\begin{bmatrix}\mathbf{P} \\ \mathbf{X}\end{bmatrix} \tag{8.61}$$

The element forces F and reaction complements R can also be written as

$$\bar{\mathbf{F}} = \begin{bmatrix}\bar{\mathbf{F}}_0 \\ \mathbf{X}\end{bmatrix} \tag{8.62}$$

From (8.61) and (8.62) it follows immediately that

$$\bar{\mathbf{F}} = \begin{bmatrix}\bar{\mathbf{F}}_0 \\ \mathbf{X}\end{bmatrix} = \left[\begin{array}{c:c}\mathbf{n}_0^{-1} & -\mathbf{n}_0^{-1}\mathbf{n}_X \\ \hdashline \mathbf{0} & \mathbf{I}\end{array}\right]\begin{bmatrix}\mathbf{P} \\ \mathbf{X}\end{bmatrix} \tag{8.63}$$

It was demonstrated in the previous section that the forces $\bar{\mathbf{F}}$ may be expressed as

$$\bar{\mathbf{F}} = \mathbf{b}_0\mathbf{P} + \mathbf{b}_X\mathbf{X} \tag{8.64}$$

Comparing Eqs. (8.63) and (8.64), we have that

$$\mathbf{b}_0 = \begin{bmatrix}\mathbf{n}_0^{-1} \\ \mathbf{0}\end{bmatrix} \tag{8.65}$$

$$\mathbf{b}_X = \begin{bmatrix}-\mathbf{n}_0^{-1}\mathbf{n}_X \\ \mathbf{I}\end{bmatrix} \tag{8.66}$$

The preceding analysis has therefore demonstrated that the $\mathbf{b}_0$ and $\mathbf{b}_X$ matrices can be generated from the equilibrium equations. The actual generation of $\mathbf{n}_0$ and $\mathbf{n}_X$ submatrices from $\bar{\mathbf{n}}$ and the matrices $\mathbf{b}_0$ and $\mathbf{b}_X$ is accomplished directly by the jordanian elimination technique applied to Eq. (8.37). In this method, an augmented matrix $[\mathbf{n}_0 \quad \mathbf{n}_X \mid \mathbf{I}]$ is formed, and the matrix is then premultiplied by $\mathbf{T}_1, \ldots, \mathbf{T}_m$ transformation matrices such that

$$\mathbf{T}_m \cdots \mathbf{T}_2\mathbf{T}_1[\mathbf{n}_0 \quad \mathbf{n}_X \mid \mathbf{I}] = [\mathbf{I} \quad \mathbf{n}_0^{-1}\mathbf{n}_X \mid \mathbf{n}_0^{-1}] \tag{8.67}$$

For practical application, it is usually preferable to form the augmented matrix as $[\mathbf{n}_0 \quad \mathbf{n}_X \mid \mathbf{P}]$ and to obtain

$$\mathbf{T}_m \cdots \mathbf{T}_2\mathbf{T}_1[\mathbf{n}_0 \quad \mathbf{n}_X \mid \mathbf{P}] = [\mathbf{I} \quad \mathbf{n}_0^{-1}\mathbf{n}_X \mid \mathbf{n}_0^{-1}\mathbf{P}] \tag{8.68}$$

This method has the advantage that the augmented matrix is of smaller size and also that it generates the matrix product $\mathbf{n}_0^{-1}\mathbf{P}$ required in

$$\mathbf{b}_0\mathbf{P} = \begin{bmatrix} \mathbf{n}_0^{-1}\mathbf{P} \\ \mathbf{0} \end{bmatrix} \tag{8.69}$$

The forces $\mathbf{X}$ selected by the jordanian elimination indicate automatically where the fictitious cuts are used on the structure. Thus, it is possible with this method to avoid the tedious task of the manual selection of redundancies, which has been customary in the past. This new technique was developed first by Denke and his associates.[76,78,80] Applications of the method will be illustrated in Secs. 8.4 and 8.5.

Once the matrices $\mathbf{b}_0$ and $\mathbf{b}_X$ are known, the redundant forces $\mathbf{X}$ can be determined from the compatibility equations (8.29*a*)

$$\mathbf{b}_X^T\bar{\mathbf{v}} = \mathbf{0} \tag{8.70}$$

where $$\bar{\mathbf{v}} = \bar{\mathbf{f}}\bar{\mathbf{F}} + \bar{\mathbf{v}}_T \tag{8.71}$$

$$\bar{\mathbf{f}} = \begin{bmatrix} \mathbf{f} & \mathbf{0} \\ \mathbf{0} & \mathbf{f}_R \end{bmatrix} \tag{8.72}$$

$$\bar{\mathbf{v}}_T = \begin{bmatrix} \mathbf{v}_T \\ \mathbf{v}_{TR} \end{bmatrix} \tag{8.73}$$

The matrix $\bar{\mathbf{f}}$ also includes the flexibility $\mathbf{f}_R$ of the reaction supports. In most practical applications, however, the supports are nonyielding (rigid), and hence $\mathbf{f}_R = \mathbf{0}$. Similarly, the thermal (or initial) deformation matrix $\bar{\mathbf{v}}_T$ includes thermal deformations of the supports represented by the submatrix $\mathbf{v}_{TR}$.

Substituting Eqs. (8.64) and (8.71) into (8.70), we have

$$\mathbf{b}_X^T\bar{\mathbf{f}}(\mathbf{b}_0\mathbf{P} + \mathbf{b}_X\mathbf{X}) + \mathbf{b}_X^T\bar{\mathbf{v}}_T = \mathbf{0} \tag{8.74}$$

Hence $\mathbf{X} = -(\mathbf{b}_X{}^T\bar{\mathbf{f}}\mathbf{b}_X)^{-1}(\mathbf{b}_X{}^T\bar{\mathbf{f}}\mathbf{b}_0\mathbf{P} + \mathbf{b}_X{}^T\bar{\mathbf{v}}_T)$ (8.75)

The redundant forces **X** can then be substituted into Eq. (8.64) to determine the element forces **F** and reaction complements **R**.

8.4 MATRIX FORCE ANALYSIS OF A PIN-JOINTED TRUSS

STATICALLY INDETERMINATE STRUCTURE; REDUNDANT REACTIONS

As a first example of the force method we shall analyze the two-dimensional truss shown in Fig. 8.8. The truss is loaded by a vertical force $P = 1{,}000$ lb at node 1. The supports are nonyielding, and all members are kept at the original temperature at which the structure was assembled with the exception of member 3, which is at a temperature $T°$ higher than other members. This example is the same we analyzed previously in Chap. 6 using the matrix displacement method. The relevant geometric data can be found in Table 6.1. As before, Young's modulus will be taken as 10×10^6 lb/in.^{2}

The first step in the force method of analysis is the formulation of the equations of equilibrium in terms of the element forces $\mathbf{F}^{(i)}$. As explained in Sec. 8.3, this may be accomplished using transformation matrices $\boldsymbol{\lambda}^{(i)}$ and $\mathbf{B}^{(i)}$. For bar elements these matrices are given by

$$\boldsymbol{\lambda}^{(i)} = \begin{bmatrix} l_{pq} & m_{pq} & 0 & 0 \\ 0 & 0 & l_{pq} & m_{pq} \end{bmatrix}$$

and $$\mathbf{B}^{(i)} = \begin{bmatrix} -1 \\ 1 \end{bmatrix}$$

To form the matrix $\mathbf{n}_F$ in the equilibrium equations we also require $\bar{\mathbf{B}}^{(i)}$, matrices

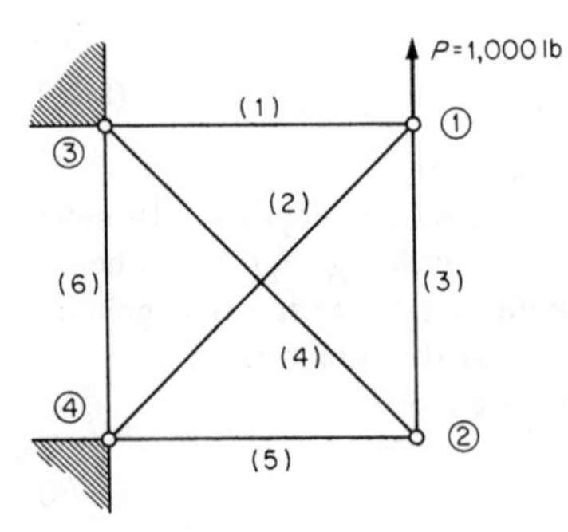

FIG. 8.8 Pin-jointed truss with redundant reactions; element and node numbering.

which are calculated from

$$\bar{\mathbf{B}}^{(i)} = (\boldsymbol{\lambda}^{(i)})^T\mathbf{B}^{(i)}$$

The transformation matrices $\boldsymbol{\lambda}^{(i)}$ have already been determined for this structure in Sec. 6.9. Hence

$$\bar{\mathbf{B}}^{(1)} = (\boldsymbol{\lambda}^{(1)})^T\mathbf{B}^{(1)} = \begin{matrix}5\\6\\1\\2\end{matrix}\begin{bmatrix}1 & 0\\0 & 0\\0 & 1\\0 & 0\end{bmatrix}\begin{bmatrix}-1\\1\end{bmatrix} = \begin{matrix}5\\6\\1\\2\end{matrix}\begin{bmatrix}-1\\0\\1\\0\end{bmatrix}$$

$$\bar{\mathbf{B}}^{(2)} = (\boldsymbol{\lambda}^{(2)})^T\mathbf{B}^{(2)} = \begin{matrix}7\\8\\1\\2\end{matrix}\begin{bmatrix}c & 0\\c & 0\\0 & c\\0 & c\end{bmatrix}\begin{bmatrix}-1\\1\end{bmatrix} = \begin{matrix}7\\8\\1\\2\end{matrix}\begin{bmatrix}-c\\-c\\c\\c\end{bmatrix}$$

$$\bar{\mathbf{B}}^{(3)} = (\boldsymbol{\lambda}^{(3)})^T\mathbf{B}^{(3)} = \begin{matrix}1\\2\\3\\4\end{matrix}\begin{bmatrix}0 & 0\\-1 & 0\\0 & 0\\0 & -1\end{bmatrix}\begin{bmatrix}-1\\1\end{bmatrix} = \begin{matrix}1\\2\\3\\4\end{matrix}\begin{bmatrix}0\\1\\0\\-1\end{bmatrix}$$

$$\bar{\mathbf{B}}^{(4)} = (\boldsymbol{\lambda}^{(4)})^T\mathbf{B}^{(4)} = \begin{matrix}5\\6\\3\\4\end{matrix}\begin{bmatrix}c & 0\\-c & 0\\0 & c\\0 & -c\end{bmatrix}\begin{bmatrix}-1\\1\end{bmatrix} = \begin{matrix}5\\6\\3\\4\end{matrix}\begin{bmatrix}-c\\c\\c\\-c\end{bmatrix}$$

$$\bar{\mathbf{B}}^{(5)} = (\boldsymbol{\lambda}^{(5)})^T\mathbf{B}^{(5)} = \begin{matrix}7\\8\\3\\4\end{matrix}\begin{bmatrix}1 & 0\\0 & 0\\0 & 1\\0 & 0\end{bmatrix}\begin{bmatrix}-1\\1\end{bmatrix} = \begin{matrix}7\\8\\3\\4\end{matrix}\begin{bmatrix}-1\\0\\1\\0\end{bmatrix}$$

$$\bar{\mathbf{B}}^{(6)} = (\boldsymbol{\lambda}^{(6)})^T\mathbf{B}^{(6)} = \begin{matrix}5\\6\\7\\8\end{matrix}\begin{bmatrix}0 & 0\\-1 & 0\\0 & 0\\0 & -1\end{bmatrix}\begin{bmatrix}-1\\1\end{bmatrix} = \begin{matrix}5\\6\\7\\8\end{matrix}\begin{bmatrix}0\\1\\0\\-1\end{bmatrix}$$

where the row numbers in $(\boldsymbol{\lambda}^{(i)})^T$ and $\bar{\mathbf{B}}^{(i)}$ indicate the directions of displacements (degrees of freedom) specified in Fig. 8.9. Also, as before, $c = \sqrt{2}/2$.

The row numbers in $\bar{\mathbf{B}}^{(i)}$ matrices identify row numbers in the matrix $\mathbf{n}_F$. Hence, there is no need to proceed with the formal derivation using $\mathbf{n}_F = \mathbf{A}^T\bar{\mathbf{B}}$, and we can set up the equilibrium equations directly as shown below.

$$\begin{matrix}1\\2\\3\\4\\5\\6\\7\\8\end{matrix}\begin{bmatrix} 1 & c & 0 & 0 & 0 & 0 \\ 0 & c & 1 & 0 & 0 & 0 \\ 0 & 0 & 0 & c & 1 & 0 \\ 0 & 0 & -1 & -c & 0 & 0 \\ -1 & 0 & 0 & -c & 0 & 0 \\ 0 & 0 & 0 & c & 0 & 1 \\ 0 & -c & 0 & 0 & -1 & 0 \\ 0 & -c & 0 & 0 & 0 & -1 \end{bmatrix} \begin{bmatrix} F_1 \\ F_2 \\ F_3 \\ F_4 \\ F_5 \\ F_6 \end{bmatrix} + \begin{bmatrix} 0 & 0 & 0 & 0 \\ 0 & 0 & 0 & 0 \\ 0 & 0 & 0 & 0 \\ 0 & 0 & 0 & 0 \\ 0 & 0 & 1 & 0 \\ 0 & 0 & 0 & 1 \\ 1 & 0 & 0 & 0 \\ 0 & 1 & 0 & 0 \end{bmatrix} \begin{bmatrix} R_1 \\ R_2 \\ R_3 \\ R_4 \end{bmatrix} = \begin{bmatrix} P_1 \\ P_2 \\ P_3 \\ P_4 \\ P_5 \\ P_6 \\ P_7 \\ P_8 \end{bmatrix}$$

The location and directions of the reaction complements $R_1, \ldots, R_4$ are indicated in Fig. 8.9.

Next, we form the augmented matrix $[\bar{\mathbf{n}} \mid \mathbf{P}]$; however, for ease of presentation we shall take $P_2 = 1$ lb and introduce the factor 10^3 only in the final solution. Thus, the augmented matrix becomes

$$\begin{matrix} \\ 1\\2\\3\\4\\5\\6\\7\\8\end{matrix}\begin{matrix} 1 & 2 & 3 & 4 & 5 & 6 & 7 & 8 & 9 & 10 & 11 \\ \hline 1 & c & 0 & 0 & 0 & 0 & 0 & 0 & 0 & 0 & 0 \\ 0 & c & 1 & 0 & 0 & 0 & 0 & 0 & 0 & 0 & 1 \\ 0 & 0 & 0 & c & 1 & 0 & 0 & 0 & 0 & 0 & 0 \\ 0 & 0 & -1 & -c & 0 & 0 & 0 & 0 & 0 & 0 & 0 \\ -1 & 0 & 0 & -c & 0 & 0 & 0 & 0 & 1 & 0 & 0 \\ 0 & 0 & 0 & c & 0 & 1 & 0 & 0 & 0 & 1 & 0 \\ 0 & -c & 0 & 0 & -1 & 0 & 1 & 0 & 0 & 0 & 0 \\ 0 & -c & 0 & 0 & 0 & -1 & 0 & 1 & 0 & 0 & 0 \end{matrix}$$

It should be noted that the column sequence in the above matrix can be changed provided the identifying column numbers are retained in the process. The transformation matrix $\mathbf{T}_1$ in the jordanian elimination technique (see Appendix A) for the first row can be applied to either column 1 or 2. Noting, however,

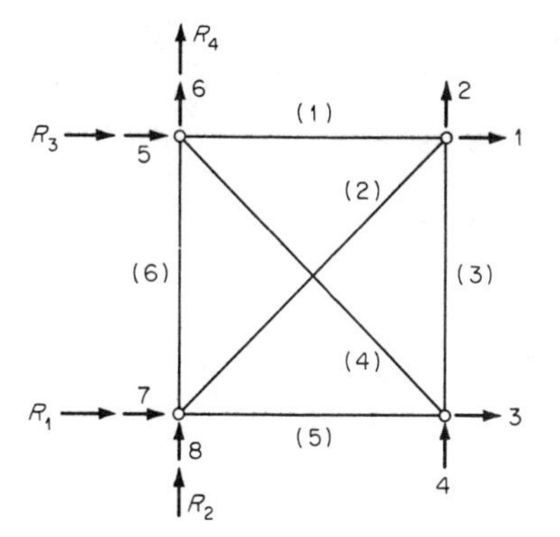

FIG. 8.9 Numbering for elements, degrees of freedom, and reaction complements.

that the column selected will correspond to the statically determinate force (either $\mathbf{F}_0$ or $\mathbf{R}_0$) in the $\bar{\mathbf{n}}$ matrix, the general rule may be established that it is preferable to select the element force which is the most significant in resisting the external force in the direction which the particular row represents. The practical rule, therefore, is to select from the columns corresponding to the element forces $\mathbf{F}$ a column with the numerically largest coefficient in a given row; however, if all coefficients are equal to zero the selection is carried out from the columns corresponding to the reaction-force complements $\mathbf{R}$. It is clear that for row 1, the largest coefficient is in column 1. Hence, applying the transformation $\mathbf{T}_1$ to the augmented matrix, we have

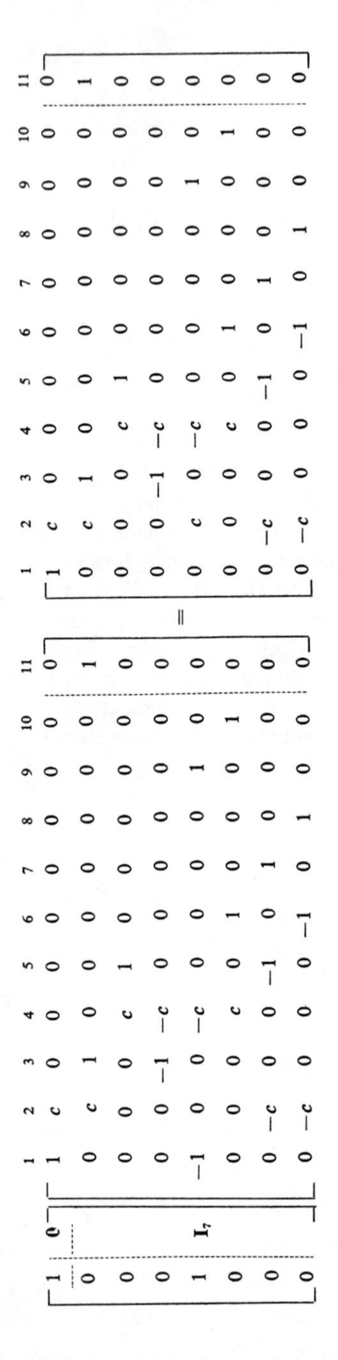

We note now that the largest coefficient in row 2 occurs in column 3. Therefore, interchanging columns 2 and 3 we can proceed with the premultiplication by the matrix $\mathbf{T}_2$. We observe that the column interchanges are not extended past column 10, that is, into the loading matrix $\mathbf{P}$, and the subsequent three multiplications in the jordanian elimination technique are as follows:

$$
\left[\begin{array}{c:c:c}
1 & 0 & \mathbf{0} \\ \hdashline
0 & 1 & \mathbf{0} \\ \hdashline
 & 0 & \\
 & 1 & \\
\mathbf{0} & 0 & \mathbf{I}_6 \\
 & 0 & \\
 & 0 & \\
 & 0 &
\end{array}\right]
\begin{array}{c}
\begin{array}{cccccccccc:c}
1 & 3 & 2 & 4 & 5 & 6 & 7 & 8 & 9 & 10 & 11
\end{array} \\
\left[\begin{array}{cccccccccc:c}
1 & 0 & c & 0 & 0 & 0 & 0 & 0 & 0 & 0 & 0 \\
0 & 1 & c & 0 & 0 & 0 & 0 & 0 & 0 & 0 & 1 \\
0 & 0 & 0 & c & 1 & 0 & 0 & 0 & 0 & 0 & 0 \\
0 & -1 & 0 & -c & 0 & 0 & 0 & 0 & 0 & 0 & 0 \\
0 & 0 & c & -c & 0 & 0 & 0 & 0 & 1 & 0 & 0 \\
0 & 0 & 0 & c & 0 & 1 & 0 & 0 & 0 & 1 & 0 \\
0 & 0 & -c & 0 & -1 & 0 & 1 & 0 & 0 & 0 & 0 \\
0 & 0 & -c & 0 & 0 & -1 & 0 & 1 & 0 & 0 & 0
\end{array}\right]
\end{array}
=
\begin{array}{c}
\begin{array}{cccccccccc:c}
1 & 3 & 2 & 4 & 5 & 6 & 7 & 8 & 9 & 10 & 11
\end{array} \\
\left[\begin{array}{cccccccccc:c}
1 & 0 & c & 0 & 0 & 0 & 0 & 0 & 0 & 0 & 0 \\
0 & 1 & c & 0 & 0 & 0 & 0 & 0 & 0 & 0 & 1 \\
0 & 0 & 0 & c & 1 & 0 & 0 & 0 & 0 & 0 & 0 \\
0 & 0 & c & -c & 0 & 0 & 0 & 0 & 0 & 0 & 1 \\
0 & 0 & c & -c & 0 & 0 & 0 & 0 & 1 & 0 & 0 \\
0 & 0 & 0 & c & 0 & 1 & 0 & 0 & 0 & 1 & 0 \\
0 & 0 & -c & 0 & -1 & 0 & 1 & 0 & 0 & 0 & 0 \\
0 & 0 & -c & 0 & 0 & -1 & 0 & 1 & 0 & 0 & 0
\end{array}\right]
\end{array}
$$

$$
\left[\begin{array}{c:c:c}
\mathbf{I}_2 & \begin{array}{c}0\\0\end{array} & \mathbf{0} \\ \hdashline
\mathbf{0} & 1 & \mathbf{0} \\ \hdashline
\mathbf{0} & \begin{array}{c}0\\0\\0\\1\\0\end{array} & \mathbf{I}_5
\end{array}\right]
\begin{array}{c}
\begin{array}{cccccccccc:c}
1 & 3 & 5 & 4 & 2 & 6 & 7 & 8 & 9 & 10 & 11
\end{array}\\
\left[\begin{array}{cccccccccc:c}
1 & 0 & 0 & 0 & c & 0 & 0 & 0 & 0 & 0 & 0 \\
0 & 1 & 0 & 0 & c & 0 & 0 & 0 & 0 & 0 & 1 \\
0 & 0 & 1 & c & 0 & 0 & 0 & 0 & 0 & 0 & 0 \\
0 & 0 & 0 & -c & c & 0 & 0 & 0 & 0 & 0 & 1 \\
0 & 0 & 0 & -c & c & 0 & 0 & 0 & 1 & 0 & 0 \\
0 & 0 & 0 & c & 0 & 1 & 0 & 0 & 0 & 1 & 0 \\
0 & 0 & -1 & 0 & -c & 0 & 1 & 0 & 0 & 0 & 0 \\
0 & 0 & 0 & 0 & -c & -1 & 0 & 1 & 0 & 0 & 0
\end{array}\right]
\end{array}
=
\begin{array}{c}
\begin{array}{cccccccccc:c}
1 & 3 & 5 & 4 & 2 & 6 & 7 & 8 & 9 & 10 & 11
\end{array}\\
\left[\begin{array}{cccccccccc:c}
1 & 0 & 0 & 0 & c & 0 & 0 & 0 & 0 & 0 & 0 \\
0 & 1 & 0 & 0 & c & 0 & 0 & 0 & 0 & 0 & 1 \\
0 & 0 & 1 & c & 0 & 0 & 0 & 0 & 0 & 0 & 0 \\
0 & 0 & 0 & -c & c & 0 & 0 & 0 & 0 & 0 & 1 \\
0 & 0 & 0 & -c & c & 0 & 0 & 0 & 1 & 0 & 0 \\
0 & 0 & 0 & c & 0 & 1 & 0 & 0 & 0 & 1 & 0 \\
0 & 0 & 0 & c & -c & 0 & 1 & 0 & 0 & 0 & 0 \\
0 & 0 & 0 & 0 & -c & -1 & 0 & 1 & 0 & 0 & 0
\end{array}\right]
\end{array}
$$

$$
\left[\begin{array}{c:c:c}
\mathbf{I}_3 & \begin{array}{c}0\\0\\1\end{array} & \mathbf{0} \\ \hdashline
\mathbf{0} & -\dfrac{1}{c} & \mathbf{0} \\ \hdashline
\mathbf{0} & \begin{array}{c}-1\\1\\1\\0\end{array} & \mathbf{I}_4
\end{array}\right]
\begin{array}{c}
\begin{array}{cccccccccc:c}
1 & 3 & 5 & 4 & 2 & 6 & 7 & 8 & 9 & 10 & 11
\end{array}\\
\left[\begin{array}{cccccccccc:c}
1 & 0 & 0 & 0 & c & 0 & 0 & 0 & 0 & 0 & 0 \\
0 & 1 & 0 & 0 & c & 0 & 0 & 0 & 0 & 0 & 1 \\
0 & 0 & 1 & c & 0 & 0 & 0 & 0 & 0 & 0 & 0 \\
0 & 0 & 0 & -c & c & 0 & 0 & 0 & 0 & 0 & 1 \\
0 & 0 & 0 & -c & c & 0 & 0 & 0 & 1 & 0 & 0 \\
0 & 0 & 0 & c & 0 & 1 & 0 & 0 & 0 & 1 & 0 \\
0 & 0 & 0 & c & -c & 0 & 1 & 0 & 0 & 0 & 0 \\
0 & 0 & 0 & 0 & -c & -1 & 0 & 1 & 0 & 0 & 0
\end{array}\right]
\end{array}
=
\begin{array}{c}
\begin{array}{cccccccccc:c}
1 & 3 & 5 & 4 & 2 & 6 & 7 & 8 & 9 & 10 & 11
\end{array}\\
\left[\begin{array}{cccccccccc:c}
1 & 0 & 0 & 0 & c & 0 & 0 & 0 & 0 & 0 & 0 \\
0 & 1 & 0 & 0 & c & 0 & 0 & 0 & 0 & 0 & 1 \\
0 & 0 & 1 & 0 & c & 0 & 0 & 0 & 0 & 0 & 1 \\
0 & 0 & 0 & 1 & -1 & 0 & 0 & 0 & 0 & 0 & -\dfrac{1}{c} \\
0 & 0 & 0 & 0 & 0 & 0 & 0 & 0 & 1 & 0 & -1 \\
0 & 0 & 0 & 0 & c & 1 & 0 & 0 & 0 & 1 & 1 \\
0 & 0 & 0 & 0 & 0 & 0 & 1 & 0 & 0 & 0 & 1 \\
0 & 0 & 0 & 0 & -c & -1 & 0 & 1 & 0 & 0 & 0
\end{array}\right]
\end{array}
$$

Interchanging columns 2 and 9, we note that the next transformation matrix $\mathbf{T}_5 = \mathbf{I}$, and consequently we may proceed directly to $\mathbf{T}_6$. Hence

$$
\left[\begin{array}{c:c:c}
 & 0 & \\
 & 0 & \\
\mathbf{I}_5 & 0 & \mathbf{0} \\
 & 0 & \\
 & 0 & \\
\hdashline
\mathbf{0} & 1 & \mathbf{0} \\
\hdashline
\mathbf{0} & 0 & \mathbf{I}_2 \\
 & 1 &
\end{array}\right]
\left[\begin{array}{cccccccccc:c}
\scriptstyle 1 & \scriptstyle 3 & \scriptstyle 5 & \scriptstyle 4 & \scriptstyle 9 & \scriptstyle 6 & \scriptstyle 7 & \scriptstyle 8 & \scriptstyle 2 & \scriptstyle 10 & \scriptstyle 11 \\
1 & 0 & 0 & 0 & 0 & 0 & 0 & 0 & c & 0 & 0 \\
0 & 1 & 0 & 0 & 0 & 0 & 0 & 0 & c & 0 & 1 \\
0 & 0 & 1 & 0 & 0 & 0 & 0 & 0 & c & 0 & 1 \\
0 & 0 & 0 & 1 & 0 & 0 & 0 & 0 & -1 & 0 & -\frac{1}{c} \\
0 & 0 & 0 & 0 & 1 & 0 & 0 & 0 & 0 & 0 & -1 \\
0 & 0 & 0 & 0 & 0 & 1 & 0 & 0 & c & 1 & 1 \\
0 & 0 & 0 & 0 & 0 & 0 & 1 & 0 & 0 & 0 & 1 \\
0 & 0 & 0 & 0 & 0 & -1 & 0 & 1 & -c & 0 & 0
\end{array}\right]
=
\left[\begin{array}{cccccccccc:c}
\scriptstyle 1 & \scriptstyle 3 & \scriptstyle 5 & \scriptstyle 4 & \scriptstyle 9 & \scriptstyle 6 & \scriptstyle 7 & \scriptstyle 8 & \scriptstyle 2 & \scriptstyle 10 & \scriptstyle 11 \\
1 & 0 & 0 & 0 & 0 & 0 & 0 & 0 & c & 0 & 0 \\
0 & 1 & 0 & 0 & 0 & 0 & 0 & 0 & c & 0 & 1 \\
0 & 0 & 1 & 0 & 0 & 0 & 0 & 0 & c & 0 & 1 \\
0 & 0 & 0 & 1 & 0 & 0 & 0 & 0 & -1 & 0 & -\frac{1}{c} \\
0 & 0 & 0 & 0 & 1 & 0 & 0 & 0 & 0 & 0 & -1 \\
0 & 0 & 0 & 0 & 0 & 1 & 0 & 0 & c & 1 & 1 \\
0 & 0 & 0 & 0 & 0 & 0 & 1 & 0 & 0 & 0 & 1 \\
0 & 0 & 0 & 0 & 0 & 0 & 0 & 1 & 0 & 1 & 1
\end{array}\right]
$$

This completes the jordanian elimination process, since it is clear that the remaining two transformation matrices $\mathbf{T}_7$ and $\mathbf{T}_8$ are unit matrices. The above equation was represented symbolically by Eq. (8.68). It should be noted that the column numbers identify the forces for the submatrices $\mathbf{n}_0$ and $\mathbf{n}_X$, which initially were combined into one rectangular matrix $\bar{\mathbf{n}}$. This, therefore, implies that forces 2 and 10 have been selected to represent redundant forces.

Using, therefore, Eqs. (8.65) and (8.66), we have

$$\mathbf{b}_0\mathbf{P} = \begin{matrix} 1 \\ 2 \\ 3 \\ 4 \\ 5 \\ 6 \\ 7 \\ 8 \\ 9 \\ 10 \end{matrix} \begin{bmatrix} 0 \\ 0 \\ 1 \\ \dfrac{-1}{c} \\ 1 \\ 1 \\ 1 \\ 1 \\ -1 \\ 0 \end{bmatrix} \times 10^3 \qquad \text{and} \qquad \mathbf{b}_X = \begin{matrix} \\ 1 \\ 2 \\ 3 \\ 4 \\ 5 \\ 6 \\ 7 \\ 8 \\ 9 \\ 10 \end{matrix} \begin{matrix} 2 & 10 \\ \begin{bmatrix} -c & 0 \\ 1 & 0 \\ -c & 0 \\ 1 & 0 \\ -c & 0 \\ -c & -1 \\ 0 & 0 \\ 0 & -1 \\ 0 & 0 \\ 0 & 1 \end{bmatrix} \end{matrix}$$

where the matrix product $\mathbf{b}_0\mathbf{P}$ has been evaluated for $P_2 = 1{,}000$ lb.

The flexibility matrix $\bar{\mathbf{f}}$ and thermal-deformation matrix $\bar{\mathbf{v}}_T$ for this example are given by

$$\bar{\mathbf{f}} = \begin{matrix} 1 & 2 & 3 & 4 & 5 & 6 & 7 & 8 & 9 & 10 \\ \lceil 2 & 4 & 2 & 4 & 2 & 2 & 0 & 0 & 0 & 0 \rfloor \end{matrix} \times 10^{-6} \text{ in./lb}$$

$$\text{and} \qquad \bar{\mathbf{v}}_T = \begin{matrix} 1 & 2 & 3 & 4 & 5 & 6 & 7 & 8 & 9 & 10 \\ \{0 & 0 & \alpha Tl & 0 & 0 & 0 & 0 & 0 & 0 & 0\} \end{matrix} \text{ in.} \qquad l = 20 \text{ in.}$$

In order to set up the compatibility equations (8.74) the following matrix products are then evaluated:

$$\mathbf{b}_X{}^T\bar{\mathbf{f}}\mathbf{b}_X = 2\begin{bmatrix} 4(1+c^2) & c \\ c & 1 \end{bmatrix} \times 10^{-6}$$

$$\mathbf{b}_X{}^T\bar{\mathbf{f}}\mathbf{b}_0\mathbf{P} = -2\begin{bmatrix} \dfrac{2+3c^2}{c} \\ 1 \end{bmatrix} \times 10^{-3}$$

$$\mathbf{b}_X{}^T\bar{\mathbf{v}}_T = \begin{bmatrix} -c\alpha Tl \\ 0 \end{bmatrix}$$

The unknown redundant forces $\mathbf{X}$ are calculated from

$$\mathbf{X} = -(\mathbf{b}_X{}^T\bar{\mathbf{f}}\mathbf{b}_X)^{-1}(\mathbf{b}_X{}^T\bar{\mathbf{f}}\mathbf{b}_0\mathbf{P} + \mathbf{b}_X{}^T\bar{\mathbf{v}}_T)$$

$$= -\frac{10^6}{2(4+3c^2)}\begin{bmatrix} 1 & -c \\ -c & 4(1+c^2) \end{bmatrix}\left\{-2\begin{bmatrix} \dfrac{2+3c^2}{c} \\ 1 \end{bmatrix} \times 10^{-3} + \begin{bmatrix} -c\alpha Tl \\ 0 \end{bmatrix}\right\}$$

$$= \frac{1}{(4+3c^2)}\begin{bmatrix} \dfrac{2(1+c^2)}{c} \\ 2+c^2 \end{bmatrix} \times 10^3 + \frac{10^6}{2(4+3c^2)}\begin{bmatrix} c \\ -c^2 \end{bmatrix}\alpha Tl$$

$$= \begin{bmatrix} \dfrac{6\sqrt{2}}{11} \\ \dfrac{5}{11} \end{bmatrix} \times 10^3 + \begin{bmatrix} \dfrac{\sqrt{2}}{22} \\ \dfrac{-1}{22} \end{bmatrix}\alpha Tl \times 10^6$$

The forces in the structure and the reaction complements are then determined from

$$\bar{\mathbf{F}} = \mathbf{b}_0\mathbf{P} + \mathbf{b}_X\mathbf{X}$$

$$= \begin{matrix} 1 \\ 2 \\ 3 \\ 4 \\ 5 \\ 6 \\ 7 \\ 8 \\ 9 \\ 10 \end{matrix}\begin{bmatrix} 0 \\ 0 \\ 1 \\ -\sqrt{2} \\ 1 \\ 1 \\ 1 \\ 1 \\ -1 \\ 0 \end{bmatrix} \times 10^3 + \begin{matrix} 1 \\ 2 \\ 3 \\ 4 \\ 5 \\ 6 \\ 7 \\ 8 \\ 9 \\ 10 \end{matrix}\begin{bmatrix} -\dfrac{\sqrt{2}}{2} & 0 \\ 1 & 0 \\ -\dfrac{\sqrt{2}}{2} & 0 \\ 1 & 0 \\ -\dfrac{\sqrt{2}}{2} & 0 \\ -\dfrac{\sqrt{2}}{2} & -1 \\ 0 & 0 \\ 0 & -1 \\ 0 & 0 \\ 0 & 1 \end{bmatrix}\left\{\begin{bmatrix} \dfrac{6\sqrt{2}}{11} \\ \dfrac{5}{11} \end{bmatrix} \times 10^3 + \begin{bmatrix} \dfrac{\sqrt{2}}{22} \\ \dfrac{-1}{22} \end{bmatrix}\alpha Tl \times 10^6\right\}$$

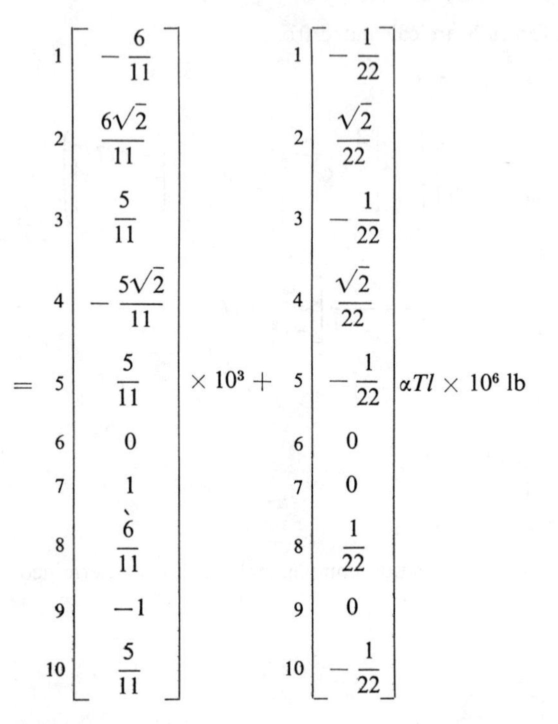

$$= \begin{matrix} 1 \\ 2 \\ 3 \\ 4 \\ 5 \\ 6 \\ 7 \\ 8 \\ 9 \\ 10 \end{matrix} \begin{bmatrix} -\frac{6}{11} \\ \frac{6\sqrt{2}}{11} \\ \frac{5}{11} \\ -\frac{5\sqrt{2}}{11} \\ \frac{5}{11} \\ 0 \\ 1 \\ \frac{6}{11} \\ -1 \\ \frac{5}{11} \end{bmatrix} \times 10^3 + \begin{matrix} 1 \\ 2 \\ 3 \\ 4 \\ 5 \\ 6 \\ 7 \\ 8 \\ 9 \\ 10 \end{matrix} \begin{bmatrix} -\frac{1}{22} \\ \frac{\sqrt{2}}{22} \\ -\frac{1}{22} \\ \frac{\sqrt{2}}{22} \\ -\frac{1}{22} \\ 0 \\ 0 \\ \frac{1}{22} \\ 0 \\ -\frac{1}{22} \end{bmatrix} \alpha T l \times 10^6 \text{ lb}$$

where the last two column matrices represent the element forces and reaction complements due to external load $P_2 = 1{,}000$ lb and the temperature respectively.

To find deflection in the direction of P_2 we may use the unit-load theorem. Hence, denoting this deflection by r_2, we have

$$r_2 = \mathbf{b}_0{}^T\bar{\mathbf{v}} = \mathbf{b}_0{}^T(\bar{\mathbf{f}}\bar{\mathbf{F}} + \mathbf{v}_T)$$
$$= \tfrac{60}{11} \times 10^{-3} + \tfrac{5}{11}\alpha T l = \tfrac{60}{11} \times 10^{-3} + \tfrac{100}{11}\alpha T \qquad \text{in.}$$

which agrees with the previous results obtained by the displacement method.

We shall consider next the application of the jordanian technique to other cases. Special consideration will be given to errors made in either the structural assembly or in the selection of support reactions to resist externally applied loads.

STATICALLY DETERMINATE REACTIONS

For this example, consider the structure just analyzed but supported only by three reactions, as shown in Fig. 8.10. We retain the same numbering scheme

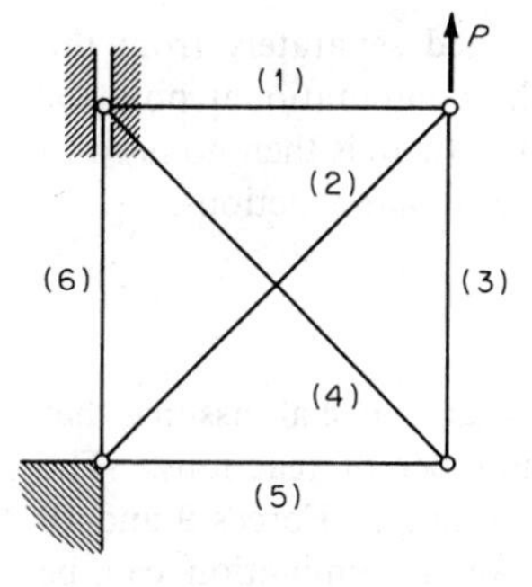

FIG. 8.10 Redundant pin-jointed truss with statically determinate reactions.

for the forces, and the augmented matrix is reduced to

$$
\begin{array}{c|ccccccccc|c}
 & 1 & 3 & 5 & 4 & 9 & 6 & 7 & 8 & 2 & 11 \\
\hline
1 & 1 & 0 & 0 & 0 & 0 & 0 & 0 & 0 & c & 0 \\
2 & 0 & 1 & 0 & 0 & 0 & 0 & 0 & 0 & c & 1 \\
3 & 0 & 0 & 1 & 0 & 0 & 0 & 0 & 0 & c & 1 \\
4 & 0 & 0 & 0 & 1 & 0 & 0 & 0 & 0 & -1 & \frac{-1}{c} \\
5 & 0 & 0 & 0 & 0 & 1 & 0 & 0 & 0 & 0 & -1 \\
6 & 0 & 0 & 0 & 0 & 0 & 1 & 0 & 0 & c & 1 \\
7 & 0 & 0 & 0 & 0 & 0 & 0 & 1 & 0 & 0 & 1 \\
8 & 0 & 0 & 0 & 0 & 0 & 0 & 0 & 1 & 0 & 1
\end{array}
$$

Using the above matrix, we obtain (for $P_2 = 1$)

$$
\mathbf{b}_0\mathbf{P} = \begin{matrix} 1 \\ 2 \\ 3 \\ 4 \\ 5 \\ 6 \\ 7 \\ 8 \\ 9 \end{matrix}
\begin{bmatrix} 0 \\ 0 \\ 1 \\ \frac{-1}{c} \\ 1 \\ 1 \\ 1 \\ 1 \\ -1 \end{bmatrix}
\qquad
\mathbf{b}_X = \begin{matrix} 1 \\ 2 \\ 3 \\ 4 \\ 5 \\ 6 \\ 7 \\ 8 \\ 9 \end{matrix}
\begin{bmatrix} -c \\ 1 \\ -c \\ 1 \\ -c \\ -c \\ 0 \\ 0 \\ 0 \end{bmatrix}
$$

Naturally, forces 7, 8, and 9 could have been obtained separately from the equations of overall equilibrium; however, from the computational point of view, it is preferable to follow the general method, since there is then no need to differentiate between statically determinate and indeterminate reactions.

INSUFFICIENT REACTIONS

As an example, consider the structure in Fig. 8.11, where we shall assume that the analyst inadvertently provided an insufficient number of reactions. The numbering scheme is the same as in the previous examples. Forces 9 and 10 (reaction complements) will be deleted. The jordanian elimination can be carried out up to and including premultiplication by $\mathbf{T}_4$, which leads to

$$
\begin{array}{c|cccccccc:c}
 & 1 & 3 & 5 & 4 & 2 & 6 & 7 & 8 & 11 \\
\hline
1 & & 0 & 0 & 0 & c & 0 & 0 & 0 & 0 \\
2 & 0 & 1 & 0 & 0 & c & 0 & 0 & 0 & 1 \\
3 & 0 & 0 & 1 & 0 & c & 0 & 0 & 0 & 1 \\
4 & 0 & 0 & 0 & 1 & -1 & 0 & 0 & 0 & \dfrac{-1}{c} \\
5 & 0 & 0 & 0 & 0 & 0 & 0 & 0 & 0 & -1 \\
6 & 0 & 0 & 0 & 0 & c & 1 & 0 & 0 & 1 \\
7 & 0 & 0 & 0 & 0 & 0 & 0 & 1 & 0 & 1 \\
8 & 0 & 0 & 0 & 0 & -c & -1 & 0 & 1 & 0
\end{array}
$$

Since the fifth row of the submatrix $\bar{\mathbf{n}}$ after four eliminations consists of zeros, the matrix $\mathbf{T}_5$ cannot be constructed, and the jordanian elimination cannot be continued past this stage. The row of zeros indicates that proper reactions were not provided for the degree of freedom 5.

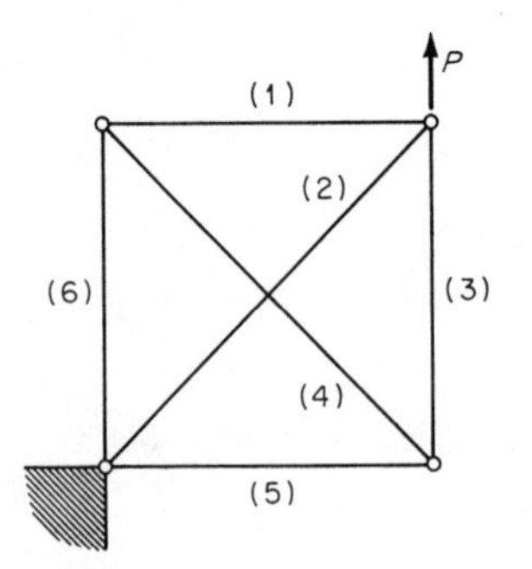

FIG. 8.11 Pin-jointed truss with insufficient reactions.

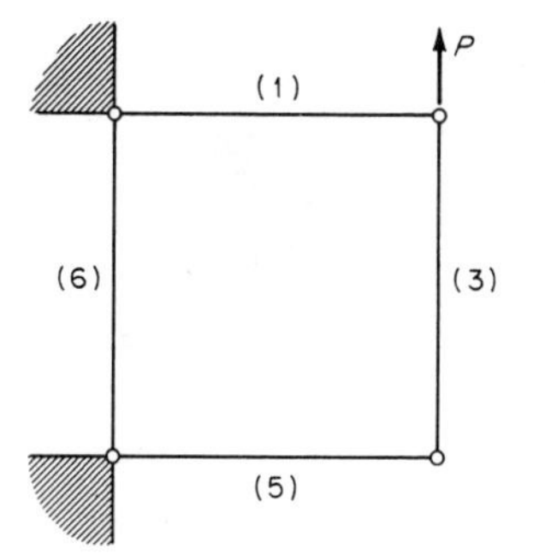

FIG. 8.12 Pin-jointed truss reduced to a mechanism.

STRUCTURE REDUCED TO A MECHANISM

Omission of the two diagonal members in the two-dimensional truss analyzed at the beginning of this section reduces the structure to a mechanism (see Fig. 8.12). Therefore, by deleting 2 and 4 it can be seen from the first example that the jordanian elimination can be carried out up to and including premultiplication by $\mathbf{T}_3$, which leads to

$$
\begin{array}{c} \\ 1 \\ 2 \\ 3 \\ 4 \\ 5 \\ 6 \\ 7 \\ 8 \end{array}
\begin{array}{c}
\begin{array}{ccccccccc} 1 & 3 & 5 & 6 & 7 & 8 & 9 & 10 & 11 \end{array} \\
\left[\begin{array}{cccccccc:c}
1 & 0 & 0 & 0 & 0 & 0 & 0 & 0 & 0 \\
0 & 1 & 0 & 0 & 0 & 0 & 0 & 0 & 1 \\
0 & 0 & 1 & 0 & 0 & 0 & 0 & 0 & 0 \\
0 & 0 & 0 & 0 & 0 & 0 & 0 & 0 & 1 \\
0 & 0 & 0 & 0 & 0 & 0 & 1 & 0 & 0 \\
0 & 0 & 0 & 1 & 0 & 0 & 0 & 1 & 0 \\
0 & 0 & 0 & 0 & 1 & 0 & 0 & 0 & 0 \\
0 & 0 & 0 & -1 & 0 & 1 & 0 & 0 & 0
\end{array}\right]
\end{array}
$$

At this stage the elimination process can no longer be continued, since the fourth row of the submatrix $\bar{\mathbf{n}}$ after three eliminations consists of zeros only. The degree of freedom 4 therefore requires either an additional structural member or an external reaction.

8.5 MATRIX FORCE ANALYSIS OF A CANTILEVER BEAM

As a second example of the force method we shall analyze a cantilever beam built-in at one end and simply supported at the other end (see Fig. 8.13). The same problem was analyzed by the displacement method in Sec. 6.10. As before, the beam length is $2l$, and the applied loading consists of forces P_3, P_4,

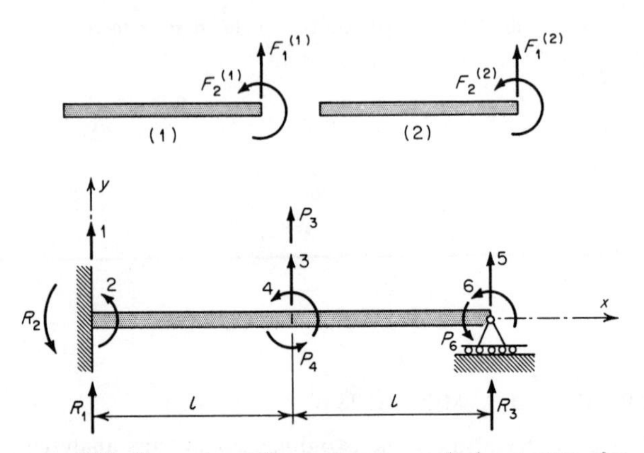

FIG. 8.13 Propped cantilever beam; numbering system for elements, degrees of freedom, and reaction complements.

and P_6. In addition to the external loads the beam is subjected to a temperature distribution varying across the beam section but constant along the length of the beam. The reaction complements are denoted by R_1, R_2, and R_3.

For the purpose of the analysis the beam is subdivided into two elements, and the selected element forces are shown in Fig. 8.13. It should be noted, however, that for beam elements other choices for the element forces are also possible, as explained in Sec. 7.2. For convenience in the subsequent analysis we introduce the following notation

$$\begin{bmatrix} F_1 \\ F_2 \\ F_3 \\ F_4 \end{bmatrix} = \begin{bmatrix} F_1^{(1)} \\ F_2^{(1)} \\ F_1^{(2)} \\ F_2^{(2)} \end{bmatrix}$$

in order to avoid repeated use of the superscripts with the element forces $\mathbf{F}$.

The transformation matrices $\mathbf{B}^{(i)}$ for the two elements are simply [see Eq. (7.23)]

$$\mathbf{B}^{(1)} = \begin{array}{c} \\ 1 \\ 2 \\ 3 \\ 4 \end{array} \begin{array}{c} \begin{array}{cc} 1 & 2 \end{array} \\ \begin{bmatrix} -1 & 0 \\ -l & -1 \\ 1 & 0 \\ 0 & 1 \end{bmatrix} \end{array} \qquad \mathbf{B}^{(2)} = \begin{array}{c} \\ 3 \\ 4 \\ 5 \\ 6 \end{array} \begin{array}{c} \begin{array}{cc} 3 & 4 \end{array} \\ \begin{bmatrix} -1 & 0 \\ -l & -1 \\ 1 & 0 \\ 0 & 1 \end{bmatrix} \end{array}$$

where the row numbers refer to the degrees of freedom (displacements) and the column numbers refer to the element forces.

Noting that the local and datum systems coincide in this example, we may proceed directly to set up equations of equilibrium using the $\mathbf{B}^{(i)}$ matrices. Hence it can be easily demonstrated that Eq. (8.36) becomes

$$\begin{array}{c} \\ 1\\2\\3\\4\\5\\6 \end{array}\overset{\begin{array}{cccc}1&2&3&4\end{array}}{\begin{bmatrix} -1 & 0 & 0 & 0 \\ -l & -1 & 0 & 0 \\ 1 & 0 & -1 & 0 \\ 0 & 1 & -l & -1 \\ 0 & 0 & 1 & 0 \\ 0 & 0 & 0 & 1 \end{bmatrix}} \begin{bmatrix} F_1 \\ F_2 \\ F_3 \\ F_4 \end{bmatrix} + \begin{array}{c} \\ 1\\2\\3\\4\\5\\6 \end{array}\overset{\begin{array}{ccc}1&2&3\end{array}}{\begin{bmatrix} 1 & 0 & 0 \\ 0 & 1 & 0 \\ 0 & 0 & 0 \\ 0 & 0 & 0 \\ 0 & 0 & 1 \\ 0 & 0 & 0 \end{bmatrix}} \begin{bmatrix} R_1 \\ R_2 \\ R_3 \end{bmatrix} = \begin{bmatrix} P_1 \\ P_2 \\ P_3 \\ P_4 \\ P_5 \\ P_6 \end{bmatrix}$$

Having obtained the equations of equilibrium, we may proceed with the jordanian elimination technique. The sequence of transformation operations on the augmented matrix from the equilibrium equations is

$$
\left[\begin{array}{c|c} -1 & \mathbf{0} \\ \hline -l & \\ 1 & \\ 0 & \mathbf{I}_5 \\ 0 & \\ 0 & \end{array}\right]
\begin{array}{c} \scriptstyle 1\;\;2\;\;3\;\;4\;\;5\;\;6\;\;7\;\;8\;\;9\;\;10 \\
\left[\begin{array}{ccccccc:ccc}
-1 & 0 & 0 & 0 & 1 & 0 & 0 & 0 & 0 & 0 \\
-l & -1 & 0 & 0 & 0 & 1 & 0 & 0 & 0 & 0 \\
1 & 0 & -1 & 0 & 0 & 0 & 0 & 1 & 0 & 0 \\
0 & 1 & -l & -1 & 0 & 0 & 0 & 0 & 1 & 0 \\
0 & 0 & 1 & 0 & 0 & 0 & 1 & 0 & 0 & 0 \\
0 & 0 & 0 & 1 & 0 & 0 & 0 & 0 & 0 & 1
\end{array}\right] \end{array}
=
\begin{array}{c} \scriptstyle 1\;\;2\;\;3\;\;4\;\;5\;\;6\;\;7\;\;8\;\;9\;\;10 \\
\left[\begin{array}{ccccccc:ccc}
1 & 0 & 0 & 0 & -1 & 0 & 0 & 0 & 0 & 0 \\
0 & -1 & 0 & 0 & -l & 1 & 0 & 0 & 0 & 0 \\
0 & 0 & -1 & 0 & 1 & 0 & 0 & 1 & 0 & 0 \\
0 & 1 & -l & -1 & 0 & 0 & 0 & 0 & 1 & 0 \\
0 & 0 & 1 & 0 & 0 & 0 & 1 & 0 & 0 & 0 \\
0 & 0 & 0 & 1 & 0 & 0 & 0 & 0 & 0 & 1
\end{array}\right] \end{array}
$$

$$
\left[\begin{array}{c|c|c} 1 & 0 & \mathbf{0} \\ \hline 0 & -1 & \mathbf{0} \\ \hline & 0 & \\ \mathbf{0} & 1 & \mathbf{I}_4 \\ & 0 & \\ & 0 & \end{array}\right]
\begin{array}{c} \scriptstyle 1\;\;2\;\;3\;\;4\;\;5\;\;6\;\;7\;\;8\;\;9\;\;10 \\
\left[\begin{array}{ccccccc:ccc}
1 & 0 & 0 & 0 & -1 & 0 & 0 & 0 & 0 & 0 \\
0 & -1 & 0 & 0 & -l & 1 & 0 & 0 & 0 & 0 \\
0 & 0 & -1 & 0 & 1 & 0 & 0 & 1 & 0 & 0 \\
0 & 1 & -l & -1 & 0 & 0 & 0 & 0 & 1 & 0 \\
0 & 0 & 1 & 0 & 0 & 0 & 1 & 0 & 0 & 0 \\
0 & 0 & 0 & 1 & 0 & 0 & 0 & 0 & 0 & 1
\end{array}\right] \end{array}
=
\begin{array}{c} \scriptstyle 1\;\;2\;\;3\;\;4\;\;5\;\;6\;\;7\;\;8\;\;9\;\;10 \\
\left[\begin{array}{ccccccc:ccc}
1 & 0 & 0 & 0 & -1 & 0 & 0 & 0 & 0 & 0 \\
0 & 1 & 0 & 0 & l & -1 & 0 & 0 & 0 & 0 \\
0 & 0 & -1 & 0 & 1 & 0 & 0 & 1 & 0 & 0 \\
0 & 0 & -l & -1 & -l & 1 & 0 & 0 & 1 & 0 \\
0 & 0 & 1 & 0 & 0 & 0 & 1 & 0 & 0 & 0 \\
0 & 0 & 0 & 1 & 0 & 0 & 0 & 0 & 0 & 1
\end{array}\right] \end{array}
$$

$$
\left[\begin{array}{c|c|c} \mathbf{I}_2 & \mathbf{0} & \mathbf{0} \\ & & \\ \hline \mathbf{0} & -1 & \mathbf{0} \\ \hline & -l & \\ \mathbf{0} & 1 & \mathbf{I}_3 \\ & 0 & \end{array}\right]
\begin{array}{c} \scriptstyle 1\;\;2\;\;3\;\;4\;\;5\;\;6\;\;7\;\;8\;\;9\;\;10 \\
\left[\begin{array}{ccccccc:ccc}
1 & 0 & 0 & 0 & -1 & 0 & 0 & 0 & 0 & 0 \\
0 & 1 & 0 & 0 & l & -1 & 0 & 0 & 0 & 0 \\
0 & 0 & -1 & 0 & 1 & 0 & 0 & 1 & 0 & 0 \\
0 & 0 & -l & -1 & -l & 1 & 0 & 0 & 1 & 0 \\
0 & 0 & 1 & 0 & 0 & 0 & 1 & 0 & 0 & 0 \\
0 & 0 & 0 & 1 & 0 & 0 & 0 & 0 & 0 & 1
\end{array}\right] \end{array}
=
\begin{array}{c} \scriptstyle 1\;\;2\;\;3\;\;4\;\;5\;\;6\;\;7\;\;8\;\;9\;\;10 \\
\left[\begin{array}{ccccccc:ccc}
1 & 0 & 0 & 0 & -1 & 0 & 0 & 0 & 0 & 0 \\
0 & 1 & 0 & 0 & l & -1 & 0 & 0 & 0 & 0 \\
0 & 0 & 1 & 0 & -1 & 0 & 0 & -1 & 0 & 0 \\
0 & 0 & 0 & -1 & -2l & 1 & 0 & -l & 1 & 0 \\
0 & 0 & 0 & 0 & 1 & 0 & 1 & 1 & 0 & 0 \\
0 & 0 & 0 & 1 & 0 & 0 & 0 & 0 & 0 & 1
\end{array}\right] \end{array}
$$

$$
\left[\begin{array}{c:c:c}
\mathbf{I}_3 & \begin{matrix}0\\0\\0\end{matrix} & \mathbf{0} \\ \hdashline
\mathbf{0} & -1 & \mathbf{0} \\ \hdashline
\mathbf{0} & \begin{matrix}0\\1\end{matrix} & \mathbf{I}_2
\end{array}\right]
\begin{array}{c}
\begin{array}{ccccccc:ccc} 1 & 2 & 3 & 4 & 5 & 6 & 7 & 8 & 9 & 10 \end{array} \\
\left[\begin{array}{ccccccc:ccc}
1 & 0 & 0 & 0 & -1 & 0 & 0 & 0 & 0 & 0 \\
0 & 1 & 0 & 0 & l & -1 & 0 & 0 & 0 & 0 \\
0 & 0 & 1 & 0 & -1 & 0 & 0 & -1 & 0 & 0 \\
0 & 0 & 0 & -1 & -2l & 1 & 0 & -l & 1 & 0 \\
0 & 0 & 0 & 0 & 1 & 0 & 1 & 1 & 0 & 0 \\
0 & 0 & 0 & 1 & 0 & 0 & 0 & 0 & 0 & 1
\end{array}\right]
\end{array}
=
\begin{array}{c}
\begin{array}{ccccccc:ccc} 1 & 2 & 3 & 4 & 5 & 6 & 7 & 8 & 9 & 10 \end{array} \\
\left[\begin{array}{ccccccc:ccc}
1 & 0 & 0 & 0 & -1 & 0 & 0 & 0 & 0 & 0 \\
0 & 1 & 0 & 0 & l & -1 & 0 & 0 & 0 & 0 \\
0 & 0 & 1 & 0 & -1 & 0 & 0 & -1 & 0 & 0 \\
0 & 0 & 0 & 1 & 2l & -1 & 0 & l & -1 & 0 \\
0 & 0 & 0 & 0 & 1 & 0 & 1 & 1 & 0 & 0 \\
0 & 0 & 0 & 0 & -2l & 1 & 0 & -l & 1 & 1
\end{array}\right]
\end{array}
$$

$$
\left[\begin{array}{c:c:c}
\mathbf{I}_4 & \begin{matrix}1\\-l\\1\\-2l\end{matrix} & \mathbf{0} \\ \hdashline
\mathbf{0} & 1 & 0 \\ \hdashline
\mathbf{0} & 2l & 1
\end{array}\right]
\begin{array}{c}
\begin{array}{ccccccc:ccc} 1 & 2 & 3 & 4 & 5 & 6 & 7 & 8 & 9 & 10 \end{array} \\
\left[\begin{array}{ccccccc:ccc}
1 & 0 & 0 & 0 & -1 & 0 & 0 & 0 & 0 & 0 \\
0 & 1 & 0 & 0 & l & -1 & 0 & 0 & 0 & 0 \\
0 & 0 & 1 & 0 & -1 & 0 & 0 & -1 & 0 & 0 \\
0 & 0 & 0 & 1 & 2l & -1 & 0 & l & -1 & 0 \\
0 & 0 & 0 & 0 & 1 & 0 & 1 & 1 & 0 & 0 \\
0 & 0 & 0 & 0 & -2l & 1 & 0 & -l & 1 & 1
\end{array}\right]
\end{array}
=
\begin{array}{c}
\begin{array}{ccccccc:ccc} 1 & 2 & 3 & 4 & 5 & 6 & 7 & 8 & 9 & 10 \end{array} \\
\left[\begin{array}{ccccccc:ccc}
1 & 0 & 0 & 0 & 0 & 0 & 1 & 1 & 0 & 0 \\
0 & 1 & 0 & 0 & 0 & -1 & -l & -l & 0 & 0 \\
0 & 0 & 1 & 0 & 0 & 0 & 1 & 0 & 0 & 0 \\
0 & 0 & 0 & 1 & 0 & -1 & -2l & -l & -1 & 0 \\
0 & 0 & 0 & 0 & 1 & 0 & 1 & 1 & 0 & 0 \\
0 & 0 & 0 & 0 & 0 & 1 & 2l & l & 1 & 1
\end{array}\right]
\end{array}
$$

$$
\left[\begin{array}{c:c}
\mathbf{I}_5 & \begin{matrix}0\\1\\0\\1\\0\end{matrix} \\ \hdashline
\mathbf{0} & 1
\end{array}\right]
\begin{array}{c}
\begin{array}{ccccccc:ccc} 1 & 2 & 3 & 4 & 5 & 6 & 7 & 8 & 9 & 10 \end{array} \\
\left[\begin{array}{ccccccc:ccc}
1 & 0 & 0 & 0 & 0 & 0 & 1 & 1 & 0 & 0 \\
0 & 1 & 0 & 0 & 0 & -1 & -l & -l & 0 & 0 \\
0 & 0 & 1 & 0 & 0 & 0 & 1 & 0 & 0 & 0 \\
0 & 0 & 0 & 1 & 0 & -1 & -2l & -l & -1 & 0 \\
0 & 0 & 0 & 0 & 1 & 0 & 1 & 1 & 0 & 0 \\
0 & 0 & 0 & 0 & 0 & 1 & 2l & l & 1 & 1
\end{array}\right]
\end{array}
=
\begin{array}{c}
\begin{array}{ccccccc:ccc} 1 & 2 & 3 & 4 & 5 & 6 & 7 & 8 & 9 & 10 \end{array} \\
\left[\begin{array}{ccccccc:ccc}
1 & 0 & 0 & 0 & 0 & 0 & 1 & 1 & 0 & 0 \\
0 & 1 & 0 & 0 & 0 & 0 & l & 0 & 1 & 1 \\
0 & 0 & 1 & 0 & 0 & 0 & 1 & 0 & 0 & 0 \\
0 & 0 & 0 & 1 & 0 & 0 & 0 & 0 & 0 & 1 \\
0 & 0 & 0 & 0 & 1 & 0 & 1 & 1 & 0 & 0 \\
0 & 0 & 0 & 0 & 0 & 1 & 2l & l & 1 & 1
\end{array}\right]
\end{array}
$$

In the preceding equations the column numbers 1 to 4 refer to the element forces, 5 to 7 refer to the reaction complements, and 8 to 10 refer to the external forces. It follows therefore from the last equation that the reaction complement R_3 (column 7) has been selected to be the redundant force. Hence, using Eqs. (8.65) and (8.66), we have

$$\mathbf{b}_0\mathbf{P} = \begin{matrix}1\\2\\3\\4\\5\\6\\7\end{matrix}\begin{bmatrix}1 & 0 & 0\\ 0 & 1 & 1\\ 0 & 0 & 0\\ 0 & 0 & 1\\ 1 & 0 & 0\\ l & 1 & 1\\ 0 & 0 & 0\end{bmatrix}\begin{bmatrix}P_3\\ P_4\\ P_6\end{bmatrix} \qquad \mathbf{b}_X = \begin{matrix}1\\2\\3\\4\\5\\6\\7\end{matrix}\begin{bmatrix}-1\\ -l\\ -1\\ 0\\ -1\\ -2l\\ 1\end{bmatrix}$$

By neglecting the effects of shear deformations the element flexibility matrices are given by [see Eq. (7.21)]

$$\mathbf{f}^{(1)} = \begin{matrix} & 1 & 2\\ 1 & \left[2l^2\right. & \left.3l\right]\\ 2 & \left[3l\right. & \left.6\right]\end{matrix}\frac{l}{6EI} \qquad \mathbf{f}^{(2)} = \begin{matrix} & 3 & 4\\ 3 & \left[2l^2\right. & \left.3l\right]\\ 4 & \left[3l\right. & \left.6\right]\end{matrix}\frac{l}{6EI}$$

while the corresponding relative thermal deformations are calculated from [see Eq. (7.22)]

$$\mathbf{v}_T^{(1)} = \begin{matrix}1\\2\end{matrix}\begin{bmatrix}-l\\ -2\end{bmatrix}\frac{\alpha l}{2I}\int_A Ty\,dA \qquad \mathbf{v}_T^{(2)} = \begin{matrix}3\\4\end{matrix}\begin{bmatrix}-l\\ -2\end{bmatrix}\frac{\alpha l}{2I}\int_A Ty\,dA$$

The flexibility matrix $\bar{\mathbf{f}}$ from (8.72) and thermal-deformation matrix $\bar{\mathbf{v}}_T$ from (8.73) are determined from

$$\bar{\mathbf{f}} = \begin{bmatrix}\mathbf{f}^{(1)} & \mathbf{0} & \mathbf{0}\\ \mathbf{0} & \mathbf{f}^{(2)} & \mathbf{0}\\ \mathbf{0} & \mathbf{0} & \mathbf{0}\end{bmatrix}$$

and
$$\bar{\mathbf{v}}_T = \begin{bmatrix}\mathbf{v}_T^{(1)}\\ \mathbf{v}_T^{(2)}\\ \mathbf{0}\end{bmatrix}$$

It can now be easily demonstrated that

$$\mathbf{b}_X{}^T\bar{\mathbf{f}}\mathbf{b}_X = \frac{8}{3}\frac{l^3}{EI}$$

$$\mathbf{b}_X{}^T\bar{\mathbf{f}}\mathbf{b}_0\mathbf{P} = -[5l \quad 9 \quad 12]\,\frac{l^2}{6EI}\begin{bmatrix} P_3 \\ P_4 \\ P_6 \end{bmatrix}$$

$$\mathbf{b}_X{}^T\bar{\mathbf{v}}_T = \frac{2\alpha l^2}{I}\int_A Ty\,dA$$

Hence $\quad \mathbf{X} = -(\mathbf{b}_X{}^T\bar{\mathbf{f}}\mathbf{b}_X)^{-1}(\mathbf{b}_X{}^T\bar{\mathbf{f}}\mathbf{b}_0\mathbf{P} + \mathbf{b}_X{}^T\bar{\mathbf{v}}_T)$

$$= [5l \quad 9 \quad 12]\,\frac{1}{16l}\begin{bmatrix} P_3 \\ P_4 \\ P_6 \end{bmatrix} - \frac{3}{4}\frac{E\alpha}{l}\int_A Ty\,dA$$

Having obtained the redundant force $\mathbf{X}$, we can calculate the element forces and reaction complements from

$$\bar{\mathbf{F}} = \mathbf{b}_0\mathbf{P} + \mathbf{b}_X\mathbf{X}$$

so that

$$\bar{\mathbf{F}} = \begin{array}{c} 1 \\ 2 \\ 3 \\ 4 \\ 5 \\ 6 \\ 7 \end{array}\begin{bmatrix} 1 & 0 & 0 \\ 0 & 1 & 1 \\ 0 & 0 & 0 \\ 0 & 0 & 1 \\ 1 & 0 & 0 \\ l & 1 & 1 \\ 0 & 0 & 0 \end{bmatrix}\begin{bmatrix} P_3 \\ P_4 \\ P_6 \end{bmatrix} + \begin{array}{c} 1 \\ 2 \\ 3 \\ 4 \\ 5 \\ 6 \\ 7 \end{array}\begin{bmatrix} -1 \\ -l \\ -1 \\ 0 \\ -1 \\ -2l \\ 1 \end{bmatrix}\left\{[5l \quad 9 \quad 12]\,\frac{1}{16l}\begin{bmatrix} P_3 \\ P_4 \\ P_6 \end{bmatrix} - \frac{3}{4}\frac{E\alpha}{l}\int_A Ty\,dA\right\}$$

$$= \begin{array}{c} 1 \\ 2 \\ 3 \\ 4 \\ 5 \\ 6 \\ 7 \end{array}\begin{bmatrix} 11l & -9 & -12 \\ -5l^2 & 7l & 4l \\ -5l & -9 & -12 \\ 0 & 0 & 16l \\ 11l & -9 & -12 \\ 6l^2 & -2l & -8l \\ 5l & 9 & 12 \end{bmatrix}\frac{1}{16l}\begin{bmatrix} P_3 \\ P_4 \\ P_6 \end{bmatrix} - \begin{array}{c} 1 \\ 2 \\ 3 \\ 4 \\ 5 \\ 6 \\ 7 \end{array}\begin{bmatrix} -1 \\ -l \\ -1 \\ 0 \\ -1 \\ -2l \\ 1 \end{bmatrix}\frac{3}{4}\frac{E\alpha}{l}\int_A Ty\,dA$$

To find the displacements in the direction of the applied forces we use the unit-load theorem, which is expressed as

$$\mathbf{r} = \mathbf{b}_0{}^T\bar{\mathbf{v}} = \mathbf{b}_0{}^T\bar{\mathbf{f}}\bar{\mathbf{F}} + \mathbf{b}_0{}^T\bar{\mathbf{v}}_T$$

Using the previously derived matrices $\mathbf{b}_0$, $\bar{\mathbf{f}}$, $\bar{\mathbf{F}}$, and $\bar{\mathbf{v}}_T$, we have that

$$\mathbf{r} = \begin{bmatrix} r_3 \\ r_4 \\ r_6 \end{bmatrix} = \begin{array}{c} 3 \\ 4 \\ 6 \end{array}\overset{\begin{array}{ccc} 3 & 4 & 6 \end{array}}{\begin{bmatrix} 7l^2 & 3l & -12l \\ 3l & 15 & -12 \\ -12l & -12 & 48 \end{bmatrix}} \frac{l}{96EI} \begin{bmatrix} P_3 \\ P_4 \\ P_6 \end{bmatrix} + \begin{array}{c} 3 \\ 4 \\ 6 \end{array}\begin{bmatrix} l \\ 1 \\ -4 \end{bmatrix} \frac{\alpha l}{8I} \int_A Ty\, dA$$

which agrees with the results obtained in Sec. 6.10 by the matrix displacement method of analysis.

8.6 COMPARISON OF THE FORCE AND DISPLACEMENT METHODS

Having developed the two alternative methods of structural analysis, we are faced with the natural question as to which method is best for practical applications. To answer this question we must consider a number of different factors.

First it should be emphasized that since the same element properties can be used for either the displacement or force methods, it is obvious that, theoretically, both methods lead to identical results, as was demonstrated in the illustrative examples in Chaps. 6 and 8. The computational path leading to the calculation of stresses and displacements is different in each method. This means that because of the different rounding-off errors and possible ill-conditioning of equations, the actual numerical results may differ slightly. For some special applications numerical solutions are obtained using both methods with different assumptions on the element stress or displacement distributions, i.e., compatible but nonequilibrium stress states for the displacement method and statically equivalent (equilibrium) but noncompatible stress states for the force method. This leads to the so called bracketing of the solution. Such solutions are particularly useful if the bracketing is small, since they provide meaningful information on the accuracy of the results.

Let us examine next the matrix operations involved in the two methods. The displacement method is based on the solution of a simple equation

$$\mathbf{P} = \mathbf{K}\mathbf{U} + \mathbf{Q} \tag{8.76}$$

relating the external forces $\mathbf{P}$ and thermal forces $\mathbf{Q}$ to the displacements $\mathbf{U}$ at the node points of the idealized structure. We have seen in Chap. 6 how the element stiffness matrices are assembled into the stiffness matrix $\mathbf{K}$ for the assembled structure. The procedure is indeed very simple and does not require any complicated programming. Once the displacements **U** have been calculated, they are used to calculate stresses in individual elements. Some difficulties may occur due to ill-conditioning of Eq. (8.76) when inverting the stiffness matrix **K**. Some of the conditioning problems have been discussed by Taig and Kerr;[318] however, constant improvements in computer technology result in increasing accuracy, such as double-precision inversion programs, and tend to eliminate ill-conditioning as a source of error.

In modern computer programs for the matrix methods, human errors in the basic input data are probably the most frequent source of errors. These are the most difficult ones to detect automatically. As a speculation for future developments we can mention the possibility of using visual displays of structural geometry from the input data. Special attention must be paid also to the design of input data sheets in order to reduce chances of erroneous entries and errors in subsequent keypunching of the input cards. One noteworthy innovation in this respect is the method introduced by Argyris,[18] whereby intermediate node points are generated automatically by the computer. This means that some of the idealization is performed by the computer, and therefore the amount of input data required from the analyst is greatly reduced.

In the force method of analysis the sequence of matrix operations required to obtain stresses and displacements is considerably more complicated than for the displacement method. A schematic flow diagram for the force analysis is presented in Fig. 8.14. This may be compared with the corresponding flow diagram for the displacement method in Fig. 6.2. We have demonstrated in this chapter that the self-equilibrating force systems can be generated automatically from the equations of equilibrium using the jordanian elimination technique. This technique allows us to use the *same* input information for the force method as for the displacement method. When the matrix force methods were first introduced, considerable difficulty was experienced in formulating the self-equilibrating force systems. The determination of the degree of redundancy and the distribution of the self-equilibrating force system was sometimes an intractable proposition for exceedingly complex structural systems. Special programs had to be written for specific structures,[29,160,267] force systems were orthogonalized to improve conditioning,[29] regularization procedures were used for cutouts,[11] and so on. The development of the automatic selection of redundancies and generation of the self-equilibrating force systems completely changed the approach to the force method of analysis. Any arbitrary structural system, no matter how complex, can now be analyzed by the force method. Furthermore, the selection procedures based on the jordanian elimination technique lead invariably to well-conditioned equations.[78]

Since the input information is identical in the two methods, it would appear at first that the choice of one or the other is largely a matter of taste and the availability of a suitable computer program. There is, however, one important consideration that has not been discussed, the number of unknown displacements

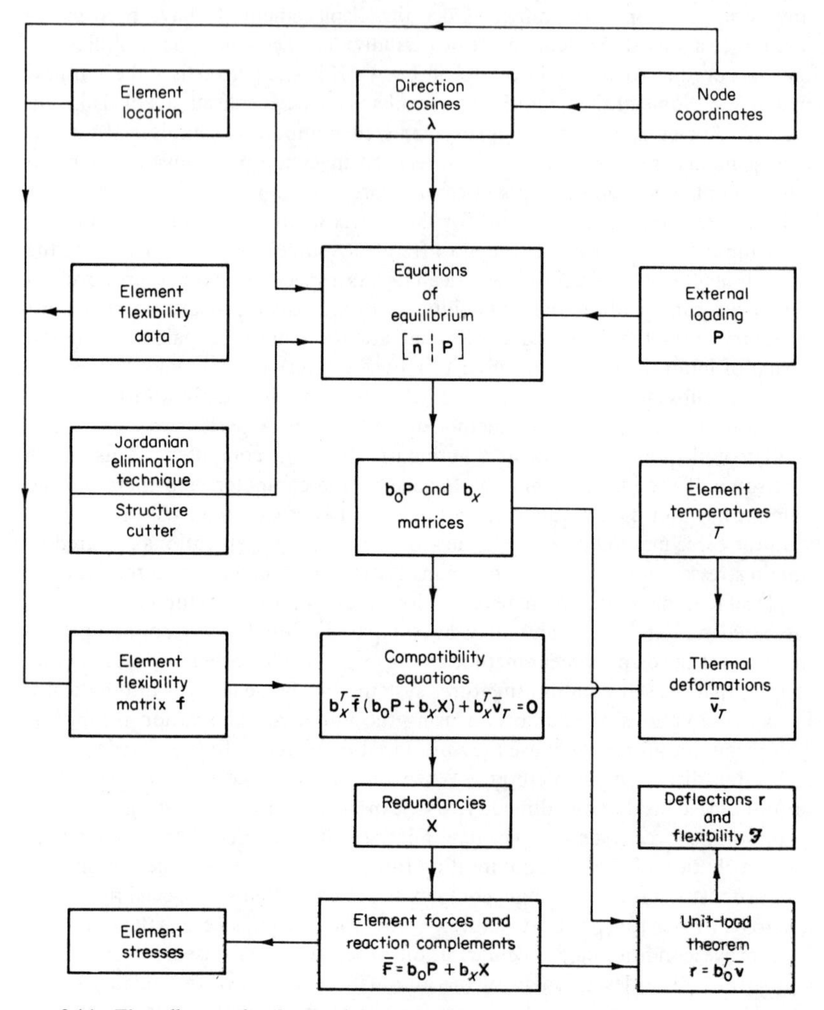

FIG. 8.14 Flow diagram for the matrix force method.

or forces and the number of structural elements. Computer programs for the displacement method have built-in limitations on the number of displacements and elements, while those for the force method have limitations on the number of node points, redundancies, and elements. Since the number of unknowns in the two methods may be widely different for the same structure, this alone may be the deciding criterion for selecting the method of analysis.

Mainly because of the simplicity of matrix operations there has been a tendency to use the displacement method for complex structural configurations. For some special structures, however, particularly if the selection of redundancies and generations of the self-equilibrating systems can be preprogrammed, the matrix force method can be used very effectively and should be simpler than the displacement method.

PROBLEMS

8.1 *a.* A redundant structure is supported in a statically determinate manner and is subjected to a set of specified displacements $\mathbf{r}$. Using the matrix force method of analysis, show that the element forces $\mathbf{F}$ are given by

$$\mathbf{F} = (\mathbf{b}_0 - \mathbf{b}_X(\mathbf{b}_X{}^T\mathbf{f}\mathbf{b}_X)^{-1}\mathbf{b}_X{}^T\mathbf{f}\mathbf{b}_0)(\mathbf{b}_0{}^T\mathbf{f}(\mathbf{b}_0 - \mathbf{b}_X(\mathbf{b}_X{}^T\mathbf{f}\mathbf{b}_X)^{-1}\mathbf{b}_X{}^T\mathbf{f}\mathbf{b}_0))^{-1}\mathbf{r}$$

where $\mathbf{b}_0$ = matrix of element forces due to unit external forces applied in directions of $\mathbf{r}$ in "cut" structure

$\mathbf{b}_X$ = matrix of element forces due to unit redundancies in "cut" structure

$\mathbf{f}$ = flexibility matrix of unassembled elements

b. Explain how the equation for the element forces $\mathbf{F}$ would be modified if the structure were statically determinate.

8.2 The element forces in a pin-jointed truss, shown in Fig. 8.15, have been determined for a single external load of 1,000 lb applied vertically down at the node 1. These forces are given by the matrix

$$\mathbf{F} = \{\overset{1}{0.442}\;\; \overset{2}{0.442}\;\; \overset{3}{0.789}\;\; \overset{4}{-0.625}\;\; \overset{5}{-0.558}\;\; \overset{6}{0}\;\; \overset{7}{1.558}\;\; \overset{8}{0.625}\;\; \overset{9}{-0.789}\;\; \overset{10}{-1.442}\;\; \overset{11}{-0.442}\} \times 10^3 \text{ lb}$$

where the row numbers refer to the element-numbering system. Apply the unit-load theorem in matrix form to determine both vertical and horizontal deflections of node 4. All cross-sectional areas are equal to 1.0 in.^{2}, and $E = 10 \times 10^6$ lb/in.^{2}

8.3 Solve the problem in Sec. 8.5 using the flexibility matrix (7.24) and thermal deformation matrix (7.25).

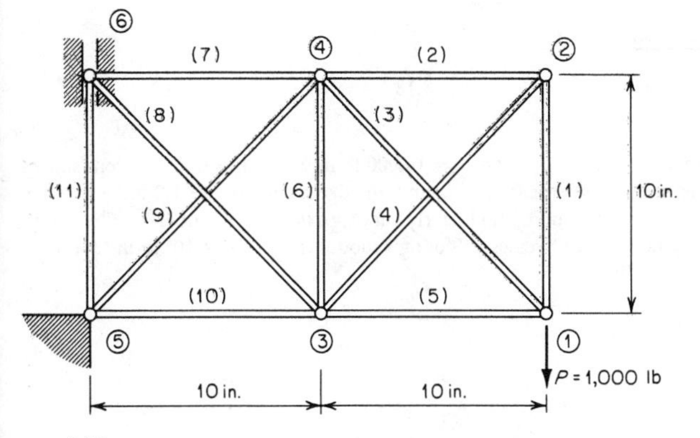

FIG. 8.15

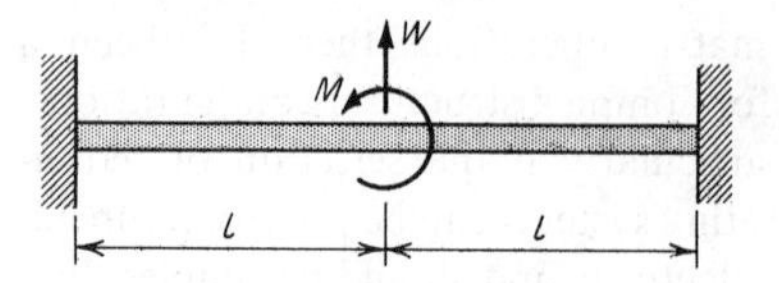

FIG. 8.16

8.4 Using the matrix force method, calculate deflection and rotation at the center of a uniform beam (Fig. 8.16) subjected to a transverse force W and moment M. The moment of inertia of the beam cross section is I, and Young's modulus is E. Neglect the effects of shear deformations.

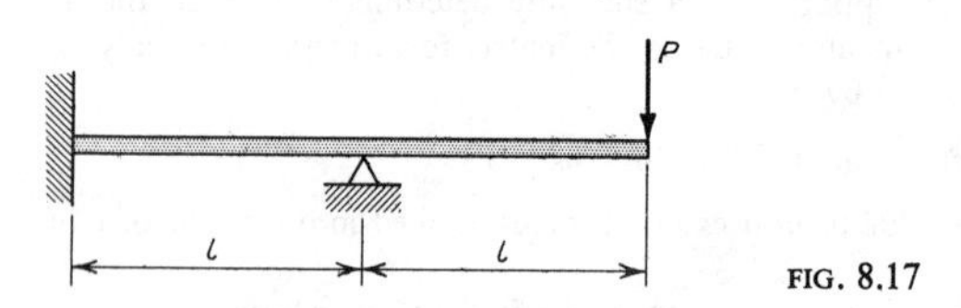

FIG. 8.17

8.5 A uniform beam of length $2l$ is built-in at $x = 0$, simply supported at $x = l$, and free at $x = 2l$ (see Fig. 8.17). The moment of inertia of the beam cross section is I, and Young's modulus of the material is E. The beam is subjected to a transverse load P at the free end. Neglecting the effects of shear deformations, calculate by the force method the beam deflection at $x = 2l$ and rotations at $x = l$ and $x = 2l$.

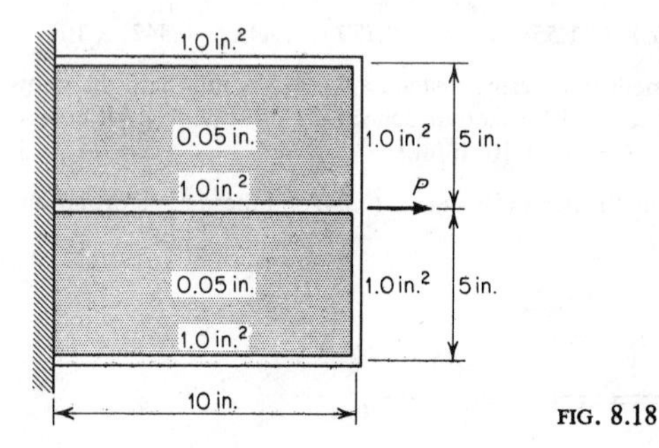

FIG. 8.18

8.6 Determine element forces due to load $P = 10{,}000$ lb in the "diffusion" problem shown in Fig. 8.18. The panel thicknesses are 0.050 in., and all edge members are 1.0 in.² Use the idealization of constant shear flow panels and linearly varying axial load members. The panels may be assumed to carry no normal stresses. Young's modulus $E = 10 \times 10^6$ lb/in.², and the shear modulus $G = 4 \times 10^6$ lb/in.²

CHAPTER 9 ANALYSIS OF SUBSTRUCTURES

In applying matrix methods of analysis to large structures, the number of structural elements very often exceeds the capacity of available computer programs, and consequently some form of structural partitioning must be employed. Structural partitioning corresponds to division of the complete structure into a number of substructures, the boundaries of which may be specified arbitrarily; however, for convenience it is preferable to make structural partitioning correspond to physical partitioning. If the stiffness or flexibility properties of each substructure are determined, the substructures can be treated as complex structural elements, and the matrix displacement or force methods of structural analysis can be formulated for the partitioned structure. Once the displacements or forces on substructure boundaries have been found, each substructure can then be analyzed separately under known substructure-boundary displacements or forces, depending on whether displacement or force methods of analysis are used. This chapter presents a general formulation of the displacement and force methods of substructure analysis.

9.1 SUBSTRUCTURE ANALYSIS BY THE MATRIX DISPLACEMENT METHOD[270]

GENERAL THEORY

In the displacement method each substructure is first analyzed separately, assuming that all common boundaries (joints) with the adjacent substructures are completely fixed; these boundaries are then relaxed simultaneously, and the actual boundary displacements are determined from the equations of equilibrium of forces at the boundary joints. Naturally, the solution for the boundary displacements involves a considerably smaller number of unknowns compared with the solution for the complete structure without partitioning. Each substructure can then be analyzed separately under known substructure loading and boundary displacements. This can be done without difficulty since the matrices involved are of a relatively small size.

The complete set of equilibrium equations for the structure regarded as a free body may be written in matrix form as

$$\mathbf{K}\mathbf{U} = \mathbf{P} - \mathbf{Q} \tag{9.1}$$

where $\mathbf{K}$ is the stiffness matrix, $\mathbf{U}$ represents a column matrix of displacements corresponding to external forces $\mathbf{P}$, and $\mathbf{Q}$ are the corresponding thermal forces calculated from some specified temperature distribution. For subsequent analysis it will be convenient to introduce the effective external loading $\bar{\mathbf{P}}$, given by

$$\bar{\mathbf{P}} = \mathbf{P} - \mathbf{Q} \tag{9.2}$$

so that Eq. (9.1) becomes

$$\mathbf{K}\mathbf{U} = \bar{\mathbf{P}} \tag{9.1a}$$

By suppressing a suitably chosen set of displacements to eliminate rigid-body displacements the matrix $\mathbf{K}$ is rendered nonsingular, and then Eq. (9.1*a*) can be solved for the unknown displacements $\mathbf{U}$.

In the following analysis the structure is divided into substructures by introducing interior boundaries. The column matrix of boundary displacements common to two or more substructures is denoted by $\mathbf{U}_b$, and the matrix of interior displacements (each of which occurs at an interior point of only one substructure) is $\mathbf{U}_i$. If the corresponding effective external forces are denoted by matrices $\bar{\mathbf{P}}_b$ and $\bar{\mathbf{P}}_i$, Eq. (9.1*a*) can be written in partitioned form as

$$\begin{bmatrix} \mathbf{K}_{bb} & \mathbf{K}_{bi} \\ \mathbf{K}_{ib} & \mathbf{K}_{ii} \end{bmatrix} \begin{bmatrix} \mathbf{U}_b \\ \mathbf{U}_i \end{bmatrix} = \begin{bmatrix} \bar{\mathbf{P}}_b \\ \bar{\mathbf{P}}_i \end{bmatrix} \tag{9.3}$$

It will now be assumed that the total displacements of the structure can be calculated from the superposition of two matrices such that

$$\mathbf{U} = \mathbf{U}^{(\alpha)} + \mathbf{U}^{(\beta)} \tag{9.4}$$

where $\mathbf{U}^{(\alpha)}$ denotes the column matrix of displacements due to $\bar{\mathbf{P}}_i$ with $\mathbf{U}_b = \mathbf{0}$, while $\mathbf{U}^{(\beta)}$ represents the necessary corrections to the displacements $\mathbf{U}^{(\alpha)}$ to allow for boundary displacements $\mathbf{U}_b$ with $\bar{\mathbf{P}}_i = \mathbf{0}$. Thus Eq. (9.4) may also be written as

$$\mathbf{U} = \begin{bmatrix} \mathbf{U}_b \\ \mathbf{U}_i \end{bmatrix} = \begin{bmatrix} \mathbf{U}_b^{(\alpha)} \\ \mathbf{U}_i^{(\alpha)} \end{bmatrix}_{\substack{\text{boundaries} \\ \text{fixed}}} + \begin{bmatrix} \mathbf{U}_b^{(\beta)} \\ \mathbf{U}_i^{(\beta)} \end{bmatrix} \tag{9.4a}$$

where the final term represents the correction due to boundary relaxation, and where, by definition,

$$\mathbf{U}_b^{(\alpha)} = \mathbf{0} \tag{9.5}$$

Similarly, corresponding to the displacements $\mathbf{U}^{(\alpha)}$ and $\mathbf{U}^{(\beta)}$ the external forces $\bar{\mathbf{P}}$ can be separated into

$$\bar{\mathbf{P}} = \bar{\mathbf{P}}^{(\alpha)} + \bar{\mathbf{P}}^{(\beta)} \tag{9.6}$$

or $$\bar{\mathbf{P}} = \begin{bmatrix} \bar{\mathbf{P}}_b \\ \bar{\mathbf{P}}_i \end{bmatrix} = \begin{bmatrix} \bar{\mathbf{P}}_b^{(\alpha)} \\ \bar{\mathbf{P}}_i^{(\alpha)} \end{bmatrix} + \begin{bmatrix} \bar{\mathbf{P}}_b^{(\beta)} \\ \bar{\mathbf{P}}_i^{(\beta)} \end{bmatrix} \tag{9.6a}$$

where, by definition,

$$\bar{\mathbf{P}}_i^{(\alpha)} = \bar{\mathbf{P}}_i \tag{9.7}$$

and $$\bar{\mathbf{P}}_i^{(\beta)} = \mathbf{0} \tag{9.8}$$

When the substructure boundaries are fixed, it can readily be shown, using Eq. (9.3), that

$$\mathbf{U}_i^{(\alpha)} = \mathbf{K}_{ii}^{-1}\bar{\mathbf{P}}_i \tag{9.9}$$

and $$\bar{\mathbf{P}}_b^{(\alpha)} = \mathbf{K}_{bi}\mathbf{K}_{ii}^{-1}\bar{\mathbf{P}}_i = \bar{\mathbf{R}}_b \tag{9.10}$$

It should be noted that $\bar{\mathbf{P}}_b^{(\alpha)}$ represents boundary reactions necessary to maintain $\mathbf{U}_b = \mathbf{0}$ when the interior forces $\bar{\mathbf{P}}_i$ are applied. When the substructure boundaries are relaxed, the displacements $\mathbf{U}^{(\beta)}$ can be determined also from Eq. (9.3), so that

$$\mathbf{U}_i^{(\beta)} = -\mathbf{K}_{ii}^{-1}\mathbf{K}_{ib}\mathbf{U}_b^{(\beta)} \tag{9.11}$$

$$\mathbf{U}_b^{(\beta)} = \mathbf{K}_b^{-1}\bar{\mathbf{P}}_b^{(\beta)} \tag{9.12}$$

where $$\mathbf{K}_b = \mathbf{K}_{bb} - \mathbf{K}_{bi}\mathbf{K}_{ii}^{-1}\mathbf{K}_{ib} \tag{9.13}$$

represents the boundary stiffness matrix. The matrix $\bar{\mathbf{P}}_b^{(\beta)}$ can be determined from Eqs. (9.6*a*) and (9.10), and hence

$$\bar{\mathbf{P}}_b^{(\beta)} = \bar{\mathbf{P}}_b - \bar{\mathbf{P}}_b^{(\alpha)} = \bar{\mathbf{P}}_b - \mathbf{K}_{bi}\mathbf{K}_{ii}^{-1}\bar{\mathbf{P}}_i = \bar{\mathbf{S}}_b \tag{9.14}$$

When the boundary displacements are set equal to zero, the substructures are completely isolated from each other, so that application of an interior force causes displacements in only one substructure. It is therefore evident that the interior displacements $\mathbf{U}_i^{(\alpha)}$ with boundaries fixed can be calculated for each substructure separately, using Eq. (9.9). The boundary displacements $\mathbf{U}_b^{(\beta)}$ are found from Eq. (9.12) involving inversion of $\mathbf{K}_b$, which is of much smaller order than the complete stiffness matrix $\mathbf{K}$.

THE SUBSTRUCTURE DISPLACEMENTS AND FORCES: BOUNDARIES FIXED

The stiffness matrix of the rth substructure, regarded as a free body, can conveniently be partitioned into

$$\mathbf{K}^{(r)} = \begin{bmatrix} \mathbf{K}_{bb}^{(r)} & \mathbf{K}_{bi}^{(r)} \\ \mathbf{K}_{ib}^{(r)} & \mathbf{K}_{ii}^{(r)} \end{bmatrix} \tag{9.15}$$

where the superscript r denotes the rth substructure and the subscripts b and i refer to the boundary and interior displacements, respectively. Naturally, because of symmetry of the stiffness matrix, $\mathbf{K}_{bi}^{(r)}$ is a transpose of $\mathbf{K}_{ib}^{(r)}$. By use of the above stiffness matrix the substructure displacements $\mathbf{U}^{(r)}$ can be related to the external forces $\bar{\mathbf{P}}^{(r)}$ by the equation

$$\mathbf{K}^{(r)}\mathbf{U}^{(r)} = \bar{\mathbf{P}}^{(r)} \tag{9.16}$$

or

$$\begin{bmatrix} \mathbf{K}_{bb}^{(r)} & \mathbf{K}_{bi}^{(r)} \\ \mathbf{K}_{ib}^{(r)} & \mathbf{K}_{ii}^{(r)} \end{bmatrix} \begin{bmatrix} \mathbf{U}_b^{(r)} \\ \mathbf{U}_i^{(r)} \end{bmatrix} = \begin{bmatrix} \bar{\mathbf{P}}_b^{(r)} \\ \bar{\mathbf{P}}_i^{(r)} \end{bmatrix} \tag{9.17}$$

When the substructure boundaries on the complete structure are fixed, the boundary fixing must be sufficient to restrain rigid-body degrees of freedom on each substructure considered separately. A typical substructure with fixed boundaries is shown in Fig. 9.1. The substructure interior displacements and

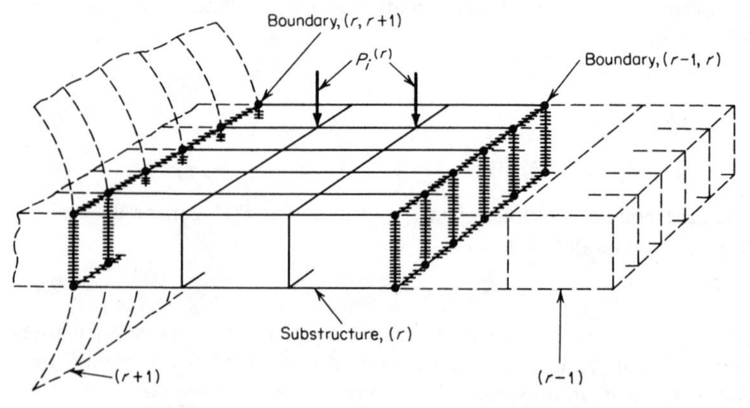

FIG. 9.1 Typical substructure with fixed boundaries.

boundary reactions due to $\bar{\mathbf{P}}_i^{(r)}$ when $\mathbf{U}_b^{(r)} = \mathbf{0}$ can be determined from Eqs. (9.9) and (9.10), and therefore

$$[\mathbf{U}_i^{(r)}]_{\text{boundaries fixed}} = (\mathbf{K}_{ii}^{(r)})^{-1}\bar{\mathbf{P}}_i^{(r)} \tag{9.18}$$

and $$\bar{\mathbf{R}}_b^{(r)} = \mathbf{K}_{bi}^{(r)}(\mathbf{K}_{ii}^{(r)})^{-1}\bar{\mathbf{P}}_i^{(r)} \tag{9.19}$$

where the matrix inversion of $\mathbf{K}_{ii}^{(r)}$ is permissible because the boundary fixing restrains all rigid-body degrees of freedom.

Before considering "matching" of displacements on common boundaries, it is necessary to evaluate the substructure stiffnesses associated with the displacements $\mathbf{U}_b^{(r)}$. To determine these stiffnesses Eq. (9.13) is applied to the rth substructure, and it follows immediately that

$$\mathbf{K}_b^{(r)} = \mathbf{K}_{bb}^{(r)} - \mathbf{K}_{bi}^{(r)}(\mathbf{K}_{ii}^{(r)})^{-1}\mathbf{K}_{ib}^{(r)} \tag{9.20}$$

which will be used subsequently to assemble the boundary-stiffness matrix $\mathbf{K}_b$ for the complete structure.

Some substructures may be analyzed more conveniently by the matrix force method. For these cases the boundary reactions $\mathbf{R}_b^{(r)}$ due to $\mathbf{P}_i^{(r)}$ for fixed boundaries can be obtained from the general analysis, while the boundary stiffness $\mathbf{K}_b^{(r)}$ can be determined from the analysis described below.

To evaluate the stiffness matrix $\mathbf{K}_b^{(r)}$ using the force method of analysis six zero displacements $\mathbf{w}^{(r)}$ are selected on the substructure boundary to restrain rigid-body degrees of freedom, and then unit loads are applied in the directions of the remaining $n - 6$ boundary displacements $\mathbf{u}^{(r)}$, where n is the total number of displacements on the boundary. The solution for displacements gives then the flexibility matrix $\mathbf{F}_{uu}^{(r)}$, relative to the fixed datum based on the selected six zero displacements which will be used to determine the stiffness matrix $\mathbf{K}_b^{(r)}$.

The boundary forces and displacements are related by the equation

$$\mathbf{K}_b^{(r)}\mathbf{U}_b^{(r)} = \mathbf{S}_b^{(r)} \tag{9.21}$$

or, in the partitioned form, by the equation

$$\begin{bmatrix} \mathbf{K}_{uu}^{(r)} & \mathbf{K}_{uw}^{(r)} \\ \mathbf{K}_{wu}^{(r)} & \mathbf{K}_{ww}^{(r)} \end{bmatrix} \begin{bmatrix} \mathbf{u}_r^{(r)} \\ \mathbf{w}^{(r)} \end{bmatrix} = \begin{bmatrix} \mathbf{S}_u^{(r)} \\ \mathbf{S}_w^{(r)} \end{bmatrix} \tag{9.21a}$$

where $\mathbf{u}_r$ represents the $n - 6$ displacements, $\mathbf{w}^{(r)}$ represents the six datum displacements, $\mathbf{S}_u^{(r)}$ and $\mathbf{S}_w^{(r)}$ are the corresponding applied forces, and

$$\mathbf{K}_{uu}^{(r)} = (\mathbf{F}_{uu}^{(r)})^{-1} \tag{9.22}$$

The column matrix for the boundary displacements $\mathbf{U}_b^{(r)}$ can be expressed as

$$\mathbf{U}_b^{(r)} = \begin{bmatrix} \mathbf{u}_r^{(r)} \\ \mathbf{w}^{(r)} \end{bmatrix} = \begin{bmatrix} \mathbf{r}^{(r)} \\ \mathbf{0} \end{bmatrix} + \begin{bmatrix} \mathbf{T}^{(r)} \\ \mathbf{I} \end{bmatrix} \mathbf{w}^{(r)} \tag{9.23}$$

where $\mathbf{r}^{(r)}$ are the displacements $\mathbf{u}_r^{(r)}$ measured relative to the fixed datum and $\mathbf{T}^{(r)}$ is a transformation matrix derived in Chap. 6. The derivation of this transformation matrix requires that at least three of the displacements in $\mathbf{w}^{(r)}$ refer to three translational displacements on the substructure boundary. Applying now virtual displacements $\delta\mathbf{w}^{(r)}$, it follows from Eq. (9.23) that

$$\delta\mathbf{u}_r^{(r)} = \mathbf{T}^{(r)}\,\delta\mathbf{w}^{(r)} \tag{9.24}$$

From the principle of virtual work, it is clear that if the virtual displacements are only those representing the rigid-body degrees of freedom, the virtual work is equal to zero. Hence

$$(\mathbf{S}_u^{(r)})^T\,\delta\mathbf{u}_r^{(r)} + (\mathbf{S}_w^{(r)})^T\,\delta\mathbf{w}^{(r)} = 0 \tag{9.25}$$

Substituting Eqs. (9.21), (9.23), and (9.24) into (9.25), and noting that $\mathbf{K}_{uw}^{(r)} = (\mathbf{K}_{wu}^{(r)})^T$, we find that

$$[\mathbf{r}^T(\mathbf{K}_{wu} + \mathbf{K}_{uu}{}^T\mathbf{T}) + \mathbf{w}^T(\mathbf{T}^T\mathbf{K}_{wu}{}^T + \mathbf{K}_{ww}{}^T + \mathbf{T}^T\mathbf{K}_{uu}{}^T\mathbf{T} + \mathbf{K}_{wu}\mathbf{T})]\,\delta\mathbf{w} = 0 \tag{9.26}$$

where for simplicity the superscripts (r) have been omitted. As Eq. (9.26) must be valid for any arbitrary values of $\mathbf{r}$ and $\mathbf{w}$, we have

$$\mathbf{K}_{wu}{}^T + \mathbf{K}_{uu}{}^T\mathbf{T} = \mathbf{0} \tag{9.27}$$

and $$\mathbf{T}^T\mathbf{K}_{wu}{}^T + \mathbf{K}_{ww}{}^T + \mathbf{T}^T\mathbf{K}_{uu}{}^T\mathbf{T} + \mathbf{K}_{wu}\mathbf{T} = \mathbf{0} \tag{9.28}$$

The stiffness submatrices $\mathbf{K}_{wu}$ and $\mathbf{K}_{ww}$ can now be determined from Eqs. (9.27) and (9.28), and the result is

$$\mathbf{K}_{wu} = -\mathbf{T}^T\mathbf{K}_{uu} \tag{9.29}$$

$$\mathbf{K}_{ww} = \mathbf{T}^T\mathbf{K}_{uu}\mathbf{T} \tag{9.30}$$

Finally, substitution of Eqs. (9.29) and (9.30) into Eq. (9.21*a*) leads to

$$\mathbf{K}_b^{(r)} = \left[\begin{array}{c|c} \mathbf{K}_{uu}^{(r)} & -\mathbf{K}_{uu}^{(r)}\mathbf{T}_{(r)} \\ \hline -(\mathbf{T}^{(r)})^T\mathbf{K}_{uu}^{(r)} & (\mathbf{T}^{(r)})^T\mathbf{K}_{uu}^{(r)}\mathbf{T}^{(r)} \end{array}\right] \tag{9.31}$$

which represents the required boundary stiffness matrix.

GENERAL SOLUTION FOR BOUNDARY DISPLACEMENTS: SUBSTRUCTURE RELAXATION

Having determined the boundary stiffnesses $\mathbf{K}_b^{(r)}$ and the reactions $\bar{\mathbf{R}}_b^{(r)}$ due to specified interior loading, we then relax all boundaries simultaneously with the

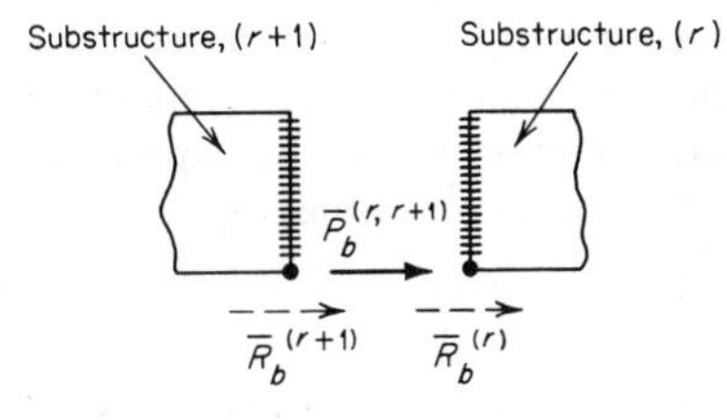

FIG. 9.2 Joint loads before boundary relaxation.

exception of a number of selected displacements serving to form a reference datum for the complete structure. When the boundaries are relaxed, the boundary reactions and any external forces applied on the boundaries will not be in balance, and therefore the boundary relaxation will induce boundary displacements of such magnitude as to satisfy equilibrium at each joint on the boundary. To calculate these boundary displacements the complete structure can be regarded as an assembly of substructures subjected to external loading

$$\bar{\mathbf{S}}_b = -\sum_r \bar{\mathbf{R}}_b^{(r)} + \bar{\mathbf{P}}_b \tag{9.32}$$

where the summation implies here addition of the corresponding boundary reactions for boundaries fixed, while $\bar{\mathbf{P}}_b$ is the loading matrix for external forces applied on the boundaries; the negative sign with $\bar{\mathbf{R}}_b^{(r)}$ is used to change the boundary reactions into externally applied forces, as indicated by Eq. (9.14). In Fig. 9.2 a typical joint on the common boundary between substructures (r) and $(r+1)$ is shown. Here a typical resultant boundary load $\bar{S}_b^{(r,r+1)}$ is given by

$$\bar{S}_b^{(r,r+1)} = -\bar{R}_b^{(r)} - \bar{R}_b^{(r+1)} + \bar{P}_b^{(r,r+1)} \tag{9.33}$$

The equations of equilibrium in terms of boundary displacements for the complete structure can now be written as

$$\mathbf{K}_b\mathbf{U}_b = \bar{\mathbf{S}}_b \tag{9.34}$$

where $\mathbf{K}_b$ is obtained by placing the submatrices $\mathbf{K}_b^{(r)}$ in their correct positions in the larger framework of the boundary stiffness matrix for the complete structure and summing all the overlapping terms. Elimination of a sufficient number of displacements to restrain rigid-body degrees of freedom for the complete structure ensures that the matrix $\mathbf{K}_b$ is nonsingular, and therefore the boundary displacements $\mathbf{U}_b$ can be determined from

$$\mathbf{U}_b = \mathbf{K}_b^{-1}\bar{\mathbf{S}}_b \tag{9.35}$$

Before proceeding to the analysis of loads and displacements on separate substructures, the displacement matrix $\mathbf{U}_b$ must be expanded into a column of substructure displacements $\mathbf{U}_b^{(r)}$, in the exact order in which they appear in Eq. (9.17). This can be obtained by a simple matrix transformation

$$\{\mathbf{U}_b^{(1)} \quad \mathbf{U}_b^{(2)} \quad \cdots \quad \mathbf{U}_b^{(r)} \quad \cdots\} = \mathbf{A}_b\mathbf{U}_b \tag{9.36}$$

where the matrix $\mathbf{A}_b$ is of the same type as the transformation matrix $\mathbf{A}$ used in Chap. 6.

When the substructure stiffness matrices $\mathbf{K}_b^{(r)}$ are assembled into the larger stiffness matrix $\mathbf{K}_b$ for the complete structure, their relative positions in this larger matrix depend on the sequence in which the individual boundary displacements are selected in Eq. (9.34). Since some of the substructures will not be physically connected, this means that their coupling stiffness matrices will be equal to zero. As the coupling matrices occur only on substructures which have common boundaries, it is therefore advantageous, when selecting a numbering system for substructures and displacements, to ensure that the component submatrices of $\mathbf{K}_b$ will occur around the principal diagonal, forming a band matrix. This arrangement may result in a considerable saving in the computing time, if special inversion programs for band matrices (also known as continuant matrices) are used to determine $\mathbf{K}_b^{-1}$.

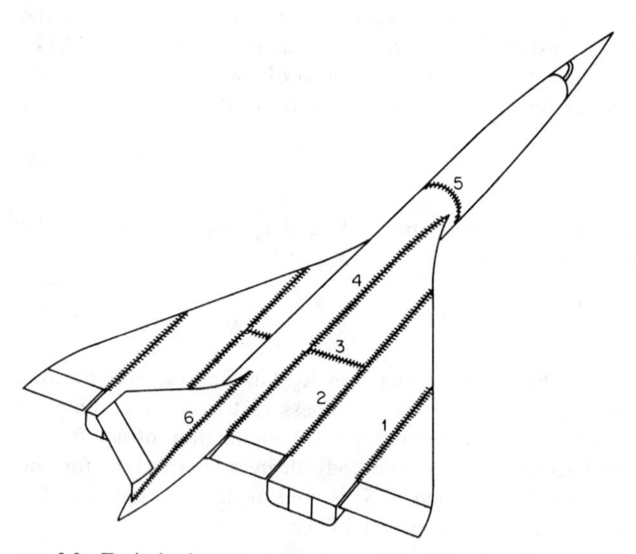

FIG. 9.3 Typical substructure arrangement for delta aircraft.

A typical substructure arrangement for a delta-wing aircraft is shown in Fig. 9.3; this arrangement results in a quintuple band matrix for $\mathbf{K}_b$

$$\mathbf{K}_b = \begin{bmatrix} [1,1] & [1,2] & \mathbf{0} & \mathbf{0} & \mathbf{0} & \mathbf{0} \\ [2,1] & [2,2] & [2,3] & [2,4] & \mathbf{0} & \mathbf{0} \\ \mathbf{0} & [3,2] & [3,3] & [3,4] & \mathbf{0} & \mathbf{0} \\ \mathbf{0} & [4,2] & [4,3] & [4,4] & [4,5] & [4,6] \\ \mathbf{0} & \mathbf{0} & \mathbf{0} & [5,4] & [5,5] & [5,6] \\ \mathbf{0} & \mathbf{0} & \mathbf{0} & [6,4] & [6,5] & [6,6] \end{bmatrix} \tag{9.37}$$

where the row and column numbers denote the substructure boundary numbers. For an aircraft whose wing has a high aspect ratio, the substructure boundaries can be selected in such a way as to ensure that each substructure will have common boundaries with not more than two adjacent substructures. A typical arrangement for these cases is shown in Fig. 9.4. Here even a greater economy of computing effort can be achieved since the boundary-stiffness matrix $\mathbf{K}_b$ would result in a triple band matrix

$$\mathbf{K}_b = \begin{bmatrix} [1,1] & [1,2] & \mathbf{0} & \mathbf{0} \\ [2,1] & [2,2] & [2,3] & \mathbf{0} \\ \mathbf{0} & [3,2] & [3,3] & [3,4] \\ \mathbf{0} & \mathbf{0} & [4,3] & [4,4] \end{bmatrix} \tag{9.38}$$

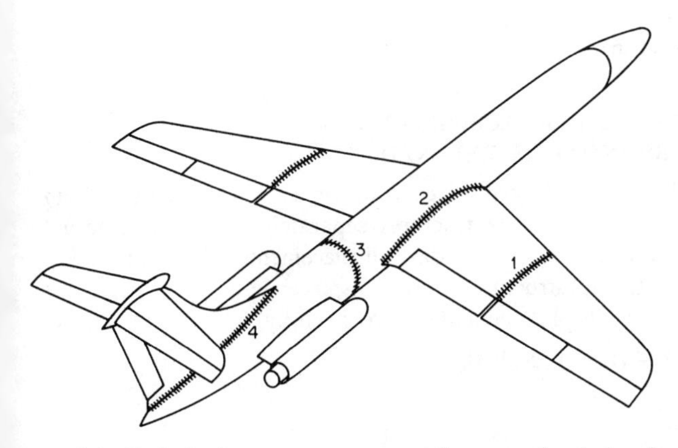

FIG. 9.4 Typical substructure arrangement for conventional aircraft.

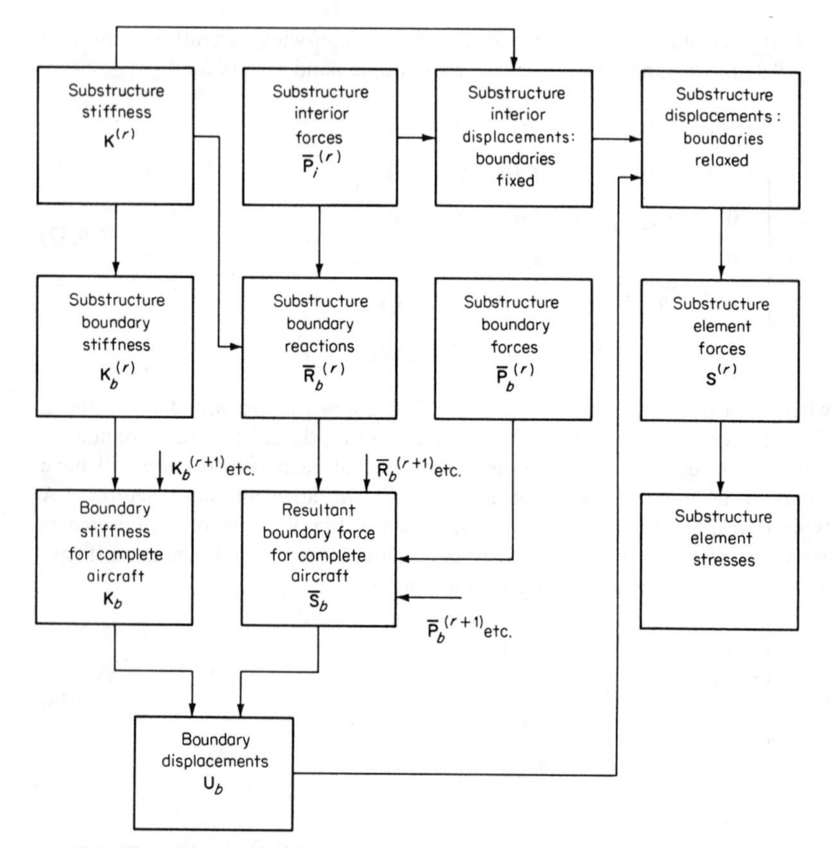

FIG. 9.5 Flow diagram of analysis.

THE SUBSTRUCTURE DISPLACEMENTS AND FORCES: BOUNDARIES RELAXED

Having determined the boundary displacements on each substructure from Eq. (9.36), we can analyze the substructures separately under the external loading $\bar{\mathbf{P}}_i^{(r)}$ together with known boundary displacements $\mathbf{U}_b^{(r)}$. From Eq. (9.17) it follows that the substructure interior displacements $\mathbf{U}_i^{(r)}$ due to the forces $\bar{\mathbf{P}}_i^{(r)}$ and boundary displacements $\mathbf{U}_b^{(r)}$ are given by

$$\mathbf{U}_i^{(r)} = (\mathbf{K}_{ii}^{(r)})^{-1}\bar{\mathbf{P}}_i^{(r)} - (\mathbf{K}_{ii}^{(r)})^{-1}\mathbf{K}_{ib}^{(r)}\mathbf{U}_b^{(r)} \tag{9.39}$$

whence

$$\begin{bmatrix} \mathbf{U}_b^{(r)} \\ \hdashline \mathbf{U}_i^{(r)} \end{bmatrix}_{\substack{\text{boundaries}\\\text{relaxed}}} = \begin{bmatrix} \mathbf{0} \\ \hdashline \mathbf{U}_i^{(r)} \end{bmatrix}_{\substack{\text{boundaries}\\\text{fixed}}} + \begin{bmatrix} \mathbf{I} \\ \hdashline -(\mathbf{K}_{ii}^{(r)})^{-1}\mathbf{K}_{ib}^{(r)} \end{bmatrix} \mathbf{U}_b^{(r)} \tag{9.40}$$

or

$$\left[\mathbf{U}^{(r)}\right]_{\substack{\text{boundaries}\\ \text{relaxed}}} = \left[\mathbf{U}^{(r)}\right]_{\substack{\text{boundaries}\\ \text{fixed}}} + \begin{bmatrix}\text{displacements due to}\\ \text{boundary relaxation}\end{bmatrix} \tag{9.40a}$$

A schematic flow diagram for the complete displacement analysis of substructures is shown in Fig. 9.5, where for simplicity only the main steps in the computation have been indicated. The diagram illustrates how the individual substructure analyses are assembled together to form boundary stiffness and boundary force matrices for the complete structure followed by the calculation of substructure boundary displacements, which are subsequently used to determine displacements and forces in each substructure independently.

9.2 SUBSTRUCTURE DISPLACEMENT ANALYSIS OF A TWO-BAY TRUSS

As an illustrative example of the substructure displacement analysis a simple two-bay pin-jointed truss (Fig. 9.6) will be analyzed. This cantilever truss is attached at one end to a rigid wall and is loaded by external forces P_1, P_2, and P_3 at the free end. The truss will be partitioned into two substructures by disconnecting the outer bay from the remainder of the structure. An exploded view of the two selected substructures is shown in Fig. 9.7. Naturally, other choices for partitioning are possible also. For example, the center vertical member can be sliced vertically into two halves to form two substructures.

The first step in the substructure displacement analysis is the determination of substructure stiffness matrices, which are obtained by combining the element stiffness matrices in the datum system into stiffness matrices for assembled substructures using the summation procedure described in Chap. 6. Since the

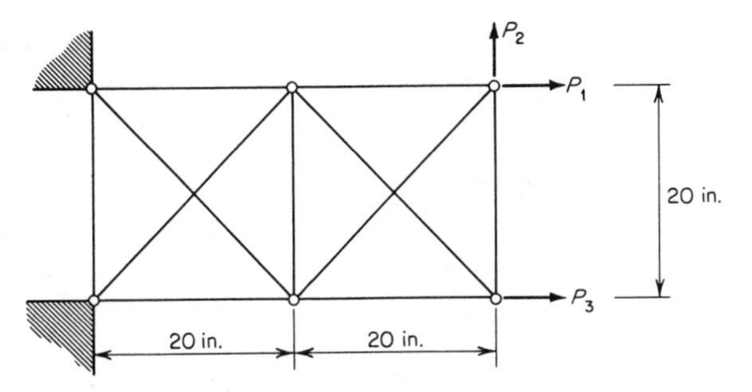

FIG. 9.6 Truss geometry and loading. Cross-sectional areas: vertical and horizontal bars 1.0 in.^{2}; diagonal bars 0.707 in.^{2} ($\sqrt{2}/2$ in.^{2}); $E = 10 \times 10^6$ lb/in.^{2}

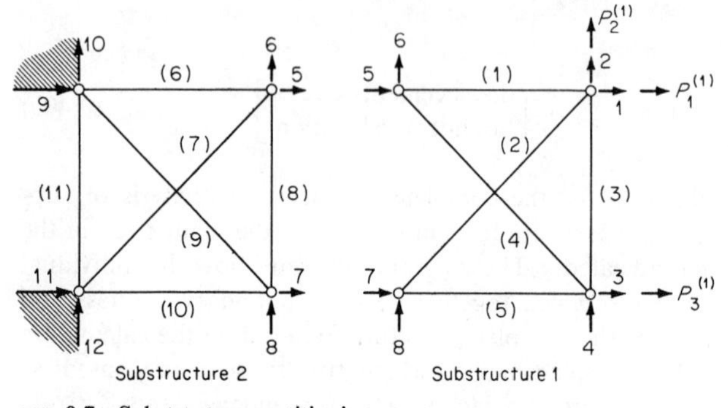

FIG. 9.7 Substructure partitioning.

present example uses the same bar elements as the example of Sec. 6.9, we can use previously derived element stiffnesses. Hence compiling the substructure stiffnesses, we obtain

$$\mathbf{K}^{(1)} = \begin{array}{c} 1 \\ 2 \\ 3 \\ 4 \\ 5 \\ 6 \\ 7 \\ 8 \end{array} \begin{bmatrix} 5 & & & & & & & \\ 1 & 5 & & & & \text{Symmetric} & & \\ 0 & 0 & 5 & & & & & \\ 0 & -4 & -1 & 5 & & & & \\ -4 & 0 & -1 & 1 & 5 & & & \\ 0 & 0 & 1 & -1 & -1 & 1 & & \\ -1 & -1 & -4 & 0 & 0 & 0 & 5 & \\ -1 & -1 & 0 & 0 & 0 & 0 & 1 & 1 \end{bmatrix} 0.125 \times 10^6$$
$$\begin{array}{cccccccc} 1 & 2 & 3 & 4 & 5 & 6 & 7 & 8 \end{array}$$

$$\mathbf{K}^{(2)} = \begin{array}{c} 5 \\ 6 \\ 7 \\ 8 \\ 9 \\ 10 \\ 11 \\ 12 \end{array} \begin{bmatrix} 5 & & & & & & & \\ 1 & 5 & & & & & & \\ 0 & 0 & 5 & & & \text{Symmetric} & & \\ 0 & -4 & -1 & 5 & & & & \\ -4 & 0 & -1 & 1 & 5 & & & \\ 0 & 0 & 1 & -1 & -1 & 5 & & \\ -1 & -1 & -4 & 0 & 0 & 0 & 5 & \\ -1 & -1 & 0 & 0 & 0 & -4 & 1 & 5 \end{bmatrix} 0.125 \times 10^6$$
$$\begin{array}{cccccccc} 5 & 6 & 7 & 8 & 9 & 10 & 11 & 12 \end{array}$$

where the row and column numbers refer to the degrees of freedom designated in Fig. 9.7. As before, the units used are pounds and inches.

To calculate boundary stiffness matrices we use Eq. (9.20). Thus for the first substructure we have

$$\mathbf{K}_b^{(1)} = \mathbf{K}_{bb}^{(1)} - \mathbf{K}_{bi}^{(1)}(\mathbf{K}_{ii}^{(1)})^{-1}\mathbf{K}_{ib}^{(1)}$$

$$= \begin{array}{c} \\ 5 \\ 6 \\ 7 \\ 8 \end{array}\!\!\begin{array}{c} \begin{array}{cccc} 5 & 6 & 7 & 8 \end{array} \\ \begin{bmatrix} 5 & -1 & 0 & 0 \\ -1 & 1 & 0 & 0 \\ 0 & 0 & 5 & 1 \\ 0 & 0 & 1 & 1 \end{bmatrix} \end{array} \frac{10^6}{8} - \begin{array}{c} \\ 5 \\ 6 \\ 7 \\ 8 \end{array}\!\!\begin{array}{c} \begin{array}{cccc} 1 & 2 & 3 & 4 \end{array} \\ \begin{bmatrix} -4 & 0 & -1 & 1 \\ 0 & 0 & 1 & -1 \\ -1 & -1 & -4 & 0 \\ -1 & -1 & 0 & 0 \end{bmatrix} \end{array} \frac{10^6}{8}$$

$$\times \left(\begin{array}{c} \\ 1 \\ 2 \\ 3 \\ 4 \end{array}\!\!\begin{array}{c} \begin{array}{cccc} 1 & 2 & 3 & 4 \end{array} \\ \begin{bmatrix} 5 & 1 & 0 & 0 \\ 1 & 5 & 0 & -4 \\ 0 & 0 & 5 & -1 \\ 0 & -4 & -1 & 5 \end{bmatrix} \end{array} \frac{10^6}{8} \right)^{-1} \begin{array}{c} \\ 1 \\ 2 \\ 3 \\ 4 \end{array}\!\!\begin{array}{c} \begin{array}{cccc} 5 & 6 & 7 & 8 \end{array} \\ \begin{bmatrix} -4 & 0 & -1 & -1 \\ 0 & 0 & -1 & -1 \\ -1 & 1 & -4 & 0 \\ 1 & -1 & 0 & 0 \end{bmatrix} \end{array} \frac{10^6}{8}$$

$$= \begin{array}{c} \\ 5 \\ 6 \\ 7 \\ 8 \end{array}\!\!\begin{array}{c} \begin{array}{cccc} 5 & 6 & 7 & 8 \end{array} \\ \begin{bmatrix} 0 & 0 & 0 & 0 \\ 0 & 1 & 0 & -1 \\ 0 & 0 & 0 & 0 \\ 0 & -1 & 0 & 1 \end{bmatrix} \end{array} \frac{10^6}{22}$$

Displacements 1 to 4 represent the interior displacements on substructure 1; however, there are no interior displacements on substructure 2, hence

$$\mathbf{K}_b^{(2)} = \mathbf{K}^{(2)}$$

Combining $\mathbf{K}_b^{(1)}$ and $\mathbf{K}_b^{(2)}$ matrices to form the boundary stiffness matrix for the entire structure and at the same time eliminating rows and columns 9 to 12, we obtain

$$\mathbf{K}_b = \begin{array}{c} \\ 5 \\ 6 \\ 7 \\ 8 \end{array}\!\!\begin{array}{c} \begin{array}{cccc} 5 & 6 & 7 & 8 \end{array} \\ \begin{bmatrix} 55 & 11 & 0 & 0 \\ 11 & 59 & 0 & -48 \\ 0 & 0 & 55 & -11 \\ 0 & -48 & -11 & 59 \end{bmatrix} \end{array} \frac{10^6}{88}$$

To determine the resultant boundary forces $\mathbf{S}_b$ we use*

$$\mathbf{S}_b = \mathbf{P}_b - \mathbf{R}_b$$
$$= \mathbf{P}_b - \mathbf{K}_{bi}^{(1)}(\mathbf{K}_{ii}^{(1)})^{-1}\mathbf{P}_i^{(1)}$$

* Bars over $\mathbf{S}_b$, $\mathbf{P}_b$, etc., are not used here because of the absence of thermal loading in this example.

Noting now that there are no external forces on the substructure boundaries we have that

$$\mathbf{P}_b = \mathbf{0}$$

Hence

$$\mathbf{S}_b = -\begin{array}{c} \\ 5\\6\\7\\8\end{array}\!\!\begin{array}{c}\begin{array}{cccc}1&2&3&4\end{array}\\ \begin{bmatrix}-4 & 0 & -1 & 1\\ 0 & 0 & 1 & -1\\ -1 & -1 & -4 & 0\\ -1 & -1 & 0 & 0\end{bmatrix}\end{array}\frac{10^6}{8}\left(\begin{array}{c} \\ 1\\2\\3\\4\end{array}\!\!\begin{array}{c}\begin{array}{cccc}1&2&3&4\end{array}\\ \begin{bmatrix}5 & 1 & 0 & 0\\ 1 & 5 & 0 & -4\\ 0 & 0 & 5 & -1\\ 0 & -4 & -1 & 5\end{bmatrix}\end{array}\frac{10^6}{8}\right)^{-1}\begin{array}{c}1\\2\\3\\4\end{array}\!\!\begin{bmatrix}P_1^{(1)}\\ P_2^{(1)}\\ P_3^{(1)}\\ 0\end{bmatrix}$$

$$= \begin{array}{c}5\\6\\7\\8\end{array}\!\!\begin{bmatrix}11 & -11 & 0 & -11\\ -1 & 5 & -1 & 6\\ 0 & 11 & 11 & 11\\ 1 & 6 & 1 & 5\end{bmatrix}\frac{1}{11}\begin{bmatrix}P_1^{(1)}\\ P_2^{(1)}\\ P_3^{(1)}\\ 0\end{bmatrix} = \begin{array}{c}5\\6\\7\\8\end{array}\!\!\begin{bmatrix}11 & -11 & 0\\ -1 & 5 & -1\\ 0 & 11 & 11\\ 1 & 6 & 1\end{bmatrix}\frac{1}{11}\begin{bmatrix}P_1^{(1)}\\ P_2^{(1)}\\ P_3^{(1)}\end{bmatrix}$$

The boundary displacements $\mathbf{U}_b$ are determined from

$$\mathbf{U}_b = (\mathbf{K}_b)^{-1}\mathbf{S}_b$$

$$= \left(\begin{array}{c} \\ 5\\6\\7\\8\end{array}\!\!\begin{array}{c}\begin{array}{cccc}5&6&7&8\end{array}\\ \begin{bmatrix}55 & 11 & 0 & 0\\ 11 & 59 & 0 & -48\\ 0 & 0 & 55 & -11\\ 0 & -48 & -11 & 59\end{bmatrix}\end{array}\frac{10^6}{88}\right)^{-1}\begin{array}{c}5\\6\\7\\8\end{array}\!\!\begin{bmatrix}11 & -11 & 0\\ -1 & 5 & -1\\ 0 & 11 & 11\\ 1 & 6 & 1\end{bmatrix}\frac{1}{11}\begin{bmatrix}P_1^{(1)}\\ P_2^{(1)}\\ P_3^{(1)}\end{bmatrix}$$

$$= \begin{array}{c} \\ 5\\6\\7\\8\end{array}\!\!\begin{array}{c}\begin{array}{cccc}5&6&7&8\end{array}\\ \begin{bmatrix}119 & -71 & -12 & -60\\ -71 & 355 & 60 & 300\\ -12 & 60 & 119 & 71\\ -60 & 300 & 71 & 355\end{bmatrix}\end{array}\frac{2\times 10^{-6}}{131}\begin{array}{c}5\\6\\7\\8\end{array}\!\!\begin{bmatrix}11 & -11 & 0\\ -1 & 5 & -1\\ 0 & 11 & 11\\ 1 & 6 & 1\end{bmatrix}\frac{1}{11}\begin{bmatrix}P_1^{(1)}\\ P_2^{(1)}\\ P_3^{(1)}\end{bmatrix}$$

$$= \begin{array}{c}5\\6\\7\\8\end{array}\!\!\begin{bmatrix}120 & -196 & -11\\ -76 & 456 & 55\\ -11 & 197 & 120\\ -55 & 461 & 76\end{bmatrix}\frac{2\times 10^{-6}}{131}\begin{bmatrix}P_1^{(1)}\\ P_2^{(1)}\\ P_3^{(1)}\end{bmatrix}$$

$$= \begin{array}{c}5\\6\\7\\8\end{array}\!\!\begin{bmatrix}1.832 & -2.992 & -0.168\\ -1.160 & 6.962 & 0.840\\ -0.168 & 3.008 & 1.832\\ -0.840 & 7.038 & 1.160\end{bmatrix}10^{-6}\begin{bmatrix}P_1^{(1)}\\ P_2^{(1)}\\ P_3^{(1)}\end{bmatrix}$$

while the interior displacements are calculated from

$$\mathbf{U}_i^{(1)} = (\mathbf{K}_{ii}^{(1)})^{-1}\mathbf{P}_i^{(1)} - (\mathbf{K}_{ii}^{(1)})^{-1}\mathbf{K}_{ib}^{(1)}\mathbf{U}_b^{(1)}$$

Now

$$(\mathbf{K}_{ii}^{(1)})^{-1}\mathbf{P}_i^{(1)} = \left\{ \begin{matrix} \\ 1\\2\\3\\4 \end{matrix} \begin{matrix} \begin{matrix} 1 & 2 & 3 & 4 \end{matrix} \\ \begin{bmatrix} 5 & 1 & 0 & 0 \\ 1 & 5 & 0 & -4 \\ 0 & 0 & 5 & -1 \\ 0 & -4 & -1 & 5 \end{bmatrix} \end{matrix} \frac{10^6}{8} \right\}^{-1} \begin{bmatrix} P_1^{(1)} \\ P_2^{(1)} \\ P_3^{(1)} \\ 0 \end{bmatrix}$$

$$= \begin{matrix} \\ 1\\2\\3\\4 \end{matrix} \begin{matrix} \begin{matrix} 1 & 2 & 3 & 4 \end{matrix} \\ \begin{bmatrix} 10 & -6 & -1 & -5 \\ -6 & 30 & 5 & 25 \\ -1 & 5 & 10 & 6 \\ -5 & 25 & 6 & 30 \end{bmatrix} \end{matrix} \frac{2 \times 10^{-6}}{11} \begin{bmatrix} P_1^{(1)} \\ P_2^{(1)} \\ P_3^{(1)} \\ 0 \end{bmatrix}$$

$$= \begin{matrix} 1\\2\\3\\4 \end{matrix} \begin{bmatrix} 10 & -6 & -1 \\ -6 & 30 & 5 \\ -1 & 5 & 10 \\ -5 & 25 & 6 \end{bmatrix} \frac{2 \times 10^{-6}}{11} \begin{bmatrix} P_1^{(1)} \\ P_2^{(1)} \\ P_3^{(1)} \end{bmatrix}$$

and

$$(\mathbf{K}_{ii}^{(1)})^{-1}\mathbf{K}_{ib}^{(1)}\mathbf{U}_b^{(1)}$$

$$= \begin{matrix} \\ 1\\2\\3\\4 \end{matrix} \begin{matrix} \begin{matrix} 1 & 2 & 3 & 4 \end{matrix} \\ \begin{bmatrix} 10 & -6 & -1 & -5 \\ -6 & 30 & 5 & 25 \\ -1 & 5 & 10 & 6 \\ -5 & 25 & 6 & 30 \end{bmatrix} \end{matrix} \frac{2 \times 10^{-6}}{11} \begin{matrix} \\ 1\\2\\3\\4 \end{matrix} \begin{matrix} \begin{matrix} 5 & 6 & 7 & 8 \end{matrix} \\ \begin{bmatrix} -4 & 0 & -1 & -1 \\ 0 & 0 & -1 & -1 \\ -1 & 1 & -4 & 0 \\ 1 & -1 & 0 & 0 \end{bmatrix} \end{matrix} \frac{10^6}{8}$$

$$\times \begin{matrix} 5\\6\\7\\8 \end{matrix} \begin{bmatrix} 120 & -196 & -11 \\ -76 & 456 & 55 \\ -11 & 197 & 120 \\ -55 & 461 & 76 \end{bmatrix} \frac{2 \times 10^{-6}}{131} \begin{bmatrix} P_1^{(1)} \\ P_2^{(1)} \\ P_3^{(1)} \end{bmatrix}$$

$$= \begin{matrix} 1\\2\\3\\4 \end{matrix} \begin{bmatrix} -1{,}341 & 2{,}151 & 100 \\ 2{,}151 & -9{,}369 & -2{,}172 \\ 100 & -2{,}172 & -1{,}341 \\ 2{,}172 & -9{,}364 & -2{,}151 \end{bmatrix} \frac{2 \times 10^{-6}}{11 \times 131} \begin{bmatrix} P_1^{(1)} \\ P_2^{(1)} \\ P_3^{(1)} \end{bmatrix}$$

Hence

$$
\mathbf{U}_i^{(1)} = \left\{ \begin{matrix} 1 \\ 2 \\ 3 \\ 4 \end{matrix} \begin{bmatrix} 10 & -6 & -1 \\ -6 & 30 & 5 \\ -1 & 5 & 10 \\ -5 & 25 & 6 \end{bmatrix} \frac{2 \times 10^{-6}}{11} \right.
$$

$$
\left. - \begin{matrix} 1 \\ 2 \\ 3 \\ 4 \end{matrix} \begin{bmatrix} -1{,}341 & 2{,}151 & 100 \\ 2{,}151 & -9{,}369 & -2{,}172 \\ 100 & -2{,}172 & -1{,}341 \\ 2{,}172 & -9{,}364 & -2{,}151 \end{bmatrix} \frac{2 \times 10^{-6}}{11 \times 131} \right\} \begin{bmatrix} P_1^{(1)} \\ P_2^{(1)} \\ P_3^{(1)} \end{bmatrix}
$$

$$
= \begin{matrix} 1 \\ 2 \\ 3 \\ 4 \end{matrix} \begin{bmatrix} 241 & -267 & -21 \\ -267 & 1{,}209 & 257 \\ -21 & 257 & 241 \\ -257 & 1{,}149 & 267 \end{bmatrix} \frac{2 \times 10^{-6}}{131} \begin{bmatrix} P_1^{(1)} \\ P_2^{(1)} \\ P_3^{(1)} \end{bmatrix}
$$

$$
= \begin{matrix} 1 \\ 2 \\ 3 \\ 4 \end{matrix} \begin{bmatrix} 3.679 & -4.076 & -0.321 \\ -4.076 & 18.458 & 3.924 \\ -0.321 & 3.924 & 3.679 \\ -3.924 & 17.542 & 4.076 \end{bmatrix} \times 10^{-6} \begin{bmatrix} P_1^{(1)} \\ P_2^{(1)} \\ P_3^{(1)} \end{bmatrix}
$$

The matrices $\mathbf{U}_b$ and $\mathbf{U}_i^{(1)}$ give all the required displacements. The element forces and stresses on individual elements can then be determined from the element stiffness matrices and the joint displacements.

9.3 SUBSTRUCTURE ANALYSIS BY THE MATRIX FORCE METHOD[278]

UNASSEMBLED SUBSTRUCTURES

For generality, it will be assumed that the complete structure is divided into N substructures. A typical substructure arrangement is shown in Fig. 9.8, where a medium-range transport-aircraft structure is divided into six structural components, wing, center fuselage, front fuselage, rear fuselage, engine pylon, and vertical stabilizer. Now, consider a typical substructure which has been detached completely from the remaining substructures (see Fig. 9.9). The external loading applied to this particular substructure r will be denoted by a column matrix $\mathbf{P}^{(r)}$, where the superscript r denotes the rth substructure. Since the disconnecting procedure is carried out by cutting all attachments on the boundaries, we must restore equilibrium and compatibility of unassembled

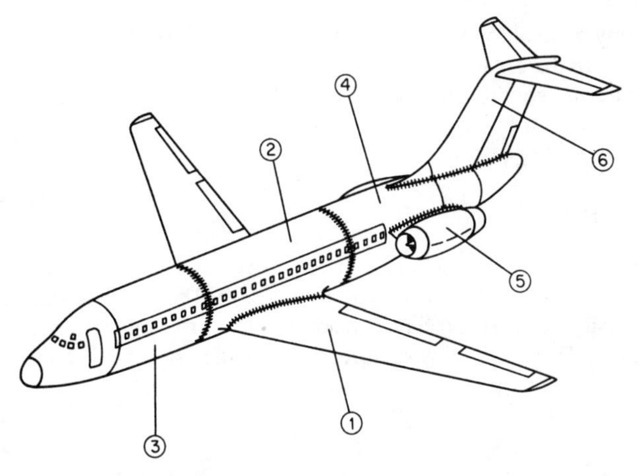

FIG. 9.8 Typical substructure arrangement.

(disconnected) substructures by applying joint internal forces to each substructure. For the rth substructure these forces will be denoted by column matrices $\mathbf{Q}^{(r)}$ and $\mathbf{F}^{(r)}$, with $\mathbf{Q}^{(r)}$ representing substructure reactions, introduced only temporarily to establish a reference datum for substructure displacements and flexibilities, and $\mathbf{F}^{(r)}$ representing all the remaining boundary forces. The forces $\mathbf{F}^{(r)}$ will be referred to as the *interaction forces*. The only restriction imposed on $\mathbf{Q}^{(r)}$ is that it must constitute a set of statically determinate reactions

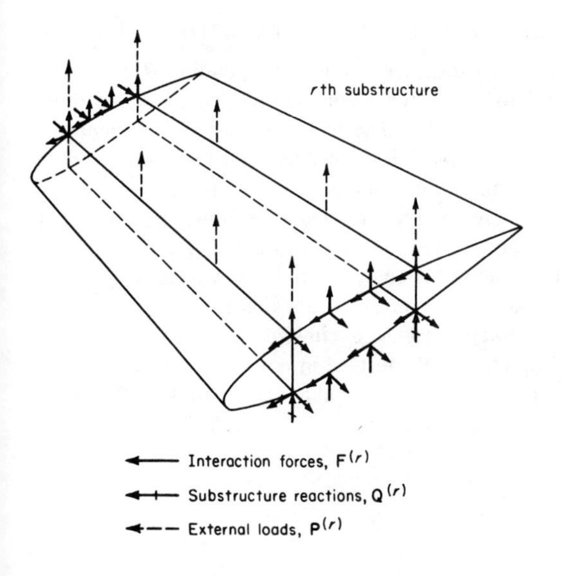

FIG. 9.9 Substructure forces.

capable of reacting any general loading. Thus for a general three-dimensional structure, there will be six forces in $\mathbf{Q}^{(r)}$.

The equation of external equilibrium for the rth substructure can be written in matrix form as

$$\mathbf{Q}^{(r)} = [\mathbf{Q}_F{}^{(r)} \quad \mathbf{Q}_P{}^{(r)}]\begin{bmatrix}\mathbf{F}^{(r)} \\ \mathbf{P}^{(r)}\end{bmatrix} \tag{9.41}$$

where $\mathbf{Q}_F{}^{(r)}$ and $\mathbf{Q}_P{}^{(r)}$ denote substructure reaction forces due to unit values of $\mathbf{F}^{(r)}$ and $\mathbf{P}^{(r)}$, respectively. The formulation of Eq. (9.41) involves only equations of statics, since the substructure reactions $\mathbf{Q}^{(r)}$ are statically determinate. Equation (9.41) for all substructures can now be assembled into a single matrix equation

$$\mathbf{Q} = [\mathbf{Q}_F \quad \mathbf{Q}_P]\begin{bmatrix}\mathbf{F} \\ \mathbf{P}\end{bmatrix} \tag{9.42}$$

where
$$\mathbf{Q} = \{\mathbf{Q}^{(1)} \quad \mathbf{Q}^{(2)} \quad \cdots \quad \mathbf{Q}^{(N)}\} \tag{9.43}$$

$$\mathbf{Q}_F = [\mathbf{Q}_F{}^{(1)} \quad \mathbf{Q}_F{}^{(2)} \quad \cdots \quad \mathbf{Q}_F{}^{(N)}] \tag{9.44}$$

$$\mathbf{Q}_P = [\mathbf{Q}_P{}^{(1)} \quad \mathbf{Q}_P{}^{(2)} \quad \cdots \quad \mathbf{Q}_P{}^{(N)}] \tag{9.45}$$

$$\mathbf{F} = \{\mathbf{F}^{(1)} \quad \mathbf{F}^{(2)} \quad \cdots \quad \mathbf{F}^{(N)}\} \tag{9.46}$$

$$\mathbf{P} = \{\mathbf{P}^{(1)} \quad \mathbf{P}^{(2)} \quad \cdots \quad \mathbf{P}^{(N)}\} \tag{9.47}$$

INTERNAL EQUILIBRIUM OF THE JOINED STRUCTURE

When the substructures are joined together, the external loading $\mathbf{P}$ is reacted by the joined-structure reactions. The force complements to these reactions will be denoted by the column matrix $\mathbf{R}$. The reaction force complement has already been introduced in Chap. 8, and it is defined as the force applied by the structure to the support. The minimum number of reactions is equal to the number of rigid-body degrees of freedom for the structure; however, additional reactions may be needed to represent redundant constraints. For example, if only one-half of a symmetric structure is analyzed under a symmetric loading, reactions across the plane of symmetry must be introduced to represent the symmetry constraint. The interaction forces $\mathbf{F}$, substructure reactions $\mathbf{Q}$, and joined-structure reaction complements $\mathbf{R}$ must be in equilibrium at each joint on the substructure boundaries. A typical joint is shown in Fig. 9.10, for which the equations of internal equilibrium are

$$F_i^{(r)} + F_j^{(r+1)} = 0 \tag{9.48}$$

$$F_{i+1}^{(r)} + F_{j+1}^{(r+1)} = 0 \tag{9.49}$$

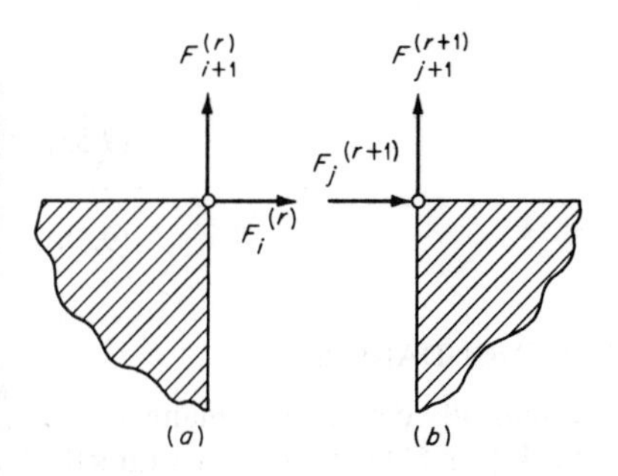

FIG. 9.10 Interaction forces at a typical joint. (*a*) *r*th substructure; (*b*) $(r+1)$st substructure.

Naturally, at some joints the equilibrium equations may contain the substructure reaction force $\mathbf{Q}$ or the joined-structure reaction force complement $\mathbf{R}$.

Equations of internal equilibrium may be formulated with reference to a common set of axes. However, in general, it is preferable to use pairs of corresponding boundary forces and formulate equilibrium equations in the direction of each pair. The internal equilibrium requires then that the sum of the two forces in every pair must be equal to zero.

The equations of internal equilibrium at the boundary joints can therefore be obtained simply from the substructure connectivity. These equations can be expressed in the form

$$[\mathbf{n}_Q \quad \mathbf{n}_R \quad \mathbf{n}_F]\begin{bmatrix}\mathbf{Q}\\ \mathbf{R}\\ \mathbf{F}\end{bmatrix} = \mathbf{0} \tag{9.50}$$

where every row in the submatrices $\mathbf{n}_Q$, $\mathbf{n}_R$, and $\mathbf{n}_F$ contains only zeros and ones. Substituting Eq. (9.42) into (9.50), in order to eliminate $\mathbf{Q}$, we have that

$$[\mathbf{n}_R \quad \bar{\mathbf{n}}_F \quad \mathbf{n}_P]\begin{bmatrix}\mathbf{R}\\ \mathbf{F}\\ \mathbf{P}\end{bmatrix} = \mathbf{0} \tag{9.51}$$

where $\quad \bar{\mathbf{n}}_F = \mathbf{n}_Q\mathbf{Q}_F + \mathbf{n}_F \qquad (9.52)$

and $\quad \mathbf{n}_P = \mathbf{n}_Q\mathbf{Q}_P \qquad (9.53)$

For subsequent analysis it is preferable to combine $\mathbf{R}$ and $\mathbf{F}$ into a single matrix

$$\bar{\mathbf{F}} = \begin{bmatrix}\mathbf{R}\\ \mathbf{F}\end{bmatrix} \tag{9.54}$$

so that Eq. (9.51) can be rewritten as

$$[\bar{\mathbf{n}} \quad \mathbf{n}_P]\begin{bmatrix}\bar{\mathbf{F}} \\ \mathbf{P}\end{bmatrix} = \mathbf{0} \tag{9.55}$$

where $$\bar{\mathbf{n}} = [\mathbf{n}_R \quad \bar{\mathbf{n}}_F] \tag{9.56}$$

SELECTION OF THE INTERACTION REDUNDANCIES

If the substructures are joined together in a statically determinate manner, the matrix $\bar{\mathbf{n}}$ is a square nonsingular matrix, and Eq. (9.55) can be solved directly to yield

$$\bar{\mathbf{F}} = -\bar{\mathbf{n}}^{-1}\mathbf{n}_P\mathbf{P} \tag{9.57}$$

If, on the other hand, the substructures are connected in a redundant manner, then $\bar{\mathbf{n}}$ is a rectangular matrix with the number of columns greater than the number of rows, and consequently no direct solution to Eq. (9.55) can be obtained. For such cases, as explained in Chap. 8, the equilibrium equations are inadequate in number to determine all the substructure boundary forces, and they must therefore be supplemented by the equations of displacement compatibility. In order to formulate the compatibility equations, we must first select from $\bar{\mathbf{F}}$ a set of redundant interaction forces and then introduce a set of structural cuts corresponding to the selected redundancies on the substructure boundaries. The cut structure becomes then a statically determinate one, as far as the boundary forces are concerned. Naturally, each substructure by itself may be highly redundant. It therefore follows that the substructure boundary forces $\bar{\mathbf{F}}$ can be separated into interaction redundancies $\mathbf{X}$ and forces $\bar{\mathbf{F}}_0$ in the statically determinate cut structure, that is,

$$\bar{\mathbf{F}} = \begin{bmatrix}\bar{\mathbf{F}}_0 \\ \mathbf{X}\end{bmatrix} \tag{9.58}$$

Using Eq. (9.58), we can modify the internal-equilibrium equations (9.55) so that*

$$[\mathbf{n}_0 \quad \mathbf{n}_X \quad \mathbf{n}_P]\begin{bmatrix}\bar{\mathbf{F}}_0 \\ \mathbf{X} \\ \mathbf{P}\end{bmatrix} = \mathbf{0} \tag{9.59}$$

where $$[\mathbf{n}_0 \quad \mathbf{n}_X] = \bar{\mathbf{n}} \tag{9.60}$$

* Note that the equilibrium equations are of different form than in Eq. (8.58). Terms with external loading $\mathbf{P}$ appear here on the left side of Eq. (9.59) instead of the right side as in Eq. (8.58).

Since the forces $\bar{\mathbf{F}}_0$ are statically determinate, it follows that $\mathbf{n}_0^{-1}$ exists and that the solution for $\bar{\mathbf{F}}_0$ is

$$\begin{aligned}\bar{\mathbf{F}}_0 &= -\mathbf{n}_0^{-1}\mathbf{n}_X\mathbf{X} - \mathbf{n}_0^{-1}\mathbf{n}_P\mathbf{P} \\ &= \mathbf{q}_X\mathbf{X} + \mathbf{q}_P\mathbf{P}\end{aligned} \tag{9.61}$$

where $$\mathbf{q}_X = -\mathbf{n}_0^{-1}\mathbf{n}_X \tag{9.62}$$

$$\mathbf{q}_P = -\mathbf{n}_0^{-1}\mathbf{n}_P \tag{9.63}$$

The solution represented by Eq. (9.61) is obtained by the jordanian elimination technique applied to Eq. (9.59). The rectangular matrix $[\bar{\mathbf{n}} \;\; \mathbf{n}_P]$ from Eq. (9.60) is premultiplied by a series of transformation matrices $\mathbf{T}_1, \ldots, \mathbf{T}_n$, where n denotes the number of rows in $\bar{\mathbf{n}}$, which change the submatrix $\mathbf{n}_0$ from $\bar{\mathbf{n}}$ into an identity matrix $\mathbf{I}$. Symbolically, this operation may be represented by the matrix equation

$$\mathbf{T}_n \quad \cdots \quad \mathbf{T}_1[\bar{\mathbf{n}} \mid \mathbf{n}_P] = [\mathbf{I} \mid \mathbf{n}_0^{-1}\mathbf{n}_X \mid \mathbf{n}_0^{-1}\mathbf{n}_P] \tag{9.64}$$

The computer program developed for this operation, frequently referred to as the *structure cutter*, ensures through the use of flexibility weighting factors associated with $\bar{\mathbf{F}}$ that the selection of the interaction redundancies $\mathbf{X}$ leads to the stiffest cut structure.[78] It should be noted that the columns of $\mathbf{n}_X$ selected from $\bar{\mathbf{n}}$ by the computer give a direct indication which boundary interaction forces have been selected as the redundants.

Combining now Eq. (9.61) with the identity $\mathbf{X} = \mathbf{X}$, it follows that

$$\begin{aligned}\bar{\mathbf{F}} = \begin{bmatrix}\bar{\mathbf{F}}_0 \\ \mathbf{X}\end{bmatrix} &= \begin{bmatrix}\mathbf{q}_X \\ \mathbf{I}\end{bmatrix}\mathbf{X} + \begin{bmatrix}\mathbf{q}_P \\ \mathbf{0}\end{bmatrix}\mathbf{P} \\ &= \mathbf{b}_X\mathbf{X} + \mathbf{b}_0\mathbf{P}\end{aligned} \tag{9.65}$$

where $$\mathbf{b}_X = \begin{bmatrix}\mathbf{q}_X \\ \mathbf{I}\end{bmatrix} \tag{9.66}$$

and $$\mathbf{b}_0 = \begin{bmatrix}\mathbf{q}_P \\ \mathbf{0}\end{bmatrix} \tag{9.67}$$

represent substructure boundary forces due to unit values of interaction redundancies $\mathbf{X}$ and external loading $\mathbf{P}$, respectively.

The information as to what columns are selected from $\bar{\mathbf{n}}$ for $\mathbf{n}_0$ and $\mathbf{n}_X$ in the structure-cutter computer program can be used to formulate two column extractor matrices $\mathbf{N}$ and $\mathbf{H}_X$ such that

$$\mathbf{n}_0 = \bar{\mathbf{n}}\mathbf{N} \tag{9.68}$$

and $$\mathbf{n}_X = \bar{\mathbf{n}}\mathbf{H}_X \tag{9.69}$$

The matrices $\mathbf{N}$ and $\mathbf{H}_X$ can be expressed symbolically for the purpose of the analysis as

$$\mathbf{N} = \begin{bmatrix} \mathbf{I} \\ \mathbf{0} \end{bmatrix} \tag{9.70}$$

$$\mathbf{H}_X = \begin{bmatrix} \mathbf{0} \\ \mathbf{I} \end{bmatrix} \tag{9.71}$$

It should be noted, however, that in practice the identity matrices of Eqs. (9.70) and (9.71) will be interspersed (see the numerical example in 9.4). The main purpose of introducing $\mathbf{N}$ and $\mathbf{H}_X$ is to use them to generate $\mathbf{b}_X$ and $\mathbf{b}_0$ directly from $\mathbf{q}_X$ and $\mathbf{q}_P$. It can be demonstrated easily that this is obtained from the equations

$$\mathbf{b}_X = \mathbf{N}\mathbf{q}_X + \mathbf{H}_X \tag{9.72}$$

and $$\mathbf{b}_0 = \mathbf{N}\mathbf{q}_P \tag{9.73}$$

COMPATIBILITY EQUATIONS

The relative displacements in the directions of the interaction forces $\mathbf{F}^{(r)}$ on each substructure can be expressed as

$$\mathbf{v}_F^{(r)} = \mathbf{D}_{FF}^{(r)}\mathbf{F}^{(r)} + \mathbf{D}_{FP}^{(r)}\mathbf{P}^{(r)} + \mathbf{e}_F^{(r)} \tag{9.74}$$

where $\mathbf{D}_{FF}^{(r)}$ and $\mathbf{D}_{FP}^{(r)}$ denote flexibility matrices for the forces $\mathbf{F}^{(r)}$ and $\mathbf{P}^{(r)}$, respectively, and $\mathbf{e}_F^{(r)}$ is a column matrix of initial displacements, e.g., due to temperature distribution. The matrices $\mathbf{D}_{FF}^{(r)}$, $\mathbf{D}_{FP}^{(r)}$, and $\mathbf{e}_F^{(r)}$ are determined here with respect to the substructure datum established by the substructure reactions $\mathbf{Q}^{(r)}$. For all substructures, Eqs. (9.74) can be combined into a single matrix equation

$$\mathbf{v}_F = \mathbf{D}_{FF}\mathbf{F} + \mathbf{D}_{FP}\mathbf{P} + \mathbf{e}_F \tag{9.75}$$

where $$\mathbf{v}_F = \{\mathbf{v}_F^{(1)} \quad \mathbf{v}_F^{(2)} \quad \cdots \quad \mathbf{v}_F^{(N)}\} \tag{9.76}$$

$$\mathbf{D}_{FF} = \lceil \mathbf{D}_{FF}^{(1)} \quad \mathbf{D}_{FF}^{(2)} \quad \cdots \quad \mathbf{D}_{FF}^{(N)} \rfloor \tag{9.77}$$

$$\mathbf{D}_{FP} = \lceil \mathbf{D}_{FP}^{(1)} \quad \mathbf{D}_{FP}^{(2)} \quad \cdots \quad \mathbf{D}_{FP}^{(N)} \rfloor \tag{9.78}$$

$$\mathbf{e}_F = \{\mathbf{e}_F^{(1)} \quad \mathbf{e}_F^{(2)} \quad \cdots \quad \mathbf{e}_F^{(N)}\} \tag{9.79}$$

For generality, it will be assumed that the joined-structure reaction supports move in the directions of $\mathbf{R}$ by some specified amounts represented by the matrix $\mathbf{e}_R$. This implies that if sinking of the supports is to be included in the analysis, the amount of sinking must be entered into $\mathbf{e}_R$ as negative values. For rigid supports $\mathbf{e}_R = \mathbf{0}$. The structural displacements in the directions of

F and **R**, relative to the substructure datum on each substructure, are now defined, and they can be used to establish the equations of compatibility of displacements on the substructure boundaries. These equations can be derived most conveniently by the applications of the unit-load theorem, which states that

$$\mathbf{b}_X{}^T\begin{bmatrix}\mathbf{e}_R\\ \mathbf{v}_F\end{bmatrix} = \mathbf{0} \tag{9.80}$$

or $$\mathbf{b}_X{}^T\bar{\mathbf{v}}_F = \mathbf{0} \tag{9.80a}$$

where $$\bar{\mathbf{v}}_F = \begin{bmatrix}\mathbf{e}_R\\ \mathbf{v}_F\end{bmatrix} \tag{9.81}$$

Upon substituting Eq. (9.75) into (9.80) it follows that

$$\mathbf{b}_X{}^T\begin{bmatrix}\mathbf{0} & \mathbf{0}\\ \mathbf{0} & \mathbf{D}_{FF}\end{bmatrix}\begin{bmatrix}\mathbf{R}\\ \mathbf{F}\end{bmatrix} + \mathbf{b}_X{}^T\begin{bmatrix}\mathbf{0}\\ \mathbf{D}_{FP}\end{bmatrix}\mathbf{P} + \mathbf{b}_X{}^T\begin{bmatrix}\mathbf{e}_R\\ \mathbf{e}_F\end{bmatrix} = \mathbf{0} \tag{9.82}$$

or $$\mathbf{b}_X{}^T\bar{\mathbf{D}}_{FF}\bar{\mathbf{F}} + \mathbf{b}_X{}^T\bar{\mathbf{D}}_{FP}\mathbf{P} + \mathbf{b}_X{}^T\bar{\mathbf{e}} = \mathbf{0} \tag{9.82a}$$

where $$\bar{\mathbf{D}}_{FF} = \begin{bmatrix}\mathbf{0} & \mathbf{0}\\ \mathbf{0} & \mathbf{D}_{FF}\end{bmatrix} \tag{9.83}$$

$$\bar{\mathbf{D}}_{FP} = \begin{bmatrix}\mathbf{0}\\ \mathbf{D}_{FP}\end{bmatrix} \tag{9.84}$$

$$\bar{\mathbf{e}} = \begin{bmatrix}\mathbf{e}_R\\ \mathbf{e}_F\end{bmatrix} \tag{9.85}$$

Using now Eqs. (9.65) and (9.82*a*), we obtain the following equation for the unknown interaction redundancies **X**

$$\mathbf{b}_X{}^T\bar{\mathbf{D}}_{FF}\mathbf{b}_X\mathbf{X} + \mathbf{b}_X{}^T\bar{\mathbf{D}}_{FF}\mathbf{b}_0\mathbf{P} + \mathbf{b}_X{}^T\bar{\mathbf{D}}_{FP}\mathbf{P} + \mathbf{b}_X{}^T\bar{\mathbf{e}} = \mathbf{0} \tag{9.86}$$

or $$\mathbf{D}_{XX}\mathbf{X} + \mathbf{D}_{XP}\mathbf{P} + \mathbf{D}_{Xe} = \mathbf{0} \tag{9.86a}$$

with $$\mathbf{D}_{XX} = \mathbf{b}_X{}^T\bar{\mathbf{D}}_{FF}\mathbf{b}_X \tag{9.87}$$

$$\mathbf{D}_{XP} = \mathbf{b}_X{}^T\bar{\mathbf{D}}_{FF}\mathbf{b}_0 + \mathbf{b}_X{}^T\bar{\mathbf{D}}_{FP} \tag{9.88}$$

$$\mathbf{D}_{Xe} = \mathbf{b}_X{}^T\bar{\mathbf{e}} \tag{9.89}$$

JOINED STRUCTURE

The solution for the redundancies **X** from Eq. (9.86*a*) is

$$\mathbf{X} = -\mathbf{D}_{XX}{}^{-1}(\mathbf{D}_{XP}\mathbf{P} + \mathbf{D}_{Xe}) \tag{9.90}$$

or $$\mathbf{X} = \mathbf{X}_P\mathbf{P} + \mathbf{X}_e\bar{\mathbf{e}} \tag{9.90a}$$

where $\qquad \mathbf{X}_P = -\mathbf{D}_{XX}{}^{-1}\mathbf{D}_{XP} \qquad (9.91)$

and $\qquad \mathbf{X}_e = -\mathbf{D}_{XX}{}^{-1}\mathbf{b}_X{}^T \qquad (9.92)$

represent the interaction redundancies due to unit values of external loads $\mathbf{P}$ and initial displacements $\bar{\mathbf{e}}$, respectively.

Substituting Eq. (9.90*a*) into (9.65), we have finally that

$$\bar{\mathbf{F}} = (\mathbf{b}_0 - \mathbf{b}_X\mathbf{D}_{XX}{}^{-1}\mathbf{D}_{XP})\mathbf{P} - \mathbf{b}_X\mathbf{D}_{XX}{}^{-1}\mathbf{D}_{Xe} \qquad (9.93)$$

or $\qquad \bar{\mathbf{F}} = \bar{\mathbf{F}}_P\mathbf{P} + \bar{\mathbf{F}}_e\bar{\mathbf{e}} \qquad (9.93a)$

where $\qquad \bar{\mathbf{F}}_P = \mathbf{b}_0 - \mathbf{b}_X\mathbf{D}_{XX}{}^{-1}\mathbf{D}_{XP} \qquad (9.94)$

and $\qquad \bar{\mathbf{F}}_e = -\mathbf{b}_X\mathbf{D}_{XX}{}^{-1}\mathbf{b}_X{}^T \qquad (9.95)$

represent the substructure boundary forces due to unit values of external loads $\mathbf{P}$ and initial displacements $\bar{\mathbf{e}}$, respectively. It should be noted that the sequence of forces in $\bar{\mathbf{F}}$ is first $\mathbf{R}$ followed by $\mathbf{F}^{(1)}, \mathbf{F}^{(2)}, \ldots, \mathbf{F}^{(r)}, \ldots, \mathbf{F}^{(N)}$. Consequently, the interaction forces for any substructure can easily be identified and extracted from the computer output. These forces are then used to determine the stress distribution in the substructures, the size of which is such that small-capacity stress-analysis computer programs can be used. The stress analysis of the substructures may, of course, be carried out either by the matrix displacement or force methods. A schematic flow diagram for the complete analysis by the present method is shown in Fig. 9.11, where for simplicity only the main steps in the computation have been indicated.

DEFLECTIONS

To determine a deflection of the joined structure a unit load (either concentrated force or moment) is applied in the direction for which the displacement is required, and for convenience this load will be assumed to coincide with one of the external loads $\mathbf{P}$. Denoting the required displacement by Δ, it follows that the virtual complementary work δW^* is simply

$$\delta W^* = 1 \times \Delta \qquad (9.96)$$

Using now Eqs. (9.75), (9.81), and (9.83) to (9.85), we find the relative displacements in the directions of $\bar{\mathbf{F}}$ due to the external loading and initial deformations to be

$$\bar{\mathbf{v}}_F = \bar{\mathbf{D}}_{FF}\bar{\mathbf{F}} + \bar{\mathbf{D}}_{FP}\mathbf{P} + \bar{\mathbf{e}} \qquad (9.97)$$

while the relative displacement in the direction of Δ is

$$v_\Delta = \bar{\mathbf{D}}_{\Delta F}\bar{\mathbf{F}} + \mathbf{D}_{\Delta P}\mathbf{P} \qquad (9.98)$$

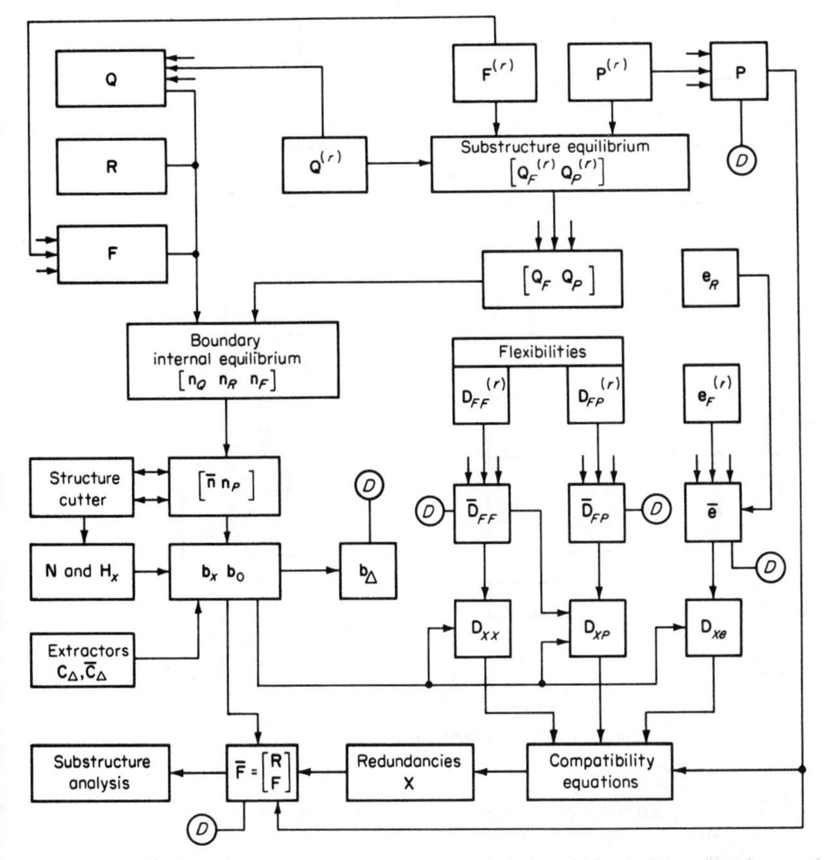

FIG. 9.11 Flow diagram for stress analysis. The encircled symbol D indicates matrices used in the deflection analysis.

where $\bar{\mathbf{D}}_{\Delta F}$ and $\mathbf{D}_{\Delta P}$ represent displacements due to unit values of $\bar{\mathbf{F}}$ and $\mathbf{P}$, respectively. The virtual complementary work can therefore be expressed, alternatively, as

$$\delta W^* = \mathbf{b}_\Delta{}^T \bar{\mathbf{v}}_F + 1 \times v_\Delta \tag{9.99}$$

where $\mathbf{b}_\Delta$ represents the interaction forces $\bar{\mathbf{F}}$ in equilibrium with the unit load in the direction Δ.

After equating Eqs. (9.96) and (9.99) and then using Eqs. (9.97) and (9.98) it is clear that the displacements on the joined structure can be calculated from

$$\Delta = (\mathbf{b}_\Delta{}^T \bar{\mathbf{D}}_{FF} + \bar{\mathbf{D}}_{\Delta F})\bar{\mathbf{F}} + (\mathbf{b}_\Delta{}^T \bar{\mathbf{D}}_{FP} + \mathbf{D}_{\Delta P})\mathbf{P} + \mathbf{b}_\Delta{}^T \bar{\mathbf{e}} \tag{9.100}$$

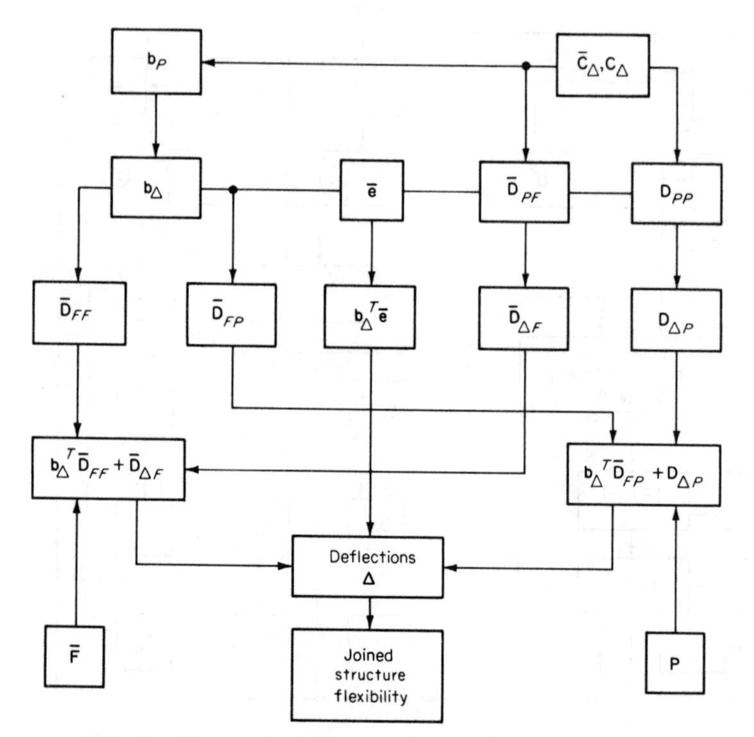

FIG. 9.12 Flow diagram for deflection analysis.

Substituting Eq. (9.93*a*) into (9.100), we can demonstrate that

$$\Delta = \mathbf{\Delta}_P\mathbf{P} + \mathbf{\Delta}_e\bar{\mathbf{e}} \tag{9.101}$$

where $$\mathbf{\Delta}_P = (\mathbf{b}_\Delta{}^T\bar{\mathbf{D}}_{FF} + \bar{\mathbf{D}}_{\Delta F})\bar{\mathbf{F}}_P + \mathbf{b}_\Delta{}^T\bar{\mathbf{D}}_{FP} + \mathbf{D}_{\Delta P} \tag{9.102}$$

and $$\mathbf{\Delta}_e = \mathbf{b}_\Delta{}^T + (\mathbf{b}_\Delta{}^T\bar{\mathbf{D}}_{FF} + \bar{\mathbf{D}}_{\Delta F})\bar{\mathbf{F}}_e \tag{9.103}$$

represent the joined-structure displacements due to unit values of $\mathbf{P}$ and $\bar{\mathbf{e}}$. Equation (9.102) can be used to determine the flexibility matrix for the directions of the applied loads $\mathbf{P}$ if $\mathbf{b}_0$ is used in place of $\mathbf{b}_\Delta$. A schematic flow diagram for the deflection analysis is shown in Fig. 9.12.

In practice, the matrices $\mathbf{b}_\Delta$, $\bar{\mathbf{D}}_{\Delta F}$, and $\mathbf{D}_{\Delta P}$ are determined from

$$\mathbf{b}_\Delta = \mathbf{b}_0\bar{\mathbf{C}}_\Delta \tag{9.104}$$

$$\bar{\mathbf{D}}_{\Delta F} = \mathbf{C}_\Delta\bar{\mathbf{D}}_{PF} \tag{9.105}$$

$$\mathbf{D}_{\Delta P} = \mathbf{C}_\Delta\mathbf{D}_{PP} \tag{9.106}$$

where $\bar{\mathbf{C}}_\Delta$ and $\mathbf{C}_\Delta$ are suitable extractor matrices. The matrices $\bar{\mathbf{D}}_{FF}$, $\bar{\mathbf{D}}_{FP}$, and $\mathbf{D}_{PP}$ are compiled from the substructure flexibility matrices of the form

$$\mathbf{D}^{(r)} = \begin{bmatrix} \mathbf{D}_{FF}{}^{(r)} & \mathbf{D}_{FP}{}^{(r)} \\ \mathbf{D}_{PF}{}^{(r)} & \mathbf{D}_{PP}{}^{(r)} \end{bmatrix} \tag{9.107}$$

9.4 SUBSTRUCTURE FORCE ANALYSIS OF A TWO-BAY TRUSS

To illustrate the substructure force analysis we shall again use the example in Sec. 9.2, as this allows a direct comparison of the two methods. As before, the structure is partitioned into two substructures by disconnecting the outer bay from the remainder of the structure. An exploded view of the two substructures with all the boundary forces **F** and **Q** is shown in Fig. 9.13. The truss is supported at one end in a statically indeterminate manner by four reactions. The reaction-force complements on the joined structure are represented by the symbols $R_1, \ldots, R_4$, and their locations are shown in Fig. 9.13.

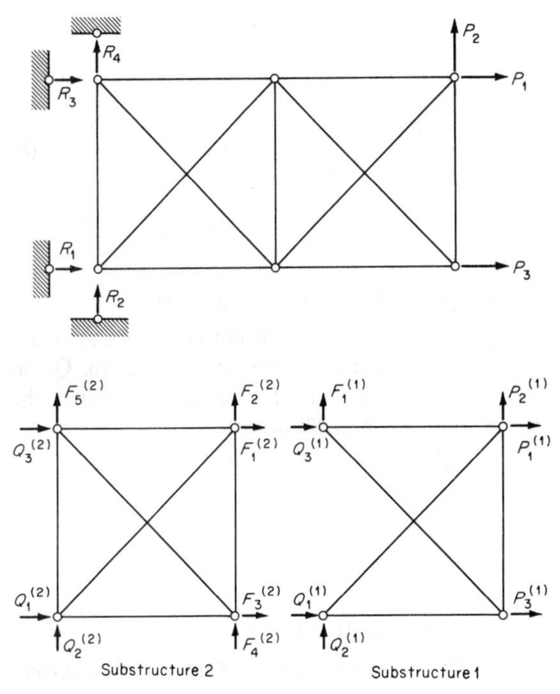

FIG. 9.13 Substructure boundary forces.

Because of the simplicity of the structure, the complete analysis has been carried out without the aid of a computer. Only the main steps in the analysis are reproduced below. The units used are pounds and inches.

EQUATIONS OF EXTERNAL EQUILIBRIUM FOR THE SUBSTRUCTURES

External equilibrium equations are set up using Eq. (9.41).

SUBSTRUCTURE 1

$$\mathbf{Q}^{(1)} = [\mathbf{Q}_F^{(1)} \quad \mathbf{Q}_P^{(1)}]\begin{bmatrix}\mathbf{F}^{(1)} \\ \mathbf{P}^{(1)}\end{bmatrix}$$

$$\begin{bmatrix} Q_1^{(1)} \\ Q_2^{(1)} \\ Q_3^{(1)} \end{bmatrix} = \left[\begin{array}{c|ccc} 0 & 0 & -1 & -1 \\ -1 & 0 & -1 & 0 \\ 0 & -1 & 1 & 0 \end{array}\right] \begin{bmatrix} F_1^{(1)} \\ \hline P_1^{(1)} \\ P_2^{(1)} \\ P_3^{(1)} \end{bmatrix} \tag{a}$$

SUBSTRUCTURE 2

$$\mathbf{Q}^{(2)} = [\mathbf{Q}_F^{(2)}][\mathbf{F}^{(2)}]$$

$$\begin{bmatrix} Q_1^{(2)} \\ Q_2^{(2)} \\ Q_3^{(2)} \end{bmatrix} = \begin{bmatrix} 0 & -1 & -1 & -1 & 0 \\ 0 & -1 & 0 & -1 & -1 \\ -1 & 1 & 0 & 1 & 0 \end{bmatrix} \begin{bmatrix} F_1^{(2)} \\ F_2^{(2)} \\ F_3^{(2)} \\ F_4^{(2)} \\ F_5^{(2)} \end{bmatrix} \tag{b}$$

COMBINED EQUATIONS OF EXTERNAL EQUILIBRIUM

It should be noted that in the present example $\mathbf{Q}_P^{(2)}$ is not used since there are no external loads $\mathbf{P}^{(2)}$ applied to the structure. The diagonal matrix $\mathbf{Q}_P$ in (9.45) therefore is reduced to only one column of submatrices. Hence the combined equations of external equilibrium (9.42) become

$$\begin{bmatrix} \mathbf{Q}^{(1)} \\ \mathbf{Q}^{(2)} \end{bmatrix} = \left[\begin{array}{cc|c} \mathbf{Q}_F^{(1)} & \mathbf{0} & \mathbf{Q}_P^{(1)} \\ \mathbf{0} & \mathbf{Q}_F^{(2)} & \mathbf{0} \end{array}\right] \begin{bmatrix} \mathbf{F}^{(1)} \\ \mathbf{F}^{(2)} \\ \hline \mathbf{P}^{(1)} \end{bmatrix} \tag{c}$$

EQUATIONS OF INTERNAL EQUILIBRIUM

It is easy to verify, using Fig. 9.13, that the equations of internal equilibrium (9.50) are

$$
\begin{bmatrix}
0 & 0 & 1 & 0 & 0 & 0 & 0 & 0 & 0 & 0 & 0 & 1 & 0 & 0 & 0 & 0 \\
0 & 0 & 0 & 0 & 0 & 0 & 0 & 0 & 0 & 0 & 1 & 0 & 1 & 0 & 0 & 0 \\
1 & 0 & 0 & 0 & 0 & 0 & 0 & 0 & 0 & 0 & 0 & 0 & 0 & 1 & 0 & 0 \\
0 & 1 & 0 & 0 & 0 & 0 & 0 & 0 & 0 & 0 & 0 & 0 & 0 & 0 & 1 & 0 \\
0 & 0 & 0 & 0 & 0 & 1 & 0 & 0 & 1 & 0 & 0 & 0 & 0 & 0 & 0 & 0 \\
0 & 0 & 0 & 0 & 0 & 0 & 0 & 0 & 0 & 1 & 0 & 0 & 0 & 0 & 0 & 1 \\
0 & 0 & 0 & 1 & 0 & 0 & 1 & 0 & 0 & 0 & 0 & 0 & 0 & 0 & 0 & 0 \\
0 & 0 & 0 & 0 & 1 & 0 & 0 & 1 & 0 & 0 & 0 & 0 & 0 & 0 & 0 & 0
\end{bmatrix}
\begin{bmatrix}
Q_1^{(1)} \\ Q_2^{(1)} \\ Q_3^{(1)} \\ Q_1^{(2)} \\ Q_2^{(2)} \\ Q_3^{(2)} \\ R_1 \\ R_2 \\ R_3 \\ R_4 \\ F_1^{(1)} \\ F_1^{(2)} \\ F_2^{(2)} \\ F_3^{(2)} \\ F_4^{(2)} \\ F_5^{(2)}
\end{bmatrix}
= \mathbf{0} \qquad (d)
$$

The equations of internal equilibrium may be described as connectivity relations defining connections between the substructures. Substituting Eq. (*c*) into (*a*), we obtain the following matrix equation [see Eqs. (9.51) or (9.55)].

$$
\begin{bmatrix}
0 & 0 & 0 & 0 & 0 & 1 & 0 & 0 & 0 & 0 & -1 & 1 & 0 \\
0 & 0 & 0 & 0 & 1 & 0 & 1 & 0 & 0 & 0 & 0 & 0 & 0 \\
0 & 0 & 0 & 0 & 0 & 0 & 0 & 1 & 0 & 0 & 0 & -1 & -1 \\
0 & 0 & 0 & 0 & -1 & 0 & 0 & 0 & 1 & 0 & 0 & -1 & 0 \\
0 & 0 & 1 & 0 & 0 & -1 & 1 & 0 & 1 & 0 & 0 & 0 & 0 \\
0 & 0 & 0 & 1 & 0 & 0 & 0 & 0 & 0 & 1 & 0 & 0 & 0 \\
1 & 0 & 0 & 0 & 0 & 0 & -1 & -1 & -1 & 0 & 0 & 0 & 0 \\
0 & 1 & 0 & 0 & 0 & 0 & -1 & 0 & -1 & -1 & 0 & 0 & 0
\end{bmatrix}
\begin{bmatrix}
R_1 \\ R_2 \\ R_3 \\ R_4 \\ F_1^{(1)} \\ F_1^{(2)} \\ F_2^{(2)} \\ F_3^{(2)} \\ F_4^{(2)} \\ F_5^{(2)} \\ P_1^{(1)} \\ P_2^{(1)} \\ P_3^{(1)}
\end{bmatrix}
= \mathbf{0} \qquad (e)
$$

FLEXIBILITY MATRICES

The flexibility matrices for the two substructures have been calculated using the standard matrix force method of analysis. It should be noted in this connection that substructure 1 is statically determinate while substructure 2 has one redundancy.

$$\mathbf{D}_{FF}^{(1)} = [22] \times 10^{-6} \tag{f}$$

$$\mathbf{D}_{FP}^{(1)} = [-2 \quad 10 \quad -2] \times 10^{-6} \tag{g}$$

$$\mathbf{D}_{PP}^{(1)} = \begin{bmatrix} 2 & -2 & 0 \\ -2 & 10 & 0 \\ 0 & 0 & 2 \end{bmatrix} \times 10^{-6} \tag{h}$$

$$\mathbf{D}_{FF}^{(2)} = \begin{bmatrix} 1.8333 & & & & \\ -1.1667 & 5.8333 & & \text{Symmetric} & \\ -0.1667 & 0.8333 & 1.8333 & & \\ -1.0000 & 5.0000 & 1.0000 & 6.0000 & \\ -0.1667 & 0.8333 & -0.1667 & 1.0000 & 1.8333 \end{bmatrix} \times 10^{-6} \tag{i}$$

The flexibility matrices $\mathbf{D}_{FP}^{(2)}$ and $\mathbf{D}_{PP}^{(2)}$ are not used since no external loads are applied to the substructure 2.

COMBINED FLEXIBILITY MATRICES

From Eqs. (9.77), (9.78), (9.83), and (9.84) it follows that

$$\bar{\mathbf{D}}_{FF} = \begin{bmatrix} \mathbf{0} & \mathbf{0} & \mathbf{0} \\ \mathbf{0} & \mathbf{D}_{FF}^{(1)} & \mathbf{0} \\ \mathbf{0} & \mathbf{0} & \mathbf{D}_{FF}^{(2)} \end{bmatrix} \tag{j}$$

$$\bar{\mathbf{D}}_{FP} = \begin{bmatrix} \mathbf{0} \\ \mathbf{D}_{FP}^{(1)} \\ \mathbf{0} \end{bmatrix} = \bar{\mathbf{D}}_{PF}^{T} \tag{k}$$

JORDANIAN ELIMINATION (MATRICES $\mathbf{b}_X$ AND $\mathbf{b}_0$)

Application of the jordanian elimination technique, represented symbolically by Eq. (9.64), to the equilibrium equations (*e*) leads directly to the generation of the matrices $\mathbf{q}_X$ and $\mathbf{q}_P$ given by Eqs. (9.62) and (9.63). A number of column interchanges were required to continue with the elimination process. The forces $\mathbf{F}_1^{(1)}$ and $\mathbf{F}_5^{(5)}$ were selected as the redundant boundary forces. The results obtained are shown by Eqs. (*l*) and (*m*), where the numbers assigned to rows and columns refer to the original column numbers in Eq. (*e*). Thus the redundant boundary forces are identified by numbers 5 and 10. Only the final steps leading to Eqs. (9.72) and (9.73) are reproduced.

$$\mathbf{b}_X = \mathbf{N}\mathbf{q}_X + \mathbf{H}_X = \begin{array}{c|cccccccc} & 6 & 7 & 8 & 9 & 3 & 4 & 1 & 2 \\ \hline 1 & 0 & 0 & 0 & 0 & 0 & 0 & 1 & 0 \\ 2 & 0 & 0 & 0 & 0 & 0 & 0 & 0 & 1 \\ 3 & 0 & 0 & 0 & 0 & 1 & 0 & 0 & 0 \\ 4 & 0 & 0 & 0 & 0 & 0 & 1 & 0 & 0 \\ 5 & 0 & 0 & 0 & 0 & 0 & 0 & 0 & 0 \\ 6 & 1 & 0 & 0 & 0 & 0 & 0 & 0 & 0 \\ 7 & 0 & 1 & 0 & 0 & 0 & 0 & 0 & 0 \\ 8 & 0 & 0 & 1 & 0 & 0 & 0 & 0 & 0 \\ 9 & 0 & 0 & 0 & 1 & 0 & 0 & 0 & \\ 10 & 0 & 0 & 0 & 0 & 0 & 0 & 0 & 0 \end{array} \; \begin{array}{c|cc} & 5 & 10 \\ \hline 6 & 0 & 0 \\ 7 & -1 & 0 \\ 8 & 0 & 0 \\ 9 & 1 & 0 \\ 3 & 0 & 0 \\ 4 & 0 & -1 \\ 1 & 0 & 0 \\ 2 & 0 & 1 \end{array} + \begin{array}{c|cc} & 5 & 10 \\ \hline 1 & 0 & 0 \\ 2 & 0 & 0 \\ 3 & 0 & \\ 4 & 0 & 0 \\ 5 & 1 & 0 \\ 6 & 0 & 0 \\ 7 & 0 & 0 \\ 8 & 0 & 0 \\ 9 & 0 & 0 \\ 10 & 0 & 1 \end{array} = \begin{array}{c|cc} & 5 & 10 \\ \hline 1 & 0 & 0 \\ 2 & 0 & 1 \\ 3 & 0 & 0 \\ 4 & 0 & -1 \\ 5 & 1 & 0 \\ 6 & 0 & 0 \\ 7 & -1 & 0 \\ 8 & 0 & 0 \\ 9 & 1 & 0 \\ 10 & 0 & 1 \end{array} \qquad (l)$$

$$\mathbf{b}_0 = \mathbf{N}\mathbf{q}_P = \begin{array}{c|cccccccc} & 6 & 7 & 8 & 9 & 3 & 4 & 1 & 2 \\ \hline 1 & 0 & 0 & 0 & 0 & 0 & 0 & 1 & 0 \\ 2 & 0 & 0 & 0 & 0 & 0 & 0 & 0 & 1 \\ 3 & 0 & 0 & 0 & 0 & 1 & 0 & 0 & 0 \\ 4 & 0 & 0 & 0 & 0 & 0 & 1 & 0 & 0 \\ 5 & 0 & 0 & 0 & 0 & 0 & 0 & 0 & 0 \\ 6 & 1 & 0 & 0 & 0 & 0 & 0 & 0 & 0 \\ 7 & 0 & 1 & 0 & 0 & 0 & 0 & 0 & 0 \\ 8 & 0 & 0 & 1 & 0 & 0 & 0 & 0 & 0 \\ 9 & 0 & 0 & 0 & 1 & 0 & 0 & 0 & 0 \\ 10 & 0 & 0 & 0 & 0 & 0 & 0 & 0 & 0 \end{array} \; \begin{array}{c|ccc} & 11 & 12 & 13 \\ \hline 6 & 1 & -1 & 0 \\ 7 & 0 & 0 & 0 \\ 8 & 0 & 1 & 1 \\ 9 & 0 & 1 & 0 \\ 3 & 1 & -2 & 0 \\ 4 & 0 & 0 & 0 \\ 1 & 0 & 2 & 1 \\ 2 & 0 & 1 & 0 \end{array} = \begin{array}{c|ccc} & 11 & 12 & 13 \\ \hline 1 & 0 & 2 & 1 \\ 2 & 0 & 1 & 0 \\ 3 & 1 & -2 & 0 \\ 4 & 0 & 0 & 0 \\ 5 & 0 & 0 & 0 \\ 6 & 1 & -1 & 0 \\ 7 & 0 & 0 & 0 \\ 8 & 0 & 1 & 1 \\ 9 & 0 & 1 & 0 \\ 10 & 0 & 0 & 0 \end{array} \qquad (m)$$

COMPATIBILITY EQUATIONS

Since there are no thermal strains, Eq. (9.86*a*) reduces to

$$\mathbf{D}_{XX}\mathbf{X} + \mathbf{D}_{XP}\mathbf{P} = \mathbf{0} \qquad (n)$$

where

$$\mathbf{D}_{XX} = \mathbf{b}_X{}^T\bar{\mathbf{D}}_{FF}\mathbf{b}_X$$

$$= \begin{bmatrix} 23.8333 & 0.1667 \\ 0.1667 & 1.8333 \end{bmatrix} \times 10^{-6} \tag{o}$$

$$\mathbf{D}_{XP} = \mathbf{b}_X{}^T\bar{\mathbf{D}}_{FF}\mathbf{b}_0 + \mathbf{b}_X{}^T\bar{\mathbf{D}}_{FP}$$

$$= \begin{bmatrix} 1.8333 & 11.0000 & 1.8333 \\ -0.1667 & 1.0000 & -0.1667 \end{bmatrix} \times 10^{-6} \tag{p}$$

Hence $\quad \mathbf{X} = -\mathbf{D}_{XX}{}^{-1}\mathbf{D}_{XP}\mathbf{P} = \mathbf{X}_P\mathbf{P}$

$$= \begin{bmatrix} 0.07634 & -0.45802 & 0.07634 \\ 0.08397 & -0.50382 & 0.08397 \end{bmatrix} \mathbf{P} \tag{q}$$

SUBSTRUCTURE BOUNDARY FORCES

The boundary forces **R** and **F** due to unit values of the external loads are calculated from Eqs. (9.91) and (9.94). Hence

$$\bar{\mathbf{F}}_P = \mathbf{b}_0 - \mathbf{b}_X\mathbf{D}_{XX}{}^{-1}\mathbf{D}_{XP} = \mathbf{b}_0 + \mathbf{b}_X\mathbf{X}_P$$

$$= \begin{array}{r} 1 \\ 2 \\ 3 \\ 4 \\ 5 \\ 6 \\ 7 \\ 8 \\ 9 \\ 10 \end{array} \left[\begin{array}{ccc} 0 & 2.0 & 1.0 \\ 0.08397 & 0.49618 & 0.08397 \\ 1.0 & -2.0 & 0 \\ -0.08397 & 0.50382 & -0.08397 \\ \hdashline 0.07634 & -0.45802 & 0.07634 \\ 1.0 & -1.0 & 0 \\ -0.07634 & 0.45802 & -0.07634 \\ 0 & 1.0 & 1.0 \\ 0.07634 & 0.54198 & 0.07634 \\ 0.08379 & -0.50382 & 0.08397 \end{array}\right] \begin{array}{l} \\ \\ R \\ \\ \\ \\ \\ F \\ \\ \\ \end{array} \tag{r}$$

The actual boundary forces due to the external loading **P** are calculated from $\bar{\mathbf{F}} = \bar{\mathbf{F}}_P\mathbf{P}$. These forces are then applied to each individual substructure in order to find stress distribution within the substructures.

DEFLECTIONS

To determine deflections in the direction of the applied forces **P** we may use $\mathbf{b}_\Delta = \mathbf{b}_0$. Hence from Eq. (9.102) deflections due to unit values of **P** are given by

$$\mathbf{\Delta}_P = (\mathbf{b}_0{}^T\bar{\mathbf{D}}_{FF} + \bar{\mathbf{D}}_{PF})\bar{\mathbf{F}}_P + \mathbf{b}_0{}^T\bar{\mathbf{D}}_{FP} + \mathbf{D}_{PP} \tag{s}$$

where in the present example

$$\mathbf{D}_{PP} = \mathbf{D}_{PP}{}^{(1)} \tag{t}$$

Using previously calculated matrices, we get

$$\mathbf{\Delta}_P = \begin{bmatrix} 3.679 & -4.076 & -0.3206 \\ -4.076 & 18.46 & 3.924 \\ -0.3206 & 3.924 & 3.679 \end{bmatrix} \times 10^{-6} \tag{u}$$

The deflections given by Eq. (*u*) agree with the results obtained using the substructure displacement analysis for the example in Sec. 9.3.

PROBLEMS

9.1 Calculate substructure flexibilities for the example in Sec. 9.4.

9.2 Using the jordanian technique applied to Eq. (*e*) in Sec. 9.4, verify that the matrices $\mathbf{b}_X$ and $\mathbf{b}_0$ are given by Eqs. (*l*) and (*m*).

CHAPTER 10 DYNAMICS OF ELASTIC SYSTEMS

So far we have discussed matrix methods of structural analysis applied to static loading; however, in many applications we require determination of stresses and displacements under dynamic loading conditions. For such cases, in addition to the previously derived structural stiffnesses or flexibilities we must also introduce inertia properties in order to describe the dynamic characteristics of the structure. This chapter gives an introduction to the general theory of structural dynamics with particular emphasis on the discrete-element representation of the actual continuous elastic systems.

10.1 FORMULATION OF THE DYNAMICAL PROBLEMS

When dynamic loading is applied to an elastic body (or structure), the elastic displacements $\mathbf{u}$ are functions not only of the coordinates but of the time as well. An infinitesimal element of volume dV at an arbitrary point is then subjected to an inertia force $-\rho\ddot{\mathbf{u}}\,dV$, where ρ is the density of the body around the point and

$$\ddot{\mathbf{u}} = \begin{bmatrix} \ddot{u}_x \\ \ddot{u}_y \\ \ddot{u}_z \end{bmatrix} \tag{10.1}$$

represents a column matrix of accelerations measured in the cartesian coordinate system. In accordance with d'Alembert's principle, a corollary of Newton's second and third laws, the equations of motion are obtained from the condition of equilibrium of the element when the inertia forces are taken into account. Since the inertia forces are proportional to the volume of the element, they constitute body forces; thus the inertia body forces, distributed throughout the elastic system, are given by

$$\mathbf{X}_{\text{inertia}} = -\rho\ddot{\mathbf{u}} \tag{10.2}$$

This indicates that the equations of motion of an elastic system can be obtained by introducing additional body forces, as given by Eq. (10.2), into the equations of static equilibrium.

We shall consider now an unconstrained body which is initially at rest. If the body is rigid and we apply an external load suddenly, all points in the body instantaneously acquire certain constant accelerations, which may be determined from the condition of equilibrium of the applied load and the inertial forces acting on the body as a whole. If the same load is applied to an elastic body, some points may have displacements relative to each other. Therefore when an external load is applied to the surface of an elastic body, the points are not set in motion all at once. The points in the immediate neighborhood of the applied load move first. This in turn produces strains in the region of the body contiguous to the loaded area, inducing stresses there which bring into play the next layer of points; these in turn transmit the motion to points in another layer, and so on, until all points participate in the motion. Thus we reach the conclusion that, in an elastic body subjected to dynamic loading, deformations must propagate with a finite velocity. Only when the body is rigid will the velocity of propagation be infinite. Since the rigid body may be considered as a physical limit (not attainable in reality) of an elastic body, we may postulate that the velocity of strain propagation is represented by an increasing function of the elastic moduli. Typical distributions of point velocities in rigid and elastic bodies are illustrated in Fig. 10.1; in a rigid body subjected to a suddenly applied load all points acquire constant velocity instantaneously without any strains, while in an elastic body points acquire velocities gradually, and this is

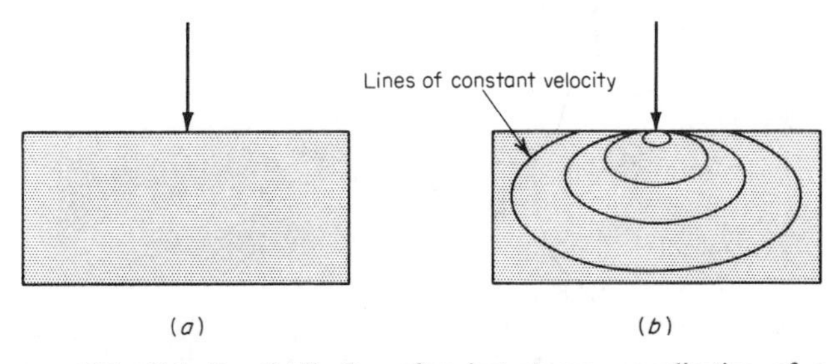

FIG. 10.1 Velocity distribution after instantaneous application of a concentrated load. (*a*) Rigid body (all points acquire constant velocity suddenly); (*b*) elastic body (points acquire velocities gradually).

accompanied by a straining action resulting from relative displacements in different parts of the elastic body.

Complete analysis of the dynamic deformations of elastic bodies presents considerable mathematical difficulties. Fortunately, in most structural problems involving dynamic loading we need not concern ourselves with the propagation of waves in elastic bodies or structures. Many dynamic problems can be approached as oscillations of the complete structure in which all the points execute periodical motion. These oscillations may be treated as being formed by superposition of similar counterrunning wave trains resulting in standing waves, which may also be regarded as the steady-state condition of the wave-propagation problem. The oscillation problems can be solved by methods independent of the theory of wave propagation. Solutions to such problems, using matrix methods for discrete-element structures, will be studied in this and in subsequent chapters.

10.2 PRINCIPLE OF VIRTUAL WORK IN DYNAMICS OF ELASTIC SYSTEMS

Consider an elastic body which undergoes a deformation under the action of dynamic loading (see Fig. 10.2). At any particular instant of time we can assume that the displacements $\mathbf{u}$ acquire virtual displacements $\delta\mathbf{u}$. The virtual displacements are infinitesimal and otherwise arbitrary but compatible with the boundary conditions on the body. Furthermore, the virtual displacements produce compatible virtual strains $\delta\boldsymbol{\epsilon}$ from which, for a known instantaneous stress distribution, the virtual strain energy δU_i for the system can be calculated at the given instant of time. The virtual work of external forces will now consist not

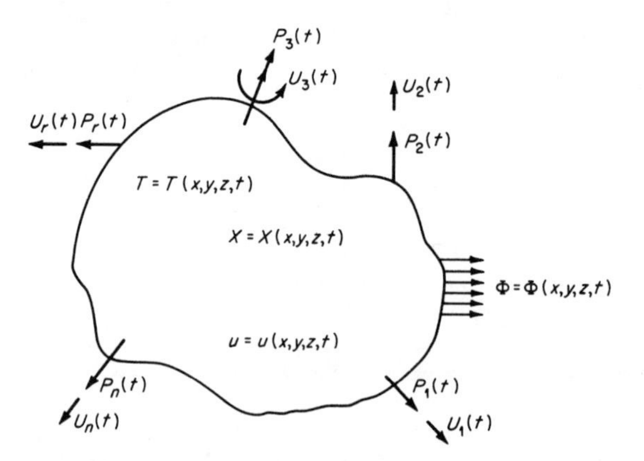

FIG. 10.2 Elastic body or structure subjected to dynamic loading.

only of work by the actual body forces and surface forces but also of work by the inertia forces represented as body forces, according to Eq. (10.2). Using therefore the results of Chap. 3, we see that the principle of virtual work generalized to include dynamic conditions may be expressed as

$$\delta U_i = \delta W - \int_v \rho\, \delta \mathbf{u}^T \ddot{\mathbf{u}}\, dV \tag{10.3}$$

where, as before,

$$\delta U_i = \int_v \delta \boldsymbol{\epsilon}^T \boldsymbol{\sigma}\, dV \tag{10.4}$$

and $$\delta W = \int_s \delta \mathbf{u}^T \boldsymbol{\Phi}\, dS + \int_v \delta \mathbf{u}^T \mathbf{X}\, dV + \delta \mathbf{U}^T \mathbf{P} \tag{10.5}$$

The third term on the right-hand side of Eq. (10.5) represents the virtual work of external forces $\mathbf{P}$ in moving through the corresponding virtual displacements $\delta\mathbf{U}$.

Equation (10.3) states that in a virtual displacement of the body from its *instantaneous* state of equilibrium, the increment of strain energy, i.e., virtual strain energy, is equal to the sum of the virtual work of all the forces, including the inertia forces. In the state of rest Eq. (10.3) takes the simpler form

$$\delta U_i = \delta W \tag{10.6}$$

which represents the principle of virtual work for static systems.

10.3 HAMILTON'S PRINCIPLE

We consider next an elastic body the deformations of which vary continously between the instant of time t_0 and t_1; thus, the displacements $\mathbf{u}$ in the body are functions of the time t. We introduce virtual displacements $\delta\mathbf{u}$ varying with time in such a way that $\delta\mathbf{u} = \mathbf{0}$ at the instants t_0 and t_1. Integrating the equation representing the principle of virtual work for dynamic systems

$$\delta U_i = \delta W - \int_v \rho\, \delta \mathbf{u}^T \ddot{\mathbf{u}}\, dV \tag{10.3}$$

over the time interval t_0 to t_1, we have

$$\int_{t_0}^{t_1} \delta U_i\, dt = \int_{t_0}^{t_1} \delta W\, dt - \int_{t_0}^{t_1} \int_v \rho\, \delta \mathbf{u}^T \ddot{\mathbf{u}}\, dV\, dt \tag{10.7}$$

The second integral on the right-hand side of Eq. (10.7) can be related to the variation of the kinetic energy K of the system. Since the kinetic energy is defined as

$$K = \frac{1}{2} \int_v \rho \dot{\mathbf{u}}^T \dot{\mathbf{u}}\, dV \tag{10.8}$$

where $$\dot{\mathbf{u}} = \begin{bmatrix} \dot{u}_x \\ \dot{u}_y \\ \dot{u}_z \end{bmatrix} \tag{10.9}$$

represents a column matrix of velocities measured in the cartesian coordinate system, it follows that

$$\delta K = \int_v \rho\, \delta\dot{\mathbf{u}}^T\, \dot{\mathbf{u}}\, dV = \int_v \frac{\partial}{\partial t} (\rho\, \delta\mathbf{u}^T\, \dot{\mathbf{u}})\, dV - \int_v \rho\, \delta\mathbf{u}^T\, \ddot{\mathbf{u}}\, dV \tag{10.10}$$

Integrating Eq. (10.10) over the time interval t_0 to t_1, we have

$$\begin{aligned} \int_t^{t_1} \delta K\, dt &= \int_{t_0}^{t_1} \left[\int_v \rho \frac{\partial}{\partial t} (\delta\mathbf{u}^T\, \dot{\mathbf{u}})\, dV \right] dt - \int_{t_0}^{t_1} \left(\int_v \rho\, \delta\mathbf{u}^T\, \ddot{\mathbf{u}}\, dV \right) dt \\ &= \int_v \left| \rho\, \delta\mathbf{u}^T\, \dot{\mathbf{u}} \right|_{t_0}^{t_1} dV - \int_{t_0}^{t_1} \left(\int_v \rho\, \delta\mathbf{u}^T\, \ddot{\mathbf{u}}\, dV \right) dt \\ &= -\int_{t_0}^{t_1} \left(\int_v \rho\, \delta\mathbf{u}^T\, \ddot{\mathbf{u}}\, dV \right) dt \end{aligned} \tag{10.11}$$

where $\left| \rho\, \delta\mathbf{u}^T\, \dot{\mathbf{u}} \right|_{t_0}^{t_1} = 0$ in view of the assumptions made that the virtual displacements $\delta\mathbf{u}$ are zero at the time instants t_0 and t_1.

Substituting Eq. (10.11) into (10.7) gives

$$\delta \int_{t_0}^{t_1} (U_i - K)\, dt = \int_{t_0}^{t_1} \delta W\, dt \tag{10.12}$$

If the loading is conservative, i.e., if the work done by external loads is independent of the load path taken and dependent only on the end points (final displacements), then Eq. (10.12) can be rewritten as

$$\delta \int_{t_0}^{t_1} (U_i - K + U_e)\, dt = 0 \tag{10.13}$$

where U_e is the potential energy of external forces. Introducing the total potential energy

$$U = U_i + U_e \tag{10.14}$$

into Eq. (10.13), it follows that

$$\delta \int_{t_0}^{t_1} (U - K)\, dt = 0 \tag{10.15}$$

Equation (10.15) is Hamilton's principle, which states that the integral $\int_{t_0}^{t_1} (U - K)\, dt$ takes a stationary value in an elastic system subjected to conservative dynamic loading. In the special case of a static loading Eq. (10.15) becomes the principle of minimum potential energy (see Chap. 3)

$$\delta_\epsilon U = 0 \qquad U = \text{minimum} \tag{3.59}$$

For rigid bodies $U_i = 0$, and Eq. (10.15) reduces to the familiar form

$$\delta \int_{t_1}^{t_2} (W + K)\, dt = 0 \tag{10.16}$$

10.4 POWER-BALANCE EQUATION

If we assume that the virtual displacements $\delta\mathbf{u}$ are identical with the actual increments of displacements $d\mathbf{u}$ and $\mathbf{u} + d\mathbf{u}$ are the actual displacements at the time instant $t + dt$, then

$$\delta\mathbf{u} = \frac{\partial \mathbf{u}}{\partial t}\, dt = \dot{\mathbf{u}}\, dt = d\mathbf{u} \tag{10.17}$$

The variation in strain energy during the time dt is given by

$$\delta U_i = \frac{\partial U_i}{\partial t}\, dt = dU_i \tag{10.18}$$

while the increment in kinetic energy is determined from

$$\begin{aligned} dK &= \int_v \rho \dot{\mathbf{u}}^T \ddot{\mathbf{u}}\, dt\, dV \\ &= \int_v \rho\, d\mathbf{u}^T\, \ddot{\mathbf{u}}\, dV \end{aligned} \tag{10.19}$$

Substituting Eqs. (10.18) and (10.19) into (10.3) and using Eqs. (10.5) and (10.17), we have

$$\frac{d}{dt}(U_i + K) = \int_s \dot{\mathbf{u}}^T \mathbf{\Phi}\, dS + \int_v \dot{\mathbf{u}}^T \mathbf{X}\, dV + \dot{\mathbf{U}}^T \mathbf{P} \tag{10.20}$$

The terms on the right of the last equation consist of a sum of products of force and velocity vectors which represents the rate of doing work, i.e., power, while the left side represents time rate of change of the sum of elastic and kinetic energies of the system. Thus Eq. (10.20) may be described as the power-balance equation for an elastic system subjected to dynamic loading.

10.5 EQUATIONS OF MOTION AND EQUILIBRIUM

CONTINUOUS SYSTEMS

All equations of elasticity theory derived for continuous elastic system can be adapted immediately to dynamic loading conditions by including inertia forces.

The equations of stress equilibrium in the x, y, and z directions are modified by the addition of the inertia body forces represented by Eq. (10.2), so that

$$
\begin{aligned}
&\frac{\partial \sigma_{xx}}{\partial x} + \frac{\partial \sigma_{xy}}{\partial y} + \frac{\partial \sigma_{xz}}{\partial z} + X_x = \rho \ddot{u}_x \\
&\frac{\partial \sigma_{yx}}{\partial x} + \frac{\partial \sigma_{yy}}{\partial y} + \frac{\partial \sigma_{yz}}{\partial z} + X_y = \rho \ddot{u}_y \\
&\frac{\partial \sigma_{zx}}{\partial x} + \frac{\partial \sigma_{zy}}{\partial y} + \frac{\partial \sigma_{zz}}{\partial z} + X_z = \rho \ddot{u}_z
\end{aligned}
\tag{10.21}
$$

while the moment equilibrium of stresses remains unchanged as

$$
\sigma_{ij} = \sigma_{ji} \tag{10.22}
$$

Stress equilibrium equations at the surface of an elastic body remain the same as in the static loading case, that is,

$$
\begin{aligned}
&l\sigma_{xx} + m\sigma_{xy} + n\sigma_{xz} = \Phi_x \\
&l\sigma_{yx} + m\sigma_{yy} + n\sigma_{yz} = \Phi_y \\
&l\sigma_{zx} + m\sigma_{zy} + n\sigma_{zz} = \Phi_z
\end{aligned}
\tag{10.23}
$$

The equilibrium equations for external forces and inertia forces can be obtained directly from Eqs. (2.64) and (2.65) by including inertia body forces. Hence

$$
\begin{aligned}
&\int_s \Phi_x \, dS + \int_v X_x \, dV + \Sigma P_x = \int_v \rho \ddot{u}_x \, dV \\
&\int_s \Phi_y \, dS + \int_v X_y \, dV + \Sigma P_y = \int_v \rho \ddot{u}_y \, dV \\
&\int_s \Phi_z \, dS + \int_v X_z \, dV + \Sigma P_z = \int_v \rho \ddot{u}_z \, dV
\end{aligned}
\tag{10.24}
$$

$$
\begin{aligned}
&\int_s (\Phi_z y - \Phi_y z) \, dS + \int_v (X_z y - X_y z) \, dV + \Sigma M_x = \int_v \rho(\ddot{u}_z y - \ddot{u}_y z) \, dV \\
&\int_s (\Phi_x z - \Phi_z x) \, dS + \int_v (X_x z - X_z x) \, dV + \Sigma M_y = \int_v \rho(\ddot{u}_x z - \ddot{u}_z x) \, dV \\
&\int_s (\Phi_y x - \Phi_x y) \, dS + \int_v (X_y x - X_x y) \, dV + \Sigma M_z = \int_v \rho(\ddot{u}_y x - \ddot{u}_x y) \, dV
\end{aligned}
\tag{10.25}
$$

Once the solution for the displacements $\mathbf{u}$ has been obtained in any particular problem, Eqs. (10.24) and (10.25) can be employed for checking purposes to ensure that the dynamic equilibrium is satisfied.

Although Eqs. (10.21) and (10.22) represent conditions of dynamic stress equilibrium, they can also be described as equations of motion of an elastic system. In fact, these equations are the governing equations for the displacements of the system. We can assume that at any instant of time Hooke's law is obeyed, and this then implies that all stresses in Eqs. (10.21) can be expressed in terms of strain components, and hence in terms of displacement derivatives; thus, we can obtain three partial differential equations for displacements. Solutions of these differential equations determine the propagation of stress waves in an elastic system subjected to some specified dynamic loading; however, such problems are beyond the scope of this text and will not be discussed here.

DISCRETE SYSTEMS

In matrix methods of structural analysis we deal with discrete quantities, concentrated forces and moments, deflections and rotations at a point, etc. Consequently, all equations of elasticity for continuous media must be reformulated as matrix equations using these discrete quantities. For static problems, the displacements $\mathbf{u}$ in a continuous structure can be related to a finite number of displacements selected at some arbitrary points on the structure. This relationship is expressed by the matrix equation

$$\mathbf{u} = \mathbf{a}\mathbf{U} \tag{10.26}$$

where
$$\mathbf{u} = \{u_x \quad u_y \quad u_z\} \tag{10.27}$$

$$\mathbf{U} = \{U_1 \quad U_2 \quad \cdots \quad U_n\} \tag{10.28}$$

$$\mathbf{a} = \mathbf{a}(x,y,z) \tag{10.29}$$

Equation (10.26) is valid only for small deflections. For large deflections, no such single relationship can be used in which the coefficients of the matrix $\mathbf{a}$ are functions of the coordinates only.

If we now use the strain-displacement equations (2.2), the total strains can be determined from

$$\mathbf{e} = \mathbf{b}\mathbf{U} \tag{10.30}$$

where
$$\mathbf{b} = \mathbf{b}(x,y,z) \tag{10.31}$$

is obtained by differentiation of the matrix $\mathbf{a}$.

In dynamic problems the simple relationship (10.26) is not valid, except in some special cases; however, if a large enough number of displacements $\mathbf{U}$

are considered, the relationship $\mathbf{u} = \mathbf{aU}$ will be a good approximation, provided $\mathbf{U}$ is determined from the dynamic equations of the system. This relationship will be employed to formulate an equivalent discrete-element system from a continuous system. To accomplish this we start with the principle of virtual work for dynamic loading

$$\delta U_i = \delta W - \int_v \rho\, \delta\mathbf{u}^T\, \ddot{\mathbf{u}}\, dV \tag{10.3}$$

The virtual displacements $\delta\mathbf{u}$ and virtual strains $\delta\boldsymbol{\epsilon}$ can be obtained from Eqs. (10.26) and (10.30). Hence

$$\delta\mathbf{u} = \mathbf{a}\,\delta\mathbf{U} \tag{10.32}$$

and $$\delta\boldsymbol{\epsilon} = \delta\mathbf{e} = \mathbf{b}\,\delta\mathbf{U} \tag{10.33}$$

After introducing Eqs. (10.32) and (10.33) into (10.3) and using the generalized Hooke's law

$$\boldsymbol{\sigma} = \boldsymbol{\varkappa}\mathbf{e} + \boldsymbol{\varkappa}_T\alpha T \tag{2.14}$$

it follows that

$$\begin{aligned}\int_v \delta\mathbf{U}^T\, \mathbf{b}^T\boldsymbol{\varkappa}\mathbf{b}\mathbf{U}\, dV + \int_v \delta\mathbf{U}^T\, \mathbf{b}^T\boldsymbol{\varkappa}_T\alpha T\, dV = \int_s \delta\mathbf{U}^T\, \mathbf{a}^T\boldsymbol{\Phi}\, dS + \int_v \delta\mathbf{U}^T\, \mathbf{a}^T\mathbf{X}\, dV \\ + \delta\mathbf{U}^T\, \mathbf{P} - \int_v \rho\, \delta\mathbf{U}^T\, \mathbf{a}^T\ddot{\mathbf{u}}\, dV\end{aligned} \tag{10.34}$$

Since the virtual displacements are arbitrary and

$$\ddot{\mathbf{u}} = \mathbf{a}\ddot{\mathbf{U}} \tag{10.35}$$

Eq. (10.34) can be rewritten as

$$\mathbf{M}\ddot{\mathbf{U}} + \mathbf{K}\mathbf{U} = \mathbf{P} - \int_v \mathbf{b}^T\boldsymbol{\varkappa}_T\alpha T\, dV + \int_s \mathbf{a}^T\boldsymbol{\Phi}\, dS + \int_v \mathbf{a}^T\mathbf{X}\, dV \tag{10.36}$$

where $$\mathbf{M} = \int_v \rho\mathbf{a}^T\mathbf{a}\, dV \tag{10.37}$$

represents mass matrix of the equivalent discrete system and

$$\mathbf{K} = \int_v \mathbf{b}^T\, \boldsymbol{\varkappa}\mathbf{b}\, dV \tag{10.38}$$

is the stiffness matrix for the displacements $\mathbf{U}$. The first term on the right of Eq. (10.36) represents a column of external concentrated loads; the second term represents equivalent concentrated forces due to temperature distribution; the third and fourth terms represent equivalent concentrated forces due to surface forces and body forces, respectively. Thus, Eq. (10.36) serves not only to determine the discrete-system mass and stiffness characteristics but also to convert distributed loading into one consisting of discrete forces.

Equation (10.36) represents equations of motion of the discrete system. These equations are coupled through the off-diagonal terms in the mass and stiffness matrices. It should be noted that mathematically the displacements **U** in the equations of motion may be regarded as the degrees of freedom of the system.

10.6 STATIC AND DYNAMIC DISPLACEMENTS IN A UNIFORM BAR

In Chap. 5 we have shown that the static displacements in a uniform bar having prescribed displacements U_1 and U_2 at the two ends are given by

$$u_x = \left[\left(1 - \frac{x}{l}\right) \quad \frac{x}{l}\right]\begin{bmatrix} U_1 \\ U_2 \end{bmatrix} \tag{10.39}$$

where l is the length of the bar. Equation (10.39) is of the form $\mathbf{u} = \mathbf{a}\mathbf{U}$, and hence

$$\mathbf{a} = \left[\left(1 - \frac{x}{l}\right) \quad \frac{x}{l}\right] \tag{10.40}$$

We shall now assume that both U_1 and U_2 will be some specified functions of time and obtain the solution for u_x in terms of U_1 and U_2 (see Fig. 10.3). We start with the equation of motion given by the first equation in (10.21). Noting that the only stress component present is $\sigma_{xx} = E\epsilon_{xx} = E\,\partial u_x/\partial x$, it follows immediately from Eq. (10.21) that the equation of motion becomes

$$c^2 \frac{\partial^2 u_x}{\partial x^2} - \ddot{u}_x = 0 \tag{10.41}$$

where $$c^2 = \frac{E}{\rho} \tag{10.42}$$

The boundary conditions are

$$u_x(0,t) = U_1(t) \tag{10.43}$$

and $$u_x(l,t) = U_2(t) \tag{10.44}$$

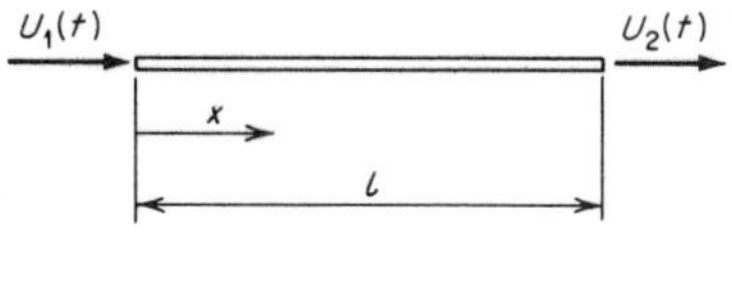

FIG. 10.3 Uniform bar with time-varying end displacements $U_1(t)$ and $U_2(t)$.

while the initial conditions will be assumed to be

$$u_x(x,0) = \dot{u}_x(x,0) = 0 \tag{10.45}$$

With use of the Laplace transforms

$$\bar{u}_x(x,p) = \int_0^\infty e^{-pt} u_x(x,t)\, dt \tag{10.46}$$

$$\bar{U}_1(p) = \int_0^\infty e^{-pt} U_1(t)\, dt \tag{10.47}$$

$$\bar{U}_2(p) = \int_0^\infty e^{-pt} U_2(t)\, dt \tag{10.48}$$

the equation of motion (10.41) and boundary conditions (10.43) and (10.44) transform into

$$c^2 \frac{d^2\bar{u}_x}{dx^2} - p^2\bar{u}_x = -pu_x(x,0) - \dot{u}_x(x,0) = 0 \tag{10.49}$$

$$\bar{u}_x(0,p) = \bar{U}_1 \tag{10.50}$$

$$\bar{u}_x(l,p) = \bar{U}_2 \tag{10.51}$$

The right side of Eq. (10.49) vanishes in view of the homogeneous initial conditions (10.45), and hence the solution for $\bar{u}_x$ can be represented by

$$\bar{u}_x = A(p) \sinh \frac{px}{c} + B(p) \cosh \frac{px}{c} \tag{10.52}$$

where the functions $A(p)$ and $B(p)$ can be determined from the boundary conditions (10.50) and (10.51). This yields

$$A(p) = \frac{\bar{U}_2 - \bar{U}_1 \cosh (pl/c)}{\sinh (pl/c)} \tag{10.53}$$

$$B(p) = \bar{U}_1 \tag{10.54}$$

Hence $$\bar{u}_x = \bar{U}_1(p) \frac{\sinh [(p/c)(l-x)]}{\sinh (pl/c)} + \bar{U}_2(p) \frac{\sinh (px/c)}{\sinh (pl/c)} \tag{10.55}$$

Noting that for functions $G(p)$ and $F(p)$

$$\mathscr{L}^{-1}\left[\frac{G(p)}{F(p)}\right] = \sum_\mu \frac{G(\mu) \exp \mu t}{\left[\dfrac{d}{dp} F(p)\right]_{p=\mu}} \tag{10.56}$$

where μ are the roots of

$$F(\mu) = 0 \tag{10.57}$$

we have

$$\mathscr{L}^{-1}\left[\frac{\sinh\{[p(l-x)]/c\}}{\sinh(pl/c)}\right] = \frac{2c}{l}\sum_{n=1}^{\infty}(-1)^{n+1}\sin\left[\frac{n\pi}{l}(l-x)\right]\sin\frac{n\pi ct}{l} \tag{10.58}$$

$$\mathscr{L}^{-1}\left[\frac{\sinh(px/c)}{\sinh(pl/c)}\right] = \frac{2c}{l}\sum_{n=1}^{\infty}(-1)^{n+1}\sin\frac{n\pi x}{l}\sin\frac{n\pi ct}{l} \tag{10.59}$$

Making use of the convolution theorem

$$\mathscr{L}^{-1}[\bar{f}_1(p)\bar{f}_2(p)] = \int_0^t f_1(\tau)f_2(t-\tau)\,dt \tag{10.60}$$

in conjunction with Eqs. (10.58) and (10.59), we can show that

$$u_x(x,t) = \frac{2c}{l}\sum_{n=1}^{\infty}(-1)^{n+1}\sin\left[\frac{n\pi}{l}(l-x)\right]\int_0^t U_1(\tau)\sin\left[\frac{n\pi c}{l}(t-\tau)\right]d\tau$$
$$+ \frac{2c}{l}\sum_{n=1}^{\infty}(-1)^{n+1}\sin\frac{n\pi x}{l}\int_0^t U_2(\tau)\sin\left[\frac{n\pi c}{l}(t-\tau)\right]d\tau \tag{10.61}$$

It is clear from the above equation that for dynamic boundary conditions the displacements $u_x(x,t)$ cannot be related to the instantaneous values of the boundary displacements $U_1(t)$ and $U_2(t)$. The relationship is in the form of an integral evaluated for times $t = 0$ to t, and this means that at any instant of time $u_x(x,t)$ depends on the previous time history of the boundary displacements. This incidentally also explains why, in general, no simple relationship of the form $\mathbf{u} = \mathbf{aU}$ can be obtained for dynamic conditions. A similar problem could be solved in which instead of specified boundary displacements we would use specified boundary forces varying with time.

The dynamic displacements given by Eq. (10.61) can also be applied to the special case of static displacements if we observe that the static condition is obtained from the dynamic one by applying the boundary displacements at an infinitesimally slow rate, so that the final values of displacements are attained at $t = \infty$. We shall therefore assume that both U_1 and U_2 are being increased exponentially according to

$$U_j(t) = U_{j,\text{static}}(1 - e^{-\beta t} - \beta t e^{-\beta t}) \qquad j = 1, 2 \tag{10.62}$$

where $\quad 0 < \beta < \epsilon \tag{10.63}$

and ϵ is an arbitrarily small constant (see Fig. 10.4). Equation (10.62) also satisfies the initial conditions $U_j(0) = \dot{U}_j(0) = 0$.

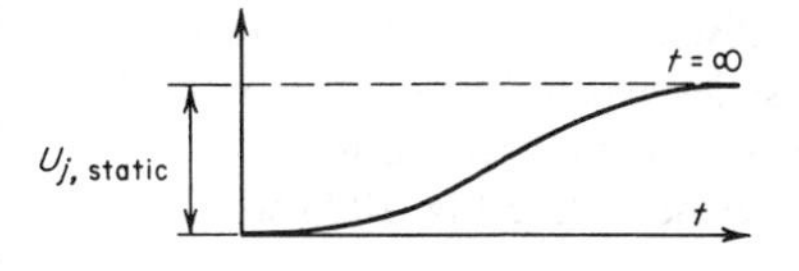

FIG. 10.4 Infinitely slowly increasing displacement representing static condition at $t = \infty$.

Introducing now the boundary displacements, as given by Eq. (10.62), into the integrals in Eq. (10.61), we have

$$\int_0^t U_j(t)\sin\left[\frac{n\pi c}{l}(t-\tau)\right]d\tau$$

$$= U_{j,\text{static}}\int_0^t (1-e^{-\beta\tau}-\beta\tau e^{-\beta\tau})\sin\left[\frac{n\pi c}{l}(t-\tau)\right]d\tau$$

$$= U_{j,\text{static}}\left[\frac{\cos[(n\pi c/l)(t-\tau)]}{n\pi c/l} - \frac{e^{-\beta\tau}}{\beta^2+n^2\pi^2c^2/l^2}\left\{-\beta\sin\left[\frac{n\pi c}{l}(t-\tau)\right] + \frac{n\pi c}{l}\cos\left[\frac{n\pi c}{l}(t-\tau)\right]\right\}\right.$$
$$- \frac{\beta\tau e^{-\beta\tau}}{\beta^2+n^2\pi^2c^2/l^2}\left\{-\beta\sin\left[\frac{n\pi c}{l}(t-\tau)\right]+\frac{n\pi c}{l}\cos\left[\frac{n\pi c}{l}(t-\tau)\right]\right\}$$
$$\left. + \frac{\beta e^{-\beta\tau}}{(\beta^2+n^2\pi^2c^2/l^2)^2}\left\{\left(\beta^2-\frac{n^2\pi^2c^2}{l^2}\right)\sin\left[\frac{n\pi c}{l}(t-\tau)\right] - 2\beta\frac{n\pi c}{l}\cos\left[\frac{n\pi c}{l}(t-\tau)\right]\right\}\right]_0^t$$

$$= U_{j,\text{static}}\left\{\frac{1}{n\pi c/l} - \frac{(n\pi c/l)e^{-\beta t}}{\beta^2+n^2\pi^2c^2/l^2} - \frac{\beta t(n\pi c/l)e^{-\beta t}}{\beta^2+n^2\pi^2c^2/l^2} - \frac{2\beta^2(n\pi c/l)e^{-\beta t}}{(\beta^2+n^2\pi^2c^2/l^2)^2}\right.$$
$$- \frac{\cos[(n\pi c/l)t]}{n\pi c/l} - \frac{\beta\sin(n\pi ct/l)}{\beta^2+n^2\pi^2c^2/l^2} + \frac{(n\pi c/l)\cos(n\pi ct/l)}{\beta^2+n^2\pi^2c^2/l^2}$$
$$\left. - \frac{\beta(\beta^2-n^2\pi^2c^2/l^2)}{(\beta^2+n^2\pi^2c^2/l^2)^2}\sin\frac{n\pi ct}{l} + \frac{2\beta^2(n\pi c/l)}{(\beta^2+n^2\pi^2c^2/l^2)^2}\cos\frac{n\pi ct}{l}\right\} \tag{10.64}$$

Because of the restriction imposed on β by Eq. (10.63) it follows from Eq. (10.64) that for $t = \infty$

$$U_{j,\text{static}}\int_0^\infty (1-e^{-\beta\tau}-\beta\tau e^{-\beta\tau})\sin\left[\frac{n\pi c}{l}(t-\tau)\right]d\tau = \frac{l}{n\pi c}U_{j,\text{static}} \tag{10.65}$$

Introducing now Eq. (10.65) into (10.61), we obtain

$$u_{x,\text{static}} = \frac{2}{\pi}U_{1,\text{static}}\sum_{n=1}^\infty \frac{(-1)^{n+1}}{n}\sin\left[\frac{n\pi}{l}(l-x)\right] + \frac{2}{\pi}U_{2,\text{static}}\sum_{n=1}^\infty\frac{(-1)^{n+1}}{n}\sin\frac{n\pi x}{l}$$
$$= \left(1-\frac{x}{l}\right)U_{1,\text{static}} + \frac{x}{l}U_{2,\text{static}} \tag{10.66}$$

which agrees with our solution for static displacements. The summations in Eq. (10.66) can be recognized as Fourier series expansions for $1 - x/l$ and x/l.

Let us suppose now that a uniform bar of length l performs harmonic forced vibrations, produced by varying end displacements (see Fig. 10.5) so that

$$U_1 = u_x(0,t) = q_1e^{i\omega t} \qquad U_2 = u_x(l,t) = q_2e^{i\omega t} \tag{10.67}$$

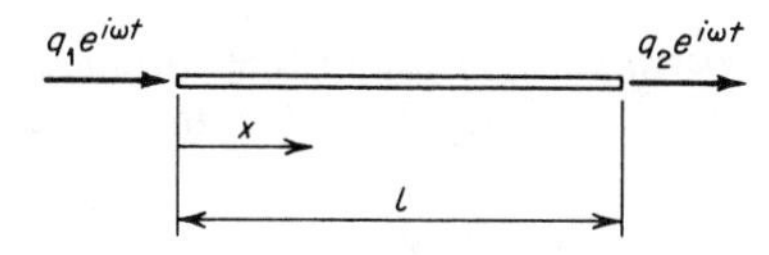

FIG. 10.5 Uniform bar with harmonic end displacements $q_1e^{i\omega t}$ and $q_2e^{i\omega t}$.

where q_1 and q_2 represent the displacement amplitudes. The displacements in the bar will be assumed to be given by

$$u_x = g(x)e^{i\omega t} \tag{10.68}$$

where $g(x)$ is a function of x only. Introducing Eq. (10.68) into (10.41), we have

$$\frac{d^2g}{dx^2} + \frac{\omega^2}{c^2}g = 0 \tag{10.69}$$

Hence $$g = A\sin\frac{\omega x}{c} + B\cos\frac{\omega x}{c} \tag{10.70}$$

The constants A and B are determined from the boundary conditions (10.67). Hence

$$A = \frac{q_2 - q_1\cos\omega l/c}{\sin\omega l/c} \tag{10.71}$$

$$B = q_1 \tag{10.72}$$

provided $\sin\dfrac{\omega l}{c} \neq 0$

or $$\frac{\omega l}{c} \neq n\pi \tag{10.73}$$

where n is an integer. Using Eqs. (10.70) to (10.72) in (10.68), we have

$$\begin{aligned} u_x &= \left[\frac{q_2 - q_1\cos(\omega l/c)}{\sin(\omega l/c)}\sin\frac{\omega x}{c} + q_1\cos\frac{\omega}{c}x\right]e^{i\omega t} \\ &= \left[\left(\cos\frac{\omega x}{c} - \cot\frac{\omega l}{c}\sin\frac{\omega x}{c}\right)\quad \operatorname{cosec}\frac{\omega l}{c}\sin\frac{\omega x}{c}\right]\begin{bmatrix} q_1 \\ q_2 \end{bmatrix}e^{i\omega t} \end{aligned} \tag{10.74}$$

When $\omega = n\pi c/l$, the forcing frequency is equal to the natural frequency of a free bar of length l, and the displacements u_x become infinite.

Equation (10.74) indicates that for harmonic boundary conditions the displacements u_x depend on the instantaneous values of the boundary displacements $u_x(0,t)$ and $u_x(l,t)$. Thus in this case the displacements at any point along the length of the bar can be related to the boundary displacement through the relationship

$$u_x = \mathbf{a}\mathbf{U} \tag{10.75}$$

where $$\mathbf{a} = \mathbf{a}(x,\omega) \tag{10.76}$$

and $\mathbf{U}$ is the column matrix of bar end displacements (10.67). This result is true for any structure or structural element performing forced or free harmonic motion; it will be used in Sec. 10.8 to determine frequency-dependent mass and stiffness matrices for a vibrating bar element.

10.7 EQUIVALENT MASSES IN MATRIX ANALYSIS

The inertia property of an idealized discrete structural system is expressed by the equivalent mass matrix *

$$\mathbf{M} = \int_v \rho \mathbf{a}^T \mathbf{a} \, dV \tag{10.37}$$

It is evident from Sec. 10.6 that for a general dynamic loading the matrix **a** does not exist, except for special cases, such as static loading or harmonic motion of the system. Even in these special cases the matrix **a** may not be exact because of limitations imposed by the idealization procedures introduced when a continuously attached structural element is replaced with an equivalent one having discrete attachments to the neighboring elements. Frequently static displacement distributions are used to determine **a**. Thus the mass representation, as given by Eq. (10.37), will of necessity be only an approximation; however, when the discrete elements selected are small, the accuracy of such representation is generally adequate for practical purposes, as can be judged from the numerical examples considered in subsequent chapters.

In this section we shall determine the equivalent mass matrix for a bar element. Equivalent mass matrices for other elements will be discussed in Chap. 11. From Fig. 10.6 it is evident that the displacements at a point distance x from node 1 are given by

$$\begin{aligned} u_{\bar{x}} &= U_1 + (U_4 - U_1)\xi \\ u_{\bar{y}} &= U_2 + (U_5 - U_2)\xi \\ u_{\bar{z}} &= U_3 + (U_6 - U_3)\xi \end{aligned} \tag{10.77}$$

$$\text{where} \qquad \xi = \frac{x}{l} \tag{10.78}$$

Alternatively, Eqs. (10.77) can be arranged in matrix form as

$$\begin{bmatrix} u_{\bar{x}} \\ u_{\bar{y}} \\ u_{\bar{z}} \end{bmatrix} = \begin{bmatrix} 1-\xi & 0 & 0 & \xi & 0 & 0 \\ 0 & 1-\xi & 0 & 0 & \xi & 0 \\ 0 & 0 & 1-\xi & 0 & 0 & \xi \end{bmatrix} \begin{bmatrix} U_1 \\ U_2 \\ U_3 \\ U_4 \\ U_5 \\ U_6 \end{bmatrix} \tag{10.79}$$

*Also referred to as the consistent mass matrix.

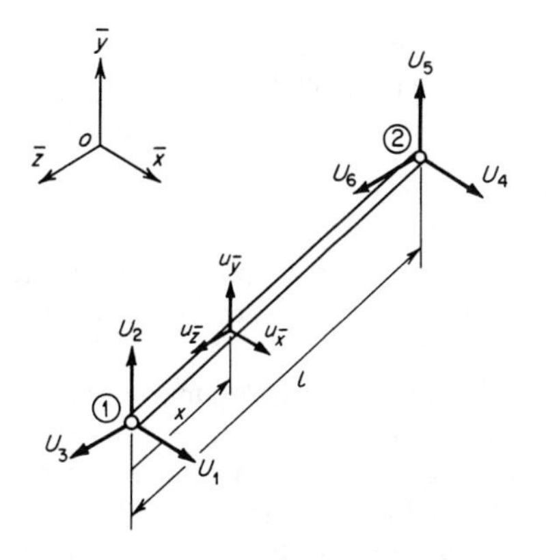

FIG. 10.6 Node displacements on a bar element in datum coordinate system.

Hence
$$\mathbf{a} = \begin{bmatrix} 1-\xi & 0 & 0 & \xi & 0 & 0 \\ 0 & 1-\xi & 0 & 0 & \xi & 0 \\ 0 & 0 & 1-\xi & 0 & 0 & \xi \end{bmatrix} \tag{10.80}$$

Introducing Eqs. (10.80) and (10.78) into (10.37) and denoting the cross-sectional area by A and the element equivalent mass matrix by $\mathbf{m}$, we have

$$\mathbf{m} = \int_v \rho \mathbf{a}^T \mathbf{a}\, dV$$

$$= \rho A l \int_0^1 \begin{bmatrix} (1-\xi)^2 & 0 & 0 & (1-\xi)\xi & 0 & 0 \\ 0 & (1-\xi)^2 & 0 & 0 & (1-\xi)\xi & 0 \\ 0 & 0 & (1-\xi)^2 & 0 & 0 & (1-\xi)\xi \\ \xi(1-\xi) & 0 & 0 & \xi^2 & 0 & 0 \\ 0 & \xi(1-\xi) & 0 & 0 & \xi^2 & 0 \\ 0 & 0 & \xi(1-\xi) & 0 & 0 & \xi^2 \end{bmatrix} d\xi$$

$$= \frac{\rho A l}{6} \begin{bmatrix} 2 & 0 & 0 & 1 & 0 & 0 \\ 0 & 2 & 0 & 0 & 1 & 0 \\ 0 & 0 & 2 & 0 & 0 & 1 \\ 1 & 0 & 0 & 2 & 0 & 0 \\ 0 & 1 & 0 & 0 & 2 & 0 \\ 0 & 0 & 1 & 0 & 0 & 2 \end{bmatrix} = \frac{\rho A l}{6} \begin{bmatrix} 2\mathbf{I}_3 & \mathbf{I}_3 \\ \mathbf{I}_3 & 2\mathbf{I}_3 \end{bmatrix} \tag{10.81}$$

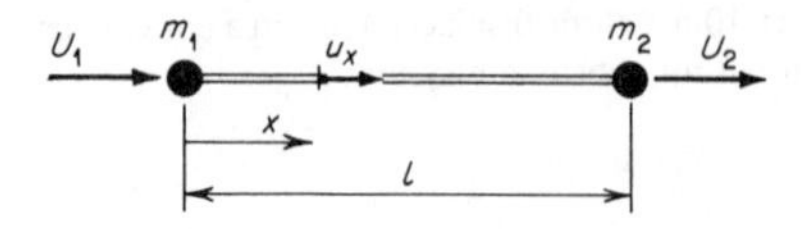

FIG. 10.7 Bar element with concentrated masses m_1 and m_2 attached to its ends.

The off-diagonal nonzero terms in **m** represent dynamic coupling between the two degrees of freedom; the coupling occurs only for degrees of freedom in the same directions, i.e., 1 and 4, 2 and 5, and 3 and 6.

We consider next a uniform bar element with two concentrated masses m_1 and m_2 attached to its ends, as shown in Fig. 10.7. Since the concentrated mass can be regarded as a region of zero volume having infinite density such that the product of density and volume is finite and equal to the mass, Eq. (10.37) can still be used to determine the equivalent mass matrix provided the integration is extended to include the concentrated masses. For simplicity we shall determine only the inertia properties in the longitudinal direction. The displacements along the length of the bar are given by

$$u_x = [(1 - \xi) \quad \xi] \begin{bmatrix} U_1 \\ U_2 \end{bmatrix} \tag{10.82}$$

Hence $$\mathbf{a} = [(1 - \xi) \quad \xi] \tag{10.83}$$

Introducing Eq. (10.83) into (10.37) gives

$$\begin{aligned} \mathbf{m} &= \int_v \rho \mathbf{a}^T \mathbf{a} \, dV \\ &= \begin{bmatrix} \int_v \rho(1 - \xi)^2 \, dV & \int_v \rho(1 - \xi)\xi \, dV \\ \int_v \rho\xi(1 - \xi) \, dV & \int_v \rho\xi^2 \, dV \end{bmatrix} \\ &= \frac{\rho A l}{6} \begin{bmatrix} 2 & 1 \\ 1 & 2 \end{bmatrix} + \begin{bmatrix} m_1 & 0 \\ 0 & m_2 \end{bmatrix} \end{aligned} \tag{10.84}$$

It follows therefore that the effect of concentrated masses at the ends of an element is accounted for by adding the actual concentrated masses to the corresponding diagonal terms in the equivalent mass matrix of the continuous system. The assembly of the complete mass matrix for a structure made up from structural elements and discrete (concentrated) masses is discussed in Chap. 11.

10.8 FREQUENCY-DEPENDENT MASS AND STIFFNESS MATRICES FOR BAR ELEMENTS[275]

It was demonstrated in Sec. 10.6 that for a bar performing harmonic motion the displacements along the length of the bar are given by

$$u_x = \left[\left(\cos\frac{\omega x}{c} - \cot\frac{\omega l}{c}\sin\frac{\omega x}{c}\right) \quad \operatorname{cosec}\frac{\omega l}{c}\sin\frac{\omega x}{c}\right]\begin{bmatrix} q_1 \\ q_2 \end{bmatrix} e^{i\omega t} \tag{10.74}$$

The strain ϵ_{xx} in the bar is obtained by differentiation of (10.74) with respect to x. Hence

$$\epsilon_{xx} = \frac{\partial u_x}{\partial x} = \frac{\omega}{c}\left[-\left(\sin\frac{\omega x}{c} + \cot\frac{\omega l}{c}\cos\frac{\omega x}{c}\right) \quad \operatorname{cosec}\frac{\omega l}{c}\cos\frac{\omega x}{c}\right]\begin{bmatrix} q_1 \\ q_2 \end{bmatrix} e^{i\omega t} \tag{10.85}$$

Hence, using the notation of Sec. 10.5, it follows that

$$\mathbf{a} = \left[\left(\cos\frac{\omega x}{c} - \cot\frac{\omega l}{c}\sin\frac{\omega x}{c}\right) \quad \operatorname{cosec}\frac{\omega l}{c}\sin\frac{\omega x}{c}\right] \tag{10.86}$$

$$\mathbf{b} = \left[\frac{-\omega}{c}\left(\sin\frac{\omega x}{c} + \cot\frac{\omega l}{c}\cos\frac{\omega x}{c}\right) \quad \frac{\omega}{c}\left(\operatorname{cosec}\frac{\omega l}{c}\cos\frac{\omega x}{c}\right)\right] \tag{10.87}$$

After substituting (10.86) into (10.37) and (10.87) into (10.38) and integrating over the volume of the bar element it can be shown that the equivalent mass and stiffness matrices are given by

$$\mathbf{m} = \frac{\rho A l}{2}\frac{c}{\omega l}\operatorname{cosec}\frac{\omega l}{c}\begin{bmatrix} \left(\frac{\omega l}{c}\operatorname{cosec}\frac{\omega l}{c} - \cos\frac{\omega l}{c}\right) & \left(1 - \frac{\omega l}{c}\cot\frac{\omega l}{c}\right) \\ \left(1 - \frac{\omega l}{c}\cot\frac{\omega l}{c}\right) & \left(\frac{\omega l}{c}\operatorname{cosec}\frac{\omega l}{c} - \cos\frac{\omega l}{c}\right) \end{bmatrix} \tag{10.88}$$

$$\mathbf{k} = \frac{AE}{2l}\frac{\omega l}{c}\operatorname{cosec}\frac{\omega l}{c}\begin{bmatrix} \left(\frac{\omega l}{c}\operatorname{cosec}\frac{\omega l}{c} + \cos\frac{\omega l}{c}\right) & -\left(1 + \frac{\omega l}{c}\cot\frac{\omega l}{c}\right) \\ -\left(1 + \frac{\omega l}{2}\cot\frac{\omega l}{c}\right) & \left(\frac{\omega l}{c}\operatorname{cosec}\frac{\omega l}{c} + \cos\frac{\omega l}{c}\right) \end{bmatrix} \tag{10.89}$$

The elements in the above matrices are functions of the circular frequency ω. In numerical calculations we usually require the natural frequencies of the system, and consequently we cannot use the frequency-dependent mass and stiffness matrices since the matrix elements are unknown initially. To avoid this difficulty the solution for displacements u_x will be assumed to be given by a series in ascending powers of the circular frequency ω, so that

$$u_x = \mathbf{aU} = \sum_{r=0}^{\infty} \omega^r \mathbf{a}_r \mathbf{U} = \sum_{r=0}^{\infty} \omega^r \mathbf{a}_r \mathbf{q} e^{i\omega t} \tag{10.90}$$

where
$$\mathbf{q} = \begin{bmatrix} q_1 \\ q_2 \end{bmatrix} \tag{10.91}$$

and the matrices $\mathbf{a}_r$ are functions of x only. Substituting Eq. (10.90) into equation of motion (10.41), we have

$$c^2 \sum_{r=0}^{\infty} \omega^r \mathbf{a}_r'' \mathbf{q} e^{i\omega t} + \omega^2 \sum_{r=0}^{\infty} \omega^r \mathbf{a}_r \mathbf{q} e^{i\omega t} = 0 \tag{10.92}$$

where primes denote differentiation with respect to x. Equating to zero coefficients of the same powers of ω in (10.92) gives

$$\mathbf{a}_0'' = \mathbf{0} \tag{10.93}$$

$$\mathbf{a}_1'' = \mathbf{0} \tag{10.94}$$

$$c^2 \mathbf{a}_2'' = -\mathbf{a}_0 \tag{10.95}$$

$$c^2 \mathbf{a}_3'' = -\mathbf{a}_1 \tag{10.96}$$

. . . .

Equations (10.93) to (10.96) can be integrated directly. The first matrix $\mathbf{a}_0$ is then used to satisfy the boundary conditions that $u_x = U_1$ at $x = 0$ and $u_x = U_2$ at $x = l$, while the remaining matrices $\mathbf{a}_1, \mathbf{a}_2, \mathbf{a}_3, \ldots$ must all vanish at $x = 0$ and l. This leads to

$$\mathbf{a}_0 = [(1 - \xi) \quad \xi] \tag{10.97}$$

$$\mathbf{a}_1 = \mathbf{0} \tag{10.98}$$

$$\mathbf{a}_2 = \frac{\rho l^2}{6E} [(2\xi - 3\xi^2 + \xi^3) \quad (\xi - \xi^3)] \tag{10.99}$$

$$\mathbf{a}_3 = \mathbf{0} \tag{10.100}$$

. . . .

Therefore the matrix $\mathbf{a}$ in Eq. (10.90) is given by

$$\mathbf{a} = \mathbf{a}_0 + \omega^2 \mathbf{a}_2 + \cdots \tag{10.101}$$

The matrix $\mathbf{a}_0$ represents the static displacement distribution due to unit values of the bar end displacements U_1 and U_2.

The distribution of strain in the bar is calculated from

$$e = \frac{\partial u_x}{\partial x} = \sum_{r=0}^{\infty} \omega^r \mathbf{a}_r' \mathbf{q} e^{i\omega t} = \sum_{r=0}^{\infty} \omega^r \mathbf{b}_r \mathbf{q} e^{i\omega t} \tag{10.102}$$

where $$\mathbf{b}_0 = \frac{d\mathbf{a}_0}{dx} = \frac{1}{l}[-1 \quad 1] \tag{10.103}$$

$$\mathbf{b}_1 = \frac{d\mathbf{a}_1}{dx} = \mathbf{0} \tag{10.104}$$

$$\mathbf{b}_2 = \frac{d\mathbf{a}_2}{dx} = \frac{\rho l}{6E}[(2 - 6\xi + 3\xi^2) \quad (1 - 3\xi^2)] \tag{10.105}$$

$$\mathbf{b}_3 = \frac{d\mathbf{a}_3}{dx} = \mathbf{0} \tag{10.106}$$

. . . .

Hence the matrix $\mathbf{b}$ in $\mathbf{e} = \mathbf{bU}$ is given by

$$\mathbf{b} = \mathbf{b}_0 + \omega^2 \mathbf{b}_2 + \cdots \tag{10.107}$$

Substituting now Eq. (10.101) into (10.37), we obtain the frequency-dependent mass matrix

$$\mathbf{m} = \mathbf{m}_0 + \omega^2 \mathbf{m}_2 + \cdots \tag{10.108}$$

where $$\mathbf{m}_0 = \frac{\rho A l}{6}\begin{bmatrix} 2 & 1 \\ 1 & 2 \end{bmatrix} \tag{10.109}$$

and $$\mathbf{m}_2 = \frac{2\rho^2 A l^3}{45E}\begin{bmatrix} 1 & \frac{7}{8} \\ \frac{7}{8} & 1 \end{bmatrix} \tag{10.110}$$

Similarly, substituting Eq. (10.107) into (10.38), we obtain the frequency-dependent stiffness matrix

$$\mathbf{k} = \mathbf{k}_0 + \omega^4 \mathbf{k}_4 + \cdots \tag{10.111}$$

where $$\mathbf{k}_0 = \frac{AE}{l}\begin{bmatrix} 1 & -1 \\ -1 & 1 \end{bmatrix} \tag{10.112}$$

and $$\mathbf{k}_4 = (\rho A l)^2 \frac{l}{45AE}\begin{bmatrix} 1 & \frac{7}{8} \\ \frac{7}{8} & 1 \end{bmatrix} \tag{10.113}$$

The matrices $\mathbf{m}_0$ and $\mathbf{k}_0$ represent the static inertia and stiffness of the bar element, while $\mathbf{m}_2$ and $\mathbf{k}_4$ and higher order terms represent dynamic corrections. Applications of $\mathbf{m}_2$ and $\mathbf{k}_4$ matrices to vibration problems will be discussed in Chap. 12.

10.9 FREQUENCY-DEPENDENT MASS AND STIFFNESS MATRICES FOR BEAM ELEMENTS[275]

The equation of motion of a beam element (see Fig. 10.8) in the transverse direction is given by

$$c^4 \frac{\partial^4 u_y}{\partial x^4} + \ddot{u}_y = 0 \tag{10.114}$$

where $$c^4 = \frac{EI}{\rho A} \tag{10.115}$$

and I is the moment of inertia of the beam cross section. For simplicity of presentation shear deformations will be neglected, but if required, they can be accounted for without any special difficulties. In addition to the transverse displacements u_y the beam element undergoing transverse vibrations will have displacements u_x, which in accordance with engineering bending theory can be calculated from

$$u_x = -\frac{\partial u_y}{\partial x} y = -l \frac{\partial u_y}{\partial x} \eta \qquad \eta = \frac{y}{l} \tag{10.116}$$

As in the case of a bar element, the displacements $\mathbf{u}$ are expanded in ascending powers of ω, so that

$$\mathbf{u} = \begin{bmatrix} u_x \\ u_y \end{bmatrix} = \left\{ \begin{bmatrix} \mathbf{a}_{0x} \\ \mathbf{a}_{0y} \end{bmatrix} + \omega \begin{bmatrix} \mathbf{a}_{1x} \\ \mathbf{a}_{1y} \end{bmatrix} + \omega^2 \begin{bmatrix} \mathbf{a}_{2x} \\ \mathbf{a}_{2y} \end{bmatrix} + \cdots \right\} \mathbf{U}$$

$$= (\mathbf{a}_0 + \omega \mathbf{a}_1 + \omega^2 \mathbf{a}_2 + \cdots)\mathbf{U} = \mathbf{a}\mathbf{q}e^{i\omega t} \tag{10.117}$$

where $$\mathbf{U} = \{U_1 \quad U_2 \quad U_3 \quad U_4\} = \{q_1 \quad q_2 \quad q_3 \quad q_4\} e^{i\omega t} \tag{10.118}$$

Hence $$u_x = \sum_{r=0}^{\infty} \omega^r \mathbf{a}_{rx} \mathbf{q} e^{i\omega t} = \mathbf{a}_x \mathbf{q} e^{i\omega t} \tag{10.119}$$

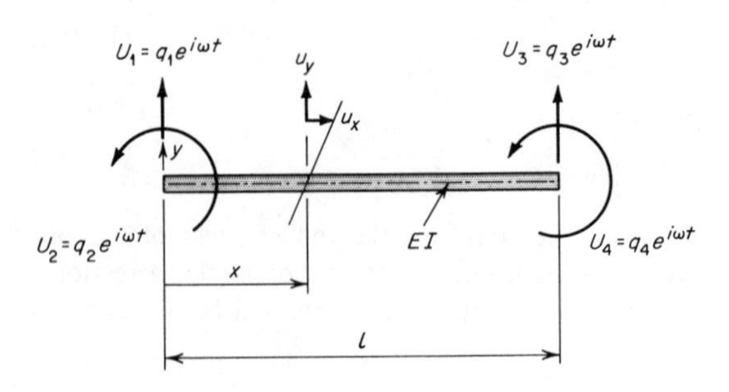

FIG. 10.8 Beam element.

and $$u_y = \sum_{r=0}^{\infty} \omega^r \mathbf{a}_{ry}\mathbf{q}e^{i\omega t} = \mathbf{a}_y\mathbf{q}e^{i\omega t} \tag{10.120}$$

Substituting Eq. (10.120) into equation of motion (10.114) gives

$$c^4 \sum_{r=0}^{\infty} \omega^r {\mathbf{a}_{ry}}^{\text{iv}}\mathbf{q}e^{i\omega t} - \omega^2 \sum_{r=0}^{\infty} \omega^r \mathbf{a}_{ry}\mathbf{q}e^{i\omega t} = 0 \tag{10.121}$$

Equating to zero coefficients of the same powers of ω in (10.121), we obtain

$${\mathbf{a}_{0y}}^{\text{iv}} = \mathbf{0} \tag{10.122}$$

$${\mathbf{a}_{1y}}^{\text{iv}} = \mathbf{0} \tag{10.123}$$

$$c^4{\mathbf{a}_{2y}}^{\text{iv}} = \mathbf{a}_{0y} \tag{10.124}$$

$$c^4{\mathbf{a}_{3y}}^{\text{iv}} = \mathbf{a}_{1y} \tag{10.125}$$

. . . .

By solving Eqs. (10.122) to (10.125) it can be demonstrated that

$$\mathbf{a}_{0y} = [(1 - 3\xi^2 + 2\xi^3) \quad (\xi - 2\xi^2 + \xi^3)l \quad (3\xi^2 - 2\xi^3) \quad (-\xi^2 + \xi^3)l] \tag{10.126}$$

$$\mathbf{a}_{1y} = \mathbf{0} \tag{10.127}$$

$$\mathbf{a}_{2y} = \frac{\rho A l^4}{2{,}520EI}[(66\xi^2 - 156\xi^3 + 105\xi^4 - 21\xi^6 + 6\xi^7) \quad (12\xi^2 - 22\xi^3 + 21\xi^5 - 14\xi^6 + 3\xi^7)l \quad (39\xi^2 - 54\xi^3 + 21\xi^6 - 6\xi^7) \quad (-9\xi^2 + 13\xi^3 - 7\xi^6 + 3\xi^7)l] \tag{10.128}$$

$$\mathbf{a}_{3y} = \mathbf{0} \tag{10.129}$$

. . . . .

The matrix $\mathbf{a}_{0y}$ represents static transverse deflection distribution due to unit values of $U_1, \ldots, U_4$. The remaining matrices in $\mathbf{a}$ are determined from Eq. (10.116). This leads to

$$\mathbf{a}_{0x} = [6(\xi - \xi^2)\eta \quad (-1 + 4\xi - 3\xi^2)l\eta \quad 6(-\xi + \xi^2)\eta \quad (2\xi - 3\xi^2)l\eta] \tag{10.130}$$

$$\mathbf{a}_{1x} = \mathbf{0} \tag{10.131}$$

$$\mathbf{a}_{2x} = \frac{\rho A l^4}{2{,}520EI}[(-132\xi + 468\xi^2 - 420\xi^3 + 126\xi^5 - 42\xi^6)\eta \quad (-24\xi + 66\xi^2 - 105\xi^4 + 84\xi^5 - 21\xi^6)l\eta \quad (-78\xi + 162\xi^2 - 126\xi^5 + 42\xi^6)\eta \quad (18\xi - 39\xi^2 + 42\xi^5 - 21\xi^6)l\eta] \tag{10.132}$$

$$\mathbf{a}_{3x} = \mathbf{0} \tag{10.133}$$

. . . . .

The strains in the beam element are derived from

$$e = \frac{\partial u_x}{\partial x} = -\frac{\partial^2 u_y}{\partial x^2} y = -l \frac{\partial^2 u_y}{\partial x^2} \eta$$
$$= (\mathbf{b}_0 + \omega \mathbf{b}_1 + \omega^2 \mathbf{b}_2 + \cdots)\mathbf{U} = \mathbf{bU} \tag{10.134}$$

Hence from Eqs. (10.126) to (10.129) and (10.134) it follows that

$$\mathbf{b}_0 = -l\eta \mathbf{a}''_{0y}$$
$$= -\frac{\eta}{l}[(-6 + 12\xi) \quad (-4 + 6\xi)l \quad (6 - 12\xi) \quad (-2 + 6\xi)l] \tag{10.135}$$

$$\mathbf{b}_1 = -l\eta \mathbf{a}''_{1y} = \mathbf{0} \tag{10.136}$$

$$\mathbf{b}_2 = -l\eta \mathbf{a}''_{2y} \tag{10.137}$$
$$= \frac{-\rho A \eta l^3}{420EI}[(22 - 156\xi + 210\xi^2 - 105\xi^4 + 42\xi^5) \quad (4 - 22\xi + 70\xi^3 - 70\xi^4$$
$$+ 21\xi^5)l \quad (13 - 54\xi + 105\xi^4 - 42\xi^5) \quad (-3 + 13\xi - 35\xi^4 + 21\xi^5)l] \tag{10.138}$$

$$\mathbf{b}_3 = -l\eta \mathbf{a}''_{3y} = \mathbf{0} \tag{10.139}$$

. . . .

The next step is to substitute the calculated series for **a** into Eq. (10.37) to determine the equivalent mass matrix **m**. For this case the mass matrix **m** is given by

$$\mathbf{m} = \mathbf{m}_0 + \omega^2 \mathbf{m}_2 + \cdots \tag{10.140}$$

where

$$\mathbf{m}_0 = \frac{\rho A l}{420}\begin{bmatrix} 156 & & \text{Symmetric} & \\ 22l & 4l^2 & & \\ 54 & 13l & 156 & \\ -13l & -3l^2 & -22l & 4l^2 \end{bmatrix} + \frac{\rho A l}{30}\left(\frac{r}{l}\right)^2 \begin{bmatrix} 36 & & \text{Symmetric} & \\ 3l & 4l^2 & & \\ -36 & -3l & 36 & \\ 3l & -l^2 & -3l & 4l^2 \end{bmatrix} \tag{10.141}$$

$$\mathbf{m}_2 = \frac{(\rho A l)^2 l^3}{EI}\begin{bmatrix} 0.729746 & & \text{Symmetric} & \\ 0.153233l & 0.0325248l^2 & & \\ 0.659142 & 0.144386l & 0.729746 & \\ -0.144386l & -0.0314082l^2 & -0.153233l & 0.0325248l^2 \end{bmatrix} \times 10^{-3}$$
$$+ \frac{(\rho A l)^2 l^3}{EI}\left(\frac{r}{l}\right)^2 \begin{bmatrix} 0.317460 & & \text{Symmetric} & \\ 0.793651l & 0.317460l^2 & & \\ -0.317460 & 0.595238l & 0.317460 & \\ -0.595238l & -0.277778l^2 & -0.793651l & 0.317460l^2 \end{bmatrix} \times 10^{-3} \tag{10.142}$$

where r is the radius of gyration of the beam cross section. The first terms in (10.141) and (10.142) represent the translational inertia of the beam element, while second terms represent the rotatory inertia.

The stiffness matrix is determined from the calculated series for $\mathbf{b}$ and Eq. (10.38). This leads to

$$\mathbf{k} = \mathbf{k}_0 + \omega^4 \mathbf{k}_4 + \cdots \tag{10.143}$$

where

$$\mathbf{k}_0 = \frac{EI}{l^3}\begin{bmatrix} 12 & & \text{Symmetric} & \\ 6l & 4l & & \\ -12 & -6l & 12 & \\ 6l & 2l^2 & -6l & 4l^2 \end{bmatrix} \tag{10.144}$$

$$\mathbf{k}_4 = (\rho A l)^2 \frac{l^3}{EI}\begin{bmatrix} 0.364872 & & \text{Symmetric} & \\ 0.0766162l & 0.0162624l^2 & & \\ 0.329571 & 0.0721933l & 0.364872 & \\ -0.0721933l & -0.0157041l^2 & -0.0766162l & 0.0162624l^2 \end{bmatrix} \times 10^{-3} \tag{10.145}$$

Examples illustrating the application of the frequency-dependent mass and stiffness matrices to cantilever beams have been given by Przemieniecki.[275]

PROBLEMS

10.1 A uniform bar is subjected to time-varying forces $P_1(t)$ and $P_2(t)$ applied at the two ends (see Fig. 10.9). Derive the analytical solution for the longitudinal displacement u_x.

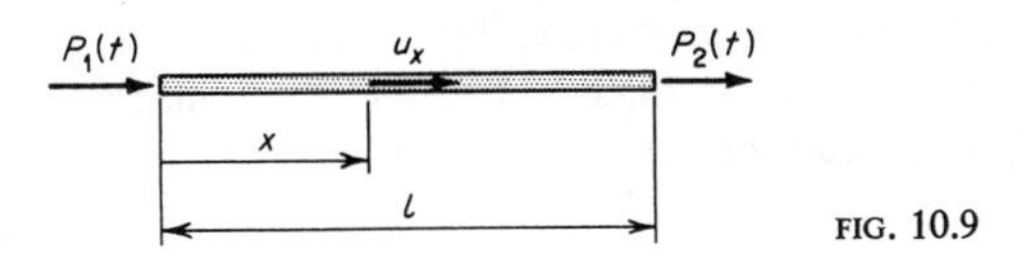

FIG. 10.9

10.2 Discuss possible methods of extending the concept of frequency-dependent mass and stiffness matrices to vibrating rectangular plates.

CHAPTER 11 INERTIA PROPERTIES OF STRUCTURAL ELEMENTS

This chapter presents detailed derivations of equivalent mass matrices for discrete elements. The underlying assumption used to derive these matrices is that the dynamic displacement distribution within each element can be adequately represented by static displacements. Calculations of mass matrices in datum coordinates, mass matrices for the assembled structure, and condensed mass matrices are also presented. The equivalent mass matrices based on static displacement distributions are determined for the following elements: pin-jointed bars, beam elements, triangular and rectangular plates with translational displacements, solid tetrahedra, solid parallelepipeds, and triangular and rectangular plates with bending displacements.

11.1 EQUIVALENT MASS MATRICES IN DATUM COORDINATE SYSTEM

From a computational point of view it is generally preferable to calculate the equivalent mass matrices for unassembled elements using the local coordinate system and then to transform these matrices into the datum system selected for the assembled structure. The element mass matrix is calculated from (see Sec. 10.5)

$$\mathbf{m} = \int_v \rho \mathbf{a}^T \mathbf{a} \, dV \tag{11.1}$$

where the matrix $\mathbf{a}$ must refer to *all* nodal displacements in local coordinate system. Thus for thin plates with negligible (zero) transverse stiffness, transverse deflections must be considered in formulating **a**. In Chap. 5 we have seen that the element displacements in local and datum systems of coordinates are related by the equation

$$\mathbf{u} = \boldsymbol{\lambda}\bar{\mathbf{u}} \tag{11.2}$$

where in the current application $\boldsymbol{\lambda}$ will always be an $n \times n$ matrix of direction cosines, n being the total number of displacements (degrees of freedom) on the element. From Eq. (11.2) it follows that

$$\ddot{\mathbf{u}} = \boldsymbol{\lambda}\ddot{\bar{\mathbf{u}}} \tag{11.3}$$

For virtual displacements

$$\delta\mathbf{u} = \boldsymbol{\lambda}\,\delta\bar{\mathbf{u}} \tag{11.4}$$

the virtual work of inertia forces must be independent of the chosen frame of reference, and therefore

$$\delta\bar{\mathbf{u}}^T\,(-\bar{\mathbf{m}}\ddot{\bar{\mathbf{u}}}) = \delta\mathbf{u}^T\,(-\mathbf{m}\ddot{\mathbf{u}}) \tag{11.5}$$

where $\bar{\mathbf{m}}$ is the equivalent mass matrix in the datum system. Introducing Eqs. (11.3) and (11.4) into (11.5) gives

$$\delta\bar{\mathbf{u}}^T\,(\bar{\mathbf{m}} - \boldsymbol{\lambda}^T\mathbf{m}\boldsymbol{\lambda})\ddot{\bar{\mathbf{u}}} = \mathbf{0} \tag{11.6}$$

Since both $\delta\bar{\mathbf{u}}$ and $\ddot{\bar{\mathbf{u}}}$ are arbitrary, it follows that

$$\bar{\mathbf{m}} = \boldsymbol{\lambda}^T\mathbf{m}\boldsymbol{\lambda} \tag{11.7}$$

Thus the same congruent transformation employed previously for element stiffnesses is used here to determine the element equivalent mass matrices in datum coordinates. Note, however, our earlier comments on the size of the $\boldsymbol{\lambda}$ matrix.

Using Eqs. (11.1) and (11.7), we have

$$\begin{aligned}\bar{\mathbf{m}} &= \int_v \rho\boldsymbol{\lambda}^T\mathbf{a}^T\mathbf{a}\boldsymbol{\lambda}\,dV \\ &= \int_v \rho\bar{\mathbf{a}}^T\bar{\mathbf{a}}\,dV \end{aligned} \tag{11.8}$$

where $\quad \bar{\mathbf{a}} = \mathbf{a}\boldsymbol{\lambda} \quad (11.9)$

represents displacement distribution for unit values of nodal displacements in the datum system of coordinates. If required, Eq. (11.8) may be used for the direct calculation of mass matrices in the datum system.

For some elements, the equivalent mass matrix is invariant with respect to the orientation and position of the coordinate axes. For example, this is true for pin-jointed bars, solid tetrahedra and parallelpipeds, and plate elements having in-plane stiffness only; however, for elements having bending stiffness, such as beam elements, the equivalent mass matrix depends on the selected frame of reference.

11.2 EQUIVALENT MASS MATRIX FOR AN ASSEMBLED STRUCTURE

The equivalent mass matrix for the complete structure made up from an assembly of idealized elements is calculated from the element mass matrices $\bar{\mathbf{m}}^{(i)}$ using the same procedure as for the calculation of the assembled stiffness matrix $\mathbf{K}$ from the element stiffnesses $\bar{\mathbf{k}}^{(i)}$. To determine the stiffness matrix we considered externally applied forces at the nodal points. To determine the mass matrix we must consider inertia forces acting on the assembled structure.

The inertia forces acting on each element are given by

$$\bar{\mathbf{S}}_I^{(i)} = -\bar{\mathbf{m}}^{(i)}\ddot{\bar{\mathbf{u}}}^{(i)} \tag{11.10}$$

where, as before, the superscript i refers to the ith element. Equations (11.10) can be combined into a single matrix equation

$$\bar{\mathbf{S}}_I = -\bar{\mathbf{m}}\ddot{\bar{\mathbf{u}}} \tag{11.11}$$

where
$$\bar{\mathbf{S}}_I = \{\mathbf{S}_I^{(1)} \quad \mathbf{S}_I^{(2)} \quad \cdots \quad \mathbf{S}_I^{(i)} \quad \cdots\} \tag{11.12}$$

$$\bar{\mathbf{m}} = \lceil \bar{\mathbf{m}}^{(1)} \quad \bar{\mathbf{m}}^{(2)} \quad \cdots \quad \bar{\mathbf{m}}^{(i)} \quad \cdots \rfloor \tag{11.13}$$

$$\ddot{\bar{\mathbf{u}}} = \{\ddot{\bar{\mathbf{u}}}^{(1)} \quad \ddot{\bar{\mathbf{u}}}^{(2)} \quad \cdots \quad \ddot{\bar{\mathbf{u}}}^{(i)} \quad \cdots\} \tag{11.14}$$

Also in Chap. 5 it was shown that

$$\bar{\mathbf{u}} = \mathbf{A}\mathbf{U} \tag{11.15}$$

and hence
$$\ddot{\bar{\mathbf{u}}} = \mathbf{A}\ddot{\mathbf{U}} \tag{11.16}$$

In addition to the equivalent mass matrices $\bar{\mathbf{m}}$, there may be actual concentrated masses placed at the nodal points. These masses will be introduced by the diagonal matrix $\bar{\mathbf{m}}_c$, the order of which is equal to the number of node displacements $\mathbf{U}$. For node points without any concentrated masses the corresponding positions in $\bar{\mathbf{m}}_c$ will be filled with zeros.

Introducing now virtual displacements

$$\delta\bar{\mathbf{u}} = \mathbf{A}\,\delta\mathbf{U} \tag{11.17}$$

and equating virtual work of inertia forces, we obtain

$$\delta\mathbf{U}^T(-\mathbf{M}\ddot{\mathbf{U}}) = \delta\bar{\mathbf{u}}^T(-\bar{\mathbf{m}}\ddot{\bar{\mathbf{u}}}) + \delta\mathbf{U}^T(-\bar{\mathbf{m}}_c\ddot{\mathbf{U}}) \tag{11.18}$$

where $\mathbf{M}$ is the equivalent mass matrix for the assembled structure. Using now Eqs. (11.16) and (11.17) in (11.18), we have

$$\delta\mathbf{U}^T(\mathbf{M} - \mathbf{A}^T\bar{\mathbf{m}}\mathbf{A} - \bar{\mathbf{m}}_c)\ddot{\mathbf{U}} = \mathbf{0} \tag{11.19}$$

and hence $\mathbf{M} = \mathbf{A}^T\bar{\mathbf{m}}\mathbf{A} + \bar{\mathbf{m}}_c$ (11.20)

It is therefore evident that the concentrated masses in $\bar{\mathbf{m}}_c$ are simply added to the corresponding diagonal terms in $\mathbf{A}^T\bar{\mathbf{m}}\mathbf{A}$. The congruent transformation $\mathbf{A}^T\bar{\mathbf{m}}\mathbf{A}$ can of course be replaced by the summation procedure, as already explained in Chap. 5 in connection with the calculation of the assembled stiffness matrix $\mathbf{K} = \mathbf{A}^T\bar{\mathbf{k}}\mathbf{A}$.

11.3 CONDENSED MASS MATRIX[133]

Because of expediency of effort or lack of reliable eigenvalue computer programs for large matrices, not all the displacements which were used in the static analysis are necessarily considered in the dynamic analysis. For example, in the conventional dynamic analysis of aircraft wing structures only the deflections normal to the wing midplane are retained. The question then arises how to formulate the equivalent mass matrix for the reduced number of degrees of freedom. This can be achieved again through the application of the virtual-work principle.

The first step is to partition the stiffness matrix $\mathbf{K}$ and the displacement matrix $\mathbf{U}$ into

$$\mathbf{K} = \begin{bmatrix} \mathbf{K}_{\text{I,I}} & \mathbf{K}_{\text{I,II}} \\ \mathbf{K}_{\text{II,I}} & \mathbf{K}_{\text{II,II}} \end{bmatrix} \tag{11.21}$$

$$\mathbf{U} = \begin{bmatrix} \mathbf{U}_{\text{I}} \\ \mathbf{U}_{\text{II}} \end{bmatrix} \tag{11.22}$$

The column matrix $\mathbf{U}_{\text{I}}$ refers to all the displacements we wish to retain as the degrees of freedom for the dynamic analysis, while $\mathbf{U}_{\text{II}}$ denotes all the remaining displacements, which, although they were used in the static analysis, will not be employed in formulating a new equivalent mass matrix. The displacements $\mathbf{U}_{\text{II}}$ may be determined from the static equilibrium equation $\mathbf{P} = \mathbf{KU}$ by assuming that the external forces $\mathbf{P}_{\text{II}}$ corresponding to the displacements $\mathbf{U}_{\text{II}}$ are all equal to zero. Hence

$$\mathbf{U}_{\text{II}} = -\mathbf{K}_{\text{II,II}}^{-1}\mathbf{K}_{\text{II,I}}\mathbf{U}_{\text{I}} \tag{11.23}$$

When we denote the equivalent mass matrix for the displacements $\mathbf{U}_{\text{I}}$ by $\mathbf{M}_c$ and introduce virtual displacements $\delta\mathbf{U}$, it follows immediately from the equivalence of virtual work of the two equivalent mass representations of the continuous system that

$$\begin{aligned} \delta\mathbf{U}_{\text{I}}^T(-\mathbf{M}_c\ddot{\mathbf{U}}_{\text{I}}) &= \delta\mathbf{U}^T(-\mathbf{M}\ddot{\mathbf{U}}) \\ &= -[\delta\mathbf{U}_{\text{I}}^T \quad \delta\mathbf{U}_{\text{II}}^T]\mathbf{M}\begin{bmatrix} \ddot{\mathbf{U}}_{\text{I}} \\ \ddot{\mathbf{U}}_{\text{II}} \end{bmatrix} \end{aligned} \tag{11.24}$$

Substituting Eq. (11.23) into (11.24), we have

$$\delta \mathbf{U}_{\mathrm{I}}^{T} \mathbf{M}_c \ddot{\mathbf{U}}_{\mathrm{I}} = \delta \mathbf{U}_{\mathrm{I}}^{T} [\mathbf{I} \quad -\mathbf{K}_{\mathrm{I,II}} \mathbf{K}_{\mathrm{II,II}}^{-1}] \mathbf{M} \begin{bmatrix} \mathbf{I} \\ -\mathbf{K}_{\mathrm{II,II}}^{-1} \mathbf{K}_{\mathrm{II,I}} \end{bmatrix} \ddot{\mathbf{U}}_I \tag{11.25}$$

Hence $\quad \mathbf{M}_c = \mathbf{A}_c^T \mathbf{M} \mathbf{A}_c \qquad (11.26)$

where $\quad \mathbf{A}_c = \begin{bmatrix} \mathbf{I} \\ -\mathbf{K}_{\mathrm{II,II}}^{-1} \mathbf{K}_{\mathrm{II,I}} \end{bmatrix} \qquad (11.27)$

The equivalent mass matrix $\mathbf{M}_c$ will be referred to as the condensed mass matrix for the reduced number of degrees of freedom. It can be seen that the condensed mass matrix is obtained in Eq. (11.26) by a congruent transformation similar to that used in determining **M** or **K**.

11.4 PIN-JOINTED BAR

The equivalent mass matrix for a pin-jointed bar was derived in Sec. 10.7 as

$$\mathbf{m} = \frac{\rho A l}{6} \begin{bmatrix} 2\mathbf{I}_3 & \mathbf{I}_3 \\ \mathbf{I}_3 & 2\mathbf{I}_3 \end{bmatrix} \tag{10.81}$$

This matrix is invariant with respect to the selected set of axes. In the special case when only motion along the length of the bar is considered (see Fig. 11.1) expression (10.81) reduces to

$$\mathbf{m} = \frac{\rho A l}{6} \begin{bmatrix} 2 & 1 \\ 1 & 2 \end{bmatrix} \tag{11.28}$$

11.5 UNIFORM BEAM

As a local coordinate system we shall select the system shown in Fig. 11.2. The origin is at node 1 with the ox axis taken along the length of the beam and with the oy and oz axes as the principal axes of the beam cross section. The matrix **U** for this element consists of twelve displacements, six deflections and six rotations, that is,

$$\mathbf{U} = \{U_1 \quad U_2 \quad \cdots \quad U_{12}\} \tag{11.29}$$

Using the engineering theory of bending and torsion and neglecting shear

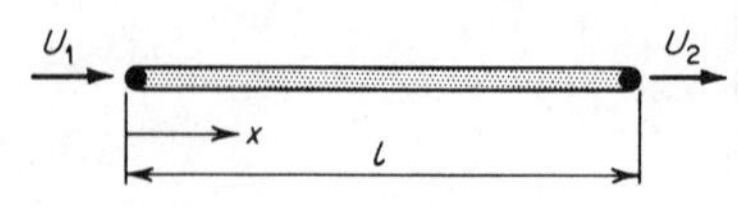

FIG. 11.1 Pin-jointed bar element.

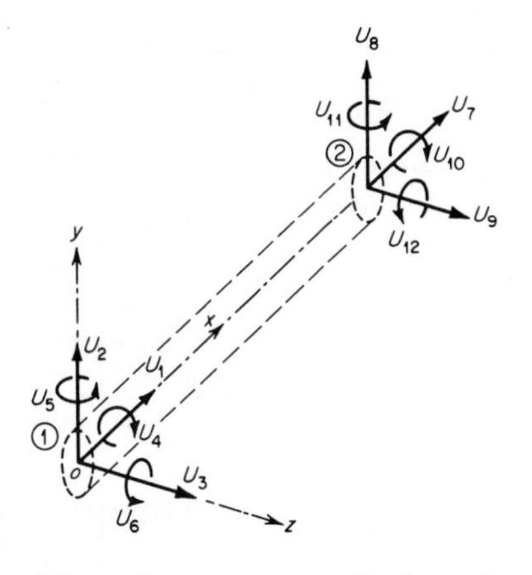

FIG. 11.2 Beam element in local coordinate system.

deformations, we can easily show that the matrix **a** in the relationship $\mathbf{u} = \mathbf{aU}$ is given by Eq. (11.30).

$$\mathbf{a}^T = \qquad (11.30)$$

	u_x	u_y	u_z
1	$1 - \xi$	0	0
2	$6(\xi - \xi^2)\eta$	$1 - 3\xi^2 + 2\xi^3$	0
3	$6(\xi - \xi^2)\zeta$	0	$1 - 3\xi^2 + 2\xi^3$
4	0	$-(1 - \xi)l\zeta$	$-(1 - \xi)l\eta$
5	$(1 - 4\xi + 3\xi^2)l\zeta$	0	$(-\xi + 2\xi^2 - \xi^3)l$
6	$(-1 + 4\xi - 3\xi^2)l\eta$	$(\xi - 2\xi^2 + \xi^3)l$	0
7	ξ	0	0
8	$6(-\xi + \xi^2)\eta$	$3\xi^2 - 2\xi^3$	0
9	$6(-\xi + \xi^2)\zeta$	0	$3\xi^2 - 2\xi^3$
10	0	$-l\xi\zeta$	$-l\xi\eta$
11	$(-2\xi + 3\xi^2)l\zeta$	0	$(\xi^2 - \xi^3)l$
12	$(2\xi - 3\xi^2)l\eta$	$(-\xi^2 + \xi^3)l$	0

The nondimensional parameters used in this equation are

$$\xi = \frac{x}{l} \qquad \eta = \frac{y}{l} \qquad \zeta = \frac{z}{l} \qquad (11.31)$$

where l is the length of the beam element. The matrix **a** in (11.30) can then be substituted into Eq. (11.1), and integration is performed over the whole volume of the element. The resulting 12×12 equivalent mass matrix is given by

$\mathbf{m} = \rho A l$ (11.32)

	1	2	3	4	5	6	7	8	9	10	11	12
1	$\frac{1}{3}$											
2	0	$\frac{13}{35} + \frac{6I_z}{5Al^2}$										
3	0	0	$\frac{13}{35} + \frac{6I_y}{5Al^2}$									
4	0	0	0	$\frac{J_x}{3A}$					Symmetric			
5	0	0	$-\frac{11l}{210} - \frac{I_y}{10Al}$	0	$\frac{l^2}{105} + \frac{2I_y}{15A}$							
6	0	$\frac{11l}{210} + \frac{I_z}{10Al}$	0	0	0	$\frac{l^2}{105} + \frac{2I_z}{15A}$						
7	$\frac{1}{6}$	0	0	0	0	0	$\frac{1}{3}$					
8	0	$\frac{9}{70} - \frac{6I_z}{5Al^2}$	0	0	0	$\frac{13l}{420} - \frac{I_z}{10Al}$	0	$\frac{13}{35} + \frac{6I_z}{5Al^2}$				
9	0	0	$\frac{9}{70} - \frac{6I_y}{5Al^2}$	0	$-\frac{13l}{420} + \frac{I_y}{10Al}$	0	0	0	$\frac{13}{35} + \frac{6I_y}{5Al^2}$			
10	0	0	0	$\frac{J_x}{6A}$	0	0	0	0	0	$\frac{J_x}{3A}$		
11	0	0	$\frac{13l}{420} - \frac{I_y}{10Al}$	0	$-\frac{l^2}{140} - \frac{I_y}{30A}$	0	0	0	$\frac{11l}{210} + \frac{I_y}{10Al}$	0	$\frac{l^2}{105} + \frac{2I_y}{15A}$	
12	0	$-\frac{13l}{420} + \frac{I_z}{10Al}$	0	0	0	$-\frac{l^2}{140} - \frac{I_z}{30A}$	0	$-\frac{11l}{210} - \frac{I_z}{10Al}$	0	0	0	$\frac{l^2}{105} + \frac{2I_z}{15A}$

where the matrix terms with the moments of inertia I_y or I_z represent rotatory inertia and the terms with the polar moment of inertia J_x represent the torsional inertia of the element.

The effect of shear deformations on a beam element can be accounted for, but this naturally leads to rather complicated expressions for the elements in the equivalent mass matrix. To simplify the presentation we shall consider transverse deflections and rotation in one plane only, as shown in Fig. 11.3. With the results of Sec. 5.6 it can be demonstrated that the matrix **a** for the beam element is given by

$$\mathbf{a} = \frac{1}{1+\Phi_s} \begin{array}{c} u_x \\ u_y \end{array} \left[\begin{array}{c|c|c|c} \overset{1}{6(\xi - \xi^2)\eta} & \overset{2}{[-1 + 4\xi - 3\xi^2 - (1-\xi)\Phi_s]l\eta} & \overset{3}{6(-\xi + \xi^2)\eta} & \overset{4}{(2\xi - 3\xi^2 - \xi\Phi_s)l\eta} \\ \hline 1 - 3\xi^2 + 2\xi^3 + (1-\xi)\Phi_s & [\xi - 2\xi^2 + \xi^3 + \frac{1}{2}(\xi - \xi^2)\Phi_s]l & 3\xi^2 - 2\xi^3 + \xi\Phi_s & [-\xi^2 + \xi^3 - \frac{1}{2}(\xi - \xi^2)\Phi_s]l \end{array} \right] \tag{11.33}$$

where
$$\Phi_s = \frac{12EI}{GA_s l^2} \tag{11.34}$$

denotes the shear-deformation parameter employed before in Chap. 5. The first and second rows in (11.33) represent displacements in the x and y directions, respectively, due to unit values of **U**. Substitution of Eq. (11.33) into (11.1) and subsequent integration yields the required mass matrix for the beam element in Fig. 11.3. This matrix is given by

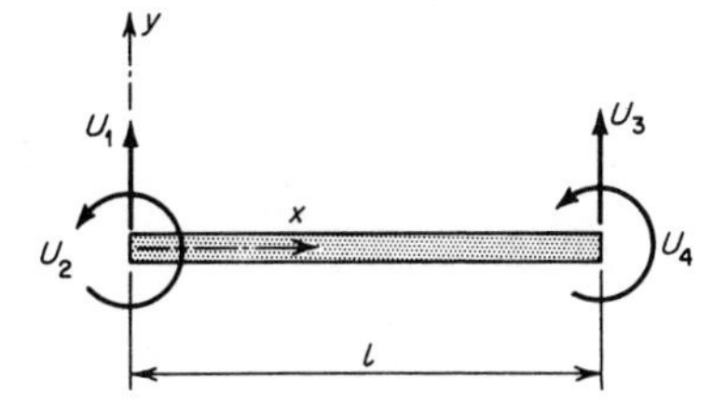

FIG. 11.3 Beam element (transverse deflections and rotations only).

$$\mathbf{m} = \frac{\rho A l}{(1+\Phi_s)^2}\begin{bmatrix} \frac{13}{35}+\frac{7}{10}\Phi_s+\frac{1}{3}\Phi_s^2 & & & \\ (\frac{11}{210}+\frac{11}{120}\Phi_s+\frac{1}{24}\Phi_s^2)l & (\frac{1}{105}+\frac{1}{60}\Phi_s+\frac{1}{120}\Phi_s^2)l^2 & & \text{Symmetric} \\ \frac{9}{70}+\frac{3}{10}\Phi_s+\frac{1}{6}\Phi_s^2 & (\frac{13}{420}+\frac{3}{40}\Phi_s+\frac{1}{24}\Phi_s^2)l & \frac{13}{35}+\frac{7}{10}\Phi_s+\frac{1}{3}\Phi_s^2 & \\ -(\frac{13}{420}+\frac{3}{40}\Phi_s+\frac{1}{24}\Phi_s^2)l & -(\frac{1}{140}+\frac{1}{60}\Phi_s+\frac{1}{120}\Phi_s^2)l^2 & -(\frac{11}{210}+\frac{11}{120}\Phi_s+\frac{1}{24}\Phi_s^2)l & (\frac{1}{105}+\frac{1}{60}\Phi_s+\frac{1}{120}\Phi_s^2)l^2 \end{bmatrix}$$

$$+\frac{\rho A l}{(1+\Phi_s)^2}\left(\frac{r}{l}\right)^2\begin{bmatrix} \frac{6}{5} & & & \\ (\frac{1}{10}-\frac{1}{2}\Phi_s)l & (\frac{2}{15}+\frac{1}{6}\Phi_s+\frac{1}{3}\Phi_s^2)l^2 & & \text{Symmetric} \\ -\frac{6}{5} & (-\frac{1}{10}+\frac{1}{2}\Phi_s)l & \frac{6}{5} & \\ (\frac{1}{10}-\frac{1}{2}\Phi_s)l & (-\frac{1}{30}-\frac{1}{6}\Phi_s+\frac{1}{6}\Phi_s^2)l^2 & (-\frac{1}{10}+\frac{1}{2}\Phi_s)l & (\frac{2}{15}+\frac{1}{6}\Phi_s+\frac{1}{3}\Phi_s^2)l^2 \end{bmatrix} \tag{11.35}$$

where an additional symbol r is introduced to represent the radius of gyration of the beam cross section. The first term in (11.35) represents the translational mass inertia, while the second term represents the rotatory inertia of the beam. Equation (11.35) was first derived independently by Archer[6,7] and by McCalley.

If the rotatory-inertia and shear deformation effects are neglected, Eq. (11.35) reduces to

$$\mathbf{m} = \frac{\rho A l}{420}\begin{bmatrix} 156 & 22l & 54 & -13l \\ 22l & 4l^2 & 13l & -3l^2 \\ 54 & 13l & 156 & -22l \\ -13l & -3l^2 & -22l & 4l^2 \end{bmatrix} \tag{11.36}$$

11.6 TRIANGULAR PLATE WITH TRANSLATIONAL DISPLACEMENTS

The in-plane displacement relationship for a triangular plate used in Chap. 5 for u_x (or u_y) measured in local coordinate system is given by

$$u_x = \frac{1}{2A_{123}}\left[\begin{array}{c|c|c} y_{32}(x-x_2) & -y_{31}(x-x_3) & y_{21}(x-x_1) \\ \quad -x_{32}(y-y_2) & \quad +x_{31}(y-y_3) & \quad -x_{21}(y-y_1) \end{array}\right]\mathbf{U} \tag{11.37}$$

where $\mathbf{U} = \{U_1 \quad U_2 \quad U_3\}$ (11.38)

represents the node displacements in either the x or in y directions. Figure 11.4 shows the displacements $\mathbf{U}$ all measured in the x direction. Thus the required matrix $\mathbf{a}$ can be obtained from (11.37), but if we had used it directly in determining the equivalent mass matrix, the resulting integration over the triangular area would have been extremely unwieldy. It is therefore preferable to introduce the triangular coordinates ξ and η shown in Fig. 11.5. These nonorthogonal coordinates are related to the rectangular coordinates x and y

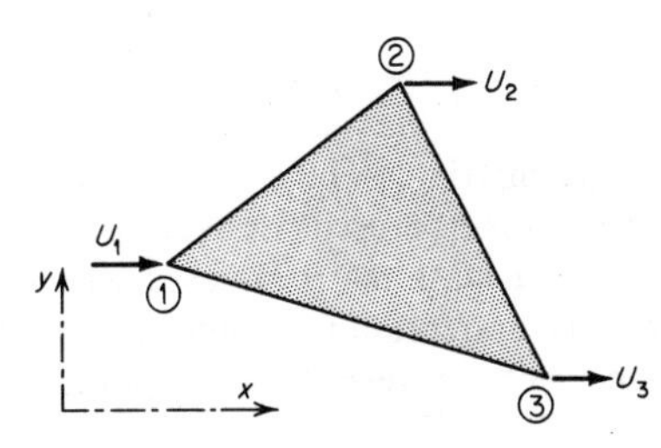

FIG. 11.4 Triangular plate element.

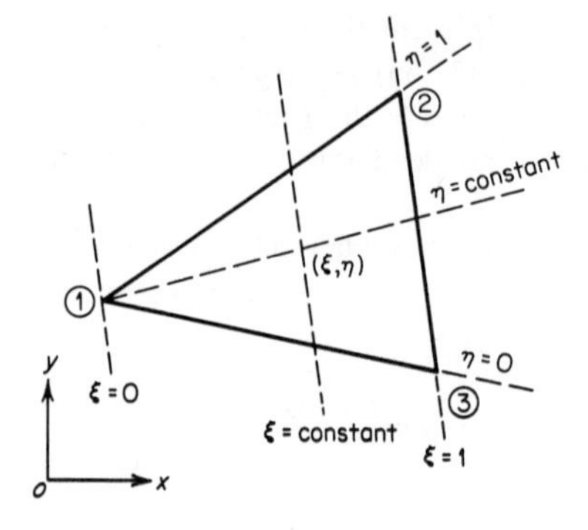

FIG. 11.5 Triangular coordinate system.

by means of the relationships

$$x = x_1 + \xi(x_{31} - \eta x_{32}) \qquad y = y_1 + \xi(y_{31} - \eta y_{32}) \tag{11.39}$$

After introducing (11.39) into (11.37) it follows immediately that

$$\mathbf{a} = [(1 - \xi) \quad \xi\eta \quad \xi(1 - \eta)] \tag{11.40}$$

Noting now that the jacobian $J(x,y)$ for the coordinate transformation is given by

$$\begin{aligned} J(x,y) &= \frac{\partial x}{\partial \xi}\frac{\partial y}{\partial \eta} - \frac{\partial x}{\partial \eta}\frac{\partial y}{\partial \xi} \\ &= -(x_{31} - \eta x_{32})y_{32}\xi + x_{32}(y_{31} - \eta y_{32})\xi \\ &= 2A_{123}\xi \end{aligned} \tag{11.41}$$

and then substituting (11.40) into (11.1), we can easily demonstrate that

$$\begin{aligned} \mathbf{m} &= \rho t \int_0^1 \int_0^1 \mathbf{a}^T \mathbf{a} \, |J(x,y)| \, d\xi \, d\eta \\ &= \frac{\rho A_{123} t}{12} \begin{bmatrix} 2 & 1 & 1 \\ 1 & 2 & 1 \\ 1 & 1 & 2 \end{bmatrix} \end{aligned} \tag{11.42}$$

where t is the thickness of the plate. The above matrix can be used either for u_x or for u_y displacements. Furthermore, it can also be used for the normal displacements, that is, u_z, but in general if such a mass matrix is used when there is a pronounced bending deformation, accurate results should not be expected since **a** is based here on linear displacements between the nodal points.

If displacements in all directions are considered, the equivalent mass matrix becomes

$$\mathbf{m}_{\text{triangle}} = \begin{bmatrix} \mathbf{m} & \mathbf{0} & \mathbf{0} \\ \mathbf{0} & \mathbf{m} & \mathbf{0} \\ \mathbf{0} & \mathbf{0} & \mathbf{m} \end{bmatrix} \tag{11.43}$$

where contrary to our previous convention the displacements are grouped in sets of three displacements in the x, y, and z directions, respectively. Matrix (11.43) can be used for triangular plates undergoing essentially translational displacements, as in the case of a triangular element in a built-up structure.

11.7 RECTANGULAR PLATE WITH TRANSLATIONAL DISPLACEMENTS

For a rectangular plate (see Fig. 11.6) the matrix $\mathbf{a}$ may be taken as

$$\mathbf{a} = [\overset{1}{(1-\xi)(1-\eta)} \quad \overset{2}{(1-\xi)\eta} \quad \overset{3}{\xi\eta} \quad \overset{4}{\xi(1-\eta)}] \tag{11.44}$$

where ξ and η represent now nondimensional rectangular coordinates. Here again we shall calculate the mass matrix for one direction only. Thus for the displacements shown in Fig. 11.6, that is, x direction,

$$\mathbf{m} = \rho V \int_0^1 \int_0^1 \mathbf{a}^T \mathbf{a} \, d\xi \, d\eta$$

$$= \frac{\rho V}{36} \begin{bmatrix} 4 & 2 & 1 & 2 \\ 2 & 4 & 2 & 1 \\ 1 & 2 & 4 & 2 \\ 2 & 1 & 2 & 4 \end{bmatrix} \tag{11.45}$$

Identical mass matrices are obtained for the y and z directions.

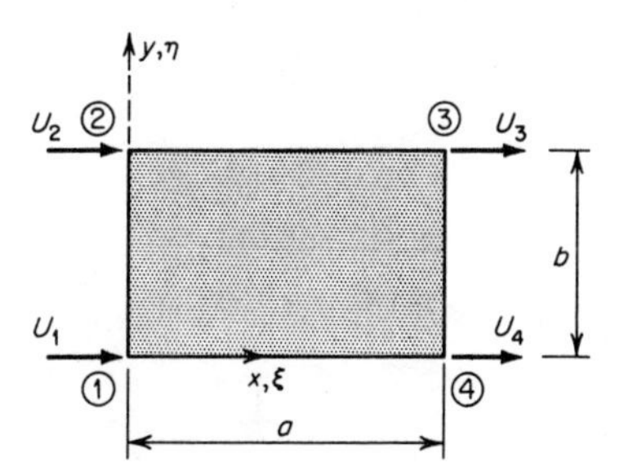

FIG. 11.6 Rectangular plate element.

11.8 SOLID TETRAHEDRON

To find the equivalent mass matrix we shall use the tetrahedral coordinates ξ, η, ζ defined in Fig. 11.7. The tetrahedral coordinates are related to the rectangular coordinates by means of the relationships

$$
\begin{aligned}
x &= x_4 - \zeta x_{41} + \xi\zeta x_{31} - \xi\eta\zeta x_{32} \\
y &= y_4 - \zeta y_{41} + \xi\zeta y_{31} - \xi\eta\zeta y_{32} \\
z &= (1 - \zeta) z_4
\end{aligned}
\qquad (11.46)
$$

where for convenience the origin of the local coordinate system is placed at node 1 and the tetrahedron side 1,2,3 lies in the xy plane. Using Eqs. (11.46) it can be demonstrated that for a linear distribution of displacements within the tetrahedron the matrix $\mathbf{a}$ in tetrahedral coordinates is given by

$$
\mathbf{a} = \begin{matrix} & 1 & 2 & 3 & 4 \\ [& (1-\xi)\zeta & \xi\eta\xi & \xi(1-\eta)\zeta & (1-\zeta) \;] \end{matrix} \qquad (11.47)
$$

where the column numbers refer to the displacements shown in Fig. 11.8.

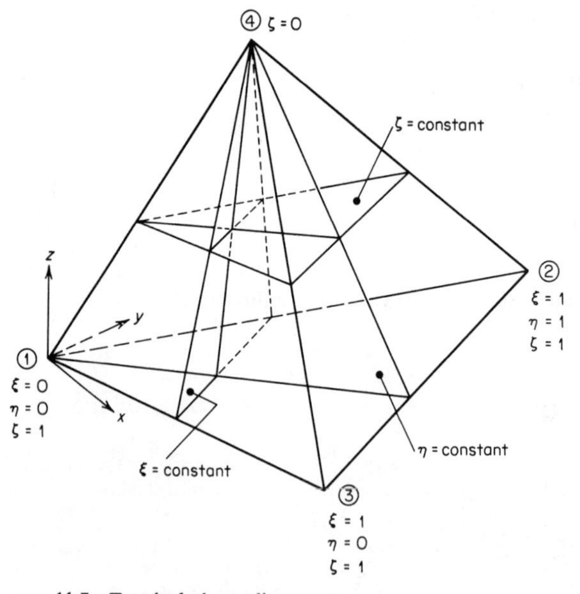

FIG. 11.7 Tetrahedral coordinate system.

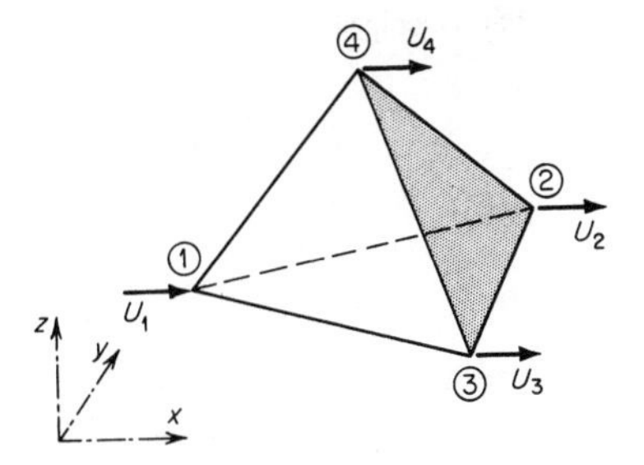

FIG. 11.8 Solid tetrahedron.

The jacobian $J(x,y,z)$ for the coordinate transformation (11.46) is given by

$$J(x,y,z) = \begin{vmatrix} \dfrac{\partial x}{\partial \xi} & \dfrac{\partial x}{\partial \eta} & \dfrac{\partial x}{\partial \zeta} \\[2ex] \dfrac{\partial y}{\partial \xi} & \dfrac{\partial y}{\partial \eta} & \dfrac{\partial y}{\partial \zeta} \\[2ex] \dfrac{\partial z}{\partial \xi} & \dfrac{\partial z}{\partial \eta} & \dfrac{\partial z}{\partial \zeta} \end{vmatrix} = -6V\xi\zeta^2 \tag{11.48}$$

where V is the volume of the element. Using, therefore, Eqs. (11.47) and (11.48) in (11.1), we have

$$\begin{aligned} \mathbf{m} &= \rho \int_0^1 \int_0^1 \int_0^1 \mathbf{a}^T \mathbf{a}\, |J(x,y,z)|\, d\xi\, d\eta\, d\zeta \\ &= \frac{\rho V}{20} \begin{bmatrix} 2 & 1 & 1 & 1 \\ 1 & 2 & 1 & 1 \\ 1 & 1 & 2 & 1 \\ 1 & 1 & 1 & 2 \end{bmatrix} \end{aligned} \tag{11.49}$$

11.9 SOLID PARALLELEPIPED

For a solid parallelepiped with linearly varying edge displacements the matrix **a** is given by

$$\begin{aligned} \mathbf{a} = [&\overset{1}{(1-\xi)(1-\eta)(1-\zeta)} \quad \overset{2}{(1-\xi)\eta(1-\zeta)} \quad \overset{3}{\xi\eta(1-\zeta)} \quad \overset{4}{\xi(1-\eta)(1-\zeta)} \\ &\overset{5}{(1-\xi)(1-\eta)\zeta} \quad \overset{6}{(1-\xi)\eta\zeta} \quad \overset{7}{\xi\eta\zeta} \quad \overset{8}{\xi(1-\eta)\zeta}] \end{aligned} \tag{11.50}$$

where $$\xi = \frac{x}{a} \qquad \eta = \frac{y}{b} \qquad \zeta = \frac{z}{c} \tag{11.51}$$

represent nondimensional coordinates measured from node 1, as shown in Fig. 11.9, and a, b, and c are the dimensions of the parallelepiped. The column numbers 1 to 8 refer here to the eight displacements for some typical direction,

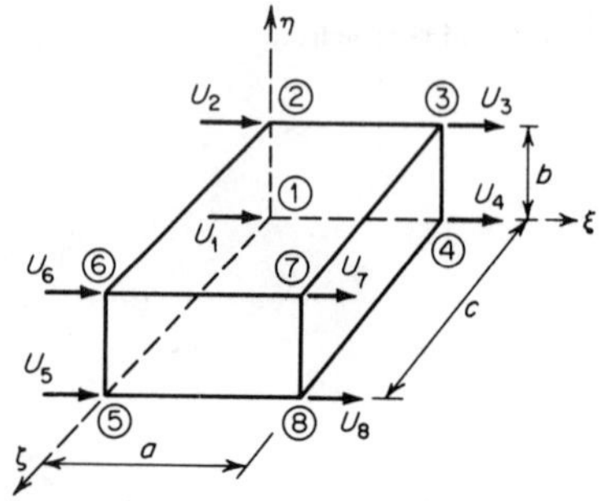

FIG. 11.9 Parallelepiped.

that is, x direction as indicated in Fig. 11.9. Substituting Eq. (11.50) into (11.1), we obtain the mass matrix in the form

$$\mathbf{m} = \frac{\rho V}{216}\begin{bmatrix} 8 & & & & & & & \\ 4 & 8 & & \text{Symmetric} & & & & \\ 2 & 4 & 8 & & & & & \\ 4 & 2 & 4 & 8 & & & & \\ 4 & 2 & 1 & 2 & 8 & & & \\ 2 & 4 & 2 & 1 & 4 & 8 & & \\ 1 & 2 & 4 & 2 & 2 & 4 & 8 & \\ 2 & 1 & 2 & 4 & 4 & 2 & 4 & 8 \end{bmatrix} \tag{11.52}$$

where $V = abc$ denotes the element volume.

11.10 TRIANGULAR PLATE WITH BENDING DISPLACEMENTS

Many different displacement distributions for triangular plate elements subjected to bending have been proposed in the technical literature, all of which could be employed to derive the equivalent mass matrix for this element. In order to illustrate the general method we shall use the displacement u_z normal to the middle plane of the plate given by Eq. (5.234), which may also be expressed in matrix form as

$$u_z = [1 \quad x \quad y \quad x^2 \quad xy \quad y^2 \quad x^3 \quad xy^2 + x^2y \quad y^3]\begin{bmatrix} c_1 \\ c_2 \\ \cdot \\ \cdot \\ \cdot \\ c_9 \end{bmatrix}$$

$$= \mathbf{dc} \tag{11.53}$$

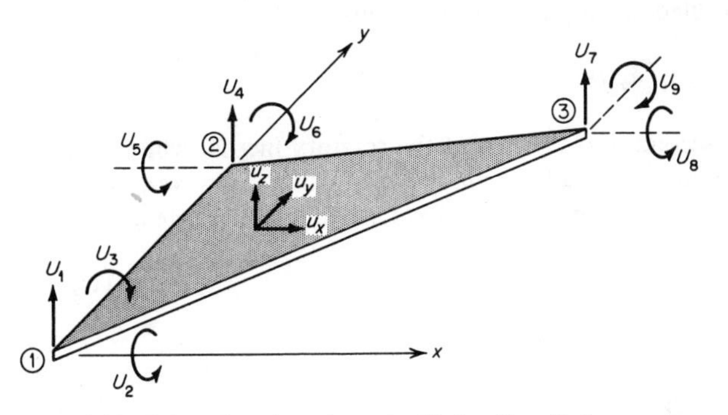

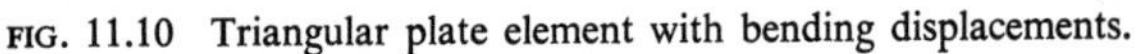
FIG. 11.10 Triangular plate element with bending displacements.

where $\mathbf{c} = \{c_1 \quad c_2 \quad \cdots \quad c_9\}$ (11.54)

and the matrix **d** has coefficients represented by functions of x and y. The directions of the coordinate axes are shown in Fig. 11.10. Although the subsequent analysis is carried out for the specific displacement distribution given by Eq. (11.53), the method is sufficiently general to allow the use of other functions in the matrix **d**.

Using the notation of Sec. 5.11, we can determine the unknown constants $c_1, \ldots, c_9$ from

$$\mathbf{U} = \mathbf{C}\mathbf{c} \tag{11.55}$$

Hence $\mathbf{c} = \mathbf{C}^{-1}\mathbf{U}$ (11.56)

and from (11.53)

$$u_z = \mathbf{d}\mathbf{C}^{-1}\mathbf{U} = \mathbf{a}_z\mathbf{U} \tag{11.57}$$

where $\mathbf{a}_z = \mathbf{d}\mathbf{C}^{-1}$ (11.58)

The displacements u_x and u_y (see Fig. 11.10), caused by the rotations of normals to the middle plane, are calculated from

$$u_x = -\frac{\partial u_z}{\partial x} z = -\frac{\partial \mathbf{d}}{\partial x} \mathbf{C}^{-1} z\mathbf{U} = \mathbf{a}_x\mathbf{U} \tag{11.59}$$

$$u_y = -\frac{\partial u_z}{\partial y} z = -\frac{\partial \mathbf{d}}{\partial y} \mathbf{C}^{-1} z\mathbf{U} = \mathbf{a}_y\mathbf{U} \tag{11.60}$$

Next, Eqs. (11.57), (11.59), and (11.60) are collected into a single matrix equation

$$\mathbf{u} = \begin{bmatrix} u_x \\ u_y \\ u_z \end{bmatrix} = \begin{bmatrix} \mathbf{a}_x \\ \mathbf{a}_y \\ \mathbf{a}_z \end{bmatrix} \mathbf{U} = \mathbf{a}\mathbf{U} \tag{11.61}$$

which can be substituted directly into (11.1). This leads to

$$\mathbf{m} = \int_v \rho \mathbf{a}_x{}^T \mathbf{a}_x \, dV + \int_v \rho \mathbf{a}_y{}^T \mathbf{a}_y \, dV + \int_v \rho \mathbf{a}_z{}^T \mathbf{a}_z \, dV \tag{11.62}$$

where the first and second terms represent the rotatory inertia, while the third term represents the translational inertia of the plate. However, for most practical calculation, the rotatory-inertia effects are usually neglected, so that

$$\mathbf{m} = \int_v \rho \mathbf{a}_z{}^T \mathbf{a}_z \, dV \tag{11.62a}$$

Substituting Eq. (11.58) into (11.62*a*), we have

$$\begin{aligned} \mathbf{m} &= \int_v \rho (\mathbf{C}^{-1})^T \mathbf{d}^T \mathbf{d} \mathbf{C}^{-1} \, dV \\ &= \rho t (\mathbf{C}^{-1})^T \iint \mathbf{d}^T \mathbf{d} \, dx \, dy \, \mathbf{C}^{-1} \end{aligned} \tag{11.63}$$

As in Sec. 5.11, the matrix $\mathbf{C}$ and its inverse* are evaluated numerically, while the integral of the matrix product $\mathbf{d}^T\mathbf{d}$, using $\mathbf{d}$ from Eq. (11.53), is determined from

$$\iint \mathbf{d}^T \mathbf{d} \, dx \, dy$$

$$= \iint \begin{bmatrix} 1 & & & & & & & & \\ x & x^2 & & & & & \text{Symmetric} & & \\ y & xy & y^2 & & & & & & \\ x^2 & x^3 & x^2y & x^4 & & & & & \\ xy & x^2y & xy^2 & x^3y & x^2y^2 & & & & \\ y^2 & xy^2 & y^3 & x^2y^2 & xy^3 & y^4 & & & \\ x^3 & x^4 & x^3y & x^5 & x^4y & x^3y^2 & x^6 & & \\ xy^2 + x^2y & x^2y^2 + x^3y & xy^3 + x^2y^2 & x^3y^2 + x^4y & x^2y^3 + x^3y^2 & xy^4 + x^2y^3 & x^4y^2 + x^5y & (xy^2 + x^2y)^2 & \\ y^3 & xy^3 & y^4 & x^2y^3 & xy^4 & y^5 & x^3y^3 & xy^5 + x^2y^4 & y^6 \end{bmatrix} dx \, dy \tag{11.64}$$

Evaluation of coefficients in $\iint \mathbf{d}^T\mathbf{d} \, dx \, dy$ involves integration of

$$I(x^m, y^n) = \iint x^m y^n \, dx \, dy \tag{11.65}$$

over the surface of the triangle. The integration in cartesian coordinates is unwieldy, and therefore it is preferable to introduce the triangular coordinate system ξ and η and also to select the oy axis to coincide with the edge 1,2 of the triangle. In addition, the origin for the cartesian coordinate system will be taken at vertex 1, as shown in Fig. 11.11. For the selected coordinate

*See the footnote on page 112.

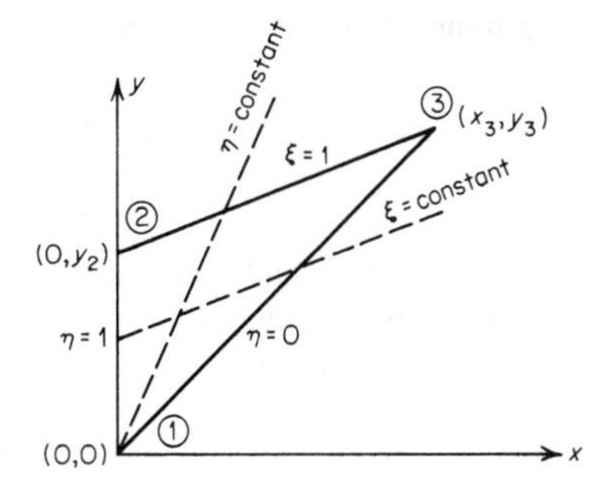

FIG. 11.11 Triangular coordinate system used to evaluate $I(x^m,y^n)$ integrals.

system we then have (see Sec. 11.6)

$$x = \xi(1 - \eta)x_3 \tag{11.66}$$

$$y = \xi[(1 - \eta)y_3 + \eta y_2] \tag{11.67}$$

while the jacobian $J(x,y)$ of the transformation is given by

$$J(x,y) = \xi x_3 y_2 \tag{11.68}$$

With the triangular coordinates the integration of (11.65) is relatively simple, since now

$$\begin{aligned} I(x^m,y^n) &= \int_0^1\int_0^1 x^m y^n\,|J(x,y)|\,d\xi\,d\eta \\ &= \int_0^1\int_0^1 \xi^{m+n+1}(1-\eta)^m[(1-\eta)y_3 + \eta y_2]^n x_3{}^{m+1} y_2\,d\xi\,d\eta \end{aligned} \tag{11.69}$$

which involves only polynomials in ξ and η, and the range of integration is from 0 to 1.

11.11 RECTANGULAR PLATE WITH BENDING DISPLACEMENTS[276]

Two different displacement distributions u_z were used in Sec. 5.12 to determine stiffness properties of rectangular plates in bending. The first distribution is such that the boundary deflections on adjacent plate elements are compatible; however, rotations of the element edges on a common boundary are not compatible, and consequently discontinuities in slopes exist across the boundaries. In the second distribution, both the deflection and slope compatibility on adjacent elements are ensured. The same two distributions will be used to derive the equivalent mass matrices for rectangular elements.

As in the case of the triangular plate element, only the translational mass matrix will be determined. The displacements at the node points on the rectangular plate are shown in Fig. 11.12. For noncompatible displacements the matrix $\mathbf{a}_z$ from

$$\mathbf{u}_z = \mathbf{a}_z \mathbf{U} \tag{11.70}$$

where $$\mathbf{U} = \{U_1 \quad U_2 \quad \cdots \quad U_{12}\} \tag{11.71}$$

is given by Eq. (5.263). Substituting (5.263) into (11.1) and performing integration over the whole volume of the rectangle, we obtain the mass matrix based on noncompatible displacements. This matrix is given by Eq. (11.72).

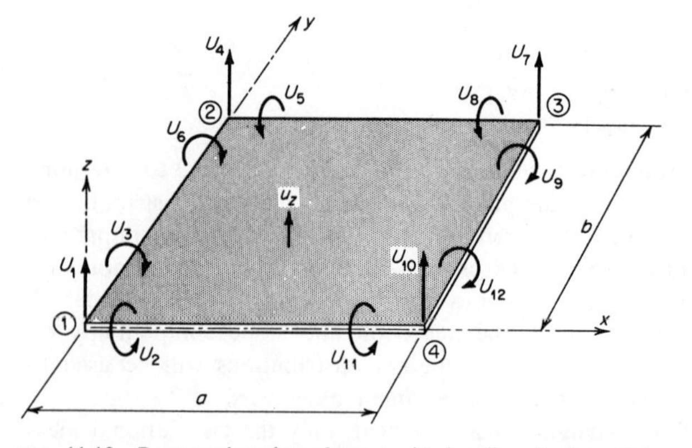

FIG. 11.12 Rectangular plate element with bending displacements.

$$
\mathbf{m} = \frac{\rho abt}{176{,}400}
\begin{bmatrix}
24{,}178 & & & & & & & & & & & \\
3{,}227b & 560b^2 & & & & & & & & & & \\
-3{,}227a & -441ab & 560a^2 & & & & & & & & & \\
8{,}582 & 1{,}918b & -1{,}393a & 24{,}178 & & & & \text{Symmetric} & & & & \\
-1{,}918b & -420b^2 & 294ab & -3{,}227b & 560b^2 & & & & & & & \\
-1{,}393a & -294ab & 280a^2 & -3{,}227a & 441ab & 560a^2 & & & & & & \\
2{,}758 & 812b & -812a & 8{,}582 & -1{,}393b & -1{,}918a & 24{,}178 & & & & & \\
-812b & -210b^2 & 196ab & -1{,}393b & 280b^2 & 294ab & -3{,}227b & 560b^2 & & & & \\
812a & 196ab & -210a^2 & 1{,}918a & -294ab & -420a^2 & 3{,}227a & -441ab & 560a^2 & & & \\
8{,}582 & 1{,}393b & -1{,}918a & 2{,}758 & -812b & -812a & 8{,}582 & -1{,}918b & 1{,}393a & 24{,}178 & & \\
1{,}393b & 280b^2 & -294ab & 812b & -210b^2 & -196ab & 1{,}918b & -420b^2 & 294ab & 3{,}227b & 560b^2 & \\
1{,}918a & 294ab & -420a^2 & 812a & -196ab & -210a^2 & 1{,}393a & -294ab & 280a^2 & 3{,}227a & 441ab & 560a^2
\end{bmatrix}
\tag{11.72}
$$

(Rows and columns numbered 1–12.)

$$
\mathbf{m} = \frac{\rho abt}{176{,}400}
\begin{bmatrix}
24{,}336 & & & & & & & & & & & \\
3{,}432b & 624b^2 & & & & & & & & & & \\
-3{,}432a & -484ab & 624a^2 & & & & & & & & & \\
8{,}424 & 2{,}028b & -1{,}188a & 24{,}336 & & & & & & & & \\
-2{,}028b & -468b^2 & 286ab & -3{,}432b & 624b^2 & & & \text{Symmetric} & & & & \\
-1{,}188a & -286ab & 216a^2 & -3{,}432a & 484ab & 624a^2 & & & & & & \\
2{,}916 & 702b & -702a & 8{,}424 & -1{,}188b & -2{,}028a & 24{,}336 & & & & & \\
-702b & -162b^2 & 169ab & -1{,}188b & 216b^2 & 286ab & -3{,}432b & 624b^2 & & & & \\
702a & 169ab & -162a^2 & 2{,}028a & -286ab & -468a^2 & 3{,}432a & -484ab & 624a^2 & & & \\
8{,}424 & 1{,}188b & -2{,}028a & 2{,}916 & -702b & -702a & 8{,}424 & -2{,}028b & 1{,}188a & 24{,}336 & & \\
1{,}188b & 216b^2 & -286ab & 702b & -162b^2 & -169ab & 2{,}028b & -468b^2 & 286ab & 3{,}432b & 624b^2 & \\
2{,}028a & 286ab & -468a^2 & 702a & -169ab & -162a^2 & 1{,}188a & -286ab & 216a^2 & 3{,}432a & 484ab & 624a^2
\end{bmatrix}
\tag{11.73}
$$

(rows and columns numbered 1–12)

Similarly, for compatible displacements the matrix $\mathbf{a}_z$ is given by Eq. (5.269), which leads to the mass matrix in Eq. (11.73). For comparison purposes both (11.72) and (11.73) are presented with a common denominator.

11.12 LUMPED-MASS REPRESENTATION

The simplest form of mathematical model for inertia properties of structural elements is the lumped-mass representation. In this idealization concentrated masses are placed at the node points in the directions of the assumed element degrees of freedom. These masses refer to translational and rotational inertia of the element. They are calculated by assuming that the material within the mean locations on either side of the specified displacement behaves like a rigid body while the remainder of the element does not participate in the motion. This assumption therefore excludes dynamic coupling between the element displacements, and the resulting element mass matrix is purely diagonal. Hence for an element with n degrees of freedom we have that

$$\mathbf{m} = \lceil m_1 \quad m_2 \quad \cdots \quad m_i \quad \cdots \quad m_n \rfloor \tag{11.74}$$

where m_i represents the lumped mass for the direction of the displacement U_i. For example, the lumped-mass representation for a bar element leads to

$$\mathbf{m} = \frac{\rho A l}{2}\begin{bmatrix} 1 & 0 \\ 0 & 1 \end{bmatrix} \tag{11.75}$$

It has been demonstrated that for a given number of degrees of freedom the lumped-mass representation is less accurate than the equivalent mass matrices derived from Eq. (11.1); however, in many practical applications we may still prefer to use the lumped-mass matrices because of the significant computational advantages derived from the fact that such matrices are diagonal.

PROBLEMS

11.1 Calculate the condensed mass matrix for displacements 3 and 5 in Prob. 6.4 (see Fig. 6.13).

11.2 Determine the equivalent mass matrix for a bar element with cross-sectional area varying linearly from A_1 to A_2. The bar length is l, and its density is ρ.

11.3 Determine the equivalent mass matrix for the curved-beam element shown in Fig. 6.14. The rotatory-inertia, shear-deformation, and axial-deformation effects are to be neglected.

11.4 Explain how the effects of rotatory inertia, shear deformations, and axial deformations can be accounted for in deriving the equivalent mass matrix in Prob. 11.3.

CHAPTER 12 VIBRATIONS OF ELASTIC SYSTEMS

In this chapter the general theory is presented for the analysis of small harmonic oscillations of elastic systems having a finite number of degrees of freedom. The harmonic oscillations may be induced in an elastic system by imposing properly selected initial displacements and then releasing these constraints, thereby causing the system to go into an oscillatory motion. This oscillatory motion is a characteristic property of the system, and it depends on the mass and stiffness distribution. In the absence of any damping forces, e.g., viscous forces proportional to velocities, the oscillatory motion will continue indefinitely, with the amplitudes of oscillations depending on the initially imposed displacements; however, if damping is present, the amplitudes will decay progressively, and if the amount of damping exceeds a certain critical value, the oscillatory character of motion will cease altogether. The oscillatory motion occurs at certain frequencies, and it follows well-defined deformation patterns, described as the *characteristic modes*. The study of such free vibrations is an important prerequisite for all dynamic-response calculations for elastic systems. In this chapter the vibration analysis is developed for both the stiffness and flexibility formulations. The orthogonality property of the vibration modes is also discussed. Several numerical examples are included to illustrate the general theory.

12.1 VIBRATION ANALYSIS BASED ON STIFFNESS

The equations of motion for an elastic system with a finite number of degrees of freedom were derived in Chap. 10 through the application of the virtual-work principle for dynamic loading. These equations are expressed in matrix notation as

$$\mathbf{M}\ddot{\mathbf{U}} + \mathbf{K}\mathbf{U} = \mathbf{P} \tag{12.1}$$

where $\mathbf{P}$ is a column matrix of equivalent forces calculated in accordance with Eq. (10.36). For aeronautical applications

$$\mathbf{P} = \mathbf{P}_d + \mathbf{P}_a \tag{12.2}$$

where $\mathbf{P}_d$ represents the disturbing forces and $\mathbf{P}_a$ represents the aerodynamic forces. The disturbing forces $\mathbf{P}_d$ may be due to a variety of causes; on aircraft structures they may represent forces on control surfaces, gust loads, landing loads, etc., while on civil engineering structures they may represent loads due to wind gusts, earthquakes, or forced vibrations. In general, these disturbing forces are represented by some specified functions of time. The aerodynamic loads $\mathbf{P}_a$, which are of great importance in aeroelastic calculations for aerospace structures, are normally expressed by the matrix equation

$$\mathbf{P}_a = \mathbf{A}_{\ddot{u}}\ddot{\mathbf{U}} + \mathbf{B}_{\dot{u}}\dot{\mathbf{U}} + \mathbf{C}_u\mathbf{U} \tag{12.3}$$

where $\mathbf{A}_{\ddot{u}}$, $\mathbf{B}_{\dot{u}}$, and $\mathbf{C}_u$ are square matrices of aerodynamic derivatives and are functions of the dynamic pressure, the Mach number, and the frequency.

In real structures we have always some energy dissipation present. For this reason damping forces must also be introduced into our discrete-element system. When the damping forces are proportional to the velocity, the damping is described as viscous. For such cases the equation of motion is written as

$$\mathbf{M}\ddot{\mathbf{U}} + \mathbf{C}\dot{\mathbf{U}} + \mathbf{K}\mathbf{U} = \mathbf{P} \tag{12.4}$$

where $\mathbf{C}$ is the damping matrix. In addition to viscous damping other forms of damping mechanism are possible, e.g., structural damping.

UNCONSTRAINED STRUCTURE

We shall consider next a completely unconstrained (free) structure undergoing free oscillations. For this case $\mathbf{P}_d = \mathbf{0}$, and $\mathbf{P}_a = \mathbf{0}$, and if we assume that the system is undamped, that is, $\mathbf{C} = \mathbf{0}$, the equation of motion becomes

$$\mathbf{M}\ddot{\mathbf{U}} + \mathbf{K}\mathbf{U} = \mathbf{0} \tag{12.5}$$

Since the free oscillations are harmonic, the displacements $\mathbf{U}$ can be written as

$$\mathbf{U} = \mathbf{q}e^{i\omega t} \tag{12.6}$$

where $\mathbf{q}$ is a column matrix of the amplitudes of the displacements $\mathbf{U}$, ω is the circular frequency of oscillations, and t is the time. Using Eq. (12.6) in (12.5) and then canceling the common factor $e^{i\omega t}$, we obtain

$$(-\omega^2\mathbf{M} + \mathbf{K})\mathbf{q} = \mathbf{0} \tag{12.7}$$

which may be regarded as the equation of motion for an undamped freely oscillating system.

Equation (12.7) has a nonzero solution for $\mathbf{q}$ provided

$$|-\omega^2\mathbf{M} + \mathbf{K}| = 0 \tag{12.8}$$

This last equation is the so-called characteristic equation from which the natural frequencies of free oscillations can be calculated. The determinant in Eq. (12.8), when expanded, yields a polynomial of nth degree in ω^2, the roots of which give the natural frequencies [eigenvalues of Eq. (12.7)]. Only for these natural frequencies will there be a nonzero solution obtained for $\mathbf{q}$ in (12.7). The number of frequencies thus obtained is equal to the number of nonzero mass coefficients on the principal diagonal of the equivalent mass matrix $\mathbf{M}$. This number includes also zero frequencies for the rigid-body degrees of freedom. For the general three-dimensional case there will be six such zero frequencies. The explanation why $\omega = 0$ is a solution of Eq. (12.5) is simple. From Eq. (12.6) for $\omega = 0$, $\mathbf{U} = \mathbf{q}$ and $\ddot{\mathbf{U}} = \mathbf{0}$. Therefore

$$\mathbf{K}\mathbf{q}_{\text{rigid body}} = \mathbf{0} \tag{12.9}$$

which is obviously satisfied by virtue of the fact that rigid-body displacements alone do not produce any elastic restoring forces in the structure.

For a given value of ω^2 determined from the characteristic equation (12.8), we can utilize Eq. (12.7) to find the amplitudes of $\mathbf{q}$. Because Eq. (12.7) represents a homogeneous set of linear equations, only the relative values or ratios of $\mathbf{q}$ can be obtained.

Returning now to Eq. (12.8), it is apparent that the roots ω^2 (eigenvalues) can be found directly only if the size of the determinant is not too large. For most problems, various numerical methods employing Eq. (12.7) are used, and they have the added advantage of producing also the "amplitudes" $\mathbf{q}$ (eigenmodes) for a given frequency. For details of these methods standard texts on matrix algebra should be consulted. For some methods the form of Eq. (12.7) may not be suitable, and we have first to premultiply the equation by $\mathbf{M}^{-1}$, so that, after changing signs,

$$(\omega^2\mathbf{I} - \mathbf{M}^{-1}\mathbf{K})\mathbf{q} = \mathbf{0} \tag{12.10}$$

which is now of the standard form for eigenvalue calculations. It should be noted that a similar transformation, where Eq. (12.7) is postmultiplied by $\mathbf{K}^{-1}/\omega^2$, is not possible because the stiffness matrix $\mathbf{K}$ for an unconstrained structure is singular.

In some practical applications some of the masses in $\mathbf{M}$ may be zero, and therefore the matrix $\mathbf{M}$ is singular. This occurs if some or all structural mass matrices are neglected and only actual concentrated masses associated with some degrees of freedom are considered. For such cases the matrices $\mathbf{M}$, $\mathbf{K}$, and $\mathbf{q}$ can be partitioned as follows:

$$\mathbf{M} = \begin{bmatrix} \mathbf{M}_c & \mathbf{0} \\ \mathbf{0} & \mathbf{0} \end{bmatrix} \tag{12.11}$$

$$\mathbf{K} = \begin{bmatrix} \mathbf{K}_{xx} & \mathbf{K}_{xy} \\ \mathbf{K}_{yx} & \mathbf{K}_{yy} \end{bmatrix} \tag{12.12}$$

$$\mathbf{q} = \begin{bmatrix} \mathbf{q}_x \\ \mathbf{q}_y \end{bmatrix} \tag{12.13}$$

where subscripts x refer to the direction of displacements in which inertia forces are present and y refers to the directions in which there are no inertia forces. Using Eqs. (12.11) to (12.13) in (12.7), we can show that

$$(\omega^2\mathbf{I} - \mathbf{M}_c^{-1}\mathbf{K}_c)\mathbf{q}_x = \mathbf{0} \tag{12.14}$$

where $\quad \mathbf{K}_c = \mathbf{K}_{xx} - \mathbf{K}_{xy}\mathbf{K}_{yy}^{-1}\mathbf{K}_{yx} \quad (12.15)$

represents the condensed stiffness matrix for the directions x.

The eigenmodes calculated from Eq. (12.10) or (12.14) can be checked by the equations of equilibrium. Since for free oscillations the inertia forces $\mathbf{P}_i$ are the only forces acting on the structure, they must be in equilibrium with themselves. Hence

$$\mathbf{P}_{iw} + \mathbf{T}^T\mathbf{P}_{iu} = \mathbf{0} \tag{12.16}$$

where $\mathbf{P}_{iw}$ represents the inertia forces in the directions of the rigid frame of reference and $\mathbf{P}_{iu}$ represents all the remaining inertia forces. Equation (12.16) can be written as

$$\mathbf{A}_0{}^T\mathbf{P}_i = \mathbf{0} \tag{12.17}$$

where $\quad \mathbf{A}_0 = \begin{bmatrix} \mathbf{I} \\ \mathbf{T} \end{bmatrix} \quad (12.18)$

and $\quad \mathbf{P}_i = \begin{bmatrix} \mathbf{P}_{iw} \\ \mathbf{P}_{iu} \end{bmatrix} = -\mathbf{M}\ddot{\mathbf{U}} \quad (12.19)$

Using Eqs. (12.6), (12.17), and (12.19) and assuming that $\omega \neq 0$, we have

$$\mathbf{A}_0{}^T\mathbf{M}\mathbf{q} = \mathbf{0} \tag{12.20}$$

Equation (12.20) can therefore be used as an independent check on eigenmodes $\mathbf{q}$ other than those corresponding to the rigid-body degrees of freedom. The rigid-body modes can be checked using Eq. (12.9).

CONSTRAINED STRUCTURE

If the structure is constrained in such a way that all rigid-body degrees of freedom are excluded, i.e., if the structure is supported in a statically determinate manner, the corresponding stiffness matrix will be nonsingular. Such a stiffness matrix is obtained from the unconstrained-structure stiffness matrix $\mathbf{K}$ by eliminating rows and columns representing the zero displacements assigned to suppress the rigid-body degrees of freedom. Denoting this new reduced matrix by $\mathbf{K}_r$ and the corresponding mass matrix by $\mathbf{M}_r$, we can write Eq. (12.7) as

$$(-\omega^2\mathbf{M}_r + \mathbf{K}_r)\mathbf{q}_r = \mathbf{0} \tag{12.21}$$

where $\mathbf{q}_r$ refers to the unconstrained degrees of freedom. Now since $|\mathbf{K}_r| \neq 0$, we may premultiply Eq. (12.21) by $(1/\omega^2)\mathbf{K}_r^{-1}$ to obtain

$$\left(\frac{1}{\omega^2}\mathbf{I} - \mathbf{K}_r^{-1}\mathbf{M}_r\right)\mathbf{q}_r = \mathbf{0} \tag{12.22}$$

or
$$\left(\frac{1}{\omega^2}\mathbf{I} - \mathbf{D}\right)\mathbf{q}_r = \mathbf{0} \tag{12.22a}$$

where
$$\mathbf{D} = \mathbf{K}_r^{-1}\mathbf{M}_r \tag{12.23}$$

is usually referred to as the *dynamical matrix*. The characteristic equation for frequencies ω for constrained vibrations is then

$$\left|\frac{1}{\omega^2}\mathbf{I} - \mathbf{D}\right| = 0 \tag{12.24}$$

It should be noted that some numerical methods applied to Eq. (12.22*a*) yield first the lowest value of ω (the highest value of $1/\omega$). This has practical advantages if only a few lowest frequencies and the corresponding modes are required.

OVERCONSTRAINED STRUCTURE

If the structure is supported in a statically indeterminate manner (with redundant supports), it may be described as being overconstrained. The characteristic frequencies and modes can still be determined for such structures from Eq. (12.22*a*) provided the additional rows and columns representing redundant constraints in $\mathbf{K}_r$ and $\mathbf{M}_r$ are eliminated.

The three types of vibrations on structures are illustrated in Fig. 12.1 for a simple beam vibrating in the transverse direction. For unconstrained vibrations the beam is oscillating freely, as would be the case of an aircraft wing in flight (see Fig. 12.1*a*). For constrained vibrations the rigid-body degrees of freedom (translation and rotation) are suppressed. This can be achieved by supporting the two end points on wedges (see Fig. 12.1*b*). There are also other combinations

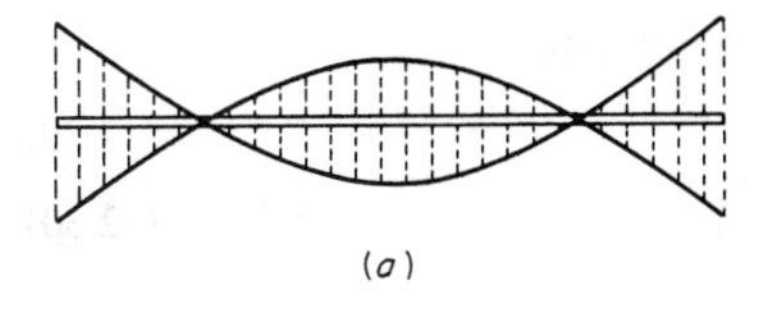

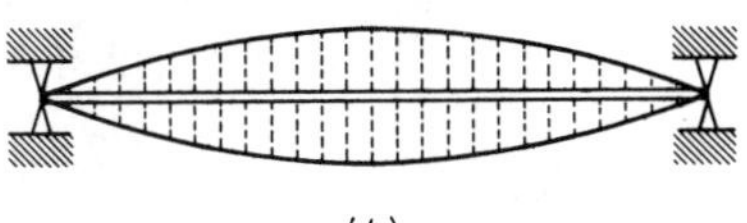

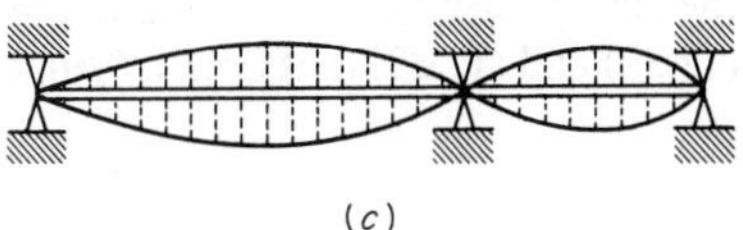

FIG. 12.1 Examples of vibrations on unconstrained, constrained, and overconstrained structures. (*a*) Unconstrained (free); (*b*) constrained; statically determinate reactions; (*c*) overconstrained; statically indeterminate reactions.

of support conditions possible for constrained vibrations, e.g., cantilever beams. To illustrate overconstrained vibrations we may take the beam on three supports (see Fig. 12.1*c*). Naturally many other combinations of support conditions are also possible, e.g., propped cantilever beams.

12.2 PROPERTIES OF THE EIGENMODES: ORTHOGONALITY RELATIONS

The modes $\mathbf{q}$ obtained from Eq. (12.7) can be separated into the rigid-body modes $\mathbf{p}_0$ and normal modes of vibration (elastic eigenmodes) $\mathbf{p}_e$ so that

$$\mathbf{p}_0 = [\mathbf{p}_1 \quad \mathbf{p}_2 \quad \cdots \quad \mathbf{p}_w] \tag{12.25}$$

and $$\mathbf{p}_e = [\mathbf{p}_{w+1} \quad \mathbf{p}_{w+2} \quad \cdots \quad \mathbf{p}_{w+m}] \tag{12.26}$$

where w represents the number of unconstrained rigid-body degrees of freedom and m represents the number of elastic eigenmodes.

If $\omega^2 \neq 0$, Eq. (12.7) can be rewritten as

$$\left(\frac{1}{\omega^2}\mathbf{K} - \mathbf{M}\right)\mathbf{q} = \mathbf{0} \tag{12.27}$$

Substitution of the elastic eigenmodes from (12.26) into (12.27) leads to

$$\begin{aligned} \omega_{w+1}^{-2}\mathbf{K}\mathbf{p}_{w+1} - \mathbf{M}\mathbf{p}_{w+1} &= \mathbf{0} \\ \omega_{w+2}^{-2}\mathbf{K}\mathbf{p}_{w+2} - \mathbf{M}\mathbf{p}_{w+2} &= \mathbf{0} \\ \cdots\cdots\cdots\cdots\cdots \\ \omega_{w+m}^{-2}\mathbf{K}\mathbf{p}_{w+m} - \mathbf{M}\mathbf{p}_{w+m} &= \mathbf{0} \end{aligned} \tag{12.28}$$

which can be combined into a single matrix equation

$$\mathbf{K}\mathbf{p}_e\mathbf{\Omega}^{-2} - \mathbf{M}\mathbf{p}_e = \mathbf{0} \tag{12.29}$$

where $\quad \mathbf{\Omega}^{-2} = \lceil \omega_{w+1}^{-2} \quad \omega_{w+2}^{-2} \quad \cdots \quad \omega_{w+m}^{-2} \rfloor \qquad (12.30)$

Premultiplying Eq. (12.29) by $\mathbf{p}_e^T$, we obtain

$$\mathbf{p}_e^T\mathbf{K}\mathbf{p}_e\mathbf{\Omega}^{-2} - \mathbf{p}_e^T\mathbf{M}\mathbf{p}_e = \mathbf{0} \tag{12.31}$$

or $\quad \mathscr{K}_e\mathbf{\Omega}^{-2} = \mathscr{M}_e \qquad (12.31a)$

where $\quad \mathscr{K}_e = \mathbf{p}_e^T\mathbf{K}\mathbf{p}_e \qquad (12.32)$

represents the generalized stiffness matrix and

$$\mathscr{M}_e = \mathbf{p}_e^T\mathbf{M}\mathbf{p}_e \tag{12.33}$$

represents the generalized mass matrix for the elastic eigenmodes. Since **K** and **M** are symmetric matrices, it follows from (12.32) and (12.33) that both $\mathscr{K}_e$ and $\mathscr{M}_e$ are also symmetric.

Writing Eq. (12.31*a*) symbolically as

$$\mathbf{A}\mathbf{D} = \mathbf{S} \tag{12.34}$$

where both **A** and **S** are symmetric and **D** is diagonal, we can find the elements of **S**, that is, $\mathscr{M}_e$, from

$$s_{ij} = \sum_{r=1}^{m} a_{ir}d_{rj} = a_{ij}d_{jj} \tag{12.35}$$

and $\quad s_{ji} = \sum_{r=1}^{m} a_{jr}d_{ri} = a_{ji}d_{ii} \qquad (12.36)$

Since $\quad s_{ij} = s_{ji} \qquad (12.37)$

we must have that

$$a_{ij}d_{jj} = a_{ji}d_{ii} \tag{12.38}$$

which is true only if either **A**, that is, $\mathscr{K}_e$, is diagonal or **D**, that is, $\mathbf{\Omega}^{-2}$, is a scalar matrix. Since **D** is not a scalar matrix, it follows that the matrix **A**, that is, $\mathscr{K}_e$, and hence **S**, that is, $\mathscr{M}_e$, must be diagonal. Hence we can note from Eqs. (12.32) and (12.33) that the elastic eigenmodes $\mathbf{p}_e$ are orthogonal with respect to either the stiffness matrix **K** or the mass matrix **M**.

It is interesting to examine Eq. (12.32). When this equation is written as

$$\tfrac{1}{2}\mathbf{p}_e^T\mathbf{K}\mathbf{p}_e = \tfrac{1}{2}\mathscr{K}_e \tag{12.39}$$

it is clear that $\mathbf{K}\mathbf{p}_e$ represents generalized elastic forces for the modes $\mathbf{p}_e$, and $\frac{1}{2}\mathbf{p}_e^T\mathbf{K}\mathbf{p}_e$ represents work done by these generalized forces. Since $\mathscr{K}_e$ is a diagonal matrix, Eq. (12.39) can be interpreted as the statement that "work of the generalized elastic forces in one mode acting over the displacements in another mode is equal to zero"; only when the generalized forces are acting over the displacements in their own mode is work not equal to zero.

We can now return to the rigid-body modes $\mathbf{p}_0$. These modes can be conveniently taken as

$$\mathbf{p}_0 = \begin{bmatrix} \mathbf{I} \\ \mathbf{T} \end{bmatrix} = \mathbf{A}_0 \tag{12.40}$$

Using Eq. (12.40) in (12.20) and noting that the equilibrium conditions must be satisfied for $\mathbf{q} = \mathbf{p}_{w+1}, \mathbf{p}_{w+2}, \ldots, \mathbf{p}_{w+m}$, we obtain

$$\mathbf{p}_0{}^T\mathbf{M}\mathbf{p}_e = \mathbf{0} \tag{12.41}$$

Equation (12.41) may be interpreted as an orthogonality condition between the rigid-body modes $\mathbf{p}_0$ and the eigenmodes $\mathbf{p}_e$ with respect to the mass matrix $\mathbf{M}$.

For the subsequent analysis the matrix product

$$\mathbf{p}_0{}^T\mathbf{M}\mathbf{p}_0 = \mathcal{M}_0 \tag{12.42}$$

will be introduced. The matrix $\mathcal{M}_0$ is symmetric since $\mathbf{M}$ is symmetric, and it represents the generalized inertia matrix of the rigid structure. It can be made into a diagonal matrix with the proper choice of the rigid frame of reference. If the displacements in $\mathbf{p}_0$ refer to the principal axes of inertia at the center of gravity, the modes $\mathbf{p}_0$ are mutually orthogonal, and

$$\mathcal{M}_0 = \lfloor M \quad M \quad M \quad I_x \quad I_y \quad I_z \rfloor \tag{12.43}$$

where M is the total mass and I_x, I_y, and I_z are the moments of inertia about the principal axes.

Combining now Eqs. (12.33), (12.41), and (12.42) into one equation, we have

$$\mathbf{p}^T\mathbf{M}\mathbf{p} = \mathcal{M} \tag{12.44}$$

where $\qquad \mathbf{p} = [\mathbf{p}_0 \quad \mathbf{p}_e] \qquad (12.45)$

and $\qquad \mathcal{M} = \lfloor \mathcal{M}_0 \quad \mathcal{M}_e \rfloor \qquad (12.46)$

Equation (12.44) represents the orthogonality condition for all rigid-body modes and elastic eigenmodes on an unconstrained structure.

For *constrained* structures all orthogonality relations are still valid provided $\mathbf{p}_0 = \mathbf{0}$ is used, which implies that no rigid-body degrees of freedom are allowed. For *overconstrained* structures $\mathbf{p}_0 = \mathbf{0}$, and the number of degrees of freedom in $\mathbf{p}_e$ is reduced, depending on the degree of overconstraint (number of redundant reactions).

12.3 VIBRATION ANALYSIS BASED ON FLEXIBILITY

UNCONSTRAINED STRUCTURE

Using Eq. (12.19), we can express the inertia forces on the structure as

$$\begin{bmatrix} \mathbf{P}_{iw} \\ \mathbf{P}_{iu} \end{bmatrix} = -\begin{bmatrix} \mathbf{M}_{ww} & \mathbf{M}_{wu} \\ \mathbf{M}_{uw} & \mathbf{M}_{uu} \end{bmatrix} \begin{bmatrix} \ddot{\mathbf{w}} \\ \ddot{\mathbf{u}} \end{bmatrix} \tag{12.47}$$

where the partitioning of the mass matrix $\mathbf{M}$ into submatrices is made to correspond to the partitioned inertia-force matrix (12.19) and the displacement matrix

$$\mathbf{U} = \begin{bmatrix} \mathbf{w} \\ \mathbf{u} \end{bmatrix} \tag{12.48}$$

The displacements $\mathbf{u}$ can be related to $\mathbf{w}$ through the equation

$$\mathbf{u} = \mathbf{r} + \mathbf{T}\mathbf{w} \tag{12.49}$$

where $\mathbf{r}$ represents the displacements $\mathbf{u}$ relative to the fixed frame of reference and $\mathbf{T}$ is a transformation matrix giving the components of $\mathbf{u}$ due to the rigid-body displacements $\mathbf{w}$. The relative displacements $\mathbf{r}$ are determined from

$$\mathbf{r} = \mathscr{F}\mathbf{P}_{iu} \tag{12.50}$$

where $\mathscr{F}$ represents the flexibility matrix calculated for the forces $\mathbf{P}_{iu}$ with $\mathbf{w} = \mathbf{0}$ as the fixed frame of reference.

Using Eqs. (12.47), (12.49), and (12.50), we have

$$\mathbf{u} - \mathbf{T}\mathbf{w} = -\mathscr{F}(\mathbf{M}_{uw}\ddot{\mathbf{w}} + \mathbf{M}_{uu}\ddot{\mathbf{u}}) \tag{12.51}$$

which may be regarded as the equation of motion of the vibrating structure. Since the motion is harmonic, the displacements $\mathbf{w}$ and $\mathbf{u}$ may be expressed as

$$\mathbf{w} = \mathbf{q}_w e^{i\omega t} \tag{12.52}$$

and $$\mathbf{u} = \mathbf{q}_u e^{i\omega t} \tag{12.53}$$

where $\mathbf{q}_w$ and $\mathbf{q}_u$ are the column matrices of the displacement amplitudes. Substituting now Eqs. (12.52) and (12.53) into the equation of motion (12.51) and then canceling the exponential factors $e^{i\omega t}$, we obtain

$$\mathbf{q}_u - \mathbf{T}\mathbf{q}_w = \omega^2\mathscr{F}(\mathbf{M}_{uw}\mathbf{q}_w + \mathbf{M}_{uu}\mathbf{q}_u) \tag{12.54}$$

To determine $\mathbf{q}_w$ in terms of $\mathbf{q}_u$ we use the equation of equilibrium

$$\mathbf{P}_{iw} + \mathbf{T}^T\mathbf{P}_{iu} = \mathbf{0} \tag{12.16}$$

in conjunction with Eqs. (12.47), (12.52), and (12.53). Thus substituting Eq. (12.47) into (12.16), we have

$$\mathbf{M}_{ww}\ddot{\mathbf{w}} + \mathbf{M}_{wu}\ddot{\mathbf{u}} + \mathbf{T}^T\mathbf{M}_{uw}\ddot{\mathbf{w}} + \mathbf{T}^T\mathbf{M}_{uu}\ddot{\mathbf{u}} = \mathbf{0} \tag{12.55}$$

which is subsequently modified by using Eqs. (12.52) and (12.53) to yield finally

$$(\mathbf{M}_{ww} + \mathbf{T}^T\mathbf{M}_{uw})\mathbf{q}_w + (\mathbf{M}_{wu} + \mathbf{T}^T\mathbf{M}_{uu})\mathbf{q}_u = \mathbf{0} \tag{12.56}$$

Hence, solving Eq. (12.56) for $\mathbf{q}_w$, we obtain

$$\begin{aligned}\mathbf{q}_w &= -(\mathbf{M}_{ww} + \mathbf{T}^T\mathbf{M}_{uw})^{-1}(\mathbf{M}_{wu} + \mathbf{T}^T\mathbf{M}_{uu})\mathbf{q}_u \\ &= \mathbf{T}_0\mathbf{q}_u\end{aligned} \tag{12.57}$$

where $$\mathbf{T}_0 = -(\mathbf{M}_{ww} + \mathbf{T}^T\mathbf{M}_{uw})^{-1}(\mathbf{M}_{wu} + \mathbf{T}^T\mathbf{M}_{uu}) \tag{12.58}$$

Substituting now Eq. (12.57) into (12.54), we obtain the equation

$$\mathbf{q}_u - \mathbf{T}\mathbf{T}_0\mathbf{q}_u = \omega^2\mathscr{F}\mathbf{M}_{uw}\mathbf{T}_0\mathbf{q}_u + \omega^2\mathscr{F}\mathbf{M}_{uu}\mathbf{q}_u \tag{12.59}$$

which can be rearranged into

$$\left[\frac{1}{\omega^2}\mathbf{I} - (\mathbf{I} - \mathbf{T}\mathbf{T}_0)^{-1}\mathscr{F}(\mathbf{M}_{uu} + \mathbf{M}_{uw}\mathbf{T}_0)\right]\mathbf{q}_u = \mathbf{0} \tag{12.60}$$

or $$\left(\frac{1}{\omega^2}\mathbf{I} - \mathbf{D}\right)\mathbf{q}_u = \mathbf{0} \tag{12.61}$$

where $$\mathbf{D} = (\mathbf{I} - \mathbf{T}\mathbf{T}_0)^{-1}\mathscr{F}(\mathbf{M}_{uu} + \mathbf{M}_{uw}\mathbf{T}_0) \tag{12.62}$$

is the dynamical matrix for the vibrating system. The condition for nonzero values for $\mathbf{q}_u$ becomes therefore

$$\left|\frac{1}{\omega^2}\mathbf{I} - \mathbf{D}\right| = 0 \tag{12.63}$$

Equation (12.63) is the characteristic equation for the frequencies ω based on the flexibility of the structural system.

Flexibility formulation of the vibration analysis does not allow for the direct determination of the rigid-body modes; however, this is not a serious restriction, since these modes can normally be obtained by inspection for simple structures, or, alternatively, they can be obtained formally from Eq. (12.40).

CONSTRAINED STRUCTURE

The problem of determining the characteristic frequencies and modes for a constrained structure is considerably easier than for a free structure. Noting that the displacements and accelerations of the fixed frame of reference are equal to zero for a constrained structure, that is,

$$\mathbf{w} = \ddot{\mathbf{w}} = \mathbf{0} \tag{12.64}$$

it follows immediately from Eq. (12.51) that

$$\mathbf{u} = -\mathscr{F}\mathbf{M}_{uu}\ddot{\mathbf{u}} \tag{12.65}$$

Substituting

$$\mathbf{u} = \mathbf{q}_u e^{i\omega t} \tag{12.66}$$

into Eq. (12.65) and canceling the exponential factor $e^{i\omega t}$, we have

$$\left(\frac{1}{\omega^2}\mathbf{I} - \mathscr{F}\mathbf{M}_{uu}\right)\mathbf{q}_u = \mathbf{0} \tag{12.67}$$

Hence the characteristic equation for the frequencies ω becomes

$$\left|\frac{1}{\omega^2}\mathbf{I} - \mathscr{F}\mathbf{M}_{uu}\right| = 0 \tag{12.68}$$

It should be noted here that Eq. (12.67) is identical with Eq. (12.22) since

$$\mathscr{F} = \mathbf{K}_r^{-1} \tag{12.69}$$

and $\qquad \mathbf{M}_{uu} = \mathbf{M}_r \qquad (12.70)$

OVERCONSTRAINED STRUCTURE

Here again $\mathbf{w} = \mathbf{0}$, and also some of the displacement $\mathbf{u}$ will be equal to zero. Equations (12.67) and (12.68) are still applicable to the case of overconstrained structure provided the appropriate flexibility and mass matrices are used. Thus the size of matrices $\mathscr{F}$ and $\mathbf{M}_{uu}$ will be reduced depending on the number of constraints for $\mathbf{u}$.

12.4 VIBRATION OF DAMPED STRUCTURAL SYSTEMS

The equation of motion for a structural system with viscous damping and without any externally applied forces is expressed by [see Eq. (12.4)]

$$\mathbf{M}\ddot{\mathbf{U}} + \mathbf{C}\dot{\mathbf{U}} + \mathbf{K}\mathbf{U} = \mathbf{0} \tag{12.71}$$

The solution to this equation may be assumed as

$$\mathbf{U} = \mathbf{q}e^{pt} \tag{12.72}$$

where p is complex. Substituting (12.72) into (12.71), we have

$$(p^2\mathbf{M} + p\mathbf{C} + \mathbf{K})\mathbf{q} = \mathbf{0} \tag{12.73}$$

which has nonzero solutions for $\mathbf{q}$ provided

$$|p^2\mathbf{M} + p\mathbf{C} + \mathbf{K}| = 0 \tag{12.74}$$

For a multi-degree-of-freedom system Eqs. (12.73) and (12.74) are inconvenient to handle. A method has been proposed by Duncan which reduces these equations to a standard form. In this method we combine the identity

$$\mathbf{M}\dot{\mathbf{U}} - \mathbf{M}\dot{\mathbf{U}} = \mathbf{0} \tag{12.75}$$

with Eq. (12.71) to get

$$\begin{bmatrix} \mathbf{0} & \mathbf{M} \\ \mathbf{M} & \mathbf{C} \end{bmatrix} \begin{bmatrix} \ddot{\mathbf{U}} \\ \dot{\mathbf{U}} \end{bmatrix} + \begin{bmatrix} -\mathbf{M} & \mathbf{0} \\ \mathbf{0} & \mathbf{K} \end{bmatrix} \begin{bmatrix} \dot{\mathbf{U}} \\ \mathbf{U} \end{bmatrix} = \begin{bmatrix} \mathbf{0} \\ \mathbf{0} \end{bmatrix} \tag{12.76}$$

Equation (12.76) can now be rewritten as

$$\mathscr{A}\dot{\mathscr{U}} + \mathscr{B}\mathscr{U} = \mathbf{0} \tag{12.76a}$$

where
$$\mathscr{A} = \begin{bmatrix} \mathbf{0} & \mathbf{M} \\ \mathbf{M} & \mathbf{C} \end{bmatrix} \tag{12.77}$$

$$\mathscr{B} = \begin{bmatrix} -\mathbf{M} & \mathbf{0} \\ \mathbf{0} & \mathbf{K} \end{bmatrix} \tag{12.78}$$

$$\mathscr{U} = \begin{bmatrix} \dot{\mathbf{U}} \\ \mathbf{U} \end{bmatrix} \quad \text{and} \quad \dot{\mathscr{U}} = \begin{bmatrix} \ddot{\mathbf{U}} \\ \dot{\mathbf{U}} \end{bmatrix} \tag{12.79}$$

Now if we let

$$\mathscr{U} = \mathbf{v}e^{pt} \tag{12.80}$$

the equation of motion (12.76*a*) can be transformed into

$$(p\mathscr{A} + \mathscr{B})\mathbf{v} = \mathbf{0} \tag{12.81}$$

This last equation is now of a standard form for which many eigenvalue computer programs are available. The only penalty we incur for this simplification is that the size of all matrices is doubled.

12.5 CRITICAL DAMPING

We shall consider next a single-degree-of-freedom system, for which the equation of motion is

$$M\ddot{U} + C\dot{U} + KU = 0 \tag{12.82}$$

we shall assume that p in (12.72) is given by

$$p = -\mu + i\omega_d \tag{12.83}$$

which when substituted into (12.74) leads to

$$(\mu^2 M - \omega_d^2 M - \mu C + K) + i(\omega_d C - 2\mu\omega_d M) = 0 \tag{12.84}$$

Hence, equating to zero the imaginary part of (12.84), we have that

$$\mu = \frac{C}{2M} \tag{12.85}$$

Similarly, after equating to zero the real part of (12.84) and then using (12.85), it follows that the damped circular frequency ω_d is obtained from

$$\begin{aligned}\omega_d^2 &= \frac{1}{M}(\mu^2 M - \mu C + K) \\ &= \frac{K}{M} - \left(\frac{C}{2M}\right)^2 \\ &= \omega^2 - \left(\frac{C}{2M}\right)^2\end{aligned} \tag{12.86}$$

where $\qquad \omega^2 = \dfrac{K}{M} \qquad\qquad (12.87)$

is the square of the circular frequency for the undamped system ($C = 0$).

When the damped circular frequency $\omega_d = 0$, the oscillatory character of the solution ceases, and the system is said to be critically damped. For this case the damping coefficient must be given by

$$C_{\text{crit}} = 2\sqrt{KM} = 2M\omega \tag{12.88}$$

For $\omega_d^2 < 0$, that is, when $C > 2M\omega$, the system is overdamped, having a nonoscillatory solution for the displacement; however, for most structural problems this condition does not occur.

The concept of critical damping is very useful in dynamic-response calculations, since it is easier to specify the amount of damping as a certain percentage of critical than it is to arrive at the numerical values of the damping coefficients in the matrix C. Details of this method are given in Chap. 13.

12.6 LONGITUDINAL VIBRATIONS OF AN UNCONSTRAINED BAR

As the first example of the application of matrix methods to a vibration problem we shall consider longitudinal vibration of a uniform bar. To simplify the calculations the bar will be idealized into two elements, as shown in Fig. 12.2; thus the idealized system contains only three displacements U_1, U_2, and U_3. The frequencies and eigenmodes for the idealized bar will be determined using both the stiffness and flexibility solutions.

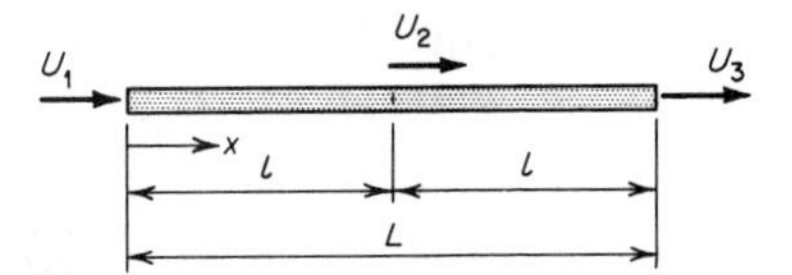

FIG. 12.2 Unconstrained bar (free-free).

STIFFNESS SOLUTION

The stiffness and mass matrices for the two-element bar shown in Fig. 12.2 are

$$\mathbf{K} = \frac{2AE}{L} \begin{bmatrix} 1 & -1 & 0 \\ -1 & 2 & -1 \\ 0 & -1 & 1 \end{bmatrix} \tag{a}$$

$$\mathbf{M} = \frac{\rho AL}{12} \begin{bmatrix} 2 & 1 & 0 \\ 1 & 4 & 1 \\ 0 & 1 & 2 \end{bmatrix} \tag{b}$$

(rows and columns numbered 1, 2, 3)

The above matrices are obtained from the stiffness and inertia properties of the bar element discussed in Secs. 5.5 and 11.4.

Substituting (*a*) and (*b*) into the equation of motion (12.7), we have

$$\left\{ -\omega^2 \frac{A\rho L}{12} \begin{bmatrix} 2 & 1 & 0 \\ 1 & 4 & 1 \\ 0 & 1 & 2 \end{bmatrix} + \frac{2AE}{L} \begin{bmatrix} 1 & -1 & 0 \\ -1 & 2 & -1 \\ 0 & -1 & 1 \end{bmatrix} \right\} \mathbf{q} = 0 \tag{c}$$

The condition for the nonzero solution for **q** is that the determinant formed by the coefficients in (*c*) must be equal to zero. Hence

$$\begin{vmatrix} 1 - 2\mu^2 & -(1 + \mu^2) & 0 \\ -(1 + \mu^2) & 2(1 - 2\mu^2) & -(1 + \mu^2) \\ 0 & -(1 + \mu^2) & 1 - 2\mu^2 \end{vmatrix} = 0 \tag{d}$$

where $\mu^2 = \dfrac{\omega^2 \rho L^2}{24E}$ (*e*)

By expanding the determinant (*d*) the following characteristic equation is obtained:

$$6\mu^2(1 - 2\mu^2)(\mu^2 - 2) = 0 \tag{f}$$

The characteristic equation yields three roots (eigenvalues)

$$\begin{aligned}
&\mu_1{}^2 = 0 \qquad &&\omega_1{}^2 = 0 \qquad &&\omega_1 = 0 \\
&\mu_2{}^2 = \tfrac{1}{2} \qquad &&\omega_2{}^2 = \frac{12E}{\rho L^2} \qquad &&\omega_2 = 3.46\sqrt{\frac{E}{\rho L^2}} \\
&\mu_3{}^2 = 2 \qquad &&\omega_3{}^2 = \frac{48E}{\rho L^2} \qquad &&\omega_3 = 6.92\sqrt{\frac{E}{\rho L^2}}
\end{aligned} \tag{g}$$

The first frequency ω_1 is the rigid-body frequency (zero), while ω_2 and ω_3 are the elastic-eigenmode frequencies (natural frequencies of vibration). The numerical values of ω_2 and ω_3 are approximately 10 percent higher than the exact values; however, had we increased the number of elements we should have improved the accuracy of our solution considerably.

For each frequency ω we can now obtain the relative values of $\mathbf{q}$ from Eq. (*c*). This determines the eigenmodes $\mathbf{p}$. It can be demonstrated easily that the displacements for the rigid-body mode are given by

$$\mathbf{p}_0 = \begin{matrix} 1 \\ 2 \\ 3 \end{matrix} \begin{bmatrix} 1 \\ 1 \\ 1 \end{bmatrix} \tag{h}$$

while those for the elastic eigenmodes are

$$\mathbf{p}_e = \begin{matrix} 1 \\ 2 \\ 3 \end{matrix} \begin{bmatrix} 1 & 1 \\ 0 & -1 \\ -1 & 1 \end{bmatrix} \tag{i}$$

The modes $\mathbf{p}_0$ and $\mathbf{p}_e$ are plotted in Fig. 12.3. The displacement distribution between the nodes on the elements is linear in accordance with the static distribution in the matrix $\mathbf{a}$ from the equation $\mathbf{u} = \mathbf{aU}$ (see Sec. 10.7).

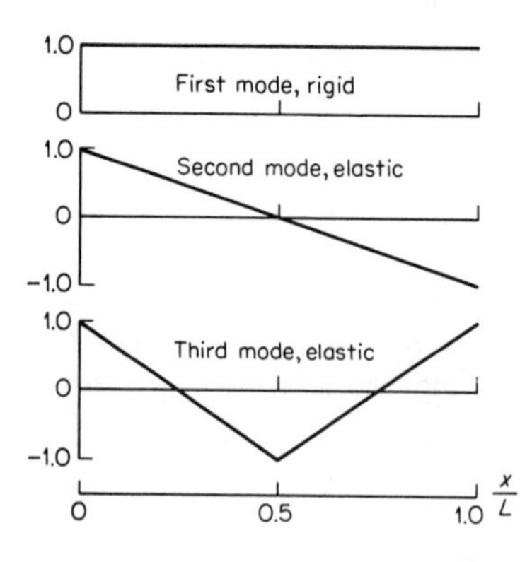

FIG. 12.3 Longitudinal vibration modes based on two-element idealization of a free-free bar.

FLEXIBILITY SOLUTION

Only one zero displacement is required here to constrain the rigid-body motion. By selecting $U_1 = 0$ to establish the rigid frame of reference, it can be shown that the flexibility matrix for the displacements U_2 and U_3 is

$$\mathscr{F} = \frac{L}{2AE} \begin{matrix} & 2 & 3 \\ 2 & 1 & 1 \\ 3 & 1 & 2 \end{matrix} \qquad \text{i.e.} \qquad \mathscr{F} = \frac{L}{2AE}\begin{bmatrix} 1 & 1 \\ 1 & 2 \end{bmatrix} \tag{j}$$

and the transformation matrix $\mathbf{T}$ is given by

$$\mathbf{T} = \begin{bmatrix} 1 \\ 1 \end{bmatrix} \tag{k}$$

where the row and column numbers refer to the degrees of freedom in Fig. 12.2.

To calculate the dynamical matrix in Eq. (12.61) we require the mass submatrices $\mathbf{M}_{ww}$, $\mathbf{M}_{wu}$, $\mathbf{M}_{uw}$, and $\mathbf{M}_{uu}$. These submatrices are obtained directly from Eq. (*b*). Hence

$$\mathbf{M}_{ww} = \frac{\rho AL}{12}[2] \tag{l}$$

$$\mathbf{M}_{wu} = \frac{\rho AL}{12}[1 \quad 0] \tag{m}$$

$$\mathbf{M}_{uw} = \frac{\rho AL}{12}\begin{bmatrix} 1 \\ 0 \end{bmatrix} \tag{n}$$

$$\mathbf{M}_{uu} = \frac{\rho AL}{12}\begin{bmatrix} 4 & 1 \\ 1 & 2 \end{bmatrix} \tag{o}$$

(Row numbers 2, 3 and column numbers 2, 3 label the matrices in (j), (k), (n), (o); row 1 and column 1 label (l), (m) as printed.)

Substituting Eqs. (*k*) to (*o*) into (12.58), we have

$$\begin{aligned} \mathbf{T}_0 &= -(\mathbf{M}_{ww} + \mathbf{T}^T\mathbf{M}_{uw})^{-1}(\mathbf{M}_{wu} + \mathbf{T}^T\mathbf{M}_{uu}) \\ &= -\left\{\frac{\rho AL}{12}[2] + [1 \quad 1]\frac{\rho AL}{12}\begin{bmatrix} 1 \\ 0 \end{bmatrix}\right\}^{-1}\left\{\frac{A\rho L}{12}[1 \quad 0] + [1 \quad 1]\frac{\rho AL}{12}\begin{bmatrix} 4 & 1 \\ 1 & 2 \end{bmatrix}\right\} \\ &= -[2 \quad 1] \end{aligned} \tag{p}$$

Having determined the transformation matrix $\mathbf{T}_0$, we can next calculate the dynamical matrix $\mathbf{D}$ from (12.62).

$$\mathbf{D} = (\mathbf{I} - \mathbf{T}\mathbf{T}_0)^{-1}\mathscr{F}(\mathbf{M}_{uu} + \mathbf{M}_{uw}\mathbf{T}_0)$$

$$= \left\{\begin{bmatrix}1 & 0\\0 & 1\end{bmatrix} + \begin{bmatrix}1\\1\end{bmatrix}[2 \quad 1]\right\}^{-1} \frac{L}{2AE}\begin{bmatrix}1 & 1\\1 & 2\end{bmatrix}\frac{\rho AL}{12}\left\{\begin{bmatrix}4 & 1\\1 & 2\end{bmatrix} - \begin{bmatrix}1\\0\end{bmatrix}[2 \quad 1]\right\}$$

$$= \frac{\rho L^2}{48E}\begin{bmatrix}1 & 0\\3 & 4\end{bmatrix} \tag{q}$$

which when substituted into the equation of motion (12.61) leads to

$$\left\{\frac{1}{\omega^2}\begin{bmatrix}1 & 0\\0 & 1\end{bmatrix} - \frac{\rho L^2}{48E}\begin{bmatrix}1 & 0\\3 & 4\end{bmatrix}\right\}\mathbf{q}_u = \mathbf{0} \tag{r}$$

Introducing again $\mu^2 = \omega^2\rho L^2/24E$, we can write the characteristic equation as

$$\begin{vmatrix}2\mu^{-2} - 1 & 0\\-3 & 2\mu^{-2} - 4\end{vmatrix} = 0 \tag{s}$$

Hence $\quad (2\mu^{-2} - 1)(2\mu^{-2} - 4) = 0 \qquad (t)$

and $\quad \mu_1{}^2 = \frac{1}{2} \qquad \mu_2{}^2 = 2 \qquad (u)$

which agrees with the previous results based on the stiffness formulation.

Substituting in turn the eigenvalues (u) into Eq. (r), we obtain

$$\mathbf{q}_u = \begin{matrix}2\\3\end{matrix}\begin{bmatrix}0 & -1\\1 & 1\end{bmatrix} \tag{v}$$

To find the eigenmode displacement in the direction of $\mathbf{w}$, that is, direction 1, we use Eq. (12.57). Hence

$$\mathbf{q}_w = \mathbf{T}_0\mathbf{q}_u$$

$$= -[2 \quad 1]\begin{bmatrix}0 & -1\\1 & 1\end{bmatrix} = [-1 \quad 1] \tag{w}$$

which may now be combined with $\mathbf{q}_u$ to yield

$$\mathbf{p}_e = \begin{matrix}1\\2\\3\end{matrix}\begin{bmatrix}-1 & 1\\0 & -1\\1 & 1\end{bmatrix} \tag{x}$$

which agrees with Eq. (i) obtained by the stiffness method except for the sign on the first column. This naturally is of no consequence, as any column can be multiplied by an arbitrary constant.

To check the calculated elastic eigenmodes we use Eq. (12.41). Noting that

$$\mathbf{p}_0 = \begin{bmatrix} \mathbf{I} \\ \mathbf{T} \end{bmatrix} = \begin{matrix} 1 \\ 2 \\ 3 \end{matrix}\begin{bmatrix} 1 \\ 1 \\ 1 \end{bmatrix} \tag{y}$$

we have

$$\mathbf{p}_0{}^T\mathbf{M}\mathbf{p}_e = [1 \quad 1 \quad 1]\,\frac{\rho AL}{12}\begin{bmatrix} 2 & 1 & 0 \\ 1 & 4 & 1 \\ 0 & 1 & 2 \end{bmatrix}\begin{bmatrix} -1 & 1 \\ 0 & -1 \\ 1 & 1 \end{bmatrix} = \mathbf{0} \tag{z}$$

which therefore satisfies the equilibrium requirement imposed by Eq. (12.41) on the inertia forces in the vibrating system.

12.7 LONGITUDINAL VIBRATIONS OF A CONSTRAINED BAR

Taking the displacement $U_1 = 0$ to indicate the built-in end condition for the two-element bar shown in Fig. 12.4, we obtain the following stiffness and mass matrices

$$\mathbf{K}_r = \frac{2AE}{L}\begin{bmatrix} 2 & -1 \\ -1 & 1 \end{bmatrix} \tag{a}$$

$$\mathbf{M}_r = \frac{\rho AL}{12}\begin{bmatrix} 4 & 1 \\ 1 & 2 \end{bmatrix} \tag{b}$$

Substituting (*a*) and (*b*) into the equation of motion (12.21), we have

$$\left\{-\omega^2\frac{\rho AL}{12}\begin{bmatrix} 4 & 1 \\ 1 & 2 \end{bmatrix} + \frac{2AE}{L}\begin{bmatrix} 2 & -1 \\ -1 & 1 \end{bmatrix}\right\}\mathbf{q}_r = \mathbf{0} \tag{c}$$

and

$$\begin{vmatrix} 2(1 - 2\mu^2) & -(1 + \mu^2) \\ -(1 + \mu^2) & 1 - 2\mu^2 \end{vmatrix} = 0 \tag{d}$$

Expanding the determinant, we obtain

$$7\mu^4 - 10\mu^2 + 1 = 0$$

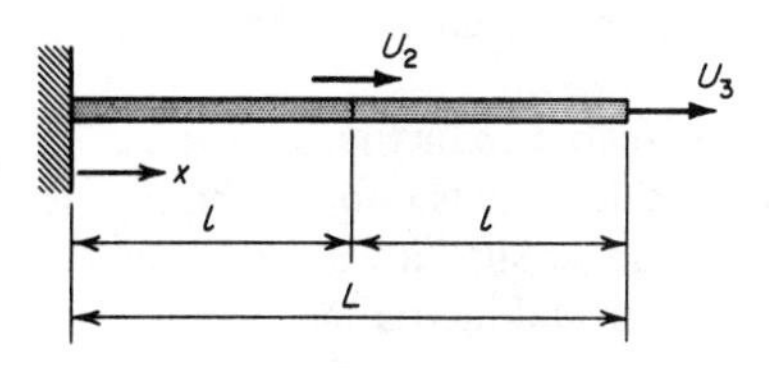

FIG. 12.4 Constrained bar (fixed-free).

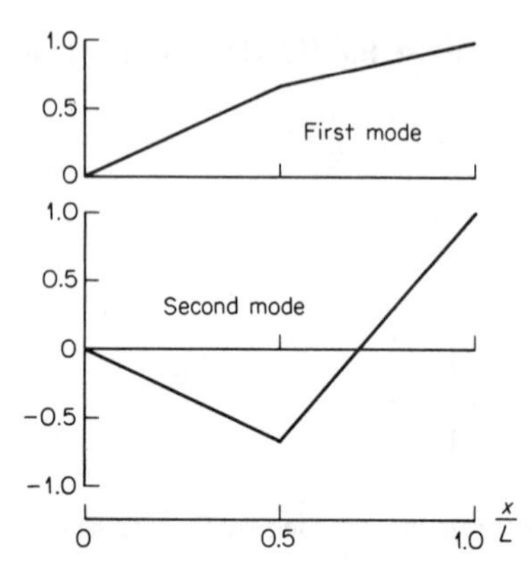

FIG. 12.5 Longitudinal vibration modes based on two-element idealization of a fixed-free bar.

Hence

$$\begin{aligned} \mu_1^2 &= \tfrac{1}{7}(5 - 3\sqrt{2}) \qquad \omega_1 = 1.6114\sqrt{E/\rho L^2} \\ \mu_2^2 &= \tfrac{1}{7}(5 + 3\sqrt{2}) \qquad \omega_2 = 5.6293\sqrt{E/\rho L^2} \end{aligned} \tag{e}$$

The calculated values for ω_1 and ω_2 are 2.6 and 19.5 percent higher than the corresponding analytical values. The elastic eigenmodes for the two calculated frequencies ω_1 and ω_2 are

$$\mathbf{p}_e = \begin{bmatrix} \dfrac{\sqrt{2}}{2} & -\dfrac{\sqrt{2}}{2} \\ 1.0 & 1.0 \end{bmatrix} \tag{f}$$

The modes $\mathbf{p}_e$ have been plotted in Fig. 12.5.

12.8 TRANSVERSE VIBRATIONS OF A FUSELAGE-WING COMBINATION

We shall calculate next natural frequencies and mode shapes of a uniform wing attached to a fuselage mass as shown in Fig. 12.6*a*. The total wing mass is distributed uniformly over the wing span of length $2L$, and it has a value of $2M_w$. The total fuselage mass has a value of $2M_F$, with vanishingly small rotational inertia. The wing is to be represented as a uniform beam with flexural stiffness EI. The effects of shear deformations and rotatory inertia are to be neglected.

ONE-ELEMENT SOLUTIONS Since the wing is symmetric about the fuselage center, we can separate the frequency and mode-shape calculations into symmetric and antisymmetric modes. As our first approximation we shall use a single-element idealization illustrated in Fig. 12.6*b*. The stiffness and inertia matrices for this

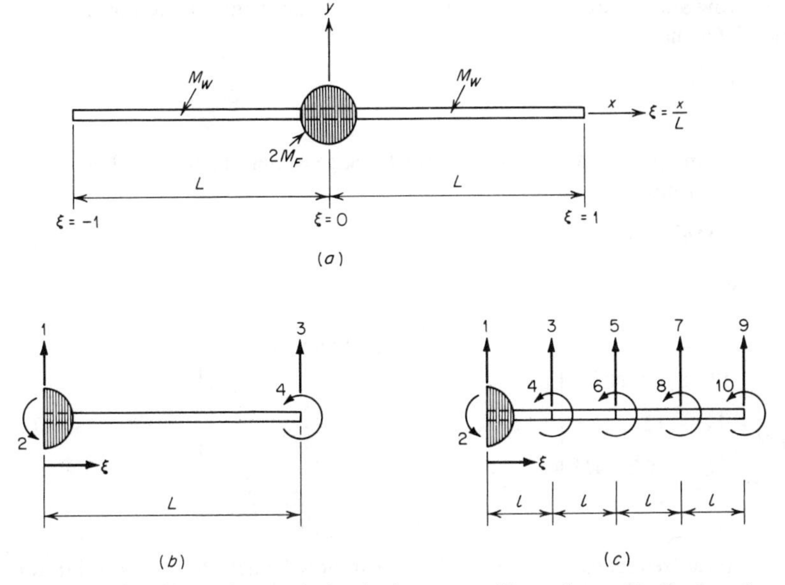

FIG. 12.6 (*a*) Uniform wing attached to fuselage mass; (*b*) one-element idealization of a uniform wing attached to fuselage mass; (*c*) four-element idealization of a uniform wing attached to fuselage mass.

case are given by

$$\mathbf{K} = \frac{EI}{L^3} \begin{array}{c} 1 \\ 2 \\ 3 \\ 4 \end{array} \begin{bmatrix} 12 & & & \\ 6L & 4L^2 & \text{Symmetric} & \\ -12 & -6L & 12 & \\ 6L & 2L^2 & -6L & 4L^2 \end{bmatrix} \tag{a}$$
$$\qquad\qquad\quad 1 \qquad 2 \qquad 3 \qquad 4$$

$$\mathbf{M} = M_w \begin{array}{c} 1 \\ \\ 2 \\ \\ 3 \\ \\ 4 \end{array} \begin{bmatrix} \frac{13}{35} + R & & & \\ & & \text{Symmetric} & \\ \frac{11}{210}L & \frac{L^2}{105} & & \\ & & & \\ \frac{9}{70} & \frac{13}{420}L & \frac{13}{35} & \\ & & & \\ \frac{-13}{420}L & \frac{-L^2}{140} & \frac{-11}{210}L & \frac{L^2}{105} \end{bmatrix} \tag{b}$$
$$\qquad\qquad\quad 1 \qquad 2 \qquad 3 \qquad 4$$

where row and column numbers refer to the displacement numbering shown in Fig. 12.6*b* and

$$R = \frac{M_F}{M_w} \tag{c}$$

For symmetric modes rotations at the fuselage center are zero, and hence the equation of motion

$$(\mathbf{K} - \omega^2\mathbf{M})\mathbf{q} = \mathbf{0} \tag{d}$$

becomes

$$\left\{ \frac{EI}{L^3}\begin{bmatrix} 12 & \text{Symmetric} & \\ -12 & 12 & \\ 6L & -6L & 4L^2 \end{bmatrix} - \omega^2 M_w \begin{bmatrix} \frac{13}{35} + R & \text{Symmetric} & \\ \frac{9}{70} & \frac{13}{35} & \\ \frac{-13L}{420} & \frac{-11L}{210} & \frac{L^2}{105} \end{bmatrix} \right\} \mathbf{q} = \mathbf{0} \tag{e}$$

The natural frequencies and mode shapes can now be determined from Eq. (*e*). Thus for $R = 0$ we obtain the following frequencies

$$\mathbf{\Omega}_{\text{symmetric}} = \sqrt{\frac{EI}{M_w L^3}}\,\lfloor 0 \quad 5.606 \quad 43.870 \rfloor \tag{f}$$

and the corresponding modes are

$$\mathbf{p} = \begin{bmatrix} 1.0 & -0.2627 & 0.05036 \\ 1.0 & 0.4293 & 0.11631 \\ 0 & \frac{1.0}{L} & \frac{1.0}{L} \end{bmatrix} \tag{g}$$

where the normalizing factors are adjusted in such a way as to make the largest numerical values in **p**, apart from the factor L, equal to unity.

The results for the frequencies when $R = 0, 1$, and 3 are presented in Table 12.1, where a comparison with the exact analytical solutions is made. The first frequency (zero) is the rigid-body frequency. The second frequency obtained for the single-element idealization is within acceptable accuracy for engineering purposes, while the accuracy of the third frequency is unacceptable. Thus in order to obtain better accuracy we must use more elements in our idealized model. Another drawback when a small number of elements is used is the inability to draw the mode shapes accurately. For example, for the single-element idealization only two deflections and two slopes (including the zero slope

TABLE 12.1 SYMMETRIC-MODE NONDIMENSIONAL FREQUENCY $c = \omega\sqrt{M_w L^3/EI}$ FOR A UNIFORM WING ATTACHED TO FUSELAGE MASS: ONE-ELEMENT IDEALIZATION

Mass ratio R	Frequency number	Exact*	One-element idealization	Percentage error
All values	1	0	0	0
0	2	5.593	5.606	+0.2
	3	30.226	43.870	+45.1
1	2	4.219	4.229	+0.2
	3	23.707	36.777	+55.1
3	2	3.822	3.835	+0.3
	3	22.677	35.575	+56.9

* Obtained from the computer solution of the exact characteristic equation.

at the fuselage center) are available. This, of course, is not sufficient for accurate graphical representation of the third mode. The symmetric mode shapes for $R = 0$, 1, and 3 based on one-element idealization are plotted in Fig. 12.7.

For antisymmetric modes deflections at the fuselage center are zero. The equations of motion for the antisymmetric case are given by

$$\left\{\frac{EI}{L^3}\begin{bmatrix} 4L^2 & \text{Symmetric} & \\ -6L & 12 & \\ 2L^2 & -6L & 4L^2 \end{bmatrix} - \omega^2 M_w \begin{bmatrix} \frac{L^2}{105} & \text{Symmetric} & \\ \frac{13L}{420} & \frac{13}{35} & \\ \frac{-L^2}{140} & \frac{-11L}{210} & \frac{L^2}{105} \end{bmatrix}\right\} \mathbf{q} = \mathbf{0} \tag{h}$$

Determination of frequencies and modes from Eq. (*h*) leads to

$$\mathbf{\Omega}_{\text{antisymmetric}} = \sqrt{\frac{EI}{M_w L^3}}\,[0 \quad 17.544 \quad 70.087] \tag{i}$$

$$\mathbf{p} = \begin{bmatrix} \frac{1.0}{L} & \frac{-0.77746}{L} & \frac{0.52322}{L} \\ 1.0 & 0.21690 & 0.09303 \\ \frac{1.0}{L} & \frac{1.0}{L} & \frac{1.0}{L} \end{bmatrix} \tag{j}$$

The antisymmetric mode shapes are plotted in Fig. 12.8. The values of frequencies obtained for the one-element idealization are compared with the exact values in Table 12.2.

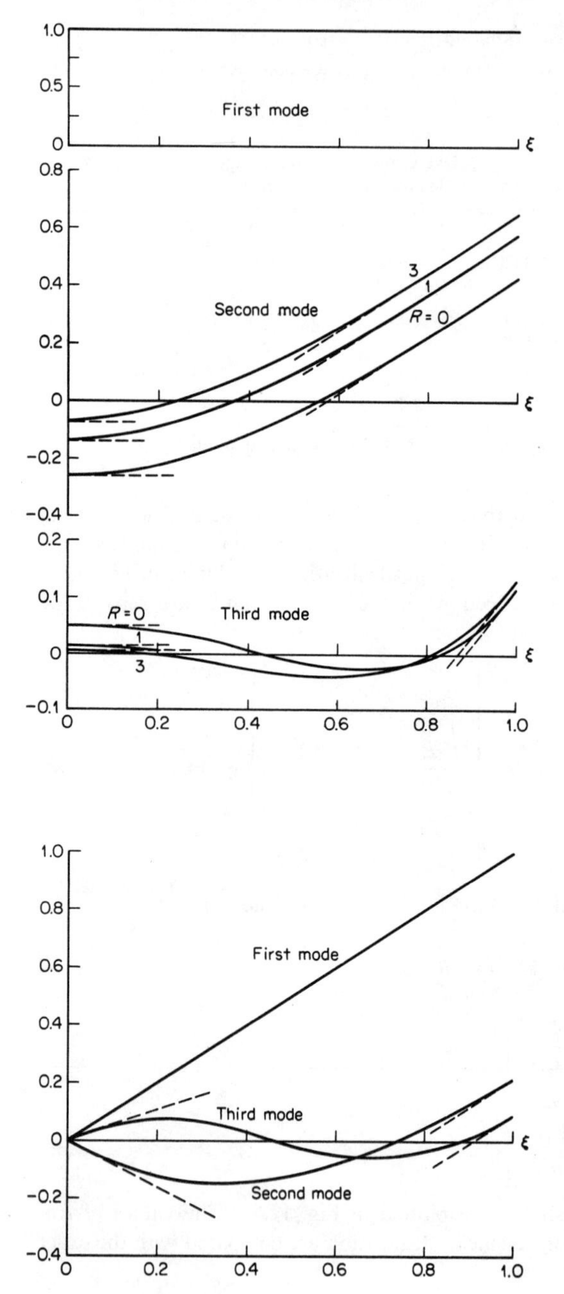

FIG. 12.7 Symmetric mode shapes based on one-element idealization of a uniform wing attached to fuselage mass; $R =$ (fuselage mass)/(wing mass).

FIG. 12.8 Antisymmetric mode shapes based on one-element idealization of a uniform wing attached to fuselage mass; rotational inertia of the fuselage is assumed to be zero.

TABLE 12.2 ANTISYMMETRIC-MODE NONDIMENSIONAL FREQUENCY $c = \omega\sqrt{M_w L^3/EI}$ FOR A UNIFORM BEAM: ONE-ELEMENT IDEALIZATION

Frequency number	Exact	One-element idealization	Percentage error
1	0	0	0
2	15.418	17.544	+13.8
3	49.965	70.087	+40.3

FOUR-ELEMENT SOLUTIONS We shall now consider a four-element idealization as illustrated in Fig. 12.6*c*. The stiffness and inertia matrices for this idealization are given by

$$
\mathbf{K} = \frac{EI}{l^3}
\begin{array}{r}
1\\2\\3\\4\\5\\6\\7\\8\\9\\10\\ \\
\end{array}
\left[\begin{array}{cccccccccc}
12 & & & & & & & & & \\
6l & 4l^2 & & & & \text{Symmetric} & & & & \\
-12 & -6l & 24 & & & & & & & \\
6l & 2l^2 & 0 & 8l^2 & & & & & & \\
0 & 0 & -12 & -6l & 24 & & & & & \\
0 & 0 & 6l & 2l^2 & 0 & 8l^2 & & & & \\
0 & 0 & 0 & 0 & -12 & -6l & 24 & & & \\
0 & 0 & 0 & 0 & 6l & 2l^2 & 0 & 8l^2 & & \\
0 & 0 & 0 & 0 & 0 & 0 & -12 & -6l & 12 & \\
0 & 0 & 0 & 0 & 0 & 0 & 6l & 2l^2 & -6l & 4l^2
\end{array}\right]
$$
$$
\begin{array}{cccccccccc}
1 & 2 & 3 & 4 & 5 & 6 & 7 & 8 & 9 & 10
\end{array}
\qquad (k)
$$

$$
\mathbf{M} = \frac{M_w}{4 \times 420}
\begin{array}{r}
1\\2\\3\\4\\5\\6\\7\\8\\9\\10\\ \\
\end{array}
\left[\begin{array}{cccccccccc}
156 + 1{,}680R & & & & & & & & & \\
22l & 4l^2 & & & & & & & & \\
54 & 13l & 312 & & & & & & & \\
-13l & -3l^2 & 0 & 8l^2 & & & \text{Symmetric} & & & \\
0 & 0 & 54 & 13l & 312 & & & & & \\
0 & 0 & -13l & -3l^2 & 0 & 8l^2 & & & & \\
0 & 0 & 0 & 0 & 54 & 13l & 312 & & & \\
0 & 0 & 0 & 0 & -13l & -3l^2 & 0 & 8l^2 & & \\
0 & 0 & 0 & 0 & 0 & 0 & 54 & 13l & 156 & \\
0 & 0 & 0 & 0 & 0 & 0 & -13l & -3l^2 & -22l & 4l^2
\end{array}\right]
$$
$$
\begin{array}{cccccccccc}
1 & 2 & 3 & 4 & 5 & 6 & 7 & 8 & 9 & 10
\end{array}
\qquad (l)
$$

TABLE 12.3 SYMMETRIC-MODE NONDIMENSIONAL FREQUENCY $c = \omega\sqrt{M_w L^3/EI}$ FOR A UNIFORM WING ATTACHED TO FUSELAGE MASS: FOUR-ELEMENT IDEALIZATION*

	$R = 0$		$R = 1$		$R = 3$	
Frequency number	Exact	Four-element idealization	Exact	Four-element idealization	Exact	Four-element idealization
1	0	0	0	0	0	0
2	5.593	5.594 (0.0)	4.219	4.219 (0.0)	3.822	3.821 (0.0)
3	30.226	30.290 (0.2)	23.707	23.731 (0.1)	22.677	22.701 (0.1)
4	74.639	75.436 (1.1)	63.457	63.934 (0.8)	62.331	62.813 (0.8)
5	138.791	140.680 (1.4)	122.722	124.344 (1.3)	121.551	123.254 (1.4)
6	222.683	248.448 (11.6)	201.725	229.768 (13.9)	200.506	228.710 (14.1)
7	326.314	392.238 (20.2)	300.442	368.131 (22.5)	299.209	366.996 (22.7)
8	449.684	603.554 (34.2)	418.894	582.223 (39.0)	417.647	581.326 (39.2)
9	592.793	954.671 (61.0)	557.081	953.153 (71.1)	555.822	953.087 (71.5)

* Values in parenthesis represent percentage errors.

where $\quad l = \dfrac{L}{4}$

For symmetric modes row and column 2, corresponding to rotation at the wing center, are removed in formulating the equations of motions for the vibrating wing. Similarly for antisymmetric modes row and column 1, corresponding to deflection at the wing center, are removed. The resulting equations were solved using an IBM 1620 eigenvalue and eigenvector computer program. The

TABLE 12.4 ANTISYMMETRIC-MODE NONDIMENSIONAL FREQUENCY $c = \omega\sqrt{M_w L^3/EI}$ FOR A UNIFORM BEAM: FOUR-ELEMENT IDEALIZATION

Frequency number	Exact	Four-element idealization	Percentage error
1	0	0	0
2	15.418	15.427	0.1
3	49.965	50.231	0.5
4	104.248	106.045	1.7
5	178.270	197.085	10.6
6	272.031	313.199	15.1
7	385.531	488.528	26.7
8	518.771	728.116	40.4
9	671.750	961.199	43.1

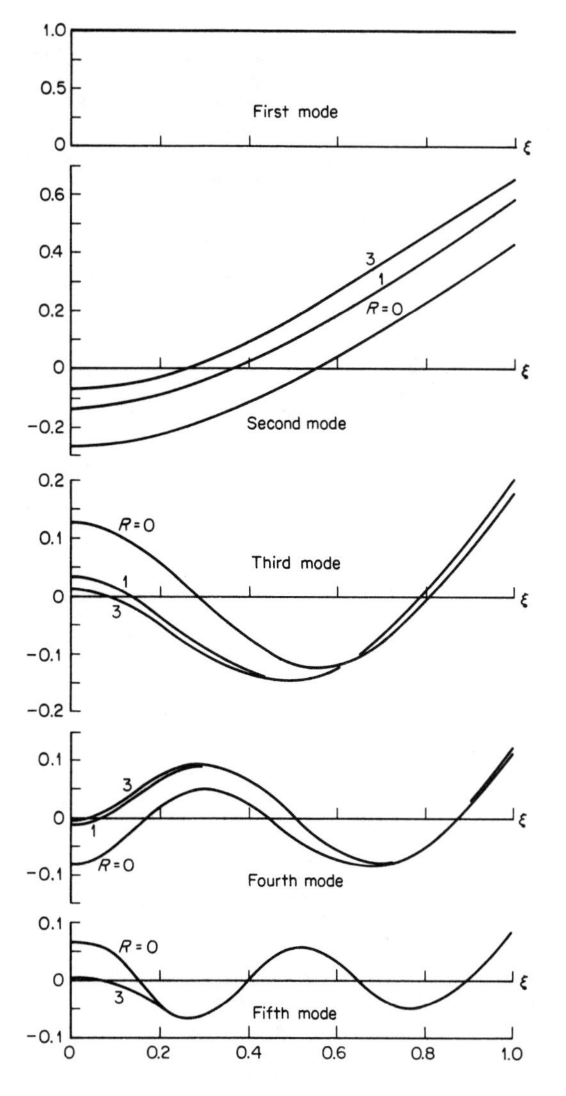

FIG. 12.9 Symmetric mode shapes based on four-element idealization of a uniform wing attached to fuselage mass; R = (fuselage mass)/(wing mass).

results for the symmetric frequencies for $R = 0$, 1, and 3 are given in Table 12.3, while for antisymmetric modes the frequencies are given in Table 12.4. The first five mode shapes calculated by the computer are illustrated in Figs. 12.9 and 12.10.

Tables 12.3 and 12.4 indicate that the accuracy of frequencies for higher modes is progressively decreased, and only the first half of all calculated frequencies may be considered satisfactory for engineering purposes.

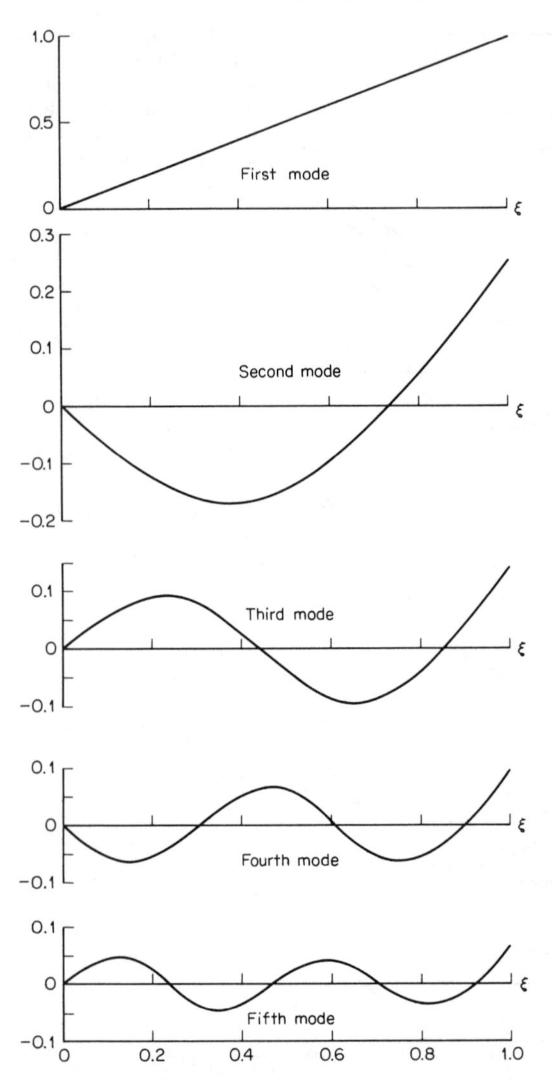

FIG. 12.10 Antisymmetric mode shapes based on four-element idealization of a uniform wing attached to fuselage mass; rotational inertia of the fuselage is assumed to be zero.

12.9 DETERMINATION OF VIBRATION FREQUENCIES FROM THE QUADRATIC MATRIX EQUATION

Substituting the frequency-dependent mass and stiffness matrices from (10.108) and (10.111) into Eq. (12.7), we obtain the equation of motion in the form of a matrix series in ascending powers of ω^2

$$[\mathbf{K}_0 - \omega^2\mathbf{M}_0 - \omega^4(\mathbf{M}_2 - \mathbf{K}_4) - \cdots]\mathbf{q} = \mathbf{0} \tag{12.89}$$

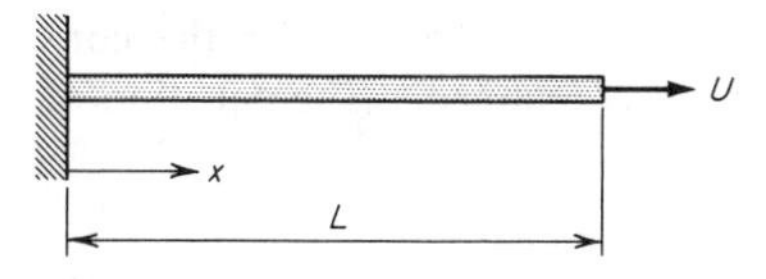

FIG. 12.11 Fixed-free bar; one-element idealization.

Since the retention of terms with ω^4 and higher presents some computational difficulties, we normally retain only the term with ω^2 when calculating vibration frequencies and modes.

To illustrate the application of the general theory we shall consider vibrations of a uniform bar built-in at one end and free at the other (see Fig. 12.11). In the equation of motion terms with ω^4 will be retained; however, in order to simplify subsequent calculations only one-element idealization will be used. For a single bar element we have the following matrices

$$\mathbf{K}_0 = \frac{AE}{L}\begin{bmatrix} 1 & -1 \\ -1 & 1 \end{bmatrix} \qquad \mathbf{M}_0 = \frac{\rho AL}{6}\begin{bmatrix} 2 & 1 \\ 1 & 2 \end{bmatrix}$$

$$\mathbf{K}_4 = \frac{\rho^2 AL^3}{45E}\begin{bmatrix} 1 & \frac{7}{8} \\ \frac{7}{8} & 1 \end{bmatrix} \qquad \mathbf{M}_2 = \frac{2\rho^2 AL^3}{45E}\begin{bmatrix} 1 & \frac{7}{8} \\ \frac{7}{8} & 1 \end{bmatrix}$$

Hence, after eliminating the row and column corresponding to the displacement at the built-in end the equation of motion becomes

$$\left(\frac{AE}{L} - \omega^2 \frac{\rho AL}{3} - \omega^4 \frac{\rho^2 AL^3}{45E}\right) q_r = 0 \tag{a}$$

while the characteristic equation is given by

$$\lambda^4 + 15\lambda^2 - 45 = 0 \tag{b}$$

$$\text{where} \qquad \lambda^2 = \frac{\omega^2 \rho L^2}{E} \tag{c}$$

The roots of (*b*) are

$$\lambda_1^2 = \tfrac{1}{2}(-15 + 9\sqrt{5}) \tag{d}$$

$$\lambda_2^2 = \tfrac{1}{2}(-15 - 9\sqrt{5}) \tag{e}$$

where only λ_1, being positive, is applicable to our problem. The eigenvalue λ_2 must be rejected because a negative value of λ^2 would require the existence of an imaginary frequency. Taking therefore λ_1^2 as the required solution, we have

$$\lambda_1 = \omega\sqrt{\frac{\rho L^2}{E}} = 1.6007 \tag{f}$$

which may be compared with the exact value of 1.5708. Hence the retention of the term with ω^4 in the equation of motion resulted in an error of only 1.9

percent. It would be interesting now to compare the results for the conventional method, where only the term with ω^2 is retained. The equation of motion for this case is

$$\left(\frac{AE}{L} - \omega^2 \frac{\rho A L}{3}\right) q_r = 0 \tag{g}$$

which leads to

$$\lambda_1 = \sqrt{3} = 1.7321 \tag{h}$$

This result differs from the exact value by 10.3 percent.

In order to demonstrate the considerable improvement in accuracy of the frequencies obtained from the quadratic equation the fixed-free bar frequencies have been calculated, using an iterative technique, the number of elements in the idealized model varying from 1 to 10. Ratios of frequencies of vibration for the fixed-free bar determined from the quadratic equation over the exact frequencies are shown in Table 12.5. For comparison, the corresponding ratios obtained from the conventional analysis are also presented. This table indicates clearly that considerable improvement in accuracy is obtained when the quadratic equations are used. In Fig. 12.12 the percentage errors are plotted against the

TABLE 12.5 RATIOS OF $\omega/\omega_{\text{exact}}$ FOR LONGITUDINAL VIBRATIONS OF A FIXED-FREE BAR (NUMBERS IN PARENTHESES REPRESENT VALUES OBTAINED FROM CONVENTIONAL ANALYSIS)

n	Frequency number									
	1	2	3	4	5	6	7	8	9	10
1	1.019 (1.103)									
2	1.002 (1.026)	1.069 (1.195)								
3	1.000 (1.012)	1.019 (1.103)	1.083 (1.200)							
4	1.000 (1.006)	1.007 (1.058)	1.041 (1.154)	1.083 (1.191)						
5	1.000 (1.004)	1.003 (1.037)	1.019 (1.103)	1.057 (1.181)	1.079 (1.182)					
6	1.000 (1.003)	1.002 (1.026)	1.010 (1.072)	1.032 (1.137)	1.069 (1.195)	1.075 (1.173)				
7	1.000 (1.002)	1.001 (1.019)	1.006 (1.053)	1.019 (1.103)	1.044 (1.161)	1.076 (1.200)	1.070 (1.166)			
8	1.000 (1.002)	1.001 (1.015)	1.003 (1.041)	1.012 (1.079)	1.029 (1.128)	1.054 (1.177)	1.080 (1.201)	1.066 (1.159)		
9	1.000 (1.001)	1.000 (1.012)	1.002 (1.032)	1.008 (1.063)	1.019 (1.103)	1.038 (1.148)	1.062 (1.188)	1.083 (1.200)	1.062 (1.154)	
10	1.000 (1.001)	1.000 (1.009)	1.002 (1.026)	1.005 (1.051)	1.013 (1.084)	1.027 (1.123)	1.046 (1.163)	1.069 (1.195)	1.084 (1.198)	1.059 (1.150)

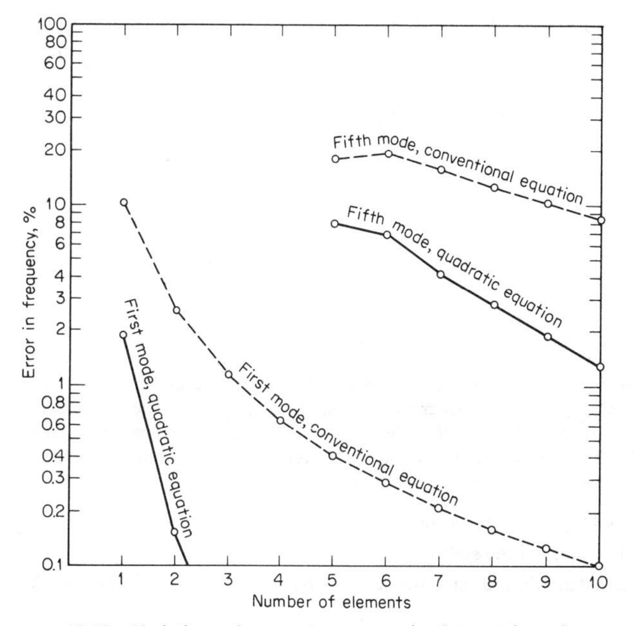

FIG. 12.12 Variation of percentage error in frequencies of a fixed-free bar calculated from quadratic and conventional equations.

number of elements for the first and fifth modes in order to show the general trends more clearly. A perusal of Table 12.5 and Fig. 12.12 reveals that the percentage error in frequencies is reduced by almost an order of magnitude when quadratic equations are used instead of the conventional eigenvalue equations. Furthermore, the rate of decrease in percentage error when the number of elements is increased is considerably greater for quadratic-equation solutions; thus the convergence to the true frequency values is much faster with quadratic equations.

PROBLEMS

12.1 Show that the natural frequency of longitudinal vibrations of a single unconstrained bar element having a linearly varying cross-sectional area is determined from the formula

$$\omega^2 = \frac{18E}{\rho l^2} \frac{(A_1 + A_2)^2}{A_1^2 + 4A_1A_2 + A_2^2}$$

where ω = circular frequency

E = Young's modulus

l = bar length

A_1, A_2 = cross-sectional areas at $x = 0$ and $x = l$, respectively

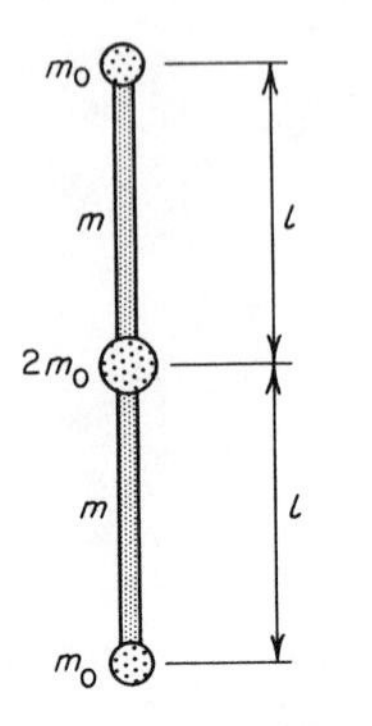

FIG. 12.13

12.2 A uniform bar has two concentrated masses attached to its ends and one mass at its center; the end masses are m_0, while the center mass is $2m_0$ (see Fig. 12.13). The total mass of the bar itself is $2m$, and its total length is $2l$. The cross-sectional area of the bar is A and its Young's modulus is E. For the purpose of analysis the bar is to be idealized into two elements, each of length l. The mass ratio $m_0/m = 1$. Determine the following:

a. The characteristic frequencies of the idealized system

b. The characteristic modes of the system

c. The generalized mass matrix and demonstrate orthogonality relations for the eigenmodes (elastic modes)

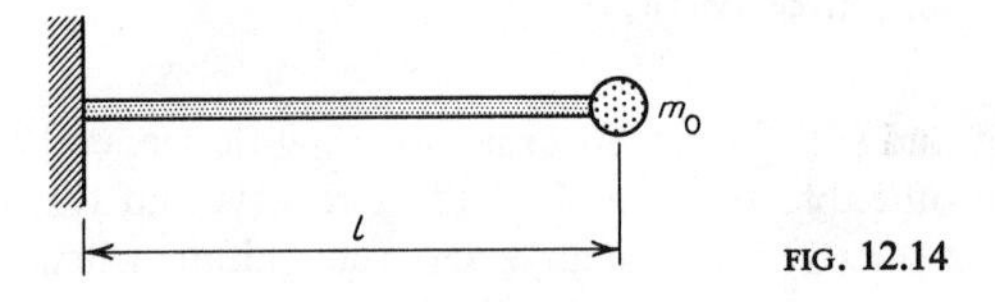

FIG. 12.14

12.3 A uniform cantilever has a concentrated mass m_0 attached to its free end, as shown in Fig. 12.14. The flexural rigidity of the beam is EI, the cross-sectional area is A, the density is ρ, and the length is l. Using only the translational and rotational degrees of freedom of the tip, determine the natural frequencies of transverse vibrations. The effects of shear deformations and rotatory inertia are to be neglected.

CHAPTER 13
DYNAMIC RESPONSE OF ELASTIC SYSTEMS

The first part of this chapter contains a discussion of the equations of motion of an elastic undamped structure with a finite number of degrees of freedom. The external loads applied to this structure are assumed to be time-dependent. Such loads may arise from a variety of causes. In aeronautical applications such loads are caused by gusts, blast-induced shock waves, rapid maneuvering, bomb release or ejection, impact forces during landing, catapulting, etc. Both unconstrained and constrained structures are considered. For unconstrained structures, e.g., an airplane in flight, the effect of rapidly applied external loads is to cause not only translational and rotational motion as a rigid body but also to induce structural vibrations. For totally constrained structures the rigid-body motion is absent, and only structural vibrations are induced. The solution of the equations of motion is accomplished by the standard technique of expressing displacements in terms of modes of vibration. Response integrals (Duhamel's integrals) for the displacements have been determined for some typical forcing functions, and the results are compiled in Table 13.1. In addition to prescribed variation of external forces, response due to forced displacements are also considered. This occurs, for example, when a building is subjected to earthquake motion.

The second part of this chapter contains a discussion of damped motion of elastic structures. Solutions to the equations of motion are presented for three different models of damping: damping proportional to mass, damping proportional to stiffness, and percentage of critical damping. Numerical examples are included to illustrate typical dynamic-response calculations.

13.1 RESPONSE OF A SINGLE-DEGREE-OF-FREEDOM SYSTEM: DUHAMEL'S INTEGRALS

Before we develop the general analysis of a multi-degree-of-freedom system, we shall consider the response of a single-degree system without damping, for which the equation of motion may be written as

$$M\ddot{U} + KU = P(t) \tag{13.1}$$

Alternatively, Eq. (13.1) may be expressed as

$$\ddot{U} + \omega^2 U = M^{-1}P(t) \tag{13.1a}$$

where $\omega^2 = K/M$. The right sides of Eqs. (13.1) and (13.1*a*) will be assumed to be known functions of time. We shall obtain solutions to Eq. (13.1*a*) for three different force-time variations, unit step function, unit impulse, and arbitrary variation of the force $P(t)$.

UNIT STEP FUNCTION

The solution of Eq. (13.1*a*) for $P(t)$ represented by a step function of amplitude P_0 (see Fig. 13.1) is given by

$$U(t) = C_1 \sin \omega t + C_2 \cos \omega t + M^{-1}\omega^{-2}P_0 \tag{13.2}$$

The first two terms with the constants C_1 and C_2 represent the complementary function, while the third term is the particular integral. Introducing the initial conditions on the displacement and velocity at $t = 0$, we can determine the constants C_1 and C_2. Hence we obtain

$$U(t) = \dot{U}(0)\omega^{-1} \sin \omega t + U(0) \cos \omega t + M^{-1}\omega^{-2}P_0(1 - \cos \omega t) \tag{13.3}$$

When the initial displacement $U(0)$ and velocity $\dot{U}(0)$ are both equal to zero and $P_0 = 1$, then

$$U(t) = A(t) = M^{-1}\omega^{-2}(1 - \cos \omega t) \tag{13.4}$$

The function $A(t)$ is defined as the *indicial admittance.* It represents the displacement response of a single-degree-of-freedom elastic system subjected to a force described by the unit step function.

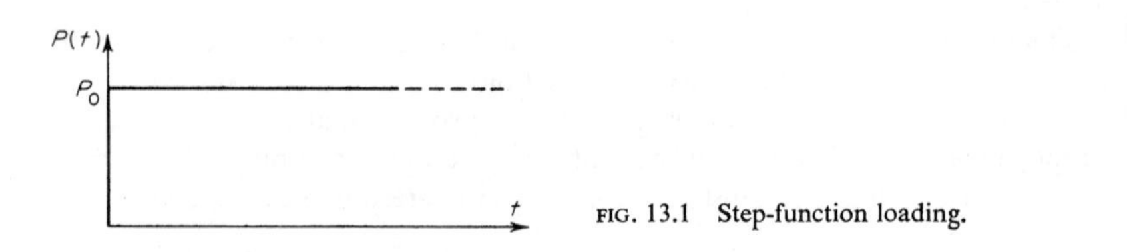

FIG. 13.1 Step-function loading.

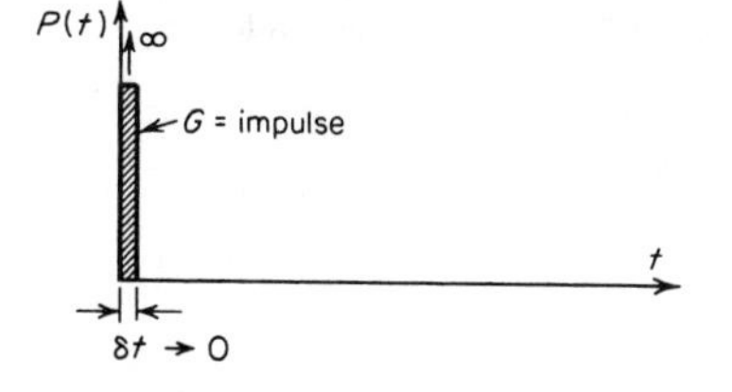

FIG. 13.2 Impulsive loading at $t = 0$.

IMPULSIVE FORCE

When an impulse G is applied to the system (see Fig. 13.2), the mass M acquires an initial velocity $\dot{U} = G/M$. Assuming that the impulse is applied at $t = 0$ and that $U(0) = 0$, we have

$$U(t) = \frac{G}{M\omega} \sin \omega t \tag{13.5}$$

Hence for a unit impulse

$$U(t) = h(t) = \frac{1}{M\omega} \sin \omega t \tag{13.6}$$

where $h(t)$ is the response due to a unit impulse. It is clear that $h(t)$ can be derived from $A(t)$ since

$$h(t) = \frac{dA(t)}{dt} \tag{13.7}$$

ARBITRARY VARIATION OF $P(t)$

An arbitrary variation of $P(t)$ may be assumed to consist of a series of step increases ΔP, as shown in Fig. 13.3. Since the system is linear, the response to the total force may be evaluated as the cumulative actions of the individual step increases. Thus using the result for a unit step increase, we have for the particular integral

$$\begin{aligned} U(t) &= P(0)A(t) + \sum_{\tau=\Delta\tau}^{\tau=t} \Delta P\, A(t-\tau) \\ &= P(0)A(t) + \sum_{\tau=\Delta\tau}^{\tau=t} \frac{\Delta P}{\Delta\tau} A(t-\tau)\, \Delta\tau \end{aligned}$$

Hence, in the limit,

$$U(t) = P(0)A(t) + \int_0^t \frac{dP}{d\tau} A(t-\tau)\, d\tau \tag{13.8}$$

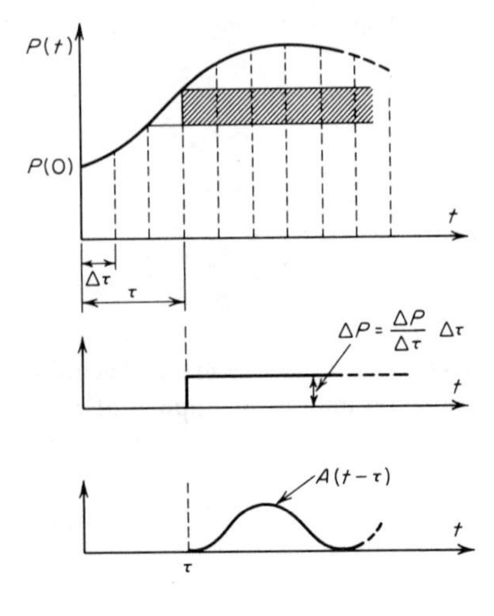

FIG. 13.3 Arbitrary function loading.

An equivalent expression may be obtained by integrating by parts

$$U(t) = P(0)A(t) + [P(\tau)A(t-\tau)]_{\tau=0}^{\tau=t} + \int_0^t P(\tau)A'(t-\tau)\,d\tau$$

$$= P(0)A(t) + P(t)A(0) - P(0)A(t) + \int_0^t P(\tau)A'(t-\tau)\,d\tau$$

$$= P(t)A(0) + \int_0^t P(\tau)A'(t-\tau)\,d\tau \tag{13.9}$$

where primes denote differentiation with respect to t. From Eq. (13.4) it follows that $A(0) = 0$, so that the particular integral becomes

$$U(t) = \int_0^t P(\tau)A'(t-\tau)\,d\tau$$

$$= M^{-1}\omega^{-1}\int_0^t P(\tau)\sin[\omega(t-\tau)]\,d\tau \tag{13.10}$$

and the complete solution for $U(t)$ is given by

$$U(t) = \dot{U}(0)\omega^{-1}\sin\omega t + U(0)\cos\omega t + M^{-1}\omega^{-1}\int_0^t P(\tau)\sin[\omega(t-\tau)]\,d\tau \tag{13.11}$$

The integral in Eq. (13.10) or (13.11) is called Duhamel's integral. This integral will be used in subsequent sections for determining response of multi-degree-of-freedom systems.

13.2 DYNAMIC RESPONSE OF AN UNCONSTRAINED (FREE) STRUCTURE

The equation of motion for an elastic undamped structure subjected to time-dependent disturbing forces $P(t)$ is

$$\mathbf{M}\ddot{\mathbf{U}} + \mathbf{K}\mathbf{U} = \mathbf{P}(t) \tag{13.12}$$

Since, in general, both $\mathbf{M}$ and $\mathbf{K}$ have nonzero off-diagonal coefficients, it is clear that the differential equations of motion represented by (13.12) are coupled and that the direct solution of these equations would be extremely unwieldy. Numerical solutions with time steps have been successfully developed for use on modern digital computers. These methods have the added advantage that the coefficients may be treated as time variables, which naturally opens up this technique to a great variety of problems. In this section only the classical method based on the so-called modal solutions will be discussed. Other methods are strictly in the realm of numerical computing techniques.

When the rigid-body modes $\mathbf{p}_0$ and elastic modes $\mathbf{p}_e$ are used to obtain a solution for $\mathbf{U}$, the equations of motion (13.12) become uncoupled, and their solutions can be easily obtained. We assume first that the displacements $\mathbf{U}$ are expressed as a linear combination of $\mathbf{p}_0$ and $\mathbf{p}_e$ with some appropriate time-dependent multipliers for the individual modes, so that

$$\mathbf{U} = \mathbf{p}\boldsymbol{\Phi} = [\mathbf{p}_0 \quad \mathbf{p}_e]\begin{bmatrix}\boldsymbol{\Phi}_0 \\ \boldsymbol{\Phi}_e\end{bmatrix} \tag{13.13}$$

where $$\boldsymbol{\Phi} = \begin{bmatrix}\boldsymbol{\Phi}_0 \\ \boldsymbol{\Phi}_e\end{bmatrix} \tag{13.14}$$

$$\boldsymbol{\Phi}_0 = \{\boldsymbol{\Phi}_1 \quad \boldsymbol{\Phi}_2 \quad \cdots \quad \boldsymbol{\Phi}_w\} \tag{13.15}$$

$$\boldsymbol{\Phi}_e = \{\boldsymbol{\Phi}_{w+1} \quad \boldsymbol{\Phi}_{w+2} \quad \cdots \quad \boldsymbol{\Phi}_{w+m}\} \tag{13.16}$$

are column matrices of unknown functions of time. Substituting Eq. (13.13) into the equation of motion (13.12), we find

$$\mathbf{M}(\mathbf{p}_0\ddot{\boldsymbol{\Phi}}_0 + \mathbf{p}_e\ddot{\boldsymbol{\Phi}}_e) + \mathbf{K}(\mathbf{p}_0\boldsymbol{\Phi}_0 + \mathbf{p}_e\boldsymbol{\Phi}_e) = \mathbf{P}(t) \tag{13.17}$$

Multiplying now Eq. (13.17) by $\mathbf{p}_0{}^T$, we have

$$\mathbf{p}_0{}^T\mathbf{M}\mathbf{p}_0\ddot{\boldsymbol{\Phi}}_0 + \mathbf{p}_0{}^T\mathbf{M}\mathbf{p}_e\ddot{\boldsymbol{\Phi}}_e + \mathbf{p}_0{}^T\mathbf{K}\mathbf{p}_0\boldsymbol{\Phi}_0 + \mathbf{p}_0{}^T\mathbf{K}\mathbf{p}_e\boldsymbol{\Phi}_e = \mathbf{p}_0{}^T\mathbf{P}(t) \tag{13.18}$$

When we use the orthogonality condition

$$\mathbf{p}_0{}^T\mathbf{M}\mathbf{p}_e = \mathbf{0} \tag{12.41}$$

and note that

$$\mathbf{p}_0{}^T\mathbf{K} = (\mathbf{K}\mathbf{p}_0)^T = \mathbf{0} \tag{13.19}$$

since the rigid body modes $\mathbf{p}_0$ do not produce any restoring forces $\mathbf{K}\mathbf{p}_0$, it follows then from Eq. (13.18) that

$$\mathbf{p}_0^T\mathbf{M}\mathbf{p}_0\ddot{\mathbf{\Phi}}_0 = \mathbf{p}_0^T\mathbf{P}(t) \tag{13.20}$$

Upon introducing the generalized force

$$\mathscr{P}_0(t) = \mathbf{p}_0^T\mathbf{P}(t) \tag{13.21}$$

and generalized mass

$$\mathscr{M}_0 = \mathbf{p}_0^T\mathbf{M}\mathbf{p}_0 \tag{13.22}$$

Eq. (13.20) becomes

$$\ddot{\mathbf{\Phi}}_0 = \mathscr{M}_0^{-1}\mathscr{P}_0(t) \tag{13.23}$$

Equation (13.23) can be integrated directly so that

$$\mathbf{\Phi}_0 = \mathscr{M}_0^{-1}\int_{\tau_2=0}^{\tau_2=t}\int_{\tau_1=0}^{\tau_1=\tau_2}\mathscr{P}_0(\tau_1)\,d\tau_1\,d\tau_2 + \mathbf{\Phi}_0(0) + \dot{\mathbf{\Phi}}_0(0)t \tag{13.24}$$

where $\mathbf{\Phi}_0(0)$ and $\dot{\mathbf{\Phi}}_0(0)$ are the integration constants, which may be determined from the initial conditions.

Multiplying now Eq. (13.17) by $\mathbf{p}_e^T$, we obtain

$$\mathbf{p}_e^T\mathbf{M}\mathbf{p}_0\ddot{\mathbf{\Phi}}_0 + \mathbf{p}_e^T\mathbf{M}\mathbf{p}_e\ddot{\mathbf{\Phi}}_e + \mathbf{p}_e^T\mathbf{K}\mathbf{p}_0\mathbf{\Phi}_0 + \mathbf{p}_e^T\mathbf{K}\mathbf{p}_e\mathbf{\Phi}_e = \mathbf{p}_e^T\mathbf{P}(t) \tag{13.25}$$

Using Eqs. (12.32), (12.33), (12.41), and (13.19), we can transform (13.25) into

$$\mathscr{M}_e\ddot{\mathbf{\Phi}}_e + \mathscr{K}_e\mathbf{\Phi}_e = \mathbf{p}_e^T\mathbf{P}(t) \tag{13.26}$$

Hence, after premultiplying by $\mathscr{M}_e^{-1}$, we obtain

$$\ddot{\mathbf{\Phi}}_e + \mathscr{M}_e^{-1}\mathscr{K}_e\mathbf{\Phi}_e = \mathscr{M}_e^{-1}\mathscr{P}_e(t) \tag{13.27}$$

where $\qquad \mathscr{P}_e(t) = \mathbf{p}_e^T\mathbf{P}(t) \qquad$ (13.28)

is the column matrix of generalized forces for the elastic modes $\mathbf{p}_e$. Noting also from Eq. (12.31*a*) that

$$\mathscr{M}_e^{-1}\mathscr{K}_e = \mathbf{\Omega}^2 \tag{13.29}$$

we see that Eq. (13.27) is transformed into

$$\ddot{\mathbf{\Phi}}_e + \mathbf{\Omega}^2\mathbf{\Phi}_e = \mathscr{M}_e^{-1}\mathscr{P}_e(t) \tag{13.30}$$

Since $\mathbf{\Omega}^2$ is a diagonal matrix, Eq. (13.30) constitutes a set of uncoupled second-order differential equations. A typical differential equation from (13.30) is

$$\ddot{\Phi}_{e_i} + \omega_{w+i}^2\Phi_{e_i} = \mathscr{M}_{e_i}^{-1}\mathscr{P}_{e_i}(t) \tag{13.31}$$

where i represents the row number in the column matrix $\mathscr{P}_e$. Solution to Eq. (13.31) can be obtained using Duhamel's integral from (13.11), so that

$$\Phi_{e_i} = \dot{\Phi}_{e_i}(0)\omega_{w+i}^{-1} \sin \omega_{w+i} t + \Phi_{e_i}(0) \cos \omega_{w+i} t + \omega_{w+i}^{-1}\mathscr{M}_{e_i}{}^{-1}\int_0^t \sin [\omega_{w+i}(t-\tau)]\mathscr{P}_{e_i}(\tau)\, d\tau \quad (13.32)$$

Equations (13.32) for $i = 1$ to m can be combined into a single matrix equation

$$\mathbf{\Phi}_e = \mathbf{\Omega}^{-1} \mathbf{sin}\,(\omega t)\dot{\mathbf{\Phi}}_e(0) + \mathbf{cos}\,(\omega t)\mathbf{\Phi}_e(0) + \mathbf{\Omega}^{-1}\mathscr{M}_e{}^{-1}\int_0^t \mathbf{sin}\,[\omega(t-\tau)]\mathscr{P}_e(\tau)\, d\tau \quad (13.33)$$

where

$$\mathbf{sin}\,(\omega t) = \lfloor \sin \omega_{w+1}t \quad \sin \omega_{w+2}t \quad \cdots \quad \sin \omega_{w+m}t \rfloor \quad (13.34)$$

$$\mathbf{cos}\,(\omega t) = \lfloor \cos \omega_{w+1}t \quad \cos \omega_{w+2}t \quad \cdots \quad \cos \omega_{w+m}t \rfloor \quad (13.35)$$

$$\mathbf{sin}\,[\omega(t-\tau)] = \lfloor \sin [\omega_{w+1}(t-\tau)] \quad \sin [\omega_{w+2}(t-\tau)] \quad \cdots \quad \sin [\omega_{w+m}(t-\tau)] \rfloor \quad (13.36)$$

Substituting Eqs. (13.24) and (13.33) into (13.13), we find the displacements $\mathbf{U}$ to be given by

$$\mathbf{U} = \mathbf{p}_0\mathbf{\Phi}_0(0) + \mathbf{p}_0\dot{\mathbf{\Phi}}_0(0)t + \mathbf{p}_e\mathbf{\Omega}^{-1} \mathbf{sin}\,(\omega t)\dot{\mathbf{\Phi}}_e(0) + \mathbf{p}_e \mathbf{cos}\,(\omega t)\mathbf{\Phi}_e(0) + \mathbf{p}_0\mathscr{M}_0{}^{-1}\int_{\tau_2=0}^{\tau_2=t}\int_{\tau_1=0}^{\tau_1=\tau_2} \mathscr{P}_0(\tau_1)\, d\tau_1\, d\tau_2 + \mathbf{p}_e\mathbf{\Omega}^{-1}\mathscr{M}_e{}^{-1}\int_0^t \mathbf{sin}\,[\omega(t-\tau)]\mathscr{P}_e(\tau)\, d\tau \quad (13.37)$$

It remains now to determine the initial values of $\mathbf{\Phi}$ and $\dot{\mathbf{\Phi}}$ at $t = 0$ in terms of the initial values of the displacements $\mathbf{U}$ and velocities $\dot{\mathbf{U}}$. From Eq. (13.13) we have

$$\mathbf{U}(0) = \mathbf{p\Phi}(0) = \mathbf{p}_0\mathbf{\Phi}_0(0) + \mathbf{p}_e\mathbf{\Phi}_e(0) \quad (13.38)$$

and $$\dot{\mathbf{U}}(0) = \mathbf{p}\dot{\mathbf{\Phi}}(0) = \mathbf{p}_0\dot{\mathbf{\Phi}}_0(0) + \mathbf{p}_e\dot{\mathbf{\Phi}}_e(0) \quad (13.39)$$

so that $$\mathbf{\Phi}(0) = \mathbf{p}^{-1}\mathbf{U}(0) \quad (13.40)$$

and $$\dot{\mathbf{\Phi}}(0) = \mathbf{p}^{-1}\dot{\mathbf{U}}(0) \quad (13.41)$$

To avoid using the inverse of the modal matrix $\mathbf{p}$ we may premultiply Eq. (13.38) by $\mathbf{p}_0{}^T\mathbf{M}$, so that

$$\mathbf{p}_0{}^T\mathbf{MU}(0) = \mathbf{p}_0{}^T\mathbf{Mp}_0\mathbf{\Phi}_0(0) + \mathbf{p}_0{}^T\mathbf{Mp}_e\mathbf{\Phi}_e(0) \quad (13.42)$$

which in view of the orthogonality relations (12.41) and (12.42) becomes

$$\mathbf{p}_0{}^T\mathbf{MU}(0) = \mathscr{M}_0\mathbf{\Phi}_0(0)$$

or $$\mathbf{\Phi}_0(0) = \mathscr{M}_0{}^{-1}\mathbf{p}_0{}^T\mathbf{MU}(0) \quad (13.43)$$

Although we still have an inversion to perform on $\mathscr{M}_0$, this is a much easier task than the calculation of $\mathbf{p}^{-1}$. The generalized mass matrix $\mathscr{M}_0$ is only a 6×6 matrix for the three-dimensional structure; furthermore, if required, this matrix can also be diagonalized by a suitable choice of reference displacements.

To obtain the initial values of $\mathbf{\Phi}_e$ we premultiply Eq. (13.38) by $\mathbf{p}_e^T\mathbf{M}$ so that

$$\mathbf{p}_e^T\mathbf{M}\mathbf{U}(0) = \mathbf{p}_e^T\mathbf{M}\mathbf{p}_0\mathbf{\Phi}_0(0) + \mathbf{p}_e^T\mathbf{M}\mathbf{p}_e\mathbf{\Phi}_e(0) \tag{13.44}$$

Using the orthogonality relations (12.33) and (12.41), we obtain

$$\mathbf{p}_e^T\mathbf{M}\mathbf{U}(0) = \mathscr{M}_e\mathbf{\Phi}_e(0)$$

or $$\mathbf{\Phi}_e(0) = \mathscr{M}_e^{-1}\mathbf{p}_e^T\mathbf{M}\mathbf{U}(0) \tag{13.45}$$

Here the calculation of the inverse of $\mathscr{M}_e$ presents no difficulty since the generalized mass $\mathscr{M}_e$ is a diagonal matrix. In an identical manner we can determine $\dot{\mathbf{\Phi}}_0(0)$ and $\dot{\mathbf{\Phi}}_e(0)$ from Eq. (13.39) and obtain

$$\dot{\mathbf{\Phi}}_0(0) = \mathscr{M}_0^{-1}\mathbf{p}_0^T\mathbf{M}\dot{\mathbf{U}}(0) \tag{13.46}$$

and $$\dot{\mathbf{\Phi}}_e(0) = \mathscr{M}_e^{-1}\mathbf{p}_e^T\mathbf{M}\dot{\mathbf{U}}(0) \tag{13.47}$$

Having found the initial values of $\mathbf{\Phi}$ and $\dot{\mathbf{\Phi}}$, we can now write the displacements $\mathbf{U}$ from (13.37) as

$$\begin{aligned}\mathbf{U} = {} & \mathbf{p}_0\mathscr{M}_0^{-1}\mathbf{p}_0^T\mathbf{M}\mathbf{U}(0) + \mathbf{p}_0\mathscr{M}_0^{-1}\mathbf{p}_0^T\mathbf{M}\dot{\mathbf{U}}(0)t + \mathbf{p}_e\mathbf{\Omega}^{-1}\,\mathbf{sin}\,(\omega t)\mathscr{M}_e^{-1}\mathbf{p}_e^T\mathbf{M}\dot{\mathbf{U}}(0)\\ & + \mathbf{p}_e\,\mathbf{cos}\,(\omega t)\mathscr{M}_e^{-1}\mathbf{p}_e^T\mathbf{M}\mathbf{U}(0) + \mathbf{p}_0\mathscr{M}_0^{-1}\int_{\tau_2=0}^{\tau_2=t}\int_{\tau_1=0}^{\tau_1=\tau_2}\mathscr{P}_0(\tau_1)\,d\tau_1\,d\tau_2\\ & + \mathbf{p}_e\mathbf{\Omega}^{-1}\mathscr{M}_e^{-1}\int_0^t \mathbf{sin}\,[\omega(t-\tau)]\mathscr{P}_e(\tau)\,d\tau\end{aligned} \tag{13.48}$$

For complicated time variations of $\mathscr{P}_0$ and $\mathscr{P}_e$ forces the integrals in (13.48) must be evaluated numerically; however, for most practical calculations the applied forces are represented by simplified functions for which the required integrals can be explicitly determined. Duhamel's integral for the last term in (13.48) has been evaluated for some typical variations of $\mathscr{P}_e$ with time, and the results are presented in Sec. 13.6.

The displacements obtained from (13.48) are used to determine the dynamic forces and stresses in the structure. Since the rigid-body components of the displacements do not contribute to the stresses, the terms premultiplied by $\mathbf{p}_0$ may be omitted when only stresses are required. Examining the form of Eq. (13.48), we may note that the solution, in accordance with (13.13), is a finite series in terms of the rigid and elastic modes and that the terms with the elastic modes are inversely proportional to the corresponding frequencies. In practical applications, we do not include all the modes, and usually only the first four or five modes are retained in the solution. Naturally no general recommendation can be made as to the exact number of modes to be used, and every solution must be examined individually to determine the relative contributions from the rejected modes.

It has been demonstrated[53] that more rapid convergence of the solution for the displacements **U** is obtained if we first determine inertia forces and then use them as the applied static loading. To determine the inertia forces we use

$$\mathbf{P}_i = -\mathbf{M}\ddot{\mathbf{U}} \tag{13.49}$$

Evaluating $\ddot{\mathbf{U}}$ from Eq. (13.48) and (13.30) gives

$$\begin{aligned}\mathbf{P}_i = {} & \mathbf{M}\mathbf{p}_e\mathbf{\Omega}\,\mathbf{sin}\,(\omega t)\mathcal{M}_e^{-1}\mathbf{p}_e^T\mathbf{M}\dot{\mathbf{U}}(0) + \mathbf{M}\mathbf{p}_e\mathbf{\Omega}^2\,\mathbf{cos}\,(\omega t)\mathcal{M}_e^{-1}\mathbf{p}_e^T\mathbf{M}\mathbf{U}(0) \\ & - \mathbf{M}\mathbf{p}_0\mathcal{M}_0^{-1}\mathcal{P}_0 - \mathbf{M}\mathbf{p}_e\mathcal{M}_e^{-1}\mathcal{P}_e + \mathbf{M}\mathbf{p}_e\mathbf{\Omega}\mathcal{M}_e^{-1}\int_0^t \mathbf{sin}\,[\omega(t-\tau)]\mathcal{P}_e(\tau)\,d\tau\end{aligned} \tag{13.50}$$

The resultant effective static loading is obtained from the sum $\mathbf{P} + \mathbf{P}_i$, which can then be used in the equilibrium equation for the matrix displacement method

$$\mathbf{K}\mathbf{U} = \mathbf{P} + \mathbf{P}_i \tag{13.51}$$

The displacements calculated from Eq. (13.51) characterize the dynamic behavior of the structure under dynamic loading $\mathbf{P}(t)$.

The term $-\mathbf{M}\mathbf{p}_0\mathcal{M}_0^{-1}\mathcal{P}_0$ in (13.50) represents the rigid-body inertia loading, while the remaining terms give the additional elastic-inertia loads on the flexible structure. Because of this separation of rigid- and flexible-body loading, fewer elastic modes $\mathbf{p}_e$ are necessary to achieve the same accuracy as compared with the direct solution for the displacements from Eq. (13.48). Normally all preliminary design calculations on aircraft and missile structures are carried out for the rigid structure; only in the subsequent design calculations are the flexibility effects included. Consequently, the use of Eqs. (13.50) and (13.51) is a logical choice for dynamic-response calculations, and, furthermore, it is consistent with present design practice.

13.3 RESPONSE RESULTING FROM IMPULSIVE FORCES

We shall assume that the system is subjected to impulsive forces at time $t = 0$. These forces (impulses) will be represented by the column matrix **G**. The initial velocities imposed on the system are therefore given by the relation

$$\mathbf{M}\dot{\mathbf{U}}(0) = \mathbf{G}$$

$$\text{or} \qquad \dot{\mathbf{U}}(0) = \mathbf{M}^{-1}\mathbf{G} \tag{13.52}$$

Substituting this expression for $\dot{\mathbf{U}}(0)$ in Eq. (13.48), setting $\mathbf{U}(0) = \mathbf{0}$, and noting that there are no forces acting on the system for $t > 0$, we obtain the following equation for the displacements caused by impulsive forces:

$$\mathbf{U} = \mathbf{p}_0 \mathscr{M}_0^{-1} \mathbf{p}_0^T \mathbf{G} t + \mathbf{p}_e \mathbf{\Omega}^{-1} \sin(\omega t) \mathscr{M}_e^{-1} \mathbf{p}_e^T \mathbf{G} \tag{13.53}$$

The first term in (13.53) represents the rigid-body motion, while the second term gives the elastic displacements of the system.

13.4 DYNAMIC RESPONSE OF A CONSTRAINED STRUCTURE

For a constrained or overconstrained structure all rigid-body degrees of freedom are suppressed. Only the elastic modes $\mathbf{p}_e$ are present, and consequently the expressions for the dynamic displacements and inertia forces for such structures are considerably simpler than those for unconstrained structures.

Using Eqs. (13.48) and (13.50) for constrained structures, we obtain expressions for the displacements and inertia forces

$$\mathbf{U} = \mathbf{p}_e \mathbf{\Omega}^{-1} \sin(\omega t) \mathscr{M}_e^{-1} \mathbf{p}_e^T \mathbf{M} \dot{\mathbf{U}}(0) + \mathbf{p}_e \cos(\omega t) \mathscr{M}_e^{-1} \mathbf{p}_e^T \mathbf{M} \mathbf{U}(0) + \mathbf{p}_e \mathbf{\Omega}^{-1} \mathscr{M}_e^{-1} \int_0^t \sin[\omega(t-\tau)] \mathscr{P}_e(\tau)\, d\tau \tag{13.54}$$

and

$$\mathbf{P}_i = \mathbf{M} \mathbf{p}_e \mathbf{\Omega} \sin(\omega t) \mathscr{M}_e^{-1} \mathbf{p}_e^T \mathbf{M} \dot{\mathbf{U}}(0) + \mathbf{M} \mathbf{p}_e \mathbf{\Omega}^2 \cos(\omega t) \mathscr{M}_e^{-1} \mathbf{p}_e^T \mathbf{M} \mathbf{U}(0) - \mathbf{M} \mathbf{p}_e \mathscr{M}_e^{-1} \mathscr{P}_e + \mathbf{M} \mathbf{p}_e \mathbf{\Omega} \mathscr{M}_e^{-1} \int_0^t \sin[\omega(t-\tau)] \mathscr{P}_e(\tau)\, d\tau \tag{13.55}$$

13.5 STEADY-STATE HARMONIC MOTION

If the applied forces are harmonic, they may be represented symbolically as $\mathbf{P} = \bar{\mathbf{P}} e^{i\omega t}$, and the steady-state solution for the displacements $\mathbf{U}$ has the form

$$\mathbf{U} = \bar{\mathbf{U}} e^{i\omega t} \tag{13.56}$$

When Eq. (13.56) is used, the equation of motion (13.12) becomes

$$(-\omega^2 \mathbf{M} + \mathbf{K}) \bar{\mathbf{U}} = \bar{\mathbf{P}} \tag{13.57}$$

and hence $$\bar{\mathbf{U}} = (-\omega^2 \mathbf{M} + \mathbf{K})^{-1} \bar{\mathbf{P}} \tag{13.58}$$

provided, of course, that $|-\omega^2 \mathbf{M} + \mathbf{K}| \neq 0$. The inversion in (13.58) can be avoided if $\mathbf{U}$ is expressed as a linear combination of the modes $\mathbf{p}_0$ and $\mathbf{p}_e$. Taking

$$\bar{\mathbf{U}} = \mathbf{p} \mathbf{\Phi} = \mathbf{p}_0 \mathbf{\Phi}_0 + \mathbf{p}_e \mathbf{\Phi}_e \tag{13.13}$$

and substituting in Eq. (13.57) gives

$$-\omega^2 \mathbf{M} \mathbf{p}_0 \mathbf{\Phi}_0 - \omega^2 \mathbf{M} \mathbf{p}_e \mathbf{\Phi}_e + \mathbf{K} \mathbf{p}_0 \mathbf{\Phi}_0 + \mathbf{K} \mathbf{p}_e \mathbf{\Phi}_e = \bar{\mathbf{P}} \tag{13.59}$$

Premultiplying Eq. (13.59) by $\mathbf{p}_0^T$ and using orthogonality relations yields

$$-\omega^2 \mathbf{p}_0^T \mathbf{M} \mathbf{p}_0 \boldsymbol{\Phi}_0 = \mathbf{p}_0^T \bar{\mathbf{P}} \tag{13.60}$$

hence $$\boldsymbol{\Phi}_0 = -\omega^{-2} \mathscr{M}_0^{-1} \mathbf{p}_0^T \bar{\mathbf{P}} \tag{13.61}$$

Similarly, premultiplying Eq. (13.59) by $\mathbf{p}_e^T$ gives

$$-\omega^2 \mathbf{p}_e^T \mathbf{M} \mathbf{p}_e \boldsymbol{\Phi}_e + \mathbf{p}_e^T \mathbf{K} \mathbf{p}_e \boldsymbol{\Phi}_e = \mathbf{p}_e^T \bar{\mathbf{P}} \tag{13.62}$$

Noting from Eq. (12.31*a*) that

$$\mathbf{p}_e^T \mathbf{K} \mathbf{p}_e = \mathscr{K}_e = \boldsymbol{\Omega}^2 \mathscr{M}_e \tag{13.63}$$

we see that Eq. (13.62) becomes

$$-\omega^2 \mathscr{M}_e \boldsymbol{\Phi}_e + \boldsymbol{\Omega}^2 \mathscr{M}_e \boldsymbol{\Phi}_e = \mathbf{p}_e^T \bar{\mathbf{P}}$$

and $$\boldsymbol{\Phi}_e = \mathscr{M}_e^{-1} (\boldsymbol{\Omega}^2 - \omega^2 \mathbf{I})^{-1} \mathbf{p}_e^T \bar{\mathbf{P}} \tag{13.64}$$

Using Eqs. (13.61) and (13.64) in (13.13), we find that the expression for the amplitudes of displacements becomes

$$\bar{\mathbf{U}} = -\omega^{-2} \mathbf{p}_0 \mathscr{M}_0^{-1} \mathbf{p}_0^T \bar{\mathbf{P}} + \mathbf{p}_e \mathscr{M}_e^{-1} (\boldsymbol{\Omega}^2 - \omega^2 \mathbf{I})^{-1} \mathbf{p}_e^T \bar{\mathbf{P}} \tag{13.65}$$

The matrix $\boldsymbol{\Omega}^2 - \omega^2 \mathbf{I}$ is diagonal, and it becomes singular whenever ω^2_{w+i} in $\boldsymbol{\Omega}^2$ is equal to ω^2. When this condition occurs, the forcing frequency ω is equal to one of the natural frequencies of the system.

13.6 DUHAMEL'S INTEGRALS FOR TYPICAL FORCING FUNCTIONS

In practical calculations the applied loading $\mathbf{P}(t)$ and hence the generalized loads $\mathscr{P}_e$ are usually approximated by simple functions for which the Duhamel's integrals can be determined exactly. For easy reference the results for most commonly employed forcing functions $\mathscr{P}_e$ are summarized in Table 13.1.

13.7 DYNAMIC RESPONSE TO FORCED DISPLACEMENTS: RESPONSE TO EARTHQUAKES

Let us suppose that the displacements $\mathbf{U}$ on a structure are partitioned into two submatrices such that

$$\mathbf{U} = \begin{bmatrix} \mathbf{U}_x \\ \mathbf{U}_y \end{bmatrix} \tag{13.66}$$

and the displacements $\mathbf{U}_y$ are forced to vary in a defined manner. Thus the displacements $\mathbf{U}_y$ are prescribed functions of time. This problem is of practical importance in calculating the response of a building to earthquake movement. The earthquake causes lateral movement and rotations of the foundation,

TABLE 13.1 RESPONSE FUNCTIONS FOR UNDAMPED SYSTEMS

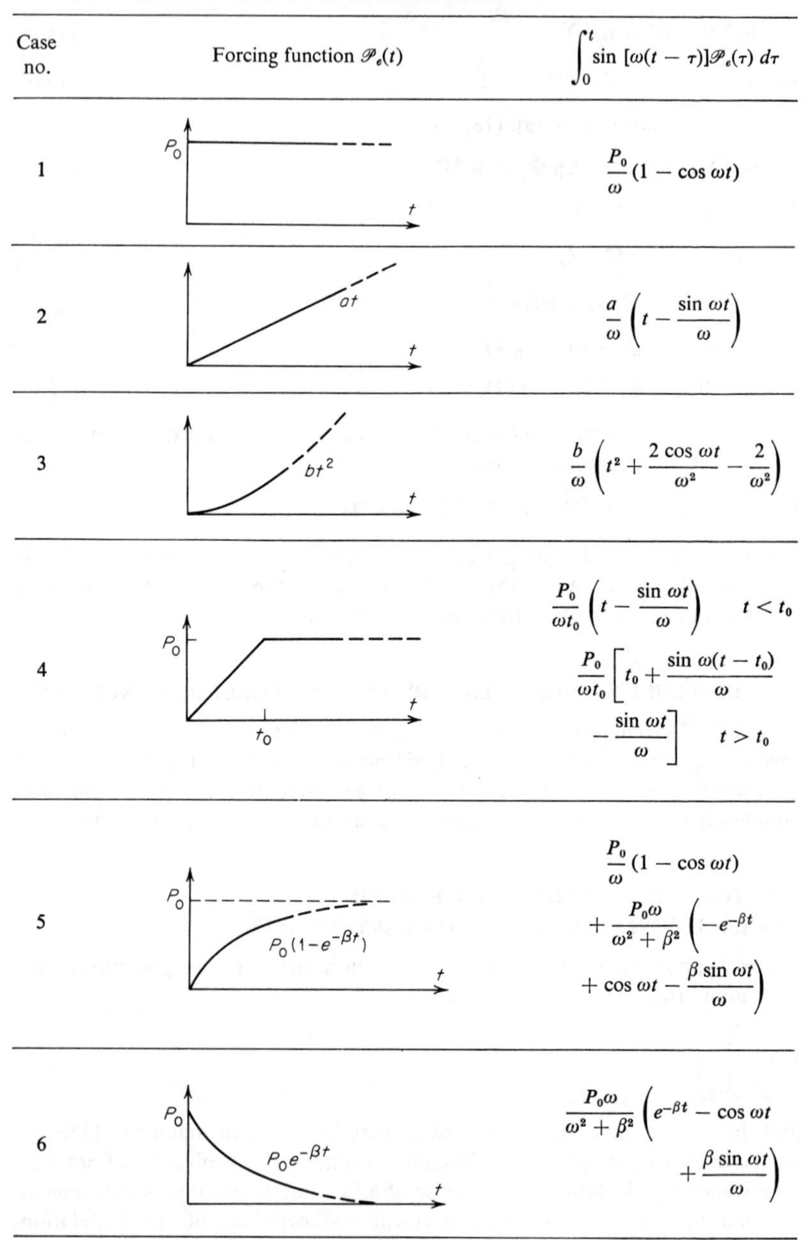

Case no.	Forcing function $\mathscr{P}_e(t)$	$\int_0^t \sin [\omega(t-\tau)]\mathscr{P}_e(\tau)\, d\tau$
1	P_0	$\dfrac{P_0}{\omega}(1-\cos \omega t)$
2	at	$\dfrac{a}{\omega}\left(t-\dfrac{\sin \omega t}{\omega}\right)$
3	bt^2	$\dfrac{b}{\omega}\left(t^2+\dfrac{2\cos \omega t}{\omega^2}-\dfrac{2}{\omega^2}\right)$
4	P_0, t_0	$\dfrac{P_0}{\omega t_0}\left(t-\dfrac{\sin \omega t}{\omega}\right) \quad t < t_0$ $\dfrac{P_0}{\omega t_0}\left[t_0+\dfrac{\sin \omega(t-t_0)}{\omega}-\dfrac{\sin \omega t}{\omega}\right] \quad t > t_0$
5	$P_0(1-e^{-\beta t})$	$\dfrac{P_0}{\omega}(1-\cos \omega t)+\dfrac{P_0\omega}{\omega^2+\beta^2}\left(-e^{-\beta t}+\cos \omega t-\dfrac{\beta \sin \omega t}{\omega}\right)$
6	$P_0e^{-\beta t}$	$\dfrac{P_0\omega}{\omega^2+\beta^2}\left(e^{-\beta t}-\cos \omega t+\dfrac{\beta \sin \omega t}{\omega}\right)$

TABLE 13.1 RESPONSE FUNCTIONS FOR UNDAMPED SYSTEMS (*Continued*)

Case no.	Forcing function $\mathscr{P}_e(t)$	$\int_0^t \sin[\omega(t-\tau)]\mathscr{P}_e(\tau)\,d\tau$
7	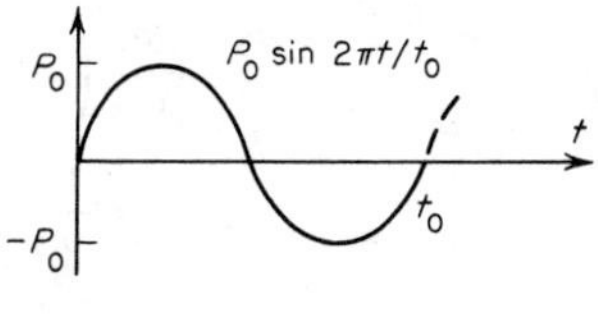	$\frac{P_0 t_0}{\omega^2 t_0^2 - 4\pi^2}\left(\omega t_0 \sin 2\pi \frac{t}{t_0} - 2\pi \sin \omega t\right)$
8	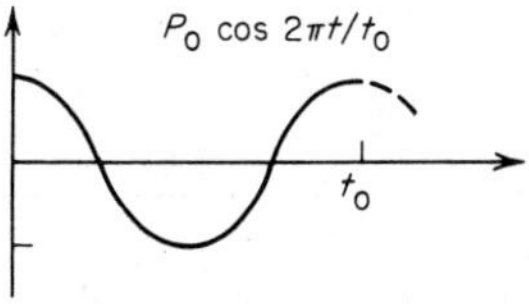	$\frac{P_0 \omega t_0^2}{\omega^2 t_0^2 - 4\pi^2} \times \left(\cos 2\pi \frac{t}{t_0} - \cos \omega t\right)$
9	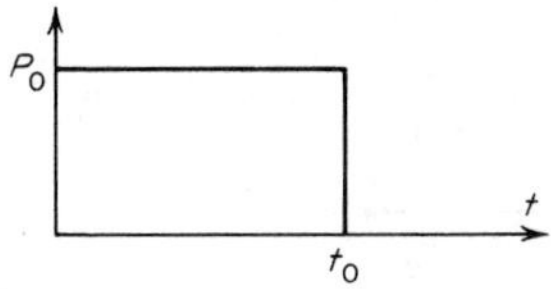	$\frac{P_0}{\omega}(1 - \cos \omega t) \qquad t < t_0$ $\frac{P_0}{\omega}[\cos \omega(t - t_0) - \cos \omega t] \qquad t > t_0$
10	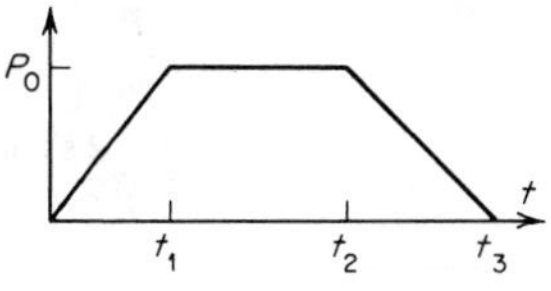	See case 4 for $t < t_2$; $t_1 = t_0$ $\frac{P_0}{\omega^2 t_1}[\omega t_1 + \sin \omega(t - t_1) - \sin \omega t] - \frac{P_0}{\omega^2(t_3 - t_2)} \times [\omega(t - t_2) - \sin \omega(t - t_2)] \qquad t_2 < t < t_3$ $\frac{P_0}{\omega}\left[\frac{\sin \omega(t - t_1)}{\omega t_1} - \frac{\sin \omega t}{\omega t_1} - \frac{\sin \omega(t - t_3)}{\omega(t_3 - t_2)} + \frac{\sin \omega(t - t_2)}{\omega(t_3 - t_2)}\right] \qquad t > t_3$

TABLE 13.1 RESPONSE FUNCTIONS FOR UNDAMPED SYSTEMS (*Continued*)

Case no.	Forcing function $\mathscr{P}_e(t)$	$\int_0^t \sin [\omega(t-\tau)]\mathscr{P}_e(\tau)\, d\tau$
11	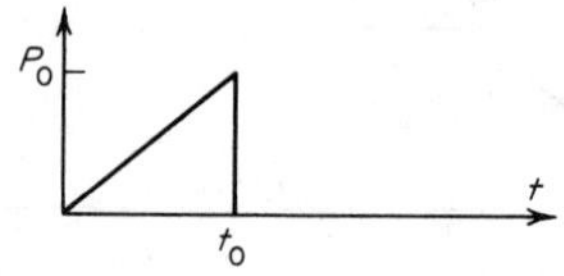	$\frac{P_0}{\omega t_0}\left(t - \frac{\sin \omega t}{\omega}\right) \quad t < t_0$ $\frac{P_0}{\omega t_0}\left[t_0 \cos \omega(t-t_0) + \frac{\sin \omega(t-t_0)}{\omega} - \frac{\sin \omega t}{\omega}\right] \quad t > t_0$
12	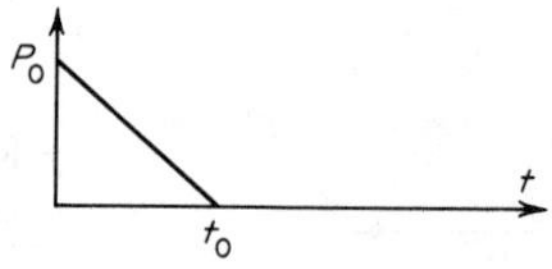	$\frac{P_0}{\omega}\left(1 - \cos \omega t - \frac{t}{t_0} + \frac{\sin \omega t}{\omega t_0}\right) \quad t < t_0$ $\frac{P_0}{\omega}\left[-\cos \omega t - \frac{\sin \omega(t-t_0)}{\omega t_0} + \frac{\sin \omega t}{\omega t_0}\right] \quad t > t_0$
13	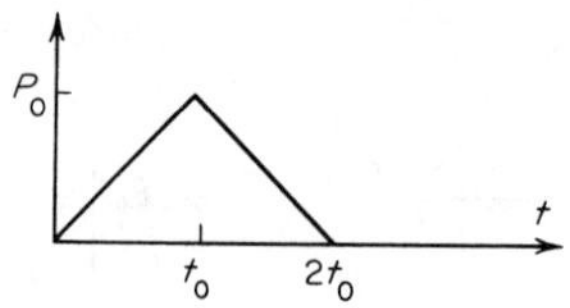	$\frac{P_0}{\omega t_0}\left(t - \frac{\sin \omega t}{\omega}\right) \quad t < t_0$ $\frac{P_0}{\omega t_0}\left[2t_0 - t + \frac{2 \sin \omega(t-t_0)}{\omega} - \frac{\sin \omega t}{\omega}\right] \quad t_0 < t < 2t_0$ $\frac{P_0}{\omega^2 t_0}[2 \sin \omega(t-t_0) - \sin \omega(t-2t_0) - \sin \omega t] \quad t > 2t_0$

TABLE 13.1 RESPONSE FUNCTIONS FOR UNDAMPED SYSTEMS (*Continued*)

Case no.	Forcing function $\mathscr{P}_e(t)$	$\int_0^t \sin [\omega(t-\tau)]\mathscr{P}_e(\tau)\,d\tau$
14	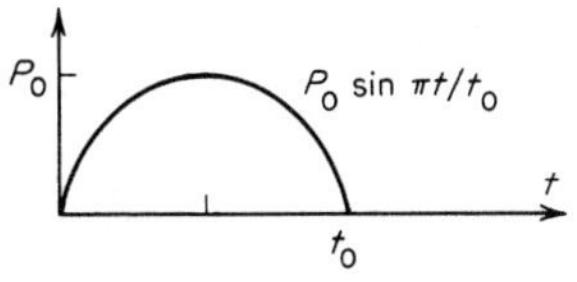 $P_0 \sin \pi t/t_0$ (graph: peak P_0, zero at t_0)	$\dfrac{P_0 t_0}{\omega^2 t_0^2 - \pi^2}\left(\omega t_0 \sin \pi \dfrac{t}{t_0} - \pi \sin \omega t\right) \quad t < t_0$ $\dfrac{-P_0 \pi t_0}{\omega^2 t_0^2 - \pi^2}[\sin \omega(t - t_0) + \sin \omega t] \quad t > t_0$
15	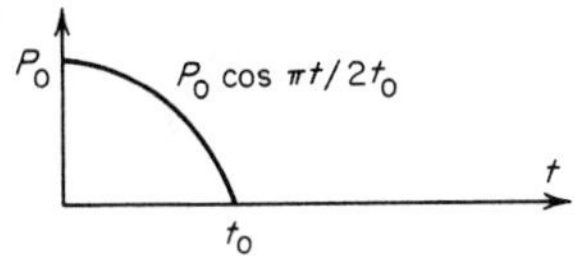 $P_0 \cos \pi t/2t_0$ (graph: starts at P_0, zero at t_0)	$\dfrac{4P_0 \omega t_0^2}{4\omega^2 t_0^2 - \pi^2}\left(\cos \dfrac{\pi t}{2t_0} - \cos \omega t\right) \quad t < t_0$ $\dfrac{-4P_0 t_0^2}{4\omega^2 t_0^2 - \pi^2}\left[\dfrac{\pi}{2t_0} \sin \omega(t - t_0) + \omega \cos \omega t\right] \quad t > t_0$
16	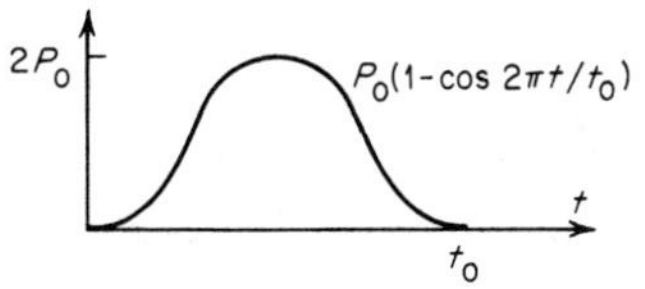 $P_0(1 - \cos 2\pi t/t_0)$ (graph: peak $2P_0$, zero at t_0)	$\dfrac{P_0}{\omega}(1 - \cos \omega t) - \dfrac{P_0 \omega t_0^2}{\omega^2 t_0^2 - 4\pi^2}\left(\cos \dfrac{2\pi t}{t_0} - \cos \omega t\right) \quad t < t_0$ $\dfrac{P_0}{\omega}\left\{\cos \omega(t - t_0) - \cos \omega t - \dfrac{\omega^2 t_0^2}{\omega^2 t_0^2 - 4\pi^2}[\cos \omega(t - t_0) - \cos \omega t]\right\} \quad t > t_0$

TABLE 13.1 RESPONSE FUNCTIONS FOR UNDAMPED SYSTEMS *(Continued)*

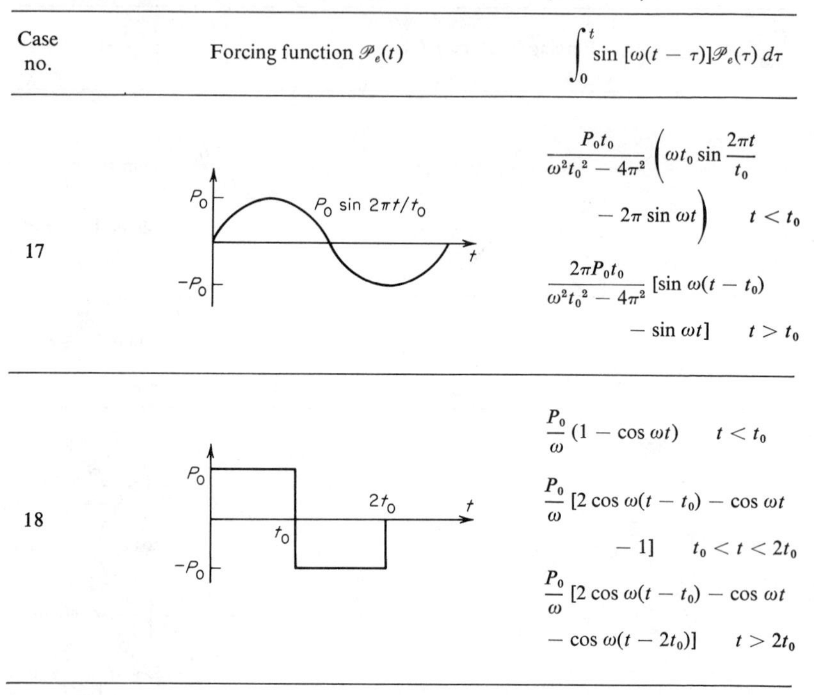

Case no.	Forcing function $\mathscr{P}_e(t)$	$\int_0^t \sin[\omega(t-\tau)]\mathscr{P}_e(\tau)\,d\tau$
17	$P_0 \sin 2\pi t/t_0$	$\frac{P_0 t_0}{\omega^2 t_0^2 - 4\pi^2}\left(\omega t_0 \sin \frac{2\pi t}{t_0} - 2\pi \sin \omega t\right) \quad t < t_0$ $\frac{2\pi P_0 t_0}{\omega^2 t_0^2 - 4\pi^2}[\sin \omega(t - t_0) - \sin \omega t] \quad t > t_0$
18		$\frac{P_0}{\omega}(1 - \cos \omega t) \quad t < t_0$ $\frac{P_0}{\omega}[2 \cos \omega(t - t_0) - \cos \omega t - 1] \quad t_0 < t < 2t_0$ $\frac{P_0}{\omega}[2 \cos \omega(t - t_0) - \cos \omega t - \cos \omega(t - 2t_0)] \quad t > 2t_0$

forcing the displacements on the structure to follow the earthquake movements. There are also many other practical applications where the structure is subjected to forced displacements rather than to applied forces.

By using the partitioning in Eq. (13.66) the equations of motion can accordingly be partitioned into

$$\begin{bmatrix} \mathbf{M}_{xx} & \mathbf{M}_{xy} \\ \mathbf{M}_{yx} & \mathbf{M}_{yy} \end{bmatrix} \begin{bmatrix} \ddot{\mathbf{U}}_x \\ \ddot{\mathbf{U}}_y \end{bmatrix} + \begin{bmatrix} \mathbf{K}_{xx} & \mathbf{K}_{xy} \\ \mathbf{K}_{yx} & \mathbf{K}_{yy} \end{bmatrix} \begin{bmatrix} \mathbf{U}_x \\ \mathbf{U}_y \end{bmatrix} = \begin{bmatrix} \mathbf{0} \\ \mathbf{P}_y \end{bmatrix} \tag{13.67}$$

where $\mathbf{P}_y$ represents the column matrix of unknown forces causing the displacements $\mathbf{U}_y$. Taking the first equation from (13.67), we have

$$\mathbf{M}_{xx}\ddot{\mathbf{U}}_x + \mathbf{M}_{xy}\ddot{\mathbf{U}}_y + \mathbf{K}_{xx}\mathbf{U}_x + \mathbf{K}_{xy}\mathbf{U}_y = \mathbf{0}$$

or $$\mathbf{M}_{xx}\ddot{\mathbf{U}}_x + \mathbf{K}_{xx}\mathbf{U}_x = \bar{\mathbf{P}}_x \tag{13.68}$$

where $$\bar{\mathbf{P}}_x = -\mathbf{M}_{xy}\ddot{\mathbf{U}}_y - \mathbf{K}_{xy}\mathbf{U}_y \tag{13.69}$$

The solution to the above modified equation of motion can be obtained in terms of the eigenmodes and frequencies of a constrained system for which

$\mathbf{U}_y = \mathbf{0}$. Assuming that $\mathbf{U}_y = \mathbf{0}$ eliminates all the rigid-body degrees of freedom and that $\mathbf{U}_x(0) = \dot{\mathbf{U}}_x(0) = \mathbf{0}$, we obtain from Eq. (13.54)

$$\mathbf{U}_x = \bar{\mathbf{p}}_e \bar{\mathbf{\Omega}}^{-1} \bar{\mathscr{M}}_e^{-1} \int_0^t \mathbf{sin}\,[\bar{\omega}(t - \tau)] \bar{\mathscr{P}}_e(\tau)\, d\tau \tag{13.70}$$

where $\bar{\mathbf{p}}_e$ denotes the eigenmodes of the constrained system and $\bar{\omega}$ the corresponding circular frequencies. Furthermore

$$\bar{\mathbf{\Omega}} = \lceil \bar{\omega}_1 \quad \bar{\omega}_2 \quad \cdots \quad \bar{\omega}_m \rfloor \tag{13.71}$$

$$\bar{\mathscr{M}}_e = \bar{\mathbf{p}}_e^T \mathbf{M}_{xx} \bar{\mathbf{p}}_e \tag{13.72}$$

$$\bar{\mathscr{P}}_e = \bar{\mathbf{p}}_e^T \bar{\mathbf{P}}_x = -\mathbf{p}_e^T (\mathbf{M}_{xy} \ddot{\mathbf{U}}_y + \mathbf{K}_{xy} \mathbf{U}_y) \tag{13.73}$$

Once the displacements $\mathbf{U}_x$ have been calculated, we can compute from Eq. (13.67) the forces $\mathbf{P}_y$ necessary to cause the required displacements $\mathbf{U}_y$.

13.8 DETERMINATION OF FREQUENCIES AND MODES OF UNCONSTRAINED (FREE) STRUCTURES USING EXPERIMENTAL DATA FOR THE CONSTRAINED STRUCTURES

The experimental difficulties of determining mode shapes and frequencies for an unconstrained structure hinge around the problem of supporting the structure during the vibration tests in a manner which will not interfere with the development of free-free modes. This, of course, is very difficult for large structures.

In this section we shall describe an analytical method of determining frequencies and mode shapes for the vibrations of an unconstrained structure using the experimental vibration data for the same structure supported rigidly on the ground. For example, an aircraft can be vibration-tested while supported rigidly at several points on the ground and the experimental results thus obtained can be applied for determining the vibrational characteristics in flight when the structure is not subjected to any external constraints.

We shall assume that the displacements on the unconstrained structure are partitioned into $\mathbf{U}_x$ and $\mathbf{U}_y$. Furthermore, we shall assume that vibration tests to determine frequencies and mode shapes can be performed while supporting the structure in such a manner that $\mathbf{U}_y = \mathbf{0}$ and that all rigid-body degrees of freedom are excluded. As an example, we may use a rocket attached to its launch pad, shown in Fig. 13.4. The displacements $\mathbf{U}_y$ will be those associated with the attachment points, while $\mathbf{U}_x$ will represent all the remaining displacements, the number of which will depend on the idealization of the structure. The equation of motion for a freely vibrating unconstrained system in which $\mathbf{U}_y \neq \mathbf{0}$ is given by

$$\begin{bmatrix} \mathbf{M}_{xx} & \mathbf{M}_{xy} \\ \mathbf{M}_{yx} & \mathbf{M}_{yy} \end{bmatrix} \begin{bmatrix} \ddot{\mathbf{U}}_x \\ \ddot{\mathbf{U}}_y \end{bmatrix} + \begin{bmatrix} \mathbf{K}_{xx} & \mathbf{K}_{xy} \\ \mathbf{K}_{yx} & \mathbf{K}_{yy} \end{bmatrix} \begin{bmatrix} \mathbf{U}_x \\ \mathbf{U}_y \end{bmatrix} = \begin{bmatrix} \mathbf{0} \\ \mathbf{0} \end{bmatrix} \tag{13.74}$$

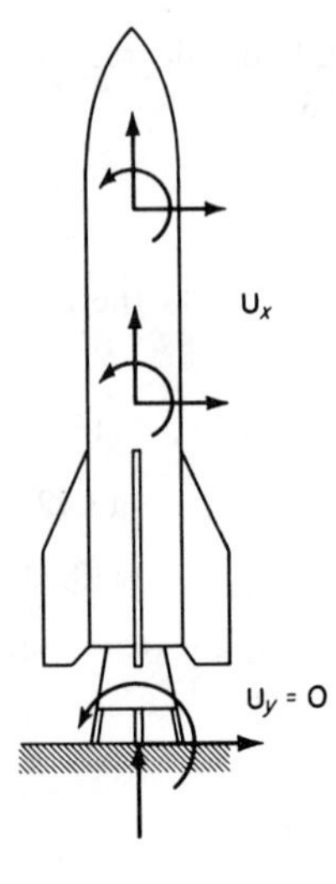

FIG. 13.4 Rocket mounted for vibration tests on a launch pad.

From the first equation in (13.74) we have

$$\mathbf{M}_{xx}\ddot{\mathbf{U}}_x + \mathbf{M}_{xy}\ddot{\mathbf{U}}_y + \mathbf{K}_{xx}\mathbf{U}_x + \mathbf{K}_{xy}\mathbf{U}_y = \mathbf{0} \tag{13.75}$$

Also, for harmonic vibrations, we have

$$\mathbf{U}_x = \mathbf{q}_x e^{i\omega t} \tag{13.76}$$

$$\mathbf{U}_y = \mathbf{q}_y e^{i\omega t} \tag{13.77}$$

and $$\ddot{\mathbf{U}}_y = -\omega^2\mathbf{q}_y e^{i\omega t} \tag{13.78}$$

where ω is the circular frequency of vibrations of the unconstrained system. Substituting now Eqs. (13.77) and (13.78) into (13.75), we obtain

$$\mathbf{M}_{xx}\ddot{\mathbf{U}}_x + \mathbf{K}_{xx}\mathbf{U}_x = (\omega^2\mathbf{M}_{xy} - \mathbf{K}_{xy})\mathbf{q}_y e^{i\omega t} \tag{13.79}$$

The right side of Eq. (13.79) represents harmonic loading, and therefore the solution (13.65) is applicable provided the modes for a constrained structure are used. Hence it follows that

$$\mathbf{U}_x = \bar{\mathbf{p}}_e \bar{\mathscr{M}}_e^{-1}(\bar{\mathbf{\Omega}}^2 - \omega^2\mathbf{I})^{-1}\bar{\mathbf{p}}_e^T(\omega^2\mathbf{M}_{xy} - \mathbf{K}_{xy})\mathbf{q}_y e^{i\omega t} = \mathbf{q}_x e^{i\omega t} \tag{13.80}$$

with $$\bar{\mathbf{\Omega}}^2 = \lceil \bar{\omega}_1^2 \quad \bar{\omega}_2^2 \quad \cdots \quad \bar{\omega}_m^2 \rfloor \tag{13.81}$$

and $$\bar{\mathscr{M}}_e = \bar{\mathbf{p}}_e^T \mathbf{M}_{xx} \bar{\mathbf{p}}_e \tag{13.82}$$

where $\bar{\mathbf{p}}_e$ is the modal matrix for the constrained system with $\mathbf{U}_y = \mathbf{0}$ and $\bar{\omega}_1, \ldots, \bar{\omega}_m$ are the associated frequencies.

Using the second equation in (13.74) and Eqs. (13.76) to (13.78), we get

$$(-\omega^2\mathbf{M}_{yx} + \mathbf{K}_{yx})\mathbf{q}_x e^{i\omega t} + (-\omega^2\mathbf{M}_{yy} + \mathbf{K}_{yy})\mathbf{q}_y e^{i\omega t} = \mathbf{0} \tag{13.83}$$

Combining Eqs. (13.80) and (13.83) into a single matrix equation and then canceling out the exponential factor $e^{i\omega t}$, we obtain

$$\left[\begin{array}{c|c} \mathbf{I} & -\bar{\mathbf{p}}_e \bar{\mathscr{M}}_e^{-1}(\bar{\mathbf{\Omega}}^2 - \omega^2\mathbf{I})^{-1}\bar{\mathbf{p}}_e^{T}(\omega^2\mathbf{M}_{xy} - \mathbf{K}_{xy}) \\ \hline -\omega^2\mathbf{M}_{yx} + \mathbf{K}_{yx} & -\omega^2\mathbf{M}_{yy} + \mathbf{K}_{yy} \end{array}\right] \left[\begin{array}{c} \mathbf{q}_x \\ \hline \mathbf{q}_y \end{array}\right] = \mathbf{0} \tag{13.84}$$

To obtain symmetry of the above equation we can premultiply the first row of submatrices by $[\bar{\mathbf{p}}_e \bar{\mathscr{M}}_e^{-1}(\bar{\mathbf{\Omega}}^2 - \omega^2\mathbf{I})^{-1}\bar{\mathbf{p}}_e^{T}]^{-1}$ so that

$$\left[\begin{array}{c|c} (\bar{\mathbf{p}}_e^{T})^{-1}(-\omega^2\mathbf{I} + \bar{\mathbf{\Omega}}^2)\bar{\mathscr{M}}_e \bar{\mathbf{p}}_e^{-1} & -\omega^2\mathbf{M}_{xy} + \mathbf{K}_{xy} \\ \hline -\omega^2\mathbf{M}_{yx} + \mathbf{K}_{yx} & -\omega^2\mathbf{M}_{yy} + \mathbf{K}_{yy} \end{array}\right] \left[\begin{array}{c} \mathbf{q}_x \\ \mathbf{q}_y \end{array}\right] = \mathbf{0}$$

or

$$\left(\begin{bmatrix} (\bar{\mathbf{p}}_e^{T})^{-1}\bar{\mathbf{\Omega}}^2\bar{\mathscr{M}}_e\bar{\mathbf{p}}_e^{-1} & \mathbf{K}_{xy} \\ \mathbf{K}_{yx} & \mathbf{K}_{yy} \end{bmatrix} - \omega^2 \begin{bmatrix} (\bar{\mathbf{p}}_e^{T})^{-1}\bar{\mathscr{M}}_e\bar{\mathbf{p}}_e^{-1} & \mathbf{M}_{xy} \\ \mathbf{M}_{yx} & \mathbf{M}_{yy} \end{bmatrix}\right) \begin{bmatrix} \mathbf{q}_x \\ \mathbf{q}_y \end{bmatrix} = \mathbf{0} \tag{13.85}$$

As a further simplification we may use

$$\bar{\mathscr{M}}_e = \bar{\mathscr{K}}_e \bar{\mathbf{\Omega}}^{-2} \tag{13.86}$$

obtained from Eq. (12.31*a*). Hence

$$\left(\begin{bmatrix} \mathbf{K}_{xx} & \mathbf{K}_{xy} \\ \mathbf{K}_{yx} & \mathbf{K}_{yy} \end{bmatrix} - \omega^2 \begin{bmatrix} \mathbf{K}_{xx}\bar{\mathbf{p}}_e\bar{\mathbf{\Omega}}^{-2}\bar{\mathbf{p}}_e^{-1} & \mathbf{M}_{xy} \\ \mathbf{M}_{yx} & \mathbf{M}_{yy} \end{bmatrix}\right) \begin{bmatrix} \mathbf{q}_x \\ \mathbf{q}_y \end{bmatrix} = \mathbf{0} \tag{13.87}$$

This equation can be used to determine the frequencies ω and modes $\{\mathbf{q}_x \quad \mathbf{q}_y\}$ for the unconstrained system. The stiffness matrices $\mathbf{K}_{xx}$, $\mathbf{K}_{xy}$, $\mathbf{K}_{yx}$, and $\mathbf{K}_{yy}$ can be obtained from static tests on the constrained structure by first determining the influence coefficients for the directions of $\mathbf{U}_x$ with $\mathbf{U}_y = \mathbf{0}$ and then using Eq. (6.106) to derive the required stiffness matrix. The modes $\bar{\mathbf{p}}_e$ and frequencies $\bar{\mathbf{\Omega}}$ are obtained from vibration tests, which allow us to determine the matrix product $\mathbf{K}_{xx}\bar{\mathbf{p}}_e\bar{\mathbf{\Omega}}^{-2}\bar{\mathbf{p}}_e^{-1}$; however, the submatrices $\mathbf{M}_{xy}$, $\mathbf{M}_{yx}$, and $\mathbf{M}_{yy}$ must be calculated, as no reliable direct experimental techniques are available to measure mass matrices. Alternatively, the structure can be supported in other locations and the vibration tests carried out to determine the remaining mass submatrices.

13.9 DYNAMIC RESPONSE OF STRUCTURAL SYSTEMS WITH DAMPING

The mathematical formulation of expressions for the damping forces in a structural system subjected to dynamic loading poses a difficult problem that still requires extensive research. Unlike mass or stiffness, damping is not necessarily an inherent property of the system. Damping forces depend not only on the oscillating system but also on the surrounding medium. For example, air has a significant effect on the oscillatory motion of the structure. The damping mechanism is customarily described as one of the following:

viscous damping
structural damping
negative damping

Viscous damping occurs when a structural system is moving in a fluid. The damping forces are then dependent on the velocities, and when the system is vibrating freely, the amplitudes of vibration decay exponentially. If the amount of damping in the system is greater than a certain critical value, the oscillatory character of motion ceases. In structural systems, however, the amount of damping present is considerably less than this critical value.

Structural damping is caused by internal friction within the material or at joints between components. The damping forces are a function of the strain in the system, and the mathematical formulation of this special case of damping is not readily amenable to structural analysis.

The so-called negative damping occurs when, instead of dissipating energy from the vibrating system, energy is added to the system. A typical example of such damping is flutter, which is defined as the dynamic instability of an elastic structure in an airstream. Flutter analysis will not be discussed in this book; however, it should be pointed out that matrix methods have also been applied to aeroelastic analysis, including flutter analysis, of aerospace vehicles.

In subsequent sections we shall discuss the formulation and solution of problems in which only viscous damping is present. For structural systems with viscous damping, the equations of motion can be written as

$$\mathbf{M}\ddot{\mathbf{U}} + \mathbf{C}\dot{\mathbf{U}} + \mathbf{K}\mathbf{U} = \mathbf{P} \tag{12.4}$$

where $\mathbf{C}$ is the symmetric matrix of damping coefficients. To solve Eq. (12.4) we may again assume that the solution for $\mathbf{U}$ is of the form

$$\mathbf{U} = \mathbf{p}_0\mathbf{\Phi}_0 + \mathbf{p}_e\mathbf{\Phi}_e \tag{13.13}$$

Hence Eq. (12.4) may be written as

$$\mathbf{M}\mathbf{p}_0\ddot{\mathbf{\Phi}}_0 + \mathbf{M}\mathbf{p}_e\ddot{\mathbf{\Phi}}_e + \mathbf{C}\mathbf{p}_0\dot{\mathbf{\Phi}}_0 + \mathbf{C}\mathbf{p}_e\dot{\mathbf{\Phi}}_e + \mathbf{K}\mathbf{p}_0\mathbf{\Phi}_0 + \mathbf{K}\mathbf{p}_e\mathbf{\Phi}_e = \mathbf{P} \tag{13.88}$$

Premultiplying this equation in turn by $\mathbf{p}_0{}^T$ and $\mathbf{p}_e{}^T$ and using orthogonality relations, we obtain two sets of equations

$$\mathscr{M}_0\ddot{\mathbf{\Phi}}_0 + \mathscr{C}_0\dot{\mathbf{\Phi}}_0 + \mathbf{p}_0{}^T\mathbf{C}\mathbf{p}_e\dot{\mathbf{\Phi}}_e = \mathscr{P}_0 \tag{13.89}$$

and $$\mathscr{M}_e\ddot{\mathbf{\Phi}}_e + \mathbf{p}_e{}^T\mathbf{C}\mathbf{p}_0\dot{\mathbf{\Phi}}_0 + \mathscr{C}_e\dot{\mathbf{\Phi}}_e + \mathbf{\Omega}^2\mathscr{M}_e\mathbf{\Phi}_e = \mathscr{P}_e \tag{13.90}$$

where $\quad \mathscr{C}_0 = \mathbf{p}_0{}^T\mathbf{C}\mathbf{p}_0$ (13.91)

and $\quad \mathscr{C}_e = \mathbf{p}_e{}^T\mathbf{C}\mathbf{p}_e$ (13.92)

represent generalized damping matrices. The two sets of equations become uncoupled from each other provided

$$\mathbf{p}_0{}^T\mathbf{C}\mathbf{p}_e = \mathbf{0} \tag{13.93}$$

Furthermore, if both $\mathscr{C}_0$ and $\mathscr{C}_e$ are diagonal, the equations in each set also become uncoupled, and they can be solved individually. Utilizing the orthogonality relations, we note that $\mathscr{C}_0$ and $\mathscr{C}_e$ become diagonal when the matrix **C** is proportional to either **M** or **K**. Alternatively, the generalized damping matrix $\mathscr{C}_e$ could be taken as a certain percentage of critical damping. All these conditions would also satisfy Eq. (13.93). Solutions for the dynamic response of elastic systems with damping matrix proportional to mass, stiffness, and critical damping will be discussed in the next three sections.

13.10 DAMPING MATRIX PROPORTIONAL TO MASS

We shall assume that the damping matrix **C** is given by

$$\mathbf{C} = 2\beta\mathbf{M} \tag{13.94}$$

where β is a constant of proportionality. When Eq. (13.94) is substituted into (13.89) and (13.90), we obtain two sets of equations

$$\ddot{\mathbf{\Phi}}_0 + 2\beta\dot{\mathbf{\Phi}}_0 = \mathscr{M}_0{}^{-1}\mathscr{P}_0 \tag{13.95}$$

and $$\ddot{\mathbf{\Phi}}_e + 2\beta\dot{\mathbf{\Phi}}_e + \mathbf{\Omega}^2\mathbf{\Phi}_e = \mathscr{M}_e{}^{-1}\mathscr{P}_e \tag{13.96}$$

The ith row from (13.95) can be written as

$$\ddot{\Phi}_{0_i} + 2\beta\dot{\Phi}_{0_i} = (\mathscr{M}_0{}^{-1}\mathscr{P}_0)_i \tag{13.97}$$

When the Laplace transform

$$\overline{\Phi}_{0_i}(p) = \int_0^\infty e^{-pt}\Phi_{0_i}(t)\,dt \tag{13.98}$$

is used, Eq. (13.97) becomes

$$\overline{\Phi}_{0_i}(p)(p^2 + 2\beta p) = [\mathscr{M}_0{}^{-1}\overline{\mathscr{P}}_0(p)]_i + p\Phi_{0_i}(0) + \dot{\Phi}_{0_i}(0) + 2\beta\Phi_{0_i}(0)$$

and hence

$$\overline{\Phi}_{0_i}(p) = \frac{[\mathscr{M}_0{}^{-1}\overline{\mathscr{P}}_0(p)]_i}{p(p + 2\beta)} + \frac{1}{p}\,\Phi_{0_i}(0) + \frac{1}{p(p + 2\beta)}\,\dot{\Phi}_{e_i}(0) \tag{13.99}$$

Noting that

$$\mathscr{L}^{-1}\left[\frac{1}{p(p + 2\beta)}\right] = \frac{1 - e^{-2\beta t}}{2\beta} \tag{13.100}$$

and using the convolution theorem

$$\mathscr{L}^{-1}[\bar{g}(p)\bar{h}(p)] = \int_0^t g(t-\tau)h(\tau)\,d\tau \tag{13.101}$$

and tables of standard Laplace transforms, we obtain from Eq. (13.99)

$$\Phi_{0_i}(t) = \int_0^t \frac{1-e^{-2\beta(t-\tau)}}{2\beta}[\mathscr{M}_0^{-1}\mathscr{P}_0(\tau)]_i\,d\tau + \Phi_{0_i}(0) + \frac{1-e^{-2\beta t}}{2\beta}\dot{\Phi}_{0_i}(0) \tag{13.102}$$

Collecting all equations (13.102) for $i = 1$ to w into a single matrix equation, we have

$$\mathbf{\Phi}_0 = \frac{\mathscr{M}_0^{-1}}{2\beta}\int_0^t (1-e^{-2\beta(t-\tau)})\mathscr{P}_0(\tau)\,d\tau + \mathbf{\Phi}_0(0) + \frac{1-e^{-2\beta t}}{2\beta}\dot{\mathbf{\Phi}}_0(0) \tag{13.103}$$

The solution to equations (13.96) is obtained in a similar manner. The ith row from (13.96) is written as

$$\ddot{\Phi}_{e_i} + 2\beta\dot{\Phi}_{e_i} + \omega_{w+i}^2\Phi_{e_i} = (\mathscr{M}_e^{-1}\mathscr{P}_e)_i \tag{13.104}$$

and the Laplace transformation is applied to this equation, leading to

$$\bar{\Phi}_{e_i}(p) = \frac{[\mathscr{M}_e^{-1}\mathscr{P}_e(p)]_i}{(p+\beta)^2 + (\omega_{w+i}^2 - \beta^2)} + \frac{p\Phi_{e_i}(0) + \dot{\Phi}_{e_i}(0) + 2\beta\Phi_{e_i}(0)}{(p+\beta)^2 + (\omega_{w+i}^2 - \beta^2)} \tag{13.105}$$

Using the convolution theorem for the first term in (13.105) and tables of standard Laplace transforms for the second term, we obtain

$$\begin{aligned}\Phi_{e_i}(t) &= (\omega_{w+i}^2 - \beta^2)^{-\frac{1}{2}}\int_0^t e^{-\beta(t-\tau)}\sin[(\omega_{w+i}^2 - \beta)^{\frac{1}{2}}(t-\tau)][\mathscr{M}_e^{-1}\mathscr{P}_e(\tau)]_i\,d\tau \\ &+ e^{-\beta t}\cos[(\omega_{w+i}^2 - \beta^2)^{\frac{1}{2}}t]\Phi_{e_i}(0) + e^{-\beta t}(\omega_{w+i}^2 - \beta^2)^{-\frac{1}{2}}\sin[(\omega_{w+i}^2 - \beta^2)^{\frac{1}{2}}t] \\ &\times [\dot{\Phi}_{e_i}(0) + \beta\Phi_{e_i}(0)]\end{aligned} \tag{13.106}$$

which can be written collectively as

$$\begin{aligned}\mathbf{\Phi}_e &= (\mathbf{\Omega}^2 - \beta^2\mathbf{I})^{-\frac{1}{2}}\mathscr{M}_e^{-1}\int_0^t e^{-\beta(t-\tau)}\,\mathbf{sin}\,[(\omega^2 - \beta^2)^{\frac{1}{2}}(t-\tau)]\mathscr{P}_e(\tau)\,d\tau \\ &+ e^{-\beta t}\,\mathbf{cos}\,[(\omega^2 - \beta^2)^{\frac{1}{2}}t]\mathbf{\Phi}_e(0) + e^{-\beta t}(\mathbf{\Omega}^2 - \beta^2\mathbf{I})^{-\frac{1}{2}}\,\mathbf{sin}\,[(\omega^2 - \beta^2)^{\frac{1}{2}}t] \\ &\times [\dot{\mathbf{\Phi}}_e(0) + \beta\mathbf{\Phi}_e(0)]\end{aligned} \tag{13.107}$$

Substituting Eqs. (13.103) and (13.107) into (13.13), we obtain an equation for the response of an elastic system with damping proportional to the mass matrix $\mathbf{M}$. Hence

$$\mathbf{U} = \frac{1}{2\beta}\mathbf{p}_0\mathcal{M}_0^{-1}\int_0^t (1 - e^{-2\beta(t-\tau)})\mathcal{P}_0(\tau)\,d\tau + \mathbf{p}_0\mathbf{\Phi}_0(0) + \frac{1}{2\beta}(1 - e^{-2\beta t})\mathbf{p}_0\dot{\mathbf{\Phi}}_0(0)$$
$$+ \mathbf{p}_e(\mathbf{\Omega}^2 - \beta^2\mathbf{I})^{-\frac{1}{2}}\mathcal{M}_e^{-1}\int_0^t e^{-\beta(t-\tau)}\,\mathbf{sin}\,[(\omega^2 - \beta^2)^{\frac{1}{2}}(t-\tau)]\mathcal{P}_e(\tau)\,d\tau$$
$$+ e^{-\beta t}\mathbf{p}_e\,\mathbf{cos}\,[(\omega^2 - \beta^2)^{\frac{1}{2}}t]\mathbf{\Phi}_e(0) + e^{-\beta t}\mathbf{p}_e(\mathbf{\Omega}^2 - \beta^2\mathbf{I})^{-\frac{1}{2}}\,\mathbf{sin}\,[(\omega^2 - \beta^2)^{\frac{1}{2}}t]$$
$$\times [\dot{\mathbf{\Phi}}_e(0) + \beta\mathbf{\Phi}_e(0)] \quad (13.108)$$

The initial values $\mathbf{\Phi}_0(0)$, $\mathbf{\Phi}_e(0)$, $\dot{\mathbf{\Phi}}_0(0)$, and $\dot{\mathbf{\Phi}}_e(0)$ are to be determined in terms of $\mathbf{U}(0)$ and $\dot{\mathbf{U}}(0)$ from Eqs. (13.43) and (13.45) to (13.47).

13.11 DAMPING MATRIX PROPORTIONAL TO STIFFNESS

For the case of damping proportional to stiffness

$$\mathbf{C} = 2\gamma\mathbf{K} \quad (13.109)$$

where γ is a constant of proportionality. Substituting Eq. (13.109) into (13.89) and (13.90) leads to

$$\ddot{\mathbf{\Phi}}_0 = \mathcal{M}_0^{-1}\mathcal{P}_0 \quad (13.110)$$

and $$\ddot{\mathbf{\Phi}}_e + 2\gamma\mathbf{\Omega}^2\dot{\mathbf{\Phi}}_e + \mathbf{\Omega}^2\mathbf{\Phi}_e = \mathcal{M}_e^{-1}\mathcal{P}_e \quad (13.111)$$

Application of Laplace transformation to Eqs. (13.110) and (13.111) results in

$$\mathbf{\Phi}_0 = \mathcal{M}_0^{-1}\int_0^t (t-\tau)\mathcal{P}_0(\tau)\,d\tau + \mathbf{\Phi}_0(0) + \dot{\mathbf{\Phi}}_0(0)t \quad (13.112)$$

$$\mathbf{\Phi}_e = (\mathbf{I} - \gamma^2\mathbf{\Omega}^2)^{-\frac{1}{2}}\mathbf{\Omega}^{-1}\mathcal{M}_e^{-1}\int_0^t \mathbf{exp}\,[-\gamma\omega^2(t-\tau)]\,\mathbf{sin}\,[\omega(1-\gamma^2\omega^2)^{\frac{1}{2}}(t-\tau)]$$
$$\times \mathcal{P}_e(\tau)\,d\tau + \mathbf{exp}\,(-\gamma\omega^2 t)\,\mathbf{cos}\,[\omega(1-\gamma^2\omega^2)^{\frac{1}{2}}t]\mathbf{\Phi}_e(0)$$
$$+ \mathbf{exp}\,(-\gamma\omega^2 t)(\mathbf{I} - \gamma^2\mathbf{\Omega}^2)^{-\frac{1}{2}}\mathbf{\Omega}^{-1}\,\mathbf{sin}\,[\omega(1-\gamma^2\omega^2)^{\frac{1}{2}}t][\dot{\mathbf{\Phi}}_e(0) + \gamma\mathbf{\Omega}^2\mathbf{\Phi}_e(0)]$$
$$(13.113)$$

The displacements $\mathbf{U}$ are then determined from

$$\mathbf{U} = \mathbf{p}_0\mathbf{\Phi}_0 + \mathbf{p}_e\mathbf{\Phi}_e$$
$$= \mathbf{p}_0\mathcal{M}_0^{-1}\int_0^t (t-\tau)\mathcal{P}_0(\tau)\,d\tau + \mathbf{p}_0\mathbf{\Phi}_0(0) + \mathbf{p}_0\dot{\mathbf{\Phi}}_0(0)t$$
$$+ \mathbf{p}_e(\mathbf{I} - \gamma^2\mathbf{\Omega}^2)^{-\frac{1}{2}}\mathbf{\Omega}^{-1}\mathcal{M}_e^{-1}\int_0^t \mathbf{exp}\,[-\gamma\omega^2(t-\tau)]\,\mathbf{sin}\,[\omega(1-\gamma^2\omega^2)^{\frac{1}{2}}(t-\tau]$$
$$\times \mathcal{P}_e(\tau)\,d\tau + \mathbf{p}_e\,\mathbf{exp}\,(-\gamma\omega^2 t)\,\mathbf{cos}\,[\omega(1-\gamma^2\omega^2)^{\frac{1}{2}}t]\mathbf{\Phi}_e(0)$$
$$+ \mathbf{p}_e\,\mathbf{exp}\,(-\gamma\omega^2 t)(\mathbf{I} - \gamma^2\mathbf{\Omega}^2)^{-\frac{1}{2}}\mathbf{\Omega}^{-1}\,\mathbf{sin}\,[\omega(1-\gamma^2\omega^2)^{\frac{1}{2}}t]$$
$$\times [\dot{\mathbf{\Phi}}_e(0) + \gamma\mathbf{\Omega}^2\mathbf{\Phi}_e(0)] \quad (13.114)$$

where

$$\mathbf{exp}\,(-\gamma\omega^2 t) = \lceil \exp(-\gamma^2_{w+1}t) \quad \exp(-\gamma\omega^2_{w+2}t) \quad \cdots \quad \exp(-\gamma\omega^2_{w+m}t)\rfloor \tag{13.115}$$

$$\begin{aligned}&\mathbf{exp}\,[-\gamma\omega^2(t-\tau)] \\ &\quad = \lceil \exp[-\gamma\omega^2_{w+1}(t-\tau)] \quad \exp[-\gamma\omega^2_{w+2}(t-\tau)] \quad \cdots \quad \exp[-\gamma\omega^2_{w+m}(t-\tau)]\rfloor\end{aligned} \tag{13.116}$$

The initial values of $\mathbf{\Phi}_0$, $\mathbf{\Phi}_e$, $\dot{\mathbf{\Phi}}_0$, and $\dot{\mathbf{\Phi}}_e$ are found, as in the previous case, from Eqs. (13.43) and (13.45) to (13.47).

Both cases $\mathbf{C} = 2\gamma\mathbf{M}$ and $\mathbf{C} = 2\gamma\mathbf{K}$ allow for the determination of damping characteristics of the system in terms of only one constant. In most practical applications, however, a single constant is inadequate to describe the damping of a multi-degree-of-freedom system. Consequently, the representation of damping to be described next appears to be more attractive because it allows the use of a number of constants equal to the number of elastic degrees of freedom.

13.12 MATRIX C PROPORTIONAL TO CRITICAL DAMPING

In this case it is more convenient first to combine Eqs. (13.89) and (13.90) into one equation, assuming that $\mathbf{p}_0{}^T\mathbf{C}\mathbf{p}_e = \mathbf{0}$, so that

$$\begin{bmatrix} \mathcal{M}_0 & \mathbf{0} \\ \mathbf{0} & \mathcal{M}_e \end{bmatrix}\begin{bmatrix} \ddot{\mathbf{\Phi}}_0 \\ \ddot{\mathbf{\Phi}}_e \end{bmatrix} + \begin{bmatrix} \mathscr{C}_0 & \mathbf{0} \\ \mathbf{0} & \mathscr{C}_e \end{bmatrix}\begin{bmatrix} \dot{\mathbf{\Phi}}_0 \\ \dot{\mathbf{\Phi}}_e \end{bmatrix} + \begin{bmatrix} \mathbf{0} & \mathbf{0} \\ \mathbf{0} & \mathbf{\Omega}^2\mathcal{M}_e \end{bmatrix}\begin{bmatrix} \mathbf{\Phi}_0 \\ \mathbf{\Phi}_e \end{bmatrix} = \begin{bmatrix} \mathscr{P}_0 \\ \mathscr{P}_e \end{bmatrix} \tag{13.117}$$

The critical damping for a single-degree-of-freedom system is given by

$$C_{\text{crit}} = 2M\omega \tag{12.88}$$

while any value of damping can be expressed as

$$C = 2\nu M\omega \tag{13.118}$$

where ν represents the ratio of the actual damping over critical damping; consequently, for structural systems ν will be less than one. By analogy we shall assume that the generalized damping may also be represented as certain fractions of the critical damping. This implies that

$$\begin{aligned}\mathscr{C} = \lceil \mathscr{C}_0 \quad \mathscr{C}_e \rfloor &= 2\boldsymbol{\nu}\mathcal{M}\lceil \mathbf{0} \quad \mathbf{\Omega}\rfloor \\ &= 2\begin{bmatrix} \mathbf{0} & \mathbf{0} \\ \mathbf{0} & \boldsymbol{\nu}_e\mathcal{M}_e\mathbf{\Omega} \end{bmatrix}\end{aligned} \tag{13.119}$$

where the diagonal matrix $\boldsymbol{\nu}$ is given by

$$\boldsymbol{\nu} = \lceil \mathbf{0} \quad \boldsymbol{\nu}_e \rfloor \tag{13.120}$$

$$\boldsymbol{\nu}_e = \lceil \nu_{w+1} \quad \nu_{w+2} \quad \cdots \quad \nu_{w+m} \rfloor \tag{13.121}$$

in which a typical term ν_{w+i} represents percentage of critical damping for the ith elastic mode $\mathbf{p}_{w+i}$.

Using Eq. (13.119) in (13.117), we obtain two sets of uncoupled equations

$$\ddot{\mathbf{\Phi}}_0 = \mathscr{M}_0^{-1}\mathscr{P}_0 \tag{13.122}$$

$$\ddot{\mathbf{\Phi}}_e + 2\boldsymbol{\nu}_e\mathbf{\Omega}\dot{\mathbf{\Phi}}_e + \mathbf{\Omega}^2\mathbf{\Phi}_e = \mathscr{M}_e^{-1}\mathscr{P}_e \tag{13.123}$$

Here again, the solutions to the above equations can be obtained using the Laplace transform technique. It can be demonstrated that the following results are obtained:

$$\mathbf{\Phi}_0 = \mathscr{M}_0^{-1}\int_0^t (t-\tau)\mathscr{P}_0(\tau)\,d\tau + \mathbf{\Phi}_0(0) + \dot{\mathbf{\Phi}}_0(0)t \tag{13.124}$$

and

$$\begin{aligned}\mathbf{\Phi}_e = (\mathbf{I} - \boldsymbol{\nu}_e^2)^{-\frac{1}{2}}\mathbf{\Omega}^{-1}\mathscr{M}_e^{-1}\int_0^t \mathbf{exp}\,[-\nu\omega(t-\tau)]\,\mathbf{sin}\,[\omega(1-\nu^2)^{\frac{1}{2}}(t-\tau)]\mathscr{P}_e(\tau)\,d\tau \\ + \mathbf{exp}\,(-\nu\omega t)\,\mathbf{cos}\,[\omega(1-\nu^2)^{\frac{1}{2}}t]\mathbf{\Phi}_e(0) \\ + \mathbf{exp}\,(-\nu\omega t)(\mathbf{I} - \boldsymbol{\nu}_e^2)^{-\frac{1}{2}}\mathbf{\Omega}^{-1}\,\mathbf{sin}\,[\omega(1-\nu^2)^{\frac{1}{2}}t][\dot{\mathbf{\Phi}}_e(0) + \boldsymbol{\nu}\mathbf{\Omega}\mathbf{\Phi}_e(0)]\end{aligned} \tag{13.125}$$

Hence the displacements are given by

$$\begin{aligned}\mathbf{U} &= \mathbf{p}_0\mathbf{\Phi}_0 + \mathbf{p}_e\mathbf{\Phi}_e \\ &= \mathbf{p}_0\mathscr{M}_0^{-1}\int_0^t (t-\tau)\mathscr{P}_0(\tau)\,d\tau + \mathbf{p}_0\mathbf{\Phi}_0(0) + \mathbf{p}_0\dot{\mathbf{\Phi}}_0(0)t \\ &+ \mathbf{p}_e(\mathbf{I} - \boldsymbol{\nu}_e^2)^{-\frac{1}{2}}\mathbf{\Omega}^{-1}\mathscr{M}_e^{-1}\int_0^t \mathbf{exp}\,[-\nu\omega(t-\tau)]\,\mathbf{sin}\,[\omega(1-\nu^2)^{\frac{1}{2}}(t-\tau)]\mathscr{P}_e(\tau)\,d\tau \\ &+ \mathbf{p}_e\,\mathbf{exp}\,(-\nu\omega t)\,\mathbf{cos}\,[\omega(1-\nu^2)^{\frac{1}{2}}t]\mathbf{\Phi}_e(0) \\ &+ \mathbf{p}_e\,\mathbf{exp}\,(-\nu\omega t)(\mathbf{I} - \boldsymbol{\nu}_e^2)^{-\frac{1}{2}}\mathbf{\Omega}^{-1}\,\mathbf{sin}\,[\omega(1-\nu^2)^{\frac{1}{2}}t][\dot{\mathbf{\Phi}}_e(0) + \boldsymbol{\nu}\mathbf{\Omega}\mathbf{\Phi}_e(0)]\end{aligned} \tag{13.126}$$

Also the same procedure as in previous cases is followed to evaluate the initial values of $\mathbf{\Phi}(0)$ and $\dot{\mathbf{\Phi}}(0)$.

To determine the damping matrix $\mathbf{C}$ we note that

$$\mathscr{C} = \mathbf{p}^T\mathbf{C}\mathbf{p} \tag{13.127}$$

where $\mathbf{p} = [\mathbf{p}_0 \quad \mathbf{p}_e]$. Consequently, premultiplying Eq. (13.127) by $(\mathbf{p}^T)^{-1}$ and postmultiplying the resulting equation by $\mathbf{p}^{-1}$, we obtain

$$\begin{aligned}\mathbf{C} &= (\mathbf{p}^T)^{-1}\mathscr{C}\mathbf{p}^{-1} \\ &= 2(\mathbf{p}^T)^{-1}\begin{bmatrix}\mathbf{0} & \mathbf{0} \\ \mathbf{0} & \boldsymbol{\nu}_e\mathscr{M}_e\mathbf{\Omega}\end{bmatrix}\mathbf{p}^{-1}\end{aligned} \tag{13.128}$$

which determines all the coefficients C_{ij} in terms of the specified "percent" damping in each mode. It should be noted, however, that knowledge of $\mathbf{C}$ is not required to determine displacements and hence stresses by the present method.

The advantages of assuming percentages of critical damping are now apparent. The coefficients in the diagonal matrix $\boldsymbol{\nu}_e$ are adjusted until a reasonable result is obtained. Although th^ number of coefficients in $\boldsymbol{\nu}_e$ is equal to the number of elastic degrees of freedom, only the first few coefficients are normally used to approximate the measured damping characteristics of the actual system.

13.13 ORTHONORMALIZATION OF THE MODAL MATRIX p

We have seen in preceding sections that the expressions for displacements due to dynamic loading require the generalized mass matrices $\mathcal{M}_0$ and $\mathcal{M}_e$. Some simplification of these expressions is possible if both $\mathcal{M}_0$ and $\mathcal{M}_e$ can be made into unit matrices. The transformations required to achieve this are quite simple. Assuming that $\mathcal{M}_0$ is a diagonal matrix (it can always be made one by a suitable choice of the frame of reference for the rigid-body displacements or by a formal matrix diagonalization) we have

$$\mathcal{M}_0 = \mathbf{p}_0{}^T\mathbf{M}\mathbf{p}_0 = \text{diagonal matrix} \tag{13.129}$$

After premultiplying by $\mathcal{M}_0{}^{-\frac{1}{2}}$ and postmultiplying by $\mathcal{M}_0{}^{-\frac{1}{2}}$ Eq. (13.129) becomes

$$\begin{aligned}\mathbf{I} = \mathcal{M}_0{}^{-\frac{1}{2}}\mathcal{M}_0\mathcal{M}_0{}^{-\frac{1}{2}} &= \mathcal{M}_0{}^{-\frac{1}{2}}\mathbf{p}_0{}^T\mathbf{M}\mathbf{p}_0\mathcal{M}_0{}^{-\frac{1}{2}} \\ &= \hat{\mathbf{p}}_0{}^T\mathbf{M}\hat{\mathbf{p}}_0\end{aligned} \tag{13.130}$$

where $$\hat{\mathbf{p}}_0 = \mathbf{p}_0\mathcal{M}_0{}^{-\frac{1}{2}} \tag{13.131}$$

is the modal matrix for the modified rigid-body modes. Such modes, on account of relationship (13.130), are called *orthonormal modes.* The transformation in (13.131) results in multiplication of each node in $\mathbf{p}_0$ by a factor equal to the reciprocal of the square root of the generalized mass in that mode.

Similarly we can show that

$$\hat{\mathbf{p}}_e{}^T\mathbf{M}\hat{\mathbf{p}}_e = \mathbf{I} \tag{13.132}$$

where $$\hat{\mathbf{p}}_e = \mathbf{p}_e\mathcal{M}_e{}^{-\frac{1}{2}} \tag{13.133}$$

is the modal matrix of the elastic orthonormal eigenmodes. We may also note that if orthonormal modes are used, then from Eq. (12.31*a*)

$$\hat{\mathcal{K}}_e = \hat{\mathbf{p}}_e{}^T\mathbf{K}\hat{\mathbf{p}}_e = \mathbf{\Omega}^2 \tag{13.134}$$

while the equilibrium equations (12.20) require

$$\hat{\mathbf{p}}_0{}^T\mathbf{M}\hat{\mathbf{p}}_e = \mathbf{0} \tag{13.135}$$

Combining Eqs. (13.130), (13.132), and (13.135), we obtain the orthogonality relationship

$$\hat{\mathbf{p}}^T\mathbf{M}\hat{\mathbf{p}} = \mathbf{I} \tag{13.136}$$

where $$\hat{\mathbf{p}} = [\hat{\mathbf{p}}_0 \quad \hat{\mathbf{p}}_e] \tag{13.137}$$

13.14 DYNAMIC RESPONSE OF AN ELASTIC ROCKET SUBJECTED TO PULSE LOADING

As an example of the dynamic-response analysis we shall consider a rocket in flight. The rocket is subjected to a sudden application of constant axial thrust of magnitude P_0 acting for the duration of time t_0, as shown in Fig. 13.5. For the purpose of matrix analysis the rocket will be idealized by two bar elements with properties based on the average properties of the rocket. Furthermore, for simplicity of presentation the two elements will be identical. The idealized structure is shown in Fig. 13.6.

The vibrational characteristics for this configuration have already been determined in the example discussed in Sec. 12.6. Using the numbering for displacements shown in Fig. 13.6, we can summarize the main results from that example

$$\mathbf{\Omega}^2 = \lceil \omega_2^2 \quad \omega_3^2 \rfloor = \frac{E}{\rho L^2} \lceil 12 \quad 48 \rfloor \tag{a}$$

$$\mathbf{p}_0 = \begin{bmatrix} 1 \\ 1 \\ 1 \end{bmatrix} \tag{b}$$

$$\mathbf{p}_e = \begin{bmatrix} 1 & 1 \\ 0 & -1 \\ -1 & 1 \end{bmatrix} \tag{c}$$

$$\mathbf{M} = \frac{\rho A L}{12} \begin{bmatrix} 2 & 1 & 0 \\ 1 & 4 & 1 \\ 0 & 1 & 2 \end{bmatrix} \tag{d}$$

We calculate first the generalized mass matrices $\mathscr{M}_0$ and $\mathscr{M}_e$.

$$\mathscr{M}_0 = \mathbf{p}_0^T \mathbf{M} \mathbf{p}_0 = [1 \quad 1 \quad 1] \frac{\rho A L}{12} \begin{bmatrix} 2 & 1 & 0 \\ 1 & 4 & 1 \\ 0 & 1 & 2 \end{bmatrix} \begin{bmatrix} 1 \\ 1 \\ 1 \end{bmatrix} = \rho A L \tag{e}$$

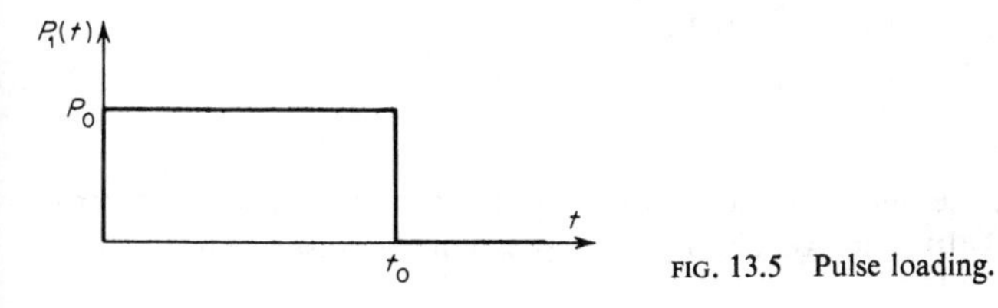

FIG. 13.5 Pulse loading.

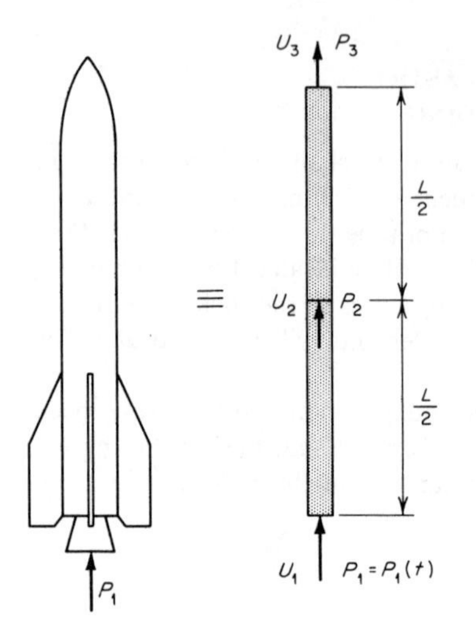

FIG. 13.6 Two-element idealization; specified thrust $P_1 = P_1(t)$.

$$\mathscr{M}_e = \mathbf{p}_e{}^T\mathbf{M}\mathbf{p}_e = \begin{bmatrix} 1 & 0 & -1 \\ 1 & -1 & 1 \end{bmatrix} \frac{\rho AL}{12} \begin{bmatrix} 2 & 1 & 0 \\ 1 & 4 & 1 \\ 0 & 1 & 2 \end{bmatrix} \begin{bmatrix} 1 & 1 \\ 0 & -1 \\ -1 & 1 \end{bmatrix} = \frac{\rho AL}{3} \begin{bmatrix} 1 & 0 \\ 0 & 1 \end{bmatrix} \tag{f}$$

Since only one force is applied to the system,

$$\mathbf{P} = \begin{bmatrix} P_1 \\ 0 \\ 0 \end{bmatrix} \tag{g}$$

where the force P_1 is represented by the rectangular pulse in Fig. 13.5. From Eqs. (*b*), (*c*), and (*g*) we can next calculate the generalized forces $\mathscr{P}_0$ and $\mathscr{P}_e$.

$$\mathscr{P}_0 = \mathbf{p}_0{}^T\mathbf{P} = [1 \quad 1 \quad 1] \begin{bmatrix} P_1 \\ 0 \\ 0 \end{bmatrix} = P_1 \tag{h}$$

and $$\mathscr{P}_e = \mathbf{p}_e{}^T\mathbf{P} = \begin{bmatrix} 1 & 0 & -1 \\ 1 & -1 & 1 \end{bmatrix} \begin{bmatrix} P_1 \\ 0 \\ 0 \end{bmatrix} = \begin{bmatrix} P_1 \\ P_1 \end{bmatrix} \tag{i}$$

Assuming that the initial displacements and velocities are all equal to zero, that is, $\mathbf{U}(0) = \dot{\mathbf{U}}(0) = \mathbf{0}$, we obtain an expression for displacements from

Eq. (13.48)

$$\mathbf{U} = \mathbf{p}_0 \mathscr{M}_0^{-1} \int_{\tau_2=0}^{\tau_2=t} \int_{\tau_1=0}^{\tau_1=\tau_2} \mathscr{P}_0(\tau_1)\, d\tau_1\, d\tau_2 + \mathbf{p}_e \mathbf{\Omega}^{-1} \mathscr{M}_e^{-1} \int_0^t \mathbf{sin}\,[\omega(t-\tau)] \mathscr{P}_e(\tau)\, d\tau \tag{j}$$

where all the necessary matrices have already been determined. For the first integral in (j) we have

$$t < t_0 \qquad \int_{\tau_2=0}^{\tau_2=t} \int_{\tau_1=0}^{\tau_1=\tau_2} \mathscr{P}_0(\tau_1)\, d\tau_1\, d\tau_2 = \int_{\tau_2=0}^{\tau_2=t} P_0 \tau_2\, d\tau_2 = \tfrac{1}{2} P_0 t^2 \tag{k}$$

$$\begin{aligned} t > t_0 \qquad \int_{\tau_2=0}^{\tau_2=t} \int_{\tau_1=0}^{\tau_1=\tau_2} \mathscr{P}_0(\tau_1)\, d\tau_1\, d\tau_2 &= \int_{\tau_2=0}^{\tau_2=t_0} \int_{\tau_1=0}^{\tau_1=\tau_2} \mathscr{P}_0(\tau_1)\, d\tau_1\, d\tau_2 \\ &\quad + \int_{\tau_2=t_0}^{\tau_2=t} \int_{\tau_1=0}^{\tau_1=\tau_2} \mathscr{P}_0(\tau_1)\, d\tau_1\, d\tau_2 \\ &= \int_{\tau_2=0}^{\tau_2=t_0} P_0 \tau_2\, d\tau_2 + \int_{\tau_2=t_0}^{\tau_2=t} P_0 t_0\, d\tau_2 \\ &= \tfrac{1}{2} P_0 t_0^2 + P_0 t_0 (t - t_0) = P_0 t_0 t - \tfrac{1}{2} P_0 t_0^2 \end{aligned} \tag{l}$$

For the second integral we have

$$\begin{aligned} \int_0^t \mathbf{sin}\,[\omega(t-\tau)] \mathscr{P}_e(\tau)\, d\tau &= \int_0^t \begin{bmatrix} \sin[\omega_2(t-\tau)] & 0 \\ 0 & \sin[\omega_3(t-\tau)] \end{bmatrix} \begin{bmatrix} P_1(\tau) \\ P_1(\tau) \end{bmatrix} d\tau \\ &= \begin{bmatrix} \int_0^t \sin[\omega_2(t-\tau)] \quad P_1(\tau)\, d\tau \\ \int_0^t \sin[\omega_3(t-\tau)] \quad P_1(\tau)\, d\tau \end{bmatrix} \end{aligned} \tag{m}$$

The integrals in Eq. (m) can be determined from case 9 in Table 13.1. Hence from Eq. (j) the displacements $\mathbf{U}$ for $0 < t < t_0$ become

$$\begin{aligned} \mathbf{U} &= \begin{bmatrix} 1 \\ 1 \\ 1 \end{bmatrix} \frac{1}{\rho AL} \frac{P_0 t^2}{2} + \begin{bmatrix} 1 & 1 \\ 0 & -1 \\ -1 & 1 \end{bmatrix} \frac{L}{12} \sqrt{\frac{3\rho}{E}} \begin{bmatrix} 2 & 0 \\ 0 & 1 \end{bmatrix} \frac{3}{\rho AL} \begin{bmatrix} 1 & 0 \\ 0 & 1 \end{bmatrix} \begin{bmatrix} \dfrac{P_0}{\omega_2}(1 - \cos\omega_2 t) \\ \dfrac{P_0}{\omega_3}(1 - \cos\omega_3 t) \end{bmatrix} \\ &= \frac{P_0 t^2}{2\rho AL} \begin{bmatrix} 1 \\ 1 \\ 1 \end{bmatrix} + \frac{P_0}{4A} \sqrt{\frac{3}{E\rho}} \begin{bmatrix} \dfrac{2}{\omega_2}(1 - \cos\omega_2 t) + \dfrac{1}{\omega_3}(1 - \cos\omega_3 t) \\ -\dfrac{1}{\omega_3}(1 - \cos\omega_3 t) \\ -\dfrac{2}{\omega_2}(1 - \cos\omega_2 t) + \dfrac{1}{\omega_3}(1 - \cos\omega_3 t) \end{bmatrix} \end{aligned} \tag{n}$$

Similarly for $t > t_0$

$$\mathbf{U} = \begin{bmatrix} 1 \\ 1 \\ 1 \end{bmatrix} \frac{1}{\rho AL} \frac{P_0 t_0}{2}(2t - t_0)$$

$$+ \begin{bmatrix} 1 & 1 \\ 0 & -1 \\ -1 & 1 \end{bmatrix} \frac{L}{12}\sqrt{\frac{3\rho}{E}} \begin{bmatrix} 2 & 0 \\ 0 & 1 \end{bmatrix} \frac{3}{\rho AL} \begin{bmatrix} 1 & 0 \\ 0 & 1 \end{bmatrix} \begin{bmatrix} \dfrac{P_0}{\omega_2}[\cos \omega_2(t - t_0) - \cos \omega_2 t] \\ \dfrac{P_0}{\omega_3}[\cos \omega_3(t - t_0) - \cos \omega_3 t] \end{bmatrix}$$

$$= \frac{P_0 t_0(2t - t_0)}{2\rho AL} \begin{bmatrix} 1 \\ 1 \\ 1 \end{bmatrix}$$

$$+ \frac{P_0}{4A}\sqrt{\frac{3}{E\rho}} \begin{bmatrix} \dfrac{2}{\omega_2}[\cos \omega_2(t - t_0) - \cos \omega_2 t] + \dfrac{1}{\omega_3}[\cos \omega_3(t - t_0) - \cos \omega_3 t] \\ -\dfrac{1}{\omega_3}[\cos \omega_3(t - t_0) - \cos \omega_3 t] \\ -\dfrac{2}{\omega_3}[\cos \omega_2(t - t_0) - \cos \omega_2 t] + \dfrac{1}{\omega_3}[\cos \omega_3(t - t_0) - \cos \omega_3 t) \end{bmatrix} \tag{o}$$

To determine the dynamic loads in the lower and upper sections of the rocket (elements 1 and 2) we use

$$F^{(1)} = 2AE\frac{U_2 - U_1}{L}$$

$$= P_0\sqrt{\frac{3E}{\rho L^2}}\left(-\frac{1 - \cos \omega_2 t}{\omega_2} - \frac{1 - \cos \omega_3 t}{\omega_3}\right) \qquad t < t_0$$

$$= P_0\sqrt{\frac{3E}{\rho L^2}}\left[-\frac{\cos \omega_2(t - t_0) - \cos \omega_2 t}{\omega_2} - \frac{\cos \omega_3(t - t_0) - \cos \omega_3 t}{\omega_3}\right] \qquad t > t_0 \tag{p}$$

and

$$F^{(2)} = 2AE\frac{U_3 - U_2}{L}$$

$$= P_0\sqrt{\frac{3E}{\rho L^2}}\left(-\frac{1 - \cos \omega_2 t}{\omega_2} + \frac{1 - \cos \omega_3 t}{\omega_3}\right) \qquad t < t_0$$

$$= P_0\sqrt{\frac{3E}{\rho L^2}}\left[-\frac{\cos \omega_2(t - t_0) - \cos \omega_2 t}{\omega_2} + \frac{\cos \omega_3(t - t_0) - \cos \omega_3 t}{\omega_3}\right] \qquad t > t_0 \tag{q}$$

13.15 RESPONSE DUE TO FORCED DISPLACEMENT AT ONE END OF A UNIFORM BAR

We shall consider problems from the previous section except that instead of applying a known force (thrust), we shall prescribe a known displacement (see Fig. 13.7). For the known displacement we shall take

$$U_1 = bt^2 \tag{a}$$

where b is a constant. The axial displacements U_2 and U_3 will be determined using the dynamic characteristics of the constrained system with $U_1 = 0$. By letting $\mathbf{U}_y = U_1$ and $\mathbf{U}_x = \{U_2 \quad U_3\}$ and taking $\mathbf{U}_x(0) = \mathbf{0}$ and $\dot{\mathbf{U}}_x(0) = \mathbf{0}$ we can use the equation

$$\mathbf{U}_x = \bar{\mathbf{p}}_e \bar{\mathbf{\Omega}}^{-1} \bar{\mathscr{M}}_e^{-1} \int_0^t \mathbf{sin}\,[\bar{\omega}(t-\tau)] \bar{\mathscr{P}}_e(\tau)\,d\tau \tag{13.70}$$

where all the required matrices have already been determined in Sec. 12.7. The relevant results are summarized below:

$$\bar{\mathbf{\Omega}} = [\bar{\omega}_1 \quad \bar{\omega}_2] = \sqrt{\frac{E}{\rho L^2}}\,[1.611 \quad 5.629] \tag{b}$$

$$\bar{\mathbf{p}}_e = \begin{bmatrix} \frac{\sqrt{2}}{2} & -\frac{\sqrt{2}}{2} \\ 1 & 1 \end{bmatrix} \tag{c}$$

$$\mathbf{M}_{xx} = \frac{\rho AL}{12}\begin{bmatrix} 4 & 1 \\ 1 & 2 \end{bmatrix} \tag{d}$$

$$\mathbf{M}_{xy} = \frac{\rho AL}{12}\begin{bmatrix} 1 \\ 0 \end{bmatrix} \tag{e}$$

$$\mathbf{K}_{xy} = \frac{2AE}{L}\begin{bmatrix} -1 \\ 0 \end{bmatrix} \tag{f}$$

The generalized mass matrix $\bar{\mathscr{M}}_e$ is evaluated as follows:

$$\bar{\mathscr{M}}_e = \bar{\mathbf{p}}_e^T \mathbf{M}_{xx} \bar{\mathbf{p}}_e = \begin{bmatrix} c & 1 \\ -c & 1 \end{bmatrix} \frac{\rho AL}{12} \begin{bmatrix} 4 & 1 \\ 1 & 2 \end{bmatrix} \begin{bmatrix} c & -c \\ 1 & 1 \end{bmatrix} = \frac{\rho AL}{6}\begin{bmatrix} 2+c & 0 \\ 0 & 2-c \end{bmatrix} \tag{g}$$

where $c = \sqrt{2}/2$. Hence

$$\bar{\mathscr{M}}_e^{-1} = \frac{12}{7\rho AL}\begin{bmatrix} 2-c & 0 \\ 0 & 2+c \end{bmatrix} \tag{h}$$

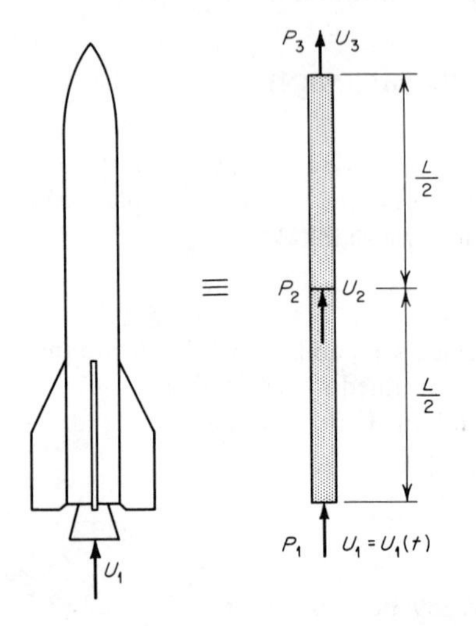

FIG. 13.7 Two-element idealization; specified displacement $U_1 = U_1(t)$.

We calculate next the equivalent force $\bar{\mathbf{P}}_x$ and the generalized force $\bar{\mathscr{P}}_e$

$$\bar{\mathbf{P}}_x = -\mathbf{M}_{xy}\ddot{\mathbf{U}}_y - \mathbf{K}_{xy}\mathbf{U}_y$$

$$= -\frac{\rho AL}{12}\begin{bmatrix}1\\0\end{bmatrix}2b - \frac{2AE}{L}\begin{bmatrix}-1\\0\end{bmatrix}bt^2 = \begin{bmatrix}-\dfrac{\rho ALb}{6} + \dfrac{2AEbt^2}{L}\\ 0\end{bmatrix} \tag{i}$$

$$\bar{\mathscr{P}}_e = \bar{\mathbf{p}}_e{}^T\bar{\mathbf{P}}_x = \begin{bmatrix}c & 1\\ -c & 1\end{bmatrix}\begin{bmatrix}-\dfrac{\rho ALb}{6} + \dfrac{2AEbt^2}{L}\\ 0\end{bmatrix}$$

$$= \begin{bmatrix}-\dfrac{c\rho ALb}{6} + \dfrac{2cAEbt^2}{L}\\ \dfrac{c\rho ALb}{6} - \dfrac{2cAEbt^2}{L}\end{bmatrix} \tag{j}$$

Evaluating first the integrals in Eq. (13.70) with the aid of Table 13.1, we have

$$\int_0^t \sin\,[\bar{\omega}(t-\tau)]\bar{\mathscr{P}}_e(\tau)\,d\tau$$

$$= \begin{bmatrix}-\dfrac{c}{6}\dfrac{\rho ALb}{\omega_1}(1-\cos\omega_1 t) + \dfrac{2cAE}{L}\dfrac{b}{\omega_1}\left(t^2 + \dfrac{2\cos\omega_1 t}{\omega_1{}^2} - \dfrac{2}{\omega_1{}^2}\right)\\ \dfrac{c}{6}\dfrac{\rho ALb}{\omega_2}(1-\cos\omega_2 t) - \dfrac{2cAE}{L}\dfrac{b}{\omega_2}\left(t^2 + \dfrac{2\cos\omega_2 t}{\omega_2{}^2} - \dfrac{2}{\omega_2{}^2}\right)\end{bmatrix} \tag{k}$$

Substituting now Eqs. (*b*), (*c*), (*g*), and (*k*) into Eq. (13.70), we obtain

$$\begin{bmatrix} U_2 \\ U_3 \end{bmatrix} = \begin{bmatrix} bt^2 - \dfrac{\rho b L^2}{E}\left(0.750 - 0.729 \cos 1.611\sqrt{\dfrac{E}{\rho L^2}}\,t - 0.021 \cos 5.629\sqrt{\dfrac{E}{\rho L^2}}\,t\right) \\ bt^2 - \dfrac{\rho b L^2}{E}\left(1.000 - 1.030 \cos 1.611\sqrt{\dfrac{E}{\rho L^2}}\,t + 0.030 \cos 5.629\sqrt{\dfrac{E}{\rho L^2}}\,t\right) \end{bmatrix} \qquad (l)$$

It may be interesting to compare the result for U_3 with the exact solution obtained by Laplace transforms operated on the differential equation for longitudinal displacements. The result is

$$\begin{aligned} U_3 &= bt^2 - \frac{\rho b L^2}{E} + \frac{\rho b L^2}{E}\frac{32}{\pi^3}\sum_{n=1}^{\infty}\frac{(-1)^{n+1}}{(2n-1)^3}\cos\left(\frac{2n-1}{2}\pi\sqrt{\frac{E}{\rho L^2}}\,t\right) \\ &= bt^2 - \frac{\rho b L^2}{E}\left(1.000 - 1.032 \cos 1.571\sqrt{\frac{E}{\rho L^2}}\,t + 0.038 \cos 4.713\sqrt{\frac{E}{\rho L^2}}\,t \cdots\right) \qquad (m) \end{aligned}$$

Comparing this result with our matrix solution, we note that even the two-element idealization leads to a very good agreement with the analytical solution.

PROBLEMS

13.1 The uniform cantilever beam of flexural rigidity EI and length l shown in Fig. 13.8*a* is subjected to a dynamic load P applied at the tip. The variation with time of load P is shown in Fig. 13.8*b*. Using only one element for the idealized system, determine the tip deflection as a function of time. Determine also the bending moment at the built-in end. The effects of shear deformations are to be neglected. The cross-sectional area of the beam is A and the density of the beam material is ρ.

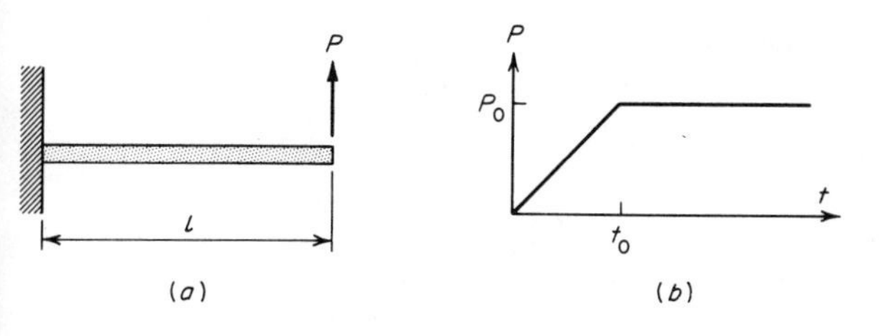

FIG. 13.8

13.2 Derive an equivalent form of Eq. (13.70) when $\mathbf{P}_x \neq \mathbf{0}$.

13.3 Solve the problem in Sec. 13.14 for a pulse loading represented by a half sine wave (case 14 in Table 13.1).

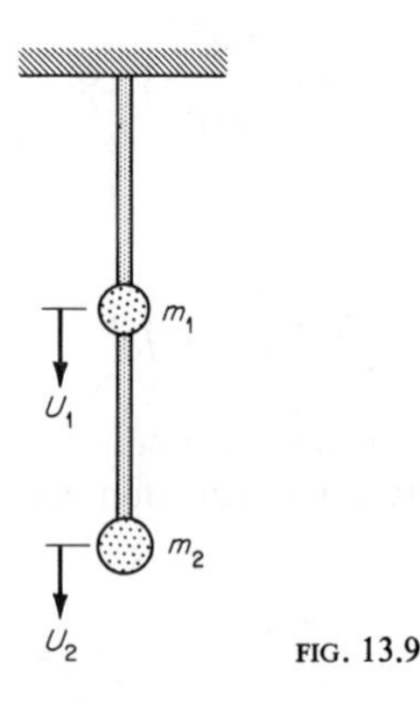

FIG. 13.9

13.4 An idealized structural system, shown in Fig. 13.9, consists of two concentrated masses m_1 and m_2. The mass matrix for this system is given by

$$\begin{bmatrix} 4 & 0 \\ 0 & 2 \end{bmatrix} \quad \text{lb-sec}^2/\text{ft}$$

The frequencies of vibration are

$$\omega_1 = 30.0 \text{ rad/sec} \qquad \text{and} \qquad \omega_2 = 60.0 \text{ rad/sec}$$

while the corresponding modal matrix for the displacements U_1 and U_2 is

$$\mathbf{p} = \begin{bmatrix} 1 & 1 \\ 2 & -1 \end{bmatrix}$$

Assuming that the system has 5 percent of critical damping in the first mode and 10 percent in the second mode, calculate coefficients in the damping matrix for this idealized system.

CHAPTER 14
STRUCTURAL SYNTHESIS

In Chap. 1 we described how a structural design is accomplished through a series of design iterations, in which a trial design is chosen, analyzed, and then modified by the designer after the examination of the numerical results. The modified structure is then reanalyzed, the analysis results examined, and the structure modified again, and so on, until a satisfactory design is achieved. Since the designer's judgment and intuition are influencing factors in the redesign process, and since only a small number of design iterations is practical, there is no guarantee that this process will evolve a design of minimum weight for all design conditions. Furthermore, such a process is very inefficient, and there is clearly a need for an automated design procedure. The requirement for the development of structural synthesis, in which the human designer is replaced by the computer, has become apparent particularly in aerospace structural designs, where weight minimization is of utmost importance. Recent advances in the fields of computer technology, linear and nonlinear optimization techniques, and matrix methods of structural analysis have provided all the necessary tools for the development of structural synthesis methods. Many papers concerned with structural synthesis have appeared in recent years. Among these perhaps the pioneering work by Gellatly and Gallagher[117,118] and also by Schmit and his associates[301–304] deserves special attention. Computer programs have been developed[119] for the determination of structural-member sizes to provide minimum weight under a set of specified load conditions and limit restrictions on stress and displacement; however, the relatively long computational times at present have precluded the application of these programs to large structural systems. This chapter is intended to provide the general concepts involved in the optimization procedures.

14.1 MATHEMATICAL FORMULATION OF THE OPTIMIZATION PROBLEM

The general optimization problem is described by the following set of functions:

$$W = W(x_1, \dots, x_n) \tag{14.1}$$

$$\begin{aligned} \psi_1 &= \psi_1(x_1, \dots, x_n) = 0 \\ &\cdots\cdots\cdots \\ \psi_m &= \psi_m(x_1, \dots, x_n) = 0 \end{aligned} \tag{14.2}$$

$$\begin{aligned} l_1 &\le \Phi_1(x_1, \dots, x_n) \le L_1 \\ &\cdots\cdots\cdots \\ l_s &\le \Phi_s(x_1, \dots, x_n) \le L_s \end{aligned} \tag{14.3}$$

The function W of the variables $x_1, \dots, x_n$ is the so called *criterion function,* which we seek to optimize. In our applications W represents the weight of the structure, and the optimization will be interpreted here as the determination of a set of design variables $x_1, \dots, x_n$ which will make W a minimum. The variables $x_1, \dots, x_n$ may be related by a set of *functional constraints* $\psi_1 = 0, \dots, \psi_m = 0$ and, in addition, they are constrained by a set of *regional constraints* Φ_i such that Φ_i must be contained between a lower limit l_i and an upper limit L_i, where $i = 1, \dots, s$. A special case of the regional constraint is a *side constraint* requiring that the variables $x_1, \dots, x_n$ all be positive.

The functional constraints $\psi_i = 0$ are not normally used in structural applications. Functional constraints would require that either some or all of the design variables be constrained by some relationship. In general, such design variables as thicknesses and areas may be varied freely; however, if, for example, aircraft machined skin panels are intended to have linear taper, the thicknesses of plate elements *within* these panels will have to be constrained accordingly to follow the linear taper. The regional constraints arise from the load conditions and the imposed limit restrictions on stress and displacement. For any given loading $\mathbf{P}$ the stress σ_i at some point in the structure is a function of the design variables $x_1, \dots, x_n$ and the loading $\mathbf{P}$, that is,

$$\sigma_i = \sigma_i(x_1, \dots, x_n, \mathbf{P}) \tag{14.4}$$

Similarly, for a representative displacement u_i we have

$$u_i = u_i(x_1, \dots, x_n, \mathbf{P}) \tag{14.5}$$

Thus, if we impose an upper and lower limit on the stress and displacement, we must have

$$\sigma^L \le \sigma_i(x_1, \dots, x_n, \mathbf{P}) \le \sigma^U \tag{14.6}$$

$$u^L \le u_i(x_1, \dots, x_n, \mathbf{P}) \le u^U \tag{14.7}$$

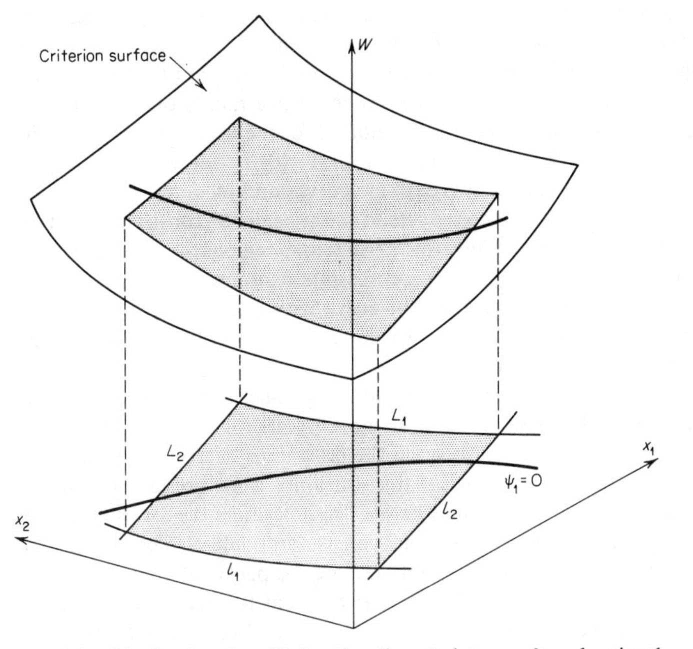

FIG. 14.1 Criterion function W, functional constraint $\psi_1 = 0$, and regional constraints $l_1 < \Phi_1 < L_1$ and $l_2 < \Phi_2 < L_2$ in three-dimensional space.

where the superscripts L and U refer respectively to the lower and upper limit values for stress or displacement. We can easily see that Eqs. (14.6) and (14.7) are of the same form as the regional constraints (14.3), since **P** may be treated here as a fixed parameter.

We describe next the geometry of the optimization problem. Since the journey into the n-dimensional space necessary for this description is not easy to visualize, we begin by discussing the three-dimensional problem first, for which the governing equations will be taken as

$$W = W(x_1,x_2) \tag{14.8}$$

$$\psi_1 = \psi_1(x_1,x_2) = 0 \tag{14.9}$$

$$l_1 \leq \Phi_1(x_1,x_2) \leq L_1 \tag{14.10}$$

$$l_2 \leq \Phi_2(x_1,x_2) \leq L_2 \tag{14.11}$$

The x_1 and x_2 axes will form a basis plane over which we shall construct the criterion function W, as shown in Fig. 14.1. The function W is therefore represented by a surface in three-dimensional space. The functional constraint

$\psi_1 = 0$ describes a curve on the basis plane. This curve can be projected upward onto the criterion surface to form a curve in space. A point moving along this space curve will rise or fall depending on the shape of the function W. The regional-constraint function $\Phi_1(x_1,x_2)$ represents a family of curves on the basis plane. Two extreme curves can be plotted for the lower and upper limits l_1 and L_1, as shown in Fig. 14.1. From Eq. (14.10) it follows that the permissible region lies between the two extreme curves and includes points on the curves. Similar curves can be constructed for the regional-constraint function Φ_2 in Eq. (14.11). By projecting the permissible regions upward onto the criterion surface we shall map out the permissible region on the surface. The lowest position of a point moving on the curve within that permissible region corresponds to the optimum, the lowest value of W subject to the constraining conditions (14.9) to (14.11).

It may be noted here that if we had an additional functional constraint ψ_2, the two constraints would intersect at one or more points on the basis plane, and only the projections of these points upon the criterion surface would represent permissible solutions. If no functional constraints were present, the whole mapped-out permissible region would be used to determine the lowest point on the criterion surface. Additional regional constraints could also be introduced. Such additional constraints would merely reduce the permissible region on the criterion surface. It should also be pointed out that the regional constraints must be such that they do not exclude each other. For example, two constraints $-10 \times 10^3 \leq \sigma_i \leq 10 \times 10^3$ and $50 \times 10^3 \leq \sigma_i \leq 80 \times 10^3$ on the stress σ_i are clearly inadmissible, since no solution for such constraints is possible.

We can now return to our original problem with n variables. We introduce $n + 1$ dimensions and the $(n+ 1)$st dimension will be used for plotting the values of W. The criterion function W in this case is a surface in our $n + 1$ space and is called a *hypersurface*. The basis plane is now n-dimensional and is referred to as a *hyperplane*. On this hyperplane we plot all functional constraints which form on it n-dimensional curves, referred to as *hypercurves*. These curves can be projected up to the hypersurface to form $(n + 1)$-dimensional hypercurves. The allowed regions formed by the regional constraints can then be mapped out on the basis hyperplane and projected up to the criterion hypersurface in the $(n + 1)$-dimensional space. The problem then reduces to the determination of the lowest value on the hypersurface within the permissible region formed by the projections of the functional and regional constraints from the basis hyperplane.

We may note from our three-dimensional example that the single functional constraint ψ_1 reduced the freedom of travel in two-dimensions within the permissible region on the basis plane to travel along the curve $\psi_1 = 0$. In the general case we have n degrees of freedom on the basis hyperplane, but as soon as we specify a functional constraint ψ_1, we lose one degree of freedom. When we

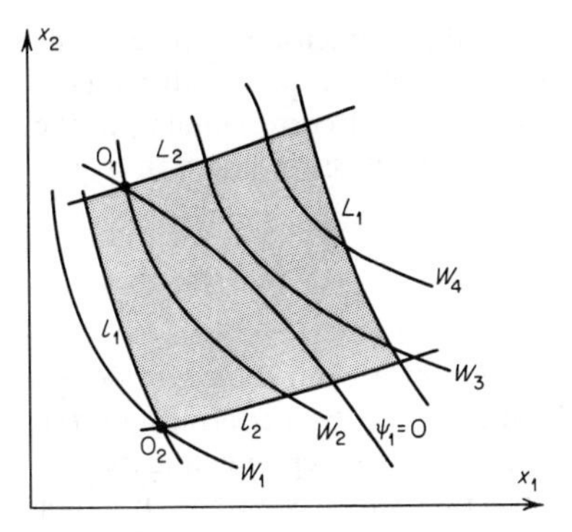

FIG. 14.2 Criterion function W, functional constraint $\psi_1 = 0$, and regional constraints $l_1 < \Phi_1 < L_1$ and $l_2 < \Phi_2 < L_2$ in two-dimensional space.

specify n functional constraints, the intersections of these functions will reduce the degrees of freedom to zero, and only a particular point, or possibly several distinct points, will represent possible solutions. Therefore, the optimization problem is said to degenerate if $m \geq n$, the constraints being too restrictive to allow for any design variations.

An alternative method of presenting the optimization problem is to use the n-dimensional space, instead of the $(n + 1)$-dimensional space, in which the values of the criterion function are plotted on the $(n + 1)$ axis. Thus the problem depicted in Fig. 14.1 would require only a two-dimensional space. This alternative representation is shown in Fig. 14.2. We note that the criterion function is plotted there as a family of curves for constant values of W; that is, the curves represent the equation

$$W(x_1,x_2) = W_i \tag{14.12}$$

where W_i is a constant. Several such curves are shown in Fig. 14.2. The functional constraint $\psi_1(x_1,x_2) = 0$ is plotted, as before, on the x_1x_2 plane, and the permissible region is mapped out on this plane by the regional constraints Φ_1 and Φ_2. In the previous representation we projected all curves from the basis onto the criterion surface $W(x_1,x_2)$ plotted on the x_3 axis. Here the process is reversed. The contours of constant values of W on the criterion surface are projected onto the basis plane.

From Fig. 14.2 it is clear that for the constraints given by Eqs. (14.9) to (14.11) the point 0_1 represents the position for the minimum value of W. If we removed the functional constraint ψ_1, any point within the quadrilateral-like region bounded by the regional constraints would represent a permissible design. If the minimum value of W is the criterion, point 0_2 gives the required design variables x_1 and x_2.

For the general case, the criterion function W is represented by a family of hypersurfaces in the n-dimensional space, each hypersurface corresponding to some constant value of W. Similarly the constraint functions ψ_i and the regional constraints Φ_i are represented by hypersurfaces in the n-dimensional space. The optimization problem then involves a search within the permissible region bounded by the constraint hypersurfaces to find a point corresponding to the lowest value of the criterion function represented by a set of criterion hypersurfaces.

14.2 STRUCTURAL OPTIMIZATION

Structural optimization is defined as the selection of a combination of design parameters which will allow the required functions to be performed by the structure at minimum weight. The design parameters describe configuration, member sizes, material properties, type of structure, etc. Naturally, it is not economically feasible at present to include all these parameters in an automated selection procedure. So far, in developing computer programs for structural optimization attention has been confined mainly to the selection of member sizes. The configuration parameters are the next logical choice for further development.

If the design parameters are the cross-sectional areas and thicknesses of each structural member, the weight of the structure is a linear function of the design parameters. For such cases, the weight can be expressed by the equation

$$W = \lambda_1 x_1 + \lambda_2 x_2 + \cdots + \lambda_n x_n \tag{14.13}$$

where $x_1, \ldots, x_n$ are the design variables (parameters) and $\lambda_1, \ldots, \lambda_n$ are constants depending upon material density and geometry of the structure. The weight function W is therefore the criterion function of the optizimation problem. In n-dimensional space this function is represented by a family of hypersurfaces, each hypersurface corresponding to some specific value of W. In three-dimensional space, when only three design variables are considered, the hypersurfaces reduce to a family of planes, as shown in Fig. 14.3.

The design variables $x_1, \ldots, x_n$ will be assumed to be independent, and therefore no functional constraints will be required. Only in special applications will the functional constraints be necessary in structural-optimization procedures.

The selection of the design variables is subject to the following type of constraints:

1. Geometric constraints: minimum and/or maximum areas or thicknesses
2. Stress constraints: maximum allowable stresses (either tensile or compressive)
3. Displacement constraints: minimum and/or maximum values

These constraints form the regional constraints, and they are applied to all load conditions for which the structure is designed. These constraints are represented by constraint hypersurfaces. Since the stresses and deflections are,

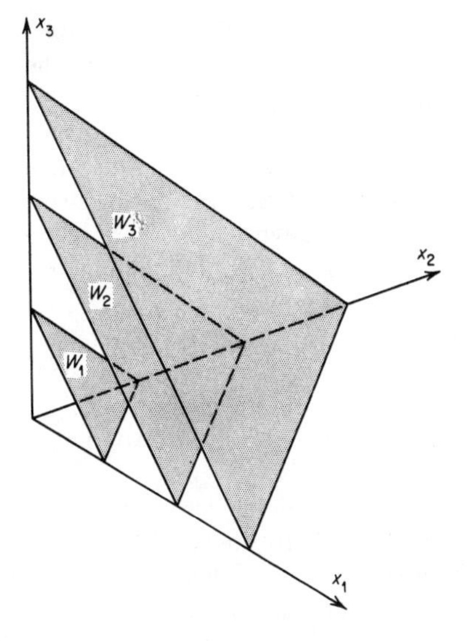

FIG. 14.3 Constant-weight planes (three-dimensional space).

in general, nonlinear functions of the design variables, the constraint hypersurfaces are nonlinear functions of $x_1, \ldots, x_n$. Typical constraint surfaces in the three-dimensional space are shown in Fig. 14.4, where additional constraints are imposed on the minimum values of x_1, x_2, and x_3. Thus in addition to the main constraint surfaces we have side constraints in the form of planes parallel to the coordinate direction. If the design point lies in the space above the

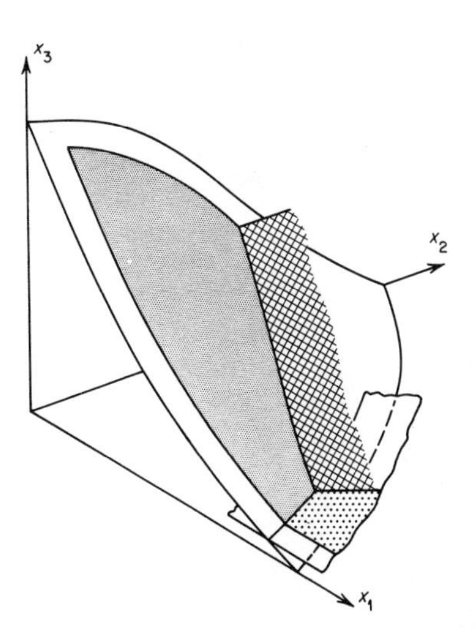

FIG. 14.4 Typical constraint surfaces in three-dimensional space (constraint surfaces $x_1 =$ const and $x_2 =$ const omitted for clarity).

constraint surfaces, the characteristic stress or displacement in the regional constraint functions will lie within the specified limits. The point where the constant-weight hypersurface touches the constraint hypersurface (or hypersurfaces) is the point for the minimum-weight design.

When constraints are imposed only on the stresses, we can use an iterative procedure to redesign the structure so that each element reaches limiting stress under at least one of the load conditions. Such a design is described as the *fully stressed design*, and, in general, it is not far removed from the minimum-weight design. The design variables for the iterative fully stressed design converge to a vertex of the n hypersurfaces representing n constraints on the stresses. Schmit[301] has demonstrated that the fully stressed design for some loading conditions may lead to an inefficient design, and for such cases the minimum-weight design should be used. To decide whether the fully stressed design is an optimum we can use a method proposed by Razani.[279]

Details of the numerical procedures for obtaining the design variables for the optimum design can be obtained from papers by Gellatly and Gallagher[117,118] and also by Schmit.[301–304] Basically these papers propose different modes of travel in the n-dimensional space to reach the minimum-weight point located on the constraining hypersurfaces. Gellatly and Gallagher use a procedure made up from three different modes: initial step (fully stressed design) used only once or several times at the beginning of the computing, followed by repeated application of the steepest-descent and side-step modes. Schmit used solely the latter two modes of travel and describes his method as the method of alternate steps. A paper by Razani[279] gives an excellent account of the mathematical formulation of both the fully stressed and minimum-weight design procedures.

CHAPTER 15 NONLINEAR STRUCTURAL ANALYSIS

Two types of nonlinearities occur in structural problems. The first type is referred to as *material nonlinearity* and is due to the nonlinearly elastic and plastic or viscoelastic behavior of the structural material. The second type is referred to as *geometric nonlinearity*, and it occurs when the deflections are large enough to cause significant changes in the geometry of the structure, so that the equations of equilibrium must be formulated for the deformed configuration.

The matrix analysis methods developed for linear structures can be extended to include the above-mentioned nonlinearities. Both the matrix displacements and force methods of analysis can be utilized for that purpose. Because of the presence of nonlinear terms, solutions to the governing matrix equations can no longer be obtained explicitly, as with linear structures, and consequently we must use iterative procedures. As a by-product of the large-deflection matrix analysis we can also formulate the eigenvalue equations for structural instability. Thus buckling loads for structures idealized into an assembly of discrete elements can be determined, an important consideration in designing lightweight structures. Furthermore, creep behavior of structures, including creep buckling, can be analyzed by matrix methods.

This chapter presents only the general principles involved in extending matrix methods into the nonlinear regime. For further details of matrix methods of analysis of nonlinear structures the extensive literature on this subject should be consulted.

15.1 MATRIX DISPLACEMENT ANALYSIS FOR LARGE DEFLECTIONS

When large deflections are present, the equations of force equilibrium must be formulated for the deformed configuration of the structure. This means that the linear relationship $\mathbf{P} = \mathbf{KU}$ between the applied forces $\mathbf{P}$ and the displacements $\mathbf{U}$ can no longer be used. To account for the effects of changes in geometry as the applied loading is increased we may obtain solutions for the displacements $\mathbf{U}$ by treating this nonlinear problem in a sequence of linear steps, each step representing a load increment. However, because of the presence of large deflections, strain-displacement equations contain nonlinear terms, which must be included in calculating the stiffness matrix $\mathbf{K}$. No discussion of these nonlinear terms in the strain-displacement equations will be given here, and for their derivation the interested reader is referred to standard texts on the nonlinear theory of elasticity.

The nonlinear terms in the strain-displacement equations modify the element stiffness matrix $\mathbf{k}$ so that

$$\mathbf{k} = \mathbf{k}_E + \mathbf{k}_G \tag{15.1}$$

where $\mathbf{k}_E$ is the standard elastic stiffness matrix calculated for the element geometry at the start of the step and $\mathbf{k}_G$ is the so-called geometrical stiffness matrix, which depends not only on the geometry but also on the initial internal forces (stresses) existing at the start of the step. Alternative names for the geometrical stiffness matrix are the incremental stiffness matrix and the initial-stress stiffness matrix. The elastic and geometrical stiffness matrices are calculated for each element and then assembled into the total stiffness matrix

$$\mathbf{K} = \mathbf{K}_E + \mathbf{K}_G \tag{15.2}$$

using the conventional procedures established for the linear analysis in Chap. 6.

The incremental displacements and internal (element) forces are calculated in the conventional manner for each load increment. If necessary, the elastic constants used in the elastic stiffness matrix can be modified for each step. Total displacements for the final values of the applied loading are then obtained by summing the incremental values. The incremental step procedure used in this application is presented symbolically in Table 15.1. It should be noted that $\mathbf{K}_G(0) = \mathbf{0}$ since the geometrical stiffness matrix is proportional to the internal forces, which are zero at the start of step 1.

The concept of incremental steps in treating the large-deflection problem by the matrix displacement method was first given by Turner et al.[330] Detailed discussion of the analysis of pin-jointed trusses, beam structures, and plates was given by Martin.[214,216] Argyris[14,18] has developed the general concept of geometrical stiffnesses extensively and applied it also to plate and shell elements. A general method of determining geometrical stiffness matrices for arbitrary

TABLE 15.1 INCREMENTAL STEP PROCEDURE FOR THE DISPLACEMENT METHOD

Step	Stiffness	Incremental displacements	Element forces
1	$\mathbf{K}_E(0) + \mathbf{K}_G(0)$	$\Delta\mathbf{U}_1$	$\mathbf{S}_1$
2	$\mathbf{K}_E(U_1) + \mathbf{K}_G(U_1)$	$\Delta\mathbf{U}_2$	$\mathbf{S}_2$
3	$\mathbf{K}_E(U_2) + \mathbf{K}_G(U_2)$	$\Delta\mathbf{U}_3$	$\mathbf{S}_3$
.	....	.	.
n	$\mathbf{K}_E(U_{n-1}) + \mathbf{K}_G(U_{n-1})$	$\Delta\mathbf{U}_n$	$\mathbf{S}_n$
Total displacements		$\mathbf{U}_n = \sum_{i=1}^{n} \Delta\mathbf{U}_i$	

structural elements has been given by Przemieniecki.[387] This method allows for a systematic inclusion of different nonlinear terms from the strain-displacement equations and is considerably simpler than the conventional techniques requiring the determination of strain energy, as described in this text.

As in any piecewise linear procedure applied to a nonlinear problem, the accuracy of the total value of displacements and internal forces increases as the number of linear steps is increased. It should be mentioned, however, that in some problems even relatively large linear steps may result in a good approximation to the true nonlinear behavior of the structure.

For the first step in Table 15.1 we have

$$\begin{aligned}\mathbf{U}_1 &= [\mathbf{K}_E(0)]^{-1}\mathbf{P} \\ &= [\mathbf{K}_E(0)]^{-1}\lambda\mathbf{P}^* \end{aligned} \tag{15.3}$$

In Eq. (15.3) the external loading $\mathbf{P}$ is expressed as

$$\mathbf{P} = \lambda\mathbf{P}^* \tag{15.4}$$

where λ is a constant and $\mathbf{P}^*$ represents the relative magnitudes of the applied forces. Since the geometrical stiffness matrix is proportional to the internal forces at the start of the loading step, it follows that

$$\mathbf{K}_G = \lambda\mathbf{K}_G^* \tag{15.5}$$

where $\mathbf{K}_G^*$ is the geometrical stiffness matrix for unit values of the applied loading ($\lambda = 1$). The elastic stiffness matrix $\mathbf{K}_E$ can be treated as a constant for quite a wide range of displacements $\mathbf{U}$. Hence we may write

$$(\mathbf{K}_E + \lambda\mathbf{K}_G^*)\mathbf{U} = \lambda\mathbf{P}^* \tag{15.6}$$

The displacements $\mathbf{U}$ may therefore be determined from

$$\mathbf{U} = (\mathbf{K}_E + \lambda\mathbf{K}_G^*)^{-1}\lambda\mathbf{P}^* \tag{15.7}$$

From the formal definition of the matrix inverse as the adjoint matrix divided by the determinant of the coefficients we note that the displacements **U** tend to infinity when

$$|\mathbf{K}_E + \lambda \mathbf{K}_G^*| = 0 \tag{15.8}$$

The lowest value of λ from Eq. (15.8) gives the classical buckling load for the idealized structure. We denote this value as λ_{crit}, and the buckling loads are found from

$$\mathbf{P}_{\text{crit}} = \lambda_{\text{crit}} \mathbf{P}^* \tag{15.9}$$

It should be noted that several loads may be applied simultaneously in **P**. Thus problems involving biaxial compression can be investigated. Several simple examples of the analysis of structural stability will be given to illustrate the application of this technique.

15.2 GEOMETRICAL STIFFNESS FOR BAR ELEMENTS

We consider next the pin-jointed bar element shown in Fig. 15.1. Under the action of applied loading the bar is displaced from its original location AB to $A'B'$. The displacements in the x and y directions respectively of the end A are u_1 and u_2 while those of the end B are u_3 and u_4. The cross-sectional area of the bar is A, its length is l, and the Young's modulus of the bar material is E.

The strain ϵ_{xx} in the direction of the bar axis must be determined from the large-deflection strain-displacement equation

$$\epsilon_{xx} = \frac{\partial u_x}{\partial x} + \frac{1}{2}\left(\frac{\partial u_y}{\partial x}\right)^2 \tag{15.10}$$

where the second term is nonlinear. The displacements u_x and u_y vary linearly along the length, and they are obtained from

$$\begin{bmatrix} u_x \\ u_y \end{bmatrix} = \begin{bmatrix} 1-\xi & 0 & \xi & 0 \\ 0 & 1-\xi & 0 & \xi \end{bmatrix} \begin{bmatrix} u_1 \\ u_2 \\ u_3 \\ u_4 \end{bmatrix} \tag{15.11}$$

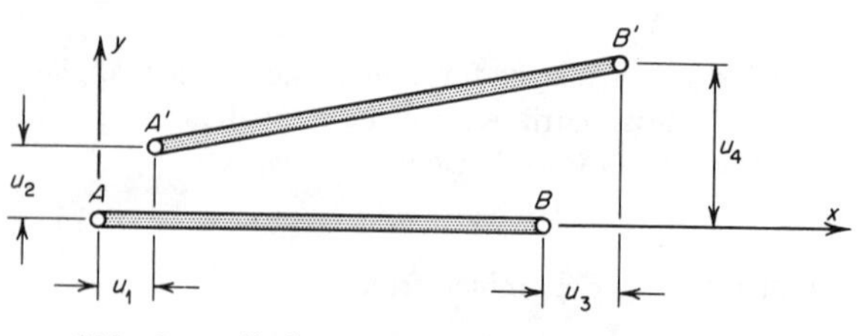

FIG. 15.1 Large displacements on a bar element.

where $\xi = x/l$. Hence

$$\frac{\partial u_x}{\partial x} = \frac{1}{l}(-u_1 + u_3) \tag{15.12}$$

and
$$\frac{\partial u_y}{\partial x} = \frac{1}{l}(-u_2 + u_4) \tag{15.13}$$

The strain energy U_i stored in the bar with a linear stress-strain law (Hookean elasticity) is determined from

$$\begin{aligned} U_i &= \frac{1}{2}\int_v E\epsilon_{xx}^2\,dV = \frac{AE}{2}\int_0^l \epsilon_{xx}^2\,dx \\ &= \frac{AE}{2}\int_0^l \left[\frac{\partial u_x}{\partial x} + \frac{1}{2}\left(\frac{\partial u_y}{\partial x}\right)^2\right]^2 dx \\ &= \frac{AE}{2}\int_0^l \left[\left(\frac{\partial u_x}{\partial x}\right)^2 + \frac{\partial u_x}{\partial x}\left(\frac{\partial u_y}{\partial x}\right)^2 + \frac{1}{4}\left(\frac{\partial u_y}{\partial x}\right)^4\right] dx \end{aligned} \tag{15.14}$$

Substituting Eqs. (15.12) and (15.13) into the strain energy expression (15.14) and then neglecting the higher-order term $(\partial u_y/\partial x)^4$, we have that

$$\begin{aligned} U_i &= \frac{AE}{2l^2}\int_0^l (u_1^2 - 2u_1u_3 + u_3^2)\,dx + \frac{AE}{2l^3}\int_0^l (u_3 - u_1)(u_2^2 - 2u_2u_4 + u_4^2)\,dx \\ &= \frac{AE}{2l}(u_1^2 - 2u_1u_3 + u_3^2) + \frac{AE}{2l^2}(u_3 - u_1)(u_2^2 - 2u_2u_4 + u_4^2) \end{aligned} \tag{15.15}$$

We may note that even for relatively large deflections the quantity $AE(u_3 - u_1)/l$ may be treated as a constant equal to the axial tensile force in the bar. Hence introducing

$$F = \frac{AE}{l}(u_3 - u_1) \tag{15.16}$$

we have that

$$U_i = \frac{AE}{2l}(u_1^2 - 2u_1u_3 + u_3^2) + \frac{F}{2l}(u_2^2 - 2u_2u_4 + u_4^2) \tag{15.17}$$

Castigliano's theorem (part I) is applicable to large deflections provided appropriate expressions for the strains are used. Hence, using Eq. (15.17), we obtain the element force-displacement relations

$$\begin{aligned} S_1 &= \frac{\partial U_i}{\partial u_1} = \frac{AE}{l}(u_1 - u_3) \qquad & S_2 &= \frac{\partial U_i}{\partial u_2} = \frac{F}{l}(u_2 - u_4) \\ S_3 &= \frac{\partial U_i}{\partial u_3} = \frac{AE}{l}(-u_1 + u_3) \qquad & S_4 &= \frac{\partial U_i}{\partial u_4} = \frac{F}{l}(-u_2 + u_4) \end{aligned} \tag{15.18}$$

Collecting Eqs. (15.18) into a single matrix equation, we obtain

$$\begin{bmatrix} S_1 \\ S_2 \\ S_3 \\ S_4 \end{bmatrix} = \frac{AE}{l}\begin{bmatrix} 1 & 0 & -1 & 0 \\ 0 & 0 & 0 & 0 \\ -1 & 0 & 1 & 0 \\ 0 & 0 & 0 & 0 \end{bmatrix}\begin{bmatrix} u_1 \\ u_2 \\ u_3 \\ u_4 \end{bmatrix} + \frac{F}{l}\begin{bmatrix} 0 & 0 & 0 & 0 \\ 0 & 1 & 0 & -1 \\ 0 & 0 & 0 & 0 \\ 0 & -1 & 0 & 1 \end{bmatrix}\begin{bmatrix} u_1 \\ u_2 \\ u_3 \\ u_4 \end{bmatrix} \tag{15.19}$$

which may be written symbolically as

$$\begin{aligned} \mathbf{S} &= (\mathbf{k}_E + \mathbf{k}_G)\mathbf{u} \\ &= \mathbf{k}\mathbf{u} \end{aligned} \tag{15.20}$$

Thus we can see clearly that the total stiffness of the bar element consists of two parts, the elastic stiffness $\mathbf{k}_E$ and the geometrical stiffness $\mathbf{k}_G$. The elastic stiffness matrix $\mathbf{k}_E$ is the same as that used in the linear analysis. The geometrical stiffness $\mathbf{k}_G$ is given by

$$\mathbf{k}_G = \frac{F}{l}\begin{bmatrix} 0 & 0 & 0 & 0 \\ 0 & 1 & 0 & -1 \\ 0 & 0 & 0 & 0 \\ 0 & -1 & 0 & 1 \end{bmatrix} \tag{15.21}$$

It should be noted that $\mathbf{k}_G$ either increases or decreases the direct stiffness coefficients (diagonal terms), depending on the sign of the force F.

15.3 GEOMETRICAL STIFFNESS FOR BEAM ELEMENTS

The displacement distribution on a beam element is given by [see Eq. (11.30)]

$$\begin{bmatrix} u_x \\ u_y \end{bmatrix} = \begin{bmatrix} \overset{1}{1-\xi} & \overset{2}{6(\xi-\xi^2)\eta} & \overset{3}{(-1+4\xi-3\xi^2)l\eta} & \overset{4}{\xi} & \overset{5}{6(-\xi+\xi^2)\eta} & \overset{6}{(2\xi-3\xi^2)l\eta} \\ 0 & 1-3\xi^2+2\xi^3 & (\xi-2\xi^2+\xi^3)l & 0 & 3\xi^2-2\xi^3 & (-\xi^2+\xi^3)l \end{bmatrix}\begin{bmatrix} u_1 \\ u_2 \\ \cdot \\ \cdot \\ \cdot \\ u_6 \end{bmatrix} \tag{15.22}$$

where $u_1, \ldots, u_6$ are the element displacements shown in Fig. 15.2. In calculating the strain energy U_i we shall neglect the contributions from the shearing strains. Thus only the normal strains ϵ_{xx} will be included. These strains for large deflections on a beam in bending are determined from

$$\epsilon_{xx} = \frac{\partial u_0}{\partial x} - \frac{\partial^2 u_y}{\partial x^2}y + \frac{1}{2}\left(\frac{\partial u_y}{\partial x}\right)^2 \tag{15.23}$$

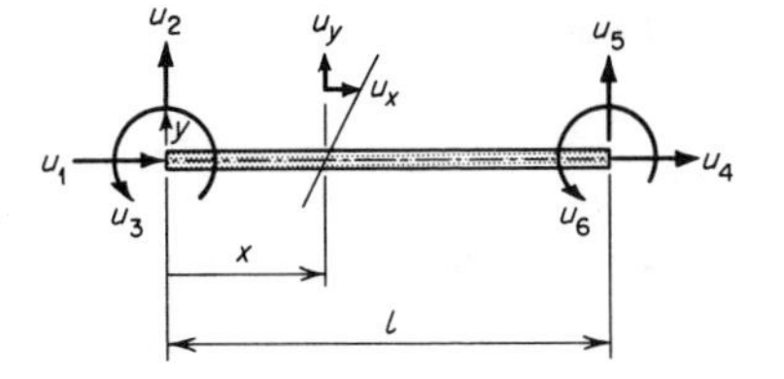

FIG. 15.2 Positive directions of displacements on a beam element.

where y is measured from the neutral axis of the beam and u_0 denotes the u_x displacement at $y = 0$. We use Eq. (15.23), and the strain energy U_i is given by

$$
\begin{aligned}
U_i &= \frac{E}{2}\int_v \epsilon_{xx}{}^2 \, dV \\
&= \frac{E}{2}\int_v \left[\frac{\partial u_0}{\partial x} - \frac{\partial^2 u_y}{\partial x^2} y + \frac{1}{2}\left(\frac{\partial u_y}{\partial x}\right)^2\right]^2 dV \\
&= \frac{E}{2}\int_{x=0}^{l}\int_A \left[\left(\frac{\partial u_0}{\partial x}\right)^2 + \left(\frac{\partial^2 u_y}{\partial x^2}\right)^2 y^2 + \frac{1}{4}\left(\frac{\partial u_y}{\partial x}\right)^4 - 2\frac{\partial u_0}{\partial x}\frac{\partial^2 u_y}{\partial x^2} y \right. \\
&\qquad \left. - \frac{\partial^2 u_y}{\partial x^2}\left(\frac{\partial u_y}{\partial x}\right)^2 y + \frac{\partial u_0}{\partial x}\left(\frac{\partial u_y}{\partial x}\right)^2\right] dx \, dA
\end{aligned}
$$

The higher-order term $\frac{1}{4}(\partial u_y/\partial x)^4$ can be neglected in the above expression. Integrating over the cross-sectional area A and noting that since y is measured from the neutral axis, all integrals of the form $\int y \, dA$ must vanish, we have that

$$
U_i = \frac{EA}{2}\int_0^l \left(\frac{\partial u_0}{\partial x}\right)^2 dx + \frac{EI}{2}\int_0^l \left(\frac{\partial^2 u_y}{\partial x^2}\right)^2 dx + \frac{EA}{2}\int_0^l \frac{\partial u_0}{\partial x}\left(\frac{\partial u_y}{\partial x}\right)^2 dx \tag{15.24}
$$

where I denotes the moment of inertia of the cross section. We may note that the first two integrals in (15.24) represent the linear strain energy while the third integral is the contribution from the nonlinear component of the strain.

From Eq. (15.22) we obtain

$$
\frac{\partial u_0}{\partial x} = \frac{1}{l}(-u_1 + u_4) \tag{15.25}
$$

$$
\frac{\partial u_y}{\partial x} = \frac{1}{l}[6(-\xi + \xi^2)u_2 + (1 - 4\xi + 3\xi^2)lu_3 + 6(\xi - \xi^2)u_5 + (-2\xi + 3\xi^2)lu_6] \tag{15.26}
$$

$$
\frac{\partial^2 u_y}{\partial x^2} = \frac{1}{l^2}[6(-1 + 2\xi)u_2 + 2(-2 + 3\xi)lu_3 + 6(1 - 2\xi)u_5 + 2(-1 + 3\xi)lu_6] \tag{15.27}
$$

Substitution of Eqs. (15.25) to (15.27) into (15.24) and integration leads to

$$
\begin{aligned}
U_i = {} & \frac{EA}{2l}(u_1^2 - 2u_1u_4 + u_4^2) \\
& + \frac{2EI}{l^3}(3u_2^2 + l^2u_3^2 + 3u_5^2 + l^2u_6^2 + 3lu_2u_3 - 6u_2u_5 + 3lu_2u_6 \\
& \qquad\qquad - 3lu_3u_5 + l^2u_3u_6 - 3lu_5u_6) \\
& + \frac{EA}{l^2}(u_4 - u_1)(\tfrac{3}{5}u_2^2 + \tfrac{1}{15}l^2u_3^2 + \tfrac{3}{5}u_5^2 + \tfrac{1}{15}l^2u_6^2 \\
& + \tfrac{1}{10}lu_2u_3 - \tfrac{6}{5}u_2u_5 + \tfrac{1}{10}lu_2u_6 - \tfrac{1}{10}lu_3u_5 - \tfrac{1}{30}l^2u_3u_6 - \tfrac{1}{10}lu_5u_6)
\end{aligned}
\tag{15.28}
$$

As in the case of the pin-jointed bar, we may introduce

$$
F = \frac{EA}{l}(u_4 - u_1) \simeq \text{const} \tag{15.29}
$$

and apply Castigliano's theorem (part I) to the strain energy expression (15.28). This results in the following element force-displacement equation

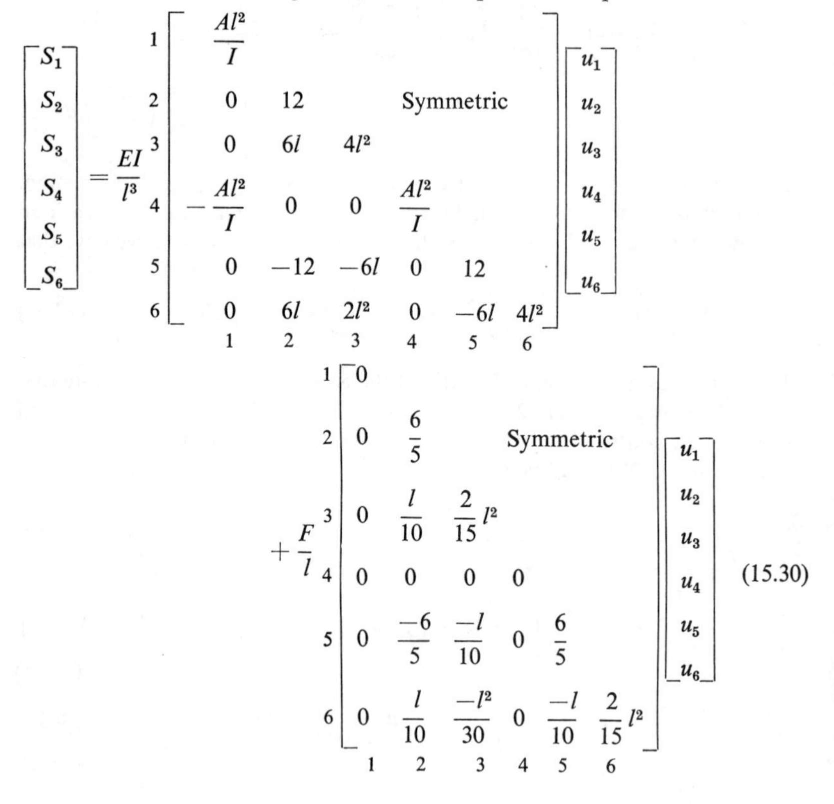

$$
\begin{bmatrix} S_1 \\ S_2 \\ S_3 \\ S_4 \\ S_5 \\ S_6 \end{bmatrix}
= \frac{EI}{l^3}
\begin{bmatrix}
\frac{Al^2}{I} & & & & & \\
0 & 12 & & \text{Symmetric} & & \\
0 & 6l & 4l^2 & & & \\
-\frac{Al^2}{I} & 0 & 0 & \frac{Al^2}{I} & & \\
0 & -12 & -6l & 0 & 12 & \\
0 & 6l & 2l^2 & 0 & -6l & 4l^2
\end{bmatrix}
\begin{bmatrix} u_1 \\ u_2 \\ u_3 \\ u_4 \\ u_5 \\ u_6 \end{bmatrix}
+ \frac{F}{l}
\begin{bmatrix}
0 & & & & & \\
0 & \frac{6}{5} & & \text{Symmetric} & & \\
0 & \frac{l}{10} & \frac{2}{15}l^2 & & & \\
0 & 0 & 0 & 0 & & \\
0 & \frac{-6}{5} & \frac{-l}{10} & 0 & \frac{6}{5} & \\
0 & \frac{l}{10} & \frac{-l^2}{30} & 0 & \frac{-l}{10} & \frac{2}{15}l^2
\end{bmatrix}
\begin{bmatrix} u_1 \\ u_2 \\ u_3 \\ u_4 \\ u_5 \\ u_6 \end{bmatrix}
\tag{15.30}
$$

which again may be written symbolically as

$$\mathbf{S} = (\mathbf{k}_E + \mathbf{k}_G)\mathbf{u}$$

The elastic stiffness matrix $\mathbf{k}_E$ is the conventional stiffness matrix obtained in Chap. 5 [see Eq. (5.121)], while the geometrical stiffness matrix is given by

$$\mathbf{k}_G = \frac{F}{l}\begin{array}{c} 1 \\ 2 \\ 3 \\ 4 \\ 5 \\ 6 \end{array}\begin{bmatrix} 0 & & & & & \\ 0 & \frac{6}{5} & & \text{Symmetric} & & \\ 0 & \frac{l}{10} & \frac{2}{15}l^2 & & & \\ 0 & 0 & 0 & 0 & & \\ 0 & \frac{-6}{5} & \frac{-l}{10} & 0 & \frac{6}{5} & \\ 0 & \frac{l}{10} & \frac{-l^2}{30} & 0 & \frac{-l}{10} & \frac{2}{15}l^2 \end{bmatrix} \tag{15.31}$$
$$\qquad\quad 1 \quad\; 2 \quad\;\; 3 \quad\;\; 4 \quad 5 \quad\; 6$$

If instead of the expression for the slope $\partial u_y/\partial x$ in (15.26) we assume that

$$\frac{\partial u_y}{\partial x} \simeq \frac{1}{l}(u_5 - u_2) \tag{15.32}$$

then the calculation of the nonlinear term in the strain energy expression (15.24) can be considerably simplified. Such an assumption implies the use of an average constant slope over the whole length of the element, and this can be justified only if the size of the element is small in relation to the overall length of the actual beam structure. When this assumption is used, the simplified geometrical stiffness matrix becomes

$$\mathbf{k}_G = \frac{F}{l}\begin{array}{c} 1 \\ 2 \\ 3 \\ 4 \\ 5 \\ 6 \end{array}\begin{bmatrix} 0 & 0 & 0 & 0 & 0 & 0 \\ 0 & 1 & 0 & 0 & -1 & 0 \\ 0 & 0 & 0 & 0 & 0 & 0 \\ 0 & 0 & 0 & 0 & 0 & 0 \\ 0 & -1 & 0 & 0 & 1 & 0 \\ 0 & 0 & 0 & 0 & 0 & 0 \end{bmatrix} \tag{15.33}$$
$$\qquad\quad 1 \quad 2 \quad 3 \quad 4 \quad 5 \quad 6$$

The above geometrical stiffness matrix is usually referred to as the *string stiffness.*

15.4 MATRIX FORCE ANALYSIS FOR LARGE DEFLECTIONS

The linear matrix force method of structural analysis can be extended to non-linear structures using the concept of fictitious forces and deformations. The idea was proposed by Denke in a report of Douglas Aircraft Company.[363] The basic concepts involved and the details of the analysis for structures idealized into bars and constant shear flow panels were presented by Warren.[335] An equivalent theory was developed independently by Lansing, Jones, and Ratner.[198] Griffin[128] and Durrett[84] analyzed pin-jointed space frameworks using the concept of fictitious forces and deformations. A detailed description of the method is also given by Argyris.[14]

The concept of the fictitious forces and deformations will be illustrated for a pin-jointed bar element. A typical element with large deflections is shown in Fig. 15.3. If the deflections are small (strictly speaking the deflections should be infinitesimal), the undeformed configuration AB is used to determine the internal force F' in the element. If the deflections are large, the deformed configuration $A'B'$ must be used to obtain the element force F. It is clear from Fig. 15.3 that if the angle θ is not too large, then

$$F = F' \tag{15.34}$$

The forces F have transverse components, and therefore in order to satisfy equilibrium in the deformed configuration we may introduce external forces P_Φ given by

$$P_\Phi \simeq F'\theta = \frac{F'(u_4 - u_2)}{l} \tag{15.35}$$

where θ is the angle of rotation of the element such that $\sin\theta \simeq \theta$ and l is the element length. Similarly, because of rotation of the element there will be an apparent contraction of the element, as indicated in Fig. 15.4. This contraction is given by

$$v_\Phi \simeq -\tfrac{1}{2}l\theta^2 = -\frac{1}{2l}(u_4 - u_2)^2 \tag{15.36}$$

The transverse forces P_Φ are described as the fictitious forces, and the contraction v_Φ is described as the fictitious deformation. In a similar manner fictitious forces and deformations can be obtained for other elements.

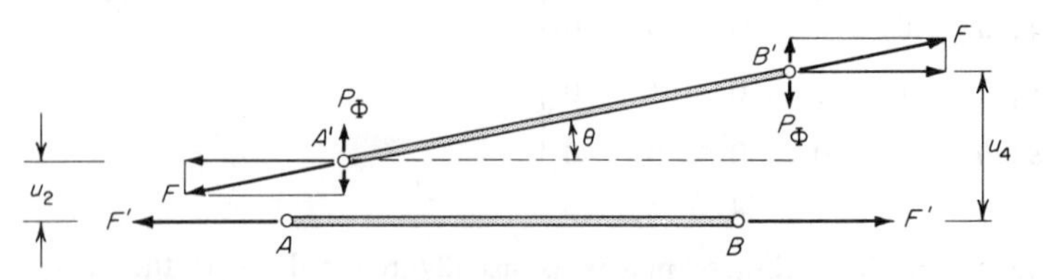

FIG. 15.3 Fictitious forces P_Φ on a pin-jointed bar element.

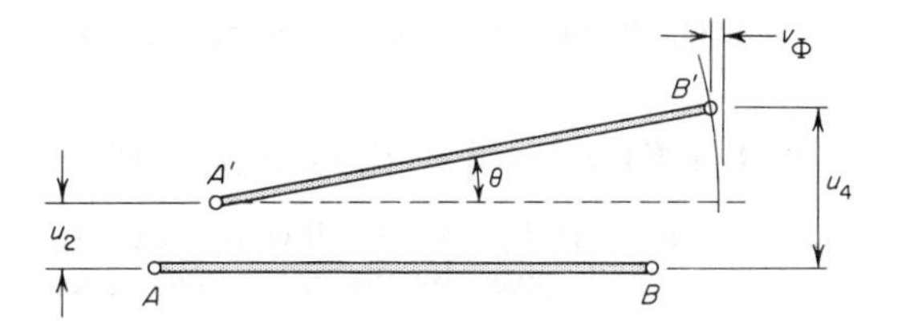

FIG. 15.4 Fictitious deformations v_Φ on a pin-jointed bar element.

The fictitious forces $\mathbf{P}_\Phi$ and fictitious deformations $\mathbf{v}_\Phi$ can be regarded as additional external loading to be applied to the undeformed structure. These fictitious forces and deformations, when applied together with the actual loading, will cause the linear-theory analysis to give the correct results for large deflections. Since both $\mathbf{P}_\Phi$ and $\mathbf{v}_\Phi$ are nonlinear functions of the displacements, the matrix analysis leads to nonlinear equations, which must be solved by iterative techniques.

The element forces and deflections can now be written as

$$\mathbf{F} = \mathbf{F}_P(\mathbf{P} + \mathbf{P}_\Phi) + \mathbf{F}_v(\mathbf{v}_T + \mathbf{v}_\Phi) \tag{15.37}$$

and
$$\mathbf{r} = \mathscr{F}_P(\mathbf{P} + \mathbf{P}_\Phi) + \mathscr{F}_v(\mathbf{v}_T + \mathbf{v}_\Phi) \tag{15.38}$$

The matrices $\mathbf{F}_P$ and $\mathbf{F}_v$ represent element forces due to unit values of external loads $\mathbf{P}$ and thermal (or initial) deformations $\mathbf{v}_T$, respectively. The matrices $\mathscr{F}_P$ and $\mathscr{F}_v$ represent structural deflections due to unit values of $\mathbf{P}$ and $\mathbf{v}_T$. The deflections $\mathbf{r}$ refer here to all displacements necessary to describe the fictitious forces $\mathbf{P}_\Phi$ and deformations $\mathbf{v}_\Phi$. The fictitious force matrix $\mathbf{P}_\Phi$ represents the resultant of all external fictitious forces acting at each joint. Consequently, the number of rows in the matrix $\mathbf{P}$ must correspond to that of $\mathbf{P}_\Phi$ even if only a few external loads are actually applied. The fictitious-deformation matrix $\mathbf{v}_\Phi$ is simply a column matrix of fictitious element deformations. The matrices $\mathbf{F}_P$, $\mathbf{F}_v$, $\mathscr{F}_P$, and $\mathscr{F}_v$ can be determined by linear force or displacement methods. The matrices $\mathbf{P}_\Phi$ and $\mathbf{v}_\Phi$ are dependent on the deflections $\mathbf{r}$, and because of this dependence Eqs. (15.37) and (15.38) are nonlinear and must be solved by iterative methods. This approach, however, is impractical, and it is preferable to eliminate the element forces $\mathbf{F}$ from Eqs. (15.37) and (15.38) and formulate equations for the deflections $\mathbf{r}$.

Since the fictitious forces $\mathbf{P}_\Phi$ are linear functions of the element forces $\mathbf{F}$, we must have that

$$\mathbf{P}_\Phi = \mathbf{\Gamma F} \tag{15.39}$$

where $\mathbf{\Gamma}$ is a matrix of fictitious forces due to unit values of the element forces $\mathbf{F}$. Substitution of Eq. (15.39) into (15.37) and solution for $\mathbf{F}$ leads to

$$\mathbf{F} = (\mathbf{I} - \mathbf{F}_P\mathbf{\Gamma})^{-1}[\mathbf{F}_P\mathbf{P} + \mathbf{F}_v(\mathbf{v}_T + \mathbf{v}_\Phi)] \tag{15.40}$$

Subsequent substitution of (15.39) and (15.40) into (15.38) results in the following equation for the deflections:

$$\mathbf{r} = \mathscr{F}_P\mathbf{P} + \mathscr{F}_P\mathbf{\Gamma}(\mathbf{I} - \mathbf{F}_P\mathbf{\Gamma})^{-1}[\mathbf{F}_P\mathbf{P} + \mathbf{F}_v(\mathbf{v}_T + \mathbf{v}_\Phi)] + \mathscr{F}_v(\mathbf{v}_T + \mathbf{v}_\Phi) \quad (15.41)$$

Warren[335] has demonstrated that elements in $\mathbf{F}_P\mathbf{\Gamma}$ are small compared with unity, and therefore the inverse in (15.41) is given by the convergent series

$$(\mathbf{I} - \mathbf{F}_P\mathbf{\Gamma})^{-1} = \mathbf{I} + \mathbf{F}_P\mathbf{\Gamma} + (\mathbf{F}_P\mathbf{\Gamma})^2 + \cdots \quad (15.42)$$

Substituting Eq. (15.42) into (15.41), we obtain

$$\begin{aligned}\mathbf{r} = \mathscr{F}_P\mathbf{P} &+ \mathscr{F}_P(\mathbf{\Gamma}\mathbf{F}_P + \mathbf{\Gamma}\mathbf{F}_P\mathbf{\Gamma}\mathbf{F}_P + \mathbf{\Gamma}\mathbf{F}_P\mathbf{\Gamma}\mathbf{F}_P\mathbf{\Gamma}\mathbf{F}_P + \cdots)\mathbf{P} \\ &+ \mathscr{F}_P(\mathbf{\Gamma}\mathbf{F}_v + \mathbf{\Gamma}\mathbf{F}_P\mathbf{\Gamma}\mathbf{F}_v + \mathbf{\Gamma}\mathbf{F}_P\mathbf{\Gamma}\mathbf{F}_P\mathbf{\Gamma}\mathbf{F}_v + \cdots)(\mathbf{v}_T + \mathbf{v}_\Phi) \\ &+ \mathscr{F}_v(\mathbf{v}_T + \mathbf{v}_\Phi) \quad (15.43)\end{aligned}$$

Introducing now

$$\mathbf{P}_{\Phi_P} = \mathbf{\Gamma}\mathbf{F}_P \quad (15.44a)$$

which is a matrix of fictitious forces resulting from unit values of external forces, and

$$\mathbf{P}_{\Phi_v} = \mathbf{\Gamma}\mathbf{F}_v \quad (15.44b)$$

which is a matrix of fictitious forces due to unit values of element deformations, we can simplify Eq. (15.43) to read

$$\begin{aligned}\mathbf{r} = \mathscr{F}_P(\mathbf{I} &+ \mathbf{P}_{\Phi_P} + \mathbf{P}_{\Phi_P}{}^2 + \cdots)\mathbf{P} \\ &+ [\mathscr{F}_v + \mathscr{F}_P(\mathbf{I} + \mathbf{P}_{\Phi_P} + \mathbf{P}_{\Phi_P}{}^2 + \cdots)\mathbf{P}_{\Phi_v}](\mathbf{v}_T + \mathbf{v}_\Phi) \quad (15.45)\end{aligned}$$

As an approximation we may delete all terms with powers of $\mathbf{P}_{\Phi_P}$ higher than the first and also the products $\mathbf{P}_{\Phi_P}\mathbf{P}_{\Phi_v}$. Hence Eq. (15.45) becomes

$$\mathbf{r} = \mathscr{F}_P(\mathbf{I} + \mathbf{P}_{\Phi_P})\mathbf{P} + (\mathscr{F}_v + \mathscr{F}_P\mathbf{P}_{\Phi_v})(\mathbf{v}_T + \mathbf{v}_\Phi) \quad (15.46)$$

We may note that in accordance with the approximations of this theory [see Eq. (15.35)] both $\mathbf{P}_{\Phi_P}\mathbf{P}$ and $\mathbf{P}_{\Phi_v}\mathbf{v}_T$ are linear functions of the deflections, and therefore we may introduce the following definitions:

$$\mathbf{P}_{\Phi_P}\mathbf{P} = \mathbf{P}_{\Phi_{Pr}}\mathbf{r} \quad (15.47)$$

and $\quad \mathbf{P}_{\Phi_v}\mathbf{v}_T = \mathbf{P}_{\Phi_{vr}}\mathbf{r} \quad (15.48)$

where $\mathbf{P}_{\Phi_{Pr}}$ is defined as the matrix of fictitious forces due to unit values of the deflections $\mathbf{r}$ when the loading $\mathbf{P}$ is applied and $\mathbf{P}_{\Phi_{vr}}$ is the matrix of fictitious forces due to unit values of the deflections $\mathbf{r}$ when the thermal (or initial) deformations $\mathbf{v}_T$ are applied.

Substituting now Eqs. (15.47) and (15.48) into (15.46), we obtain

$$[\mathbf{I} - \mathscr{F}_P(\mathbf{P}_{\Phi_{Pr}} + \mathbf{P}_{\Phi_{vr}})]\mathbf{r} = \mathscr{F}_P\mathbf{P} + \mathscr{F}_v\mathbf{v}_T + (\mathscr{F}_v + \mathscr{F}_P\mathbf{P}_{\Phi_v})\mathbf{v}_\Phi \tag{15.49}$$

Hence, solving for **r**, we have

$$\mathbf{r} = [\mathbf{I} - \mathscr{F}_P(\mathbf{P}_{\Phi_{Pr}} + \mathbf{P}_{\Phi_{vr}})]^{-1}[\mathscr{F}_P\mathbf{P} + \mathscr{F}_v\mathbf{v}_T + (\mathscr{F}_v + \mathscr{F}_P\mathbf{P}_{\Phi_v})\mathbf{v}_\Phi] \tag{15.50}$$

Equation (15.50) is nonlinear in **r** since both $\mathbf{P}_{\Phi_v}$ and $\mathbf{v}_\Phi$ are functions of **r**. The elements of matrix $\mathbf{P}_{\Phi_v}$ are proportional to **r**, while those of $\mathbf{v}_\Phi$ are proportional to squares of the differences of displacements. Equation (15.50) can be solved by iteration and is much more convenient to handle than Eqs. (15.37) and (15.38), where we had not only the unknown deflections **r** but also the element forces **F**. Once the deflections **r** have been calculated, the element forces can be found from Eq. (15.37).

From Eq. (15.49) we obtain the linear stability equation

$$[\mathbf{I} - \mathscr{F}_P(\mathbf{P}_{\Phi_{Pr}} + \mathbf{P}_{\Phi_{vr}})]\mathbf{r} = \mathbf{0} \tag{15.51}$$

In solving for the eigenvalues of this equation it is convenient to introduce

$$\mathbf{P}_{\Phi_{Pr}} = \lambda_P\mathbf{P}^*_{\Phi_{Pr}} \tag{15.52}$$

and $$\mathbf{P}_{\Phi_{vr}} = \lambda_v\mathbf{P}^*_{\Phi_{vr}} \tag{15.53}$$

where λ_P and λ_v are constants, while $\mathbf{P}^*_{\Phi_{Pr}}$ are the fictitious forces due to unit values of the deflections **r** when a reference loading is applied and $\mathbf{P}^*_{\Phi_{vr}}$ are the fictitious forces due to unit values of deflections when some reference thermal (or initial) deformations are applied, i.e., deformations caused by some reference temperatures. Hence the eigenvalue equation (15.51) may be rewritten as

$$(\mathbf{I} - \lambda_P\mathscr{F}_P\mathbf{P}^*_{\Phi_{Pr}} - \lambda_v\mathscr{F}_P\mathbf{P}^*_{\Phi_{vr}})\mathbf{r} = \mathbf{0} \tag{15.54}$$

The determinant

$$|\mathbf{I} - \lambda_P\mathscr{F}_P\mathbf{P}^*_{\Phi_{Pr}} - \lambda_v\mathscr{F}_P\mathbf{P}^*_{\Phi_{vr}}| = 0 \tag{15.55}$$

is therefore the stability criterion formulated by the matrix force method. The smallest values of λ_P and λ_v from Eq. (15.55) give a functional relationship specifying the stability boundary for the combined mechanical applied loading **P** and thermal loading $\mathbf{v}_T$. This includes two special cases, buckling due to **P** when temperature is zero and buckling due to the temperature distribution alone when $\mathbf{P} = \mathbf{0}$.

15.5 INELASTIC ANALYSIS AND CREEP

The incremental step procedure used for the nonlinear matrix displacement analysis is ideally suited for structures with inelastic material properties.

Applications of the matrix displacement method to elastoplastic problems are discussed by Gallagher, Padlog, and Bijlaard,[109] and also by Besseling[44] and Pope.[258,259] Extension of the method to plastic-hinge design is discussed by Livesley.[208]

The matrix force method can be utilized for the analysis of inelastic structures by suitably modifying the element relative displacement-force equations. These equations are of the form

$$\mathbf{v} = \mathbf{f}\mathbf{F} + \mathbf{v}_T + \mathbf{v}_I \tag{15.56}$$

where both the initial deformations $\mathbf{v}_I$ and the flexibility $\mathbf{f}$ are functions of the element forces $\mathbf{F}$. The analysis is normally carried out for the stress-strain curve idealized into a series of straight lines in order to simplify computer programming of Eq. (15.56). Because of the nonlinearities an iterative method of solution must be used, and essentially the problem involves the determination of the initial deformations $\mathbf{v}_I$ and the tangent flexibilities $\mathbf{f}$ for each element. Details of such an analysis including the description of a computer program are given by Warren[335] and Argyris.[14]

The analysis of creep and creep buckling can be conveniently carried out by matrix methods. Since creep is a time-dependent process, the matrix analysis for creep is based on a stepwise approach utilizing increments of time. The deformations due to creep from the preceding time increment can be considered as initial deformations $\mathbf{v}_I$ in the time increment for which the stress distribution is being calculated. This stress distribution can then be used to determine the additional creep deformations which will occur at the end of the time increment. These additional creep deformations are then utilized to evaluate initial deformations for the next time increment and so on. If nonlinear matrix analysis is used, an iterative cycle can be introduced within each time increment. The success of the creep analysis, however, depends largely on an accurate knowledge of the creep law and the establishment of a valid relationship for cumulative creep under conditions of varying stress and temperature.

15.6 STABILITY ANALYSIS OF A SIMPLE TRUSS

DISPLACEMENT METHOD

As an example of the stability analysis by matrix methods we shall consider the simple truss shown in Fig. 15.5. The cross-sectional areas of the truss members are A, and Young's modulus is E. The external loading consists of a vertical force P applied at node 1. We shall use first the displacement method of analysis to determine the load P which will cause the truss to become unstable in its own plane. Subsequently the same problem will be solved by the matrix force method using the concept of fictitious forces.

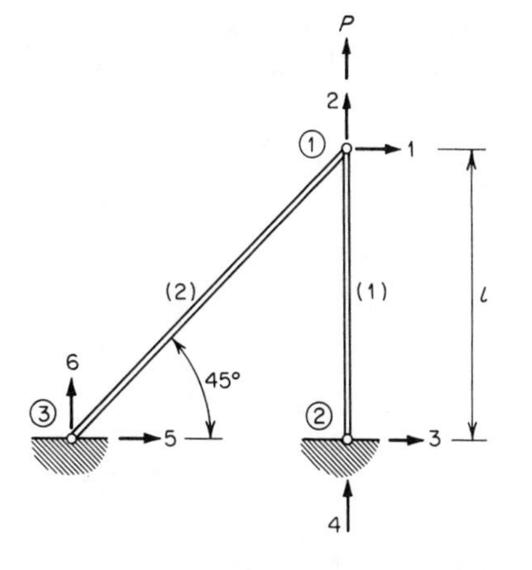

FIG. 15.5 Truss geometry and loading.

The elastic stiffness matrices for elements 1 and 2 are

$$\mathbf{k}_E^{(1)} = \frac{AE}{l} \begin{array}{c} \\ 1 \\ 2 \\ 3 \\ 4 \end{array} \begin{array}{c} \begin{array}{cccc} 1 & 2 & 3 & 4 \end{array} \\ \begin{bmatrix} 0 & 0 & 0 & 0 \\ 0 & 1 & 0 & -1 \\ 0 & 0 & 0 & 0 \\ 0 & -1 & 0 & 1 \end{bmatrix} \end{array} \tag{15.57}$$

$$\mathbf{k}_E^{(2)} = \frac{\sqrt{2}}{4} \frac{AE}{l} \begin{array}{c} \\ 5 \\ 6 \\ 1 \\ 2 \end{array} \begin{array}{c} \begin{array}{cccc} 5 & 6 & 1 & 2 \end{array} \\ \begin{bmatrix} 1 & 1 & -1 & -1 \\ 1 & 1 & -1 & -1 \\ -1 & -1 & 1 & 1 \\ -1 & -1 & 1 & 1 \end{bmatrix} \end{array} \tag{15.58}$$

where the row and column numbers refer to the displacements as shown in Fig. 15.5. It can be demonstrated easily that the element forces obtained from the conventional linear analysis are given by

$$F^{(1)} = P \qquad \text{and} \qquad F^{(2)} = 0 \tag{15.59}$$

Hence it follows from Eqs. (15.21) and (15.59) that the geometrical stiffness matrices are given by

$$\mathbf{k}_G^{(1)} = \frac{P}{l} \begin{array}{c} \\ 1 \\ 2 \\ 3 \\ 4 \end{array} \begin{array}{c} \begin{array}{cccc} 1 & 2 & 3 & 4 \end{array} \\ \begin{bmatrix} 1 & 0 & -1 & 0 \\ 0 & 0 & 0 & 0 \\ -1 & 0 & 1 & 0 \\ 0 & 0 & 0 & 0 \end{bmatrix} \end{array} \tag{15.60}$$

and $\quad \mathbf{k}_G^{(2)} = \mathbf{0}$ $\hfill (15.61)$

The element stiffness matrices can now be assembled to form $\mathbf{K}_E$ and $\mathbf{K}_G$ for the entire structure. Hence after deleting the rows and columns corresponding to displacements 3 to 6, we obtain

$$\mathbf{K}_E = \frac{\sqrt{2}}{4}\frac{AE}{l}\begin{matrix} & 1 & 2 \\ 1 & 1 & 1 \\ 2 & 1 & 1+2\sqrt{2} \end{matrix} \tag{15.62}$$

$$\mathbf{K}_G = \frac{P}{l}\begin{matrix} & 1 & 2 \\ 1 & 1 & 0 \\ 2 & 0 & 0 \end{matrix} \tag{15.63}$$

whence
$$\mathbf{K}_G^* = \frac{1}{l}\begin{matrix} & 1 & 2 \\ 1 & 1 & 0 \\ 2 & 0 & 0 \end{matrix} \tag{15.64}$$

Substitution of Eqs. (15.62) and (15.64) into the stability determinant (15.8) results in

$$\begin{vmatrix} \frac{\sqrt{2}}{4}\frac{AE}{l} + \frac{\lambda}{l} & \frac{\sqrt{2}}{4}\frac{AE}{l} \\ \frac{\sqrt{2}}{4}\frac{AE}{l} & \frac{\sqrt{2}}{4}\frac{AE}{l}(1+2\sqrt{2}) \end{vmatrix} = 0 \tag{15.65}$$

from which we obtain the buckling load

$$P_{\text{crit}} = \lambda_{\text{crit}} \times 1 = -\frac{2\sqrt{2}-1}{7}AE \tag{15.66}$$

This agrees with the exact result obtained by Timoshenko and Gere. The negative value indicates that the buckling load is in the direction opposite to that shown in Fig. 15.5.

FORCE METHOD

Since there is no thermal loading applied to the truss, $\mathbf{v}_T = \mathbf{0}$, and the stability criterion from Eq. (15.55) becomes

$$|\mathbf{I} - \lambda_P \mathscr{F}_P \mathbf{P}^*_{\Phi_{Pr}}| = 0 \tag{15.67}$$

The flexibility matrix $\mathscr{F}_P$ can be obtained by the force method, but since we have already calculated the corresponding stiffness $\mathbf{K}_E$, it will be more expedient to use

$$\begin{aligned} \mathscr{F}_P &= \mathbf{K}_E^{-1} \\ &= \left\{\frac{\sqrt{2}}{4}\frac{AE}{l}\begin{bmatrix} 1 & 1 \\ 1 & 1+2\sqrt{2} \end{bmatrix}\right\}^{-1} = \frac{l}{AE}\begin{bmatrix} 1+2\sqrt{2} & -1 \\ -1 & 1 \end{bmatrix} \end{aligned} \tag{15.68}$$

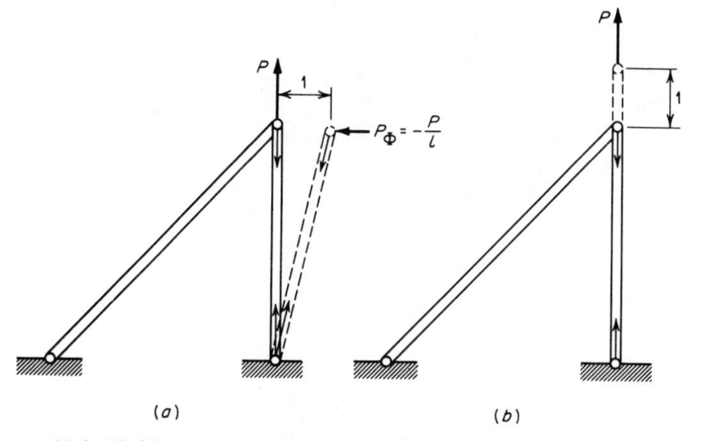

FIG. 15.6 Fictitious forces P_Φ on a two-bar truss with a vertical force P. (*a*) $r_1 = 1$; (*b*) $r_2 = 1$.

We also note from the previous section that the linear analysis gives $F^{(1)} = P$ and $F^{(2)} = 0$. Therefore with the aid of Fig. 15.6 it is easy to see that the matrix of fictitious forces due to unit values of the deflections when the loading is applied is given by

$$\mathbf{P}_{\Phi_{Pr}} = \begin{bmatrix} -\dfrac{P}{l} & 0 \\ 0 & 0 \end{bmatrix} \tag{15.69}$$

whence

$$\mathbf{P}^*_{\Phi_{Pr}} = \begin{bmatrix} \dfrac{-1}{l} & 0 \\ 0 & 0 \end{bmatrix} \tag{15.70}$$

Multiplying out the product $\mathscr{F}_P \mathbf{P}^*_{\Phi_{Pr}}$ and then substituting into the stability determinant (15.67), we obtain

$$\begin{vmatrix} 1 + \dfrac{\lambda_P}{AE}(1 + 2\sqrt{2}) & 0 \\ \dfrac{-\lambda_P}{AE} & 1 \end{vmatrix} = 0 \tag{15.71}$$

The root of this determinant is

$$\lambda_P = -\frac{AE}{1 + 2\sqrt{2}} = -\frac{2\sqrt{2} - 1}{7} AE \tag{15.72}$$

which agrees with the result obtained by the displacement method.

15.7 STABILITY ANALYSIS OF A COLUMN

As a second example we shall consider stability of a fixed-pinned column shown in Fig. 15.7. The column length is L, the flexural rigidity is EI, and the cross-sectional area is A. For the purpose of the analysis the column will be idealized into two elements. Only the displacement method of analysis will be illustrated.

Using the numbering scheme for deflections in Fig. 15.7, we obtain the following stiffness matrices from Eq. (15.30):

$$\mathbf{k}_E^{(1)} = \frac{EI}{l^3}\begin{array}{c}1\\2\\3\\4\\5\\6\end{array}\begin{bmatrix} \phi & & & & & \\ 0 & 12 & & \text{Symmetric} & & \\ 0 & 6l & 4l^2 & & & \\ -\phi & 0 & 0 & \phi & & \\ 0 & -12 & -6l & 0 & 12 & \\ 0 & 6l & 2l^2 & 0 & -6l & 4l^2 \end{bmatrix} \tag{15.73}$$
$$\begin{matrix} 1 & 2 & 3 & 4 & 5 & 6 \end{matrix}$$

$$\mathbf{k}_E^{(2)} = \frac{EI}{l^3}\begin{array}{c}4\\5\\6\\7\\8\\9\end{array}\begin{bmatrix} \phi & & & & & \\ 0 & 12 & & \text{Symmetric} & & \\ 0 & 6l & 4l^2 & & & \\ -\phi & 0 & 0 & \phi & & \\ 0 & -12 & -6l & 0 & 12 & \\ 0 & 6l & 2l^2 & 0 & -6l & 4l^2 \end{bmatrix} \tag{15.74}$$
$$\begin{matrix} 4 & 5 & 6 & 7 & 8 & 9 \end{matrix}$$

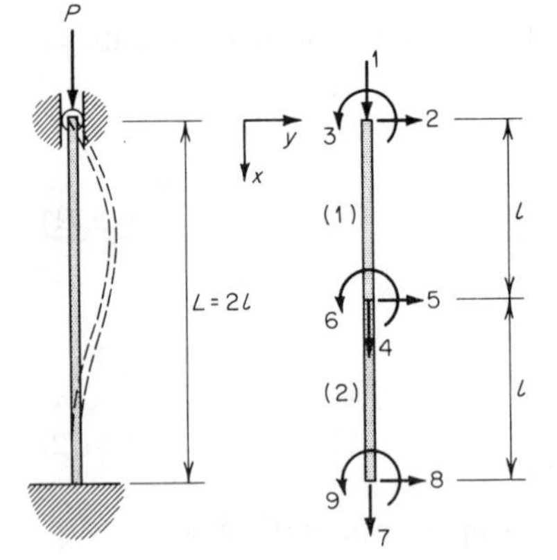

FIG. 15.7 Column with one end fixed and the other pinned.

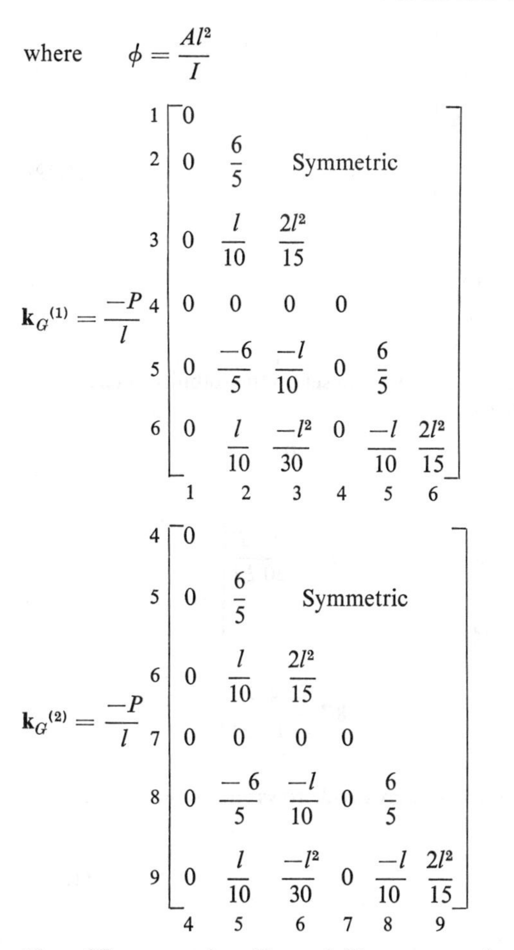

where $\phi = \dfrac{Al^2}{I}$ (15.75)

$$
\mathbf{k}_G^{(1)} = \frac{-P}{l}
\begin{array}{c}1\\2\\3\\4\\5\\6\end{array}
\begin{bmatrix}
0 & & & & & \\
0 & \frac{6}{5} & & \text{Symmetric} & & \\
0 & \frac{l}{10} & \frac{2l^2}{15} & & & \\
0 & 0 & 0 & 0 & & \\
0 & \frac{-6}{5} & \frac{-l}{10} & 0 & \frac{6}{5} & \\
0 & \frac{l}{10} & \frac{-l^2}{30} & 0 & \frac{-l}{10} & \frac{2l^2}{15}
\end{bmatrix}
\quad \text{columns } 1\ 2\ 3\ 4\ 5\ 6 \tag{15.76}
$$

$$
\mathbf{k}_G^{(2)} = \frac{-P}{l}
\begin{array}{c}4\\5\\6\\7\\8\\9\end{array}
\begin{bmatrix}
0 & & & & & \\
0 & \frac{6}{5} & & \text{Symmetric} & & \\
0 & \frac{l}{10} & \frac{2l^2}{15} & & & \\
0 & 0 & 0 & 0 & & \\
0 & \frac{-6}{5} & \frac{-l}{10} & 0 & \frac{6}{5} & \\
0 & \frac{l}{10} & \frac{-l^2}{30} & 0 & \frac{-l}{10} & \frac{2l^2}{15}
\end{bmatrix}
\quad \text{columns } 4\ 5\ 6\ 7\ 8\ 9 \tag{15.77}
$$

The stiffness matrices $\mathbf{K}_E$ and $\mathbf{K}_G$ can now be assembled from the element stiffnesses. Eliminating rows and columns 2, 7, 8, and 9 corresponding to zero displacements on the column, we obtain, after some rearrangement,

$$
\mathbf{K}_E = \frac{EI}{l^3}
\begin{array}{c}1\\4\\3\\5\\6\end{array}
\begin{bmatrix}
\phi & -\phi & 0 & 0 & 0 \\
-\phi & 2\phi & 0 & 0 & 0 \\
0 & 0 & 4l^2 & -6l & 2l^2 \\
0 & 0 & -6l & 24 & 0 \\
0 & 0 & 2l^2 & 0 & 8l^2
\end{bmatrix}
\quad \text{columns } 1\ 4\ 3\ 5\ 6 \tag{15.78}
$$

$$\mathbf{K}_G = \frac{-P}{l}\begin{array}{c} 1 \\ 4 \\ 3 \\ 5 \\ 6 \\ \\ \end{array}\!\!\begin{array}{c}\left[\begin{array}{ccccc} 0 & 0 & 0 & 0 & 0 \\ 0 & 0 & 0 & 0 & 0 \\ 0 & 0 & \frac{2}{15}l^2 & \frac{-l}{10} & \frac{-l^2}{30} \\ 0 & 0 & \frac{-l}{10} & \frac{12}{5} & 0 \\ 0 & 0 & \frac{-l^2}{30} & 0 & \frac{4}{15}l^2 \end{array}\right] \\ \begin{array}{ccccc} 1 & 4 & 3 & 5 & 6 \end{array}\end{array} \tag{15.79}$$

Noting that $\mathbf{K}_G^*$ is equal to $\mathbf{K}_G$ for $P = 1$, we can set up the stability determinant $|\,\mathbf{K}_E + \lambda\mathbf{K}_G^*\,| = 0$. This leads to

$$\begin{array}{c} 1 \\ 4 \\ 3 \\ 5 \\ 6 \\ \\ \end{array}\!\!\begin{array}{c}\left|\begin{array}{ccccc} \phi & -\phi & 0 & 0 & 0 \\ -\phi & 2\phi & 0 & 0 & 0 \\ 0 & 0 & 4l^2 - \frac{2}{15}\frac{\lambda l^4}{EI} & -6l + \frac{1}{10}\frac{\lambda l^3}{EI} & 2l^2 + \frac{1}{30}\frac{\lambda l^4}{EI} \\ 0 & 0 & -6l + \frac{1}{10}\frac{\lambda l^3}{EI} & 24 - \frac{12}{5}\frac{\lambda l^2}{EI} & 0 \\ 0 & 0 & 2l^2 + \frac{1}{30}\frac{\lambda l^4}{EI} & 0 & 8l^2 - \frac{4}{15}\frac{\lambda l^4}{EI} \end{array}\right| \\ \begin{array}{ccccc} 1 & 4 & 3 & 5 & 6 \end{array}\end{array} = 0 \tag{15.80}$$

To simplify subsequent calculations we may divide rows and columns 3 and 6 by l and introduce

$$\mu = \frac{\lambda l^2}{EI} \tag{15.81}$$

so that the determinant becomes

$$\left|\begin{array}{ccccc} \phi & -\phi & 0 & 0 & 0 \\ -\phi & 2\phi & 0 & 0 & 0 \\ 0 & 0 & 2\left(2 - \frac{\mu}{15}\right) & -6 + \frac{\mu}{10} & 2 + \frac{\mu}{30} \\ 0 & 0 & -6 + \frac{\mu}{10} & 12\left(2 - \frac{\mu}{5}\right) & 0 \\ 0 & 0 & 2 + \frac{\mu}{30} & 0 & 4\left(2 - \frac{\mu}{15}\right) \end{array}\right| = 0 \tag{15.82}$$

Expanding this determinant, we obtain a cubic equation in μ

$$3\mu^3 - 220\mu^2 + 3{,}840\mu - 14{,}400 = 0 \tag{15.83}$$

The lowest root of this equation is

$$\mu = 5.1772 \tag{15.84}$$

This value is only 2.6 percent higher than the exact value 5.0475. Thus we have demonstrated that even with only two elements excellent accuracy is obtained for the buckling load. It should also be noted that the effect of shear deformations on the buckling load can be investigated by modifying the elastic and geometrical stiffness.

15.8 INFLUENCE OF A CONSTANT AXIAL FORCE ON TRANSVERSE VIBRATIONS OF BEAMS

As a further example of the application of the matrix displacement analysis for large deflections we shall determine the influence of an axial load in a beam column (Fig. 15.8) on the frequency of transverse vibrations. Both ends are pinned, and the column is subjected to an axial load P. The positive load P corresponds to tensile axial load in the beam. The flexural rigidity of the beam is EI, the beam length is l, the cross-sectional area is A, and the density of the material is ρ. For simplicity of presentation of the analysis the beam will be idealized into only one element. Naturally, in any practical application more elements would have to be used for this problem. Since the transverse and longitudinal displacements are uncoupled, we may use only transverse deflections and rotations in setting up the equations of motion.

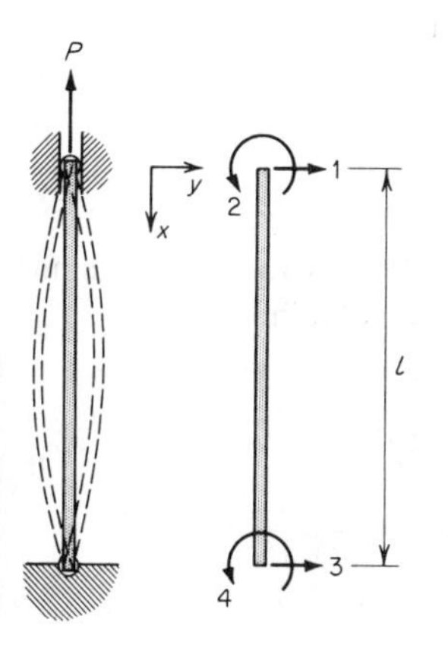

FIG. 15.8 Vibrations of a beam column with both ends pinned.

Using the numbering system indicated in Fig. 15.8, we obtain the following stiffness matrices:

$$\mathbf{K}_E = \frac{EI}{l^3}\begin{bmatrix} 12 & 6l & -12 & 6l \\ 6l & 4l^2 & -6l & 2l^2 \\ -12 & -6l & 12 & -6l \\ 6l & 2l^2 & -6l & 4l^2 \end{bmatrix} \tag{15.85}$$

$$\mathbf{K}_G = \frac{P}{l}\begin{bmatrix} \frac{6}{5} & \frac{l}{10} & \frac{-6}{5} & \frac{l}{10} \\ \frac{l}{10} & \frac{2}{15}l^2 & \frac{-l}{10} & \frac{-l^2}{30} \\ \frac{-6}{5} & \frac{-l}{10} & \frac{6}{5} & \frac{-l}{10} \\ \frac{l}{10} & \frac{-l^2}{30} & \frac{-l}{10} & \frac{2}{15}l^2 \end{bmatrix} \tag{15.86}$$

The mass matrix for the corresponding degrees of freedom is

$$\mathbf{M} = \frac{\rho A l}{420}\begin{bmatrix} 156 & 22l & 54 & -13l \\ 22l & 4l^2 & 13l & -3l^2 \\ 54 & 13l & 156 & -22l \\ -13l & -3l^2 & -22l & 4l^2 \end{bmatrix} \tag{15.87}$$

(Rows and columns of each matrix are numbered 1, 2, 3, 4.)

For large deflections the equation of motion of a freely vibrating system is obtained by replacing $\mathbf{K}$ with $\mathbf{K}_E + \mathbf{K}_G$. Hence we have

$$(-\omega^2\mathbf{M} + \mathbf{K}_E + \mathbf{K}_G)\mathbf{q} = \mathbf{0} \tag{15.88}$$

Substituting Eqs. (15.85) to (15.87) into (15.88) and deleting rows and columns corresponding to the transverse deflections 1 and 3, we obtain

$$\left\{-\omega^2\frac{\rho A l}{420}\begin{bmatrix} 4l^2 & -3l^2 \\ -3l^2 & 4l^2 \end{bmatrix} + \frac{EI}{l^3}\begin{bmatrix} 4l^2 & 2l^2 \\ 2l^2 & 4l^2 \end{bmatrix} + \frac{P}{l}\begin{bmatrix} \frac{2}{15}l^2 & \frac{-l^2}{30} \\ \frac{-l^2}{30} & \frac{2}{15}l^2 \end{bmatrix}\right\}\begin{bmatrix} q_2 \\ q_4 \end{bmatrix} = \mathbf{0} \tag{15.89}$$

(Rows and columns of each matrix in Eq. (15.89) are numbered 2, 4.)

Next the characteristic equation

$$|-\omega^2\mathbf{M} + \mathbf{K}_E + \mathbf{K}_G| = 0 \tag{15.90}$$

can be formulated from Eq. (15.89), and this leads to

$$\begin{vmatrix} 4 + \frac{2}{15}\mu - \frac{1}{105}\Theta^2 & 2 - \frac{1}{30}\mu + \frac{1}{140}\Theta^2 \\ 2 - \frac{1}{30}\mu + \frac{1}{140}\Theta^2 & 4 + \frac{2}{15}\mu - \frac{1}{105}\Theta^2 \end{vmatrix} = 0 \tag{15.91}$$

where $$\mu = \frac{Pl^2}{EI} \tag{15.92}$$

and $$\Theta^2 = \omega^2 \frac{\rho A l^4}{EI} \tag{15.93}$$

Evaluating the determinant (15.91), we get

$$(2 + \tfrac{1}{6}\mu - \tfrac{1}{60}\Theta^2)(6 + \tfrac{1}{10}\mu - \tfrac{1}{420}\Theta^2) = 0 \tag{15.94}$$

from which two eigenvalues are obtained

$$\Theta_1{}^2 = 120(1 + \tfrac{1}{12}\mu) \tag{15.95}$$

$$\Theta_2{}^2 = 2{,}520(1 + \tfrac{1}{60}\mu) \tag{15.96}$$

Hence the two natural frequencies derived in this simple analysis are

$$\omega_1 = \left(\frac{EI}{\rho A l^4}\right)^{\frac{1}{2}} \Theta_1 = 10.954\left(\frac{EI}{\rho A l^4}\right)^{\frac{1}{2}}\left(1 + \frac{Pl^2}{12EI}\right)^{\frac{1}{2}} \tag{15.97}$$

and $$\omega_2 = \left(\frac{EI}{\rho A l^4}\right)^{\frac{1}{2}} \Theta_2 = 50.200\left(\frac{EI}{\rho A l^4}\right)^{\frac{1}{2}}\left(1 + \frac{Pl^2}{60EI}\right)^{\frac{1}{2}} \tag{15.98}$$

From Eqs. (15.97) and (15.98) it is clear that the frequencies increase with increasing tensile load P and decrease with increasing compressive load (when P is negative). The influence of the axial load is greater on the lowest frequency than on the higher frequency. When the axial load is

$$P = -12\frac{EI}{l^2} \tag{15.99}$$

the first frequency ω_1 is reduced to zero. This condition corresponds to buckling of the column. It may be noted that the value of the buckling load given by Eq. (15.99) is 21.6 percent higher than the exact value π^2EI/l^2. However, had we used more elements, as in the example in Sec. 15.7, we should have obtained much better accuracy. The frequencies given by Eqs. (15.97) and (15.98) are 11.0 and 26.7 percent higher than the exact analytical values for the first and second frequency of a beam with both ends pinned and with $P = 0$.

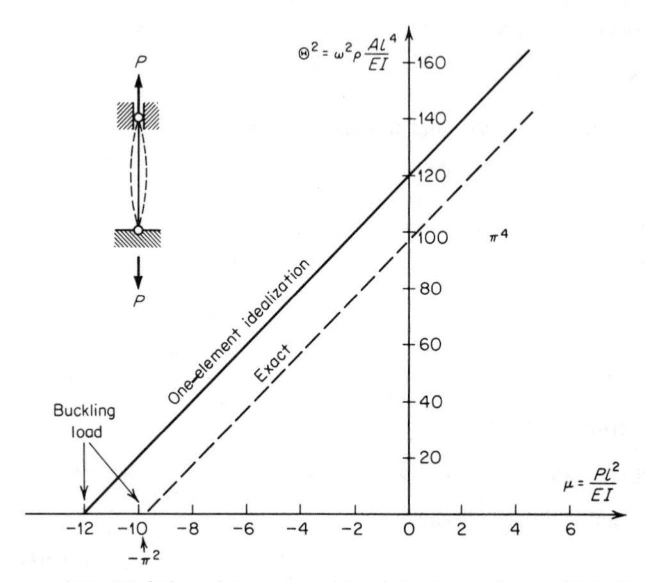

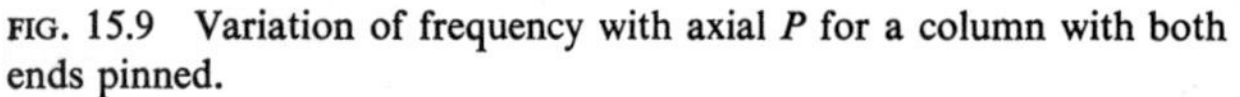

FIG. 15.9 Variation of frequency with axial P for a column with both ends pinned.

A plot of Eq. (15.95) is shown in Fig. 15.9. This demonstrates clearly that frequency measurements for varying axial loads can be used in nondestructive testing to predict buckling loads.

PROBLEMS

15.1 The truss shown in Fig. 15.5 is subjected to both horizontal and vertical loads applied at the free node point. Determine the functional relationship between the two loads for instability using (*a*) matrix displacement method and (*b*) matrix force method.

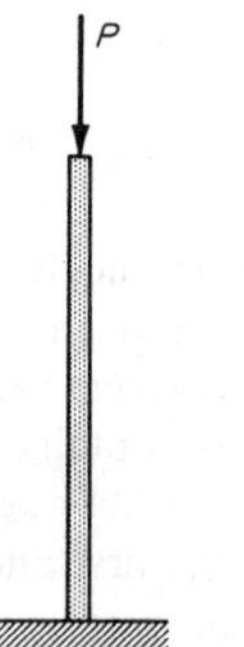

FIG. 15.10

15.2 Determine the natural frequency of vibration of a uniform column free at one end and built-in at the other and subjected to a constant axial load P, as shown in Fig. 15.10. The effects of shear deformations must be included in the derivation. Use matrix formulation based on one element only.

15.3 Derive the *string stiffness* for a beam element given by Eq. (15.33).

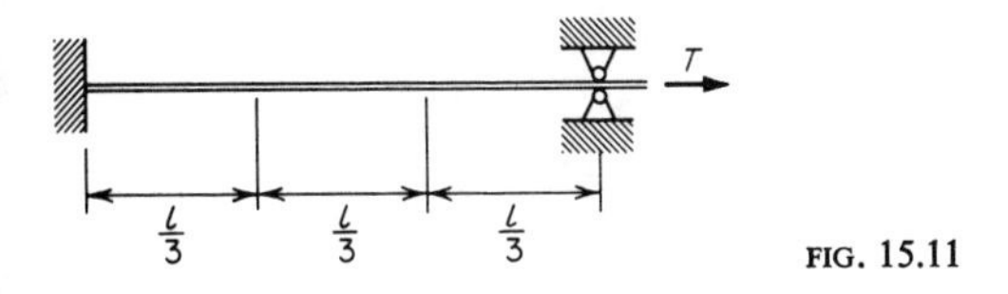

FIG. 15.11

15.4 Vibrations of cables having no bending stiffness can be analyzed by introducing the geometrical string stiffness of bar element. Determine natural frequencies of vibration of a pre-tensioned cable idealized into three elements, as shown in Fig. 15.11. The amount of tension is T, and the weight of the cable per unit length is m.

APPENDIX A MATRIX ALGEBRA

A.1 INTRODUCTION

Matrix algebra represents mathematical operations performed on a group of algebraic or numerical quantities in such a way that a single symbol suffices to denote the whole group. These groups are represented by matrices, which may be thought of as a type of algebraic shorthand notation. The mathematical operations performed on matrices must be explicitly defined, and since these operations may be defined in a number of ways, different matrix algebras may be formulated. For linear structural analysis the conventional matrix operations are normally used. Only for nonlinear structural analysis are special matrix operations required which are not found in the conventional matrix algebra.

When matrix algebra is used in structural analysis, the organizational properties of matrices allow for systematic compilation of the required data, and the structural analysis itself can then be defined as a sequence of matrix operations which can be programmed directly for a digital computer. Since in any structural analysis the matrices are used only as a mathematical tool, there is no need to know much of the pure mathematical properties of matrices but only some of the simple operations in which they can be used. This appendix is therefore limited to the rudimentary elements of matrix algebra and is provided here with a view to convenient reference and not as an exhaustive treatment of the subject.

The various types of matrices occurring in structural analysis are described, in addition to the fundamental matrix operations comprising addition, subtraction, transposition, multiplication, and inversion. Matrix inversion is associated with the solution of simultaneous linear algebraic equations; it is usually obtained on a digital computer if the matrix size is large. However, in order to understand the mechanism of matrix inversion, which only superficially resembles division in ordinary algebra, several inversion techniques are described. Some of these techniques are in fact used in computer programs. The problem of finding the characteristic roots (eigenvalues) of matrices is treated in only a cursory manner. For this and other details of the theory of matrices standard textbooks on the subject may be consulted.

A.2 MATRIX NOTATION AND DEFINITIONS

MATRICES

A matrix is defined as a rectangular array of symbols or numerical quantities arranged in rows and columns. The array is enclosed in brackets, and thus if there are m rows and n columns, the matrix can be represented by

$$\mathbf{A} = \begin{bmatrix} a_{11} & a_{12} & a_{13} & \cdots & a_{1j} & \cdots & a_{1n} \\ a_{21} & a_{22} & a_{23} & \cdots & a_{2j} & \cdots & a_{2n} \\ a_{31} & a_{32} & a_{33} & \cdots & a_{3j} & \cdots & a_{3n} \\ \cdots & \cdots & \cdots & \cdots & \cdots & \cdots & \cdots \\ a_{i1} & a_{i2} & a_{i3} & \cdots & a_{ij} & \cdots & a_{in} \\ \cdots & \cdots & \cdots & \cdots & \cdots & \cdots & \cdots \\ a_{m1} & a_{m2} & a_{m3} & \cdots & a_{mj} & \cdots & a_{mn} \end{bmatrix} \tag{A.1}$$

where the typical element a_{ij} has two subscripts, of which the first denotes the row (ith) and the second denotes the column (jth) which the element occupies in the matrix. A matrix with m rows and n columns is defined as a matrix of order $m \times n$, or simply an $m \times n$ matrix. It should be noted, however, that the number of rows is always specified first. In Eq. (A.1) the symbol $\mathbf{A}$ stands for the whole array of m rows and n columns and it is usually printed in boldface type.

A matrix should not be confused with a determinant, which is also formed from an array of symbols or numerical quantities. A determinant must always be square; i.e., the number of rows must be equal to the number of columns, and it can be evaluated since the determinant signifies a certain relationship among its elements. On the other hand, a matrix merely represents an array and it does not imply a relationship among the elements.

ROW AND COLUMN MATRICES

If $m = 1$, the matrix $\mathbf{A}$ reduces to the single row

$$\mathbf{A} = [a_{11} \quad a_{12} \quad a_{13} \quad \cdots \quad a_{1j} \quad \cdots \quad a_{1n}] \tag{A.2}$$

which is called a *row matrix*. Similarly, if $n = 1$, the matrix $\mathbf{A}$ reduces to the single column

$$\mathbf{A} = \begin{bmatrix} a_{11} \\ a_{21} \\ \cdot \\ \cdot \\ \cdot \\ a_{m1} \end{bmatrix} = \{a_{11} \quad a_{21} \quad \cdots \quad a_{m1}\} \tag{A.3}$$

which is called a *column matrix*. To save space, however, column matrices may be written horizontally and enclosed in braces, as shown above.

The order of a row matrix is $1 \times n$, and the order of a column matrix is $m \times 1$.

NULL MATRIX (ZERO MATRIX)

When all the elements of a matrix are equal to zero, the matrix is called *null* or *zero* and is indicated by $\mathbf{0}$. A null matrix serves the same function in matrix algebra as zero does in ordinary algebra.

SQUARE MATRIX

If $m = n$, the matrix $\mathbf{A}$ reduces to the square array

$$\mathbf{A} = \begin{bmatrix} a_{11} & a_{12} & \cdots & a_{1n} \\ a_{21} & a_{22} & \cdots & a_{2n} \\ \cdots & \cdots & \cdots & \cdots \\ a_{n1} & a_{n2} & \cdots & a_{nn} \end{bmatrix} \tag{A.4}$$

which is called a square matrix. Square matrices occupy a special place in matrix algebra since only square matrices have reciprocals (provided they also satisfy certain relationships among the elements). There are several important types of square matrices which will now be discussed.

DIAGONAL MATRIX

A diagonal matrix is one which has zero elements everywhere outside the principal diagonal, defined as the diagonal running from the upper left to the lower right corner of the array. It follows therefore that for a diagonal matrix $a_{ij} = 0$ when $i \neq j$ and not all a_{ii} are zero. A typical diagonal matrix may be represented by

$$\mathbf{A} = \begin{bmatrix} a_{11} & 0 & \cdots & 0 \\ 0 & a_{22} & \cdots & 0 \\ \cdots & \cdots & \cdots & \cdots \\ 0 & 0 & \cdots & a_{nn} \end{bmatrix} \tag{A.5}$$

or simply, in order to save space, as

$$\mathbf{A} = \lceil a_{11} \quad a_{22} \quad \cdots \quad a_{nn} \rfloor \tag{A.6}$$

The principal diagonal in a square matrix is usually referred to as *the diagonal.*

UNIT MATRIX (IDENTITY MATRIX)

A unit matrix is one which has unit elements on the principal diagonal and zeros elsewhere. It is usually denoted by the symbol **I**. Thus a unit matrix of order 3×3 can be written as

$$\mathbf{I} = \begin{bmatrix} 1 & 0 & 0 \\ 0 & 1 & 0 \\ 0 & 0 & 1 \end{bmatrix} = \lceil 1 \quad 1 \quad 1 \rfloor \tag{A.7}$$

The unit matrix serves the same function in matrix algebra as unity does in ordinary algebra. The order of the unit matrix is sometimes indicated by the subscript n, and thus $\mathbf{I}_n$ denotes a unit matrix of order $n \times n$, for example,

$$\mathbf{I}_3 = \lceil 1 \quad 1 \quad 1 \rfloor \tag{A.8}$$

SCALAR MATRIX

If **A** is a diagonal matrix in which $a_{ii} = a$ for all i, the matrix is called a *scalar matrix.* Thus a typical scalar matrix may be represented by

$$\mathbf{A} = \lceil a \quad a \quad \cdots \quad a \rfloor \tag{A.9}$$

BAND MATRIX

If the nonzero elements in a square matrix are grouped around the principal diagonal, forming a diagonal band of elements, the matrix is called a *band matrix.* A typical example of the band matrix is the flexibility matrix for unassembled structural elements in which the linearly varying axial force is reacted by a constant shear flow. This flexibility matrix is of the form

$$\mathbf{A} = \begin{bmatrix} a_{11} & a_{12} & 0 & 0 & \cdots & 0 & 0 \\ a_{21} & a_{22} & 0 & 0 & \cdots & 0 & 0 \\ 0 & 0 & a_{33} & a_{34} & \cdots & 0 & 0 \\ 0 & 0 & a_{43} & a_{44} & \cdots & 0 & 0 \\ \cdots & \cdots & \cdots & \cdots & \cdots & \cdots & \cdots \\ 0 & 0 & 0 & 0 & \cdots & a_{n-1,n-1} & a_{n-1,n} \\ 0 & 0 & 0 & 0 & \cdots & a_{n,n-1} & a_{n,n} \end{bmatrix} \tag{A.10}$$

TRIANGULAR MATRIX

If all the elements on one side of the principal diagonal of a square matrix are zero, the matrix is called a *triangular matrix.* There are two types of triangular matrices, an upper triangular matrix $\mathbf{U}$, whose elements below the principal diagonal are all zero, and a lower triangular matrix $\mathbf{L}$, whose elements above the diagonal are all zero. In the special cases when the elements on the principal diagonal are all equal to unity, the other nonzero elements being arbitrary, attention is drawn to this fact by denoting $\mathbf{U}$ and $\mathbf{L}$ by $\mathbf{U}(1)$ and $\mathbf{L}(1)$, respectively, and referring to these special matrices as *unit upper* and *unit lower triangular matrices.* A typical lower triangular matrix of order $n \times n$ is

$$\mathbf{L} = \begin{bmatrix} a_{11} & 0 & 0 & \cdots & 0 \\ a_{21} & a_{22} & 0 & \cdots & 0 \\ a_{31} & a_{32} & a_{33} & \cdots & 0 \\ \cdots & \cdots & \cdots & \cdots & \cdots \\ a_{n1} & a_{n2} & a_{n3} & \cdots & a_{nn} \end{bmatrix} \tag{A.11}$$

SYMMETRIC AND SKEW-SYMMETRIC MATRICES

A symmetric matrix is a square matrix whose elements are symmetrical about the principal diagonal. Thus if the matrix $\mathbf{A}$ is symmetric, then $a_{ij} = a_{ji}$.

A skew-symmetric matrix is a square matrix in which the elements symmetrically located with respect to the principal diagonal are opposite in sign, and those located on the principal diagonal are zero. Thus for a skew-symmetric matrix $a_{ij} = -a_{ji}$ and $a_{ii} = 0$. If $a_{ij} = -a_{ji}$ but the elements on the principal diagonal are not all equal to zero, the matrix is called a *skew matrix.*

A.3 MATRIX PARTITIONING: SUBMATRICES

The array of elements in a matrix may be divided into smaller arrays by horizontal and vertical lines. Such a matrix is then referred to as a *partitioned matrix*, and the smaller arrays are called *submatrices*. For example, a square matrix of order 3 may be partitioned into four submatrices as shown below:

$$\mathbf{A} = \left[\begin{array}{cc:c} a_{11} & a_{12} & a_{13} \\ a_{21} & a_{22} & a_{23} \\ \hdashline a_{31} & a_{32} & a_{33} \end{array}\right] = \begin{bmatrix} \mathbf{A}_{11} & \mathbf{A}_{12} \\ \mathbf{A}_{21} & \mathbf{A}_{22} \end{bmatrix} \tag{A.12}$$

where $$\mathbf{A}_{11} = \begin{bmatrix} a_{11} & a_{12} \\ a_{21} & a_{22} \end{bmatrix} \tag{A.13}$$

$$\mathbf{A}_{12} = \{a_{13} \quad a_{23}\} \tag{A.14}$$

$$\mathbf{A}_{21} = [a_{31} \quad a_{32}] \tag{A.15}$$

$$\mathbf{A}_{22} = [a_{33}] \tag{A.16}$$

Other partitioning arrangements are also possible with the above matrix.

Provided the general rules for matrix operations (addition, subtraction, etc.) are observed, the submatrices can be treated as if they were ordinary matrix elements. For certain types of matrices in structural analysis, the matrix partitioning has a corresponding *physical* counterpart in structural partitioning, a fact which greatly simplifies the arrangement of submatrices. For example, flexibility matrices are usually partitioned so as to distinguish the location on the structure to which the flexibility submatrices are referred. Stiffness matrices can be partitioned in such a way that a large number of the stiffness submatrices are null matrices, which can be very useful in simplifying subsequent matrix operations in the analysis.

A.4 EQUALITY, ADDITION, AND SUBTRACTION OF MATRICES

Matrices of the same order are equal if each element of one is equal to the corresponding element of the other. Thus if

$$\mathbf{A} = \mathbf{B} \tag{A.17}$$

it then follows that

$$a_{ij} = b_{ij} \tag{A.18}$$

for all values of i and j.

If the corresponding elements in matrices **A** and **B**, of the same order, are added algebraically, the resulting elements form a third matrix which is the sum of the first two, that is,

$$\mathbf{A} + \mathbf{B} = \mathbf{C} \tag{A.19}$$

where $$c_{ij} = a_{ij} + b_{ij} \tag{A.20}$$

For example,

$$\begin{bmatrix} 3 & 2 \\ 1 & 7 \\ 3 & 5 \end{bmatrix} + \begin{bmatrix} 0 & -1 \\ 7 & 2 \\ 6 & 3 \end{bmatrix} = \begin{bmatrix} 3 & 1 \\ 8 & 9 \\ 9 & 8 \end{bmatrix} \tag{A.21}$$

Similarly, if the elements in matrices **A** and **B** are algebraically subtracted, the resulting elements form a third matrix which is the difference of the first two, that is,

$$\mathbf{A} - \mathbf{B} = \mathbf{C} \tag{A.22}$$

where $c_{ij} = a_{ij} - b_{ij}$ (A.23)

Taking again Eq. (A.21) as an example,

$$\begin{bmatrix} 3 & 2 \\ 1 & 7 \\ 3 & 5 \end{bmatrix} - \begin{bmatrix} 0 & -1 \\ 7 & 2 \\ 6 & 3 \end{bmatrix} = \begin{bmatrix} 3 & 3 \\ -6 & 5 \\ -3 & 2 \end{bmatrix} \tag{A.24}$$

From the above definitions it is clear that equality, addition, and subtraction of matrices are meaningful only among matrices of the same order. Furthermore, it can be seen that the commutative and associative laws of ordinary algebra are also applicable to the addition of matrices. That is,

$$\mathbf{A} + \mathbf{B} = \mathbf{B} + \mathbf{A} \qquad \text{commutative law} \tag{A.25}$$

and $(\mathbf{A} + \mathbf{B}) + \mathbf{C} = \mathbf{A} + (\mathbf{B} + \mathbf{C})$ associative law (A.26)

It should also be noted that a null matrix can be defined as the difference of two equal matrices. Thus if

$$\mathbf{A} = \mathbf{B} \tag{A.27}$$

it follows that

$$\mathbf{A} - \mathbf{B} = \mathbf{0} \tag{A.28}$$

where the right-hand side represents the null matrix of the same order as **A** and **B**.

A.5 MATRIX TRANSPOSITION

The transposed matrix (or simply the transpose) is formed from the matrix **A** by interchanging all rows for the corresponding columns. Thus the transpose of a row matrix is a column matrix, and vice versa, while the transpose of a symmetric matrix, including a diagonal matrix, is the same matrix. The transposition of a matrix will be denoted here by the superscript T. Thus $\mathbf{A}^T$

represents the transpose of **A**. Other symbols used to denote the transpose of **A** are $\tilde{\mathbf{A}}$ and $\mathbf{A}'$.

For example, if

$$\mathbf{A} = \begin{bmatrix} a_{11} & a_{12} & a_{13} \\ a_{21} & a_{22} & a_{23} \end{bmatrix} \tag{A.29}$$

then $$\mathbf{A}^T = \begin{bmatrix} a_{11} & a_{21} \\ a_{12} & a_{22} \\ a_{13} & a_{23} \end{bmatrix} \tag{A.30}$$

Thus if the order of the matrix **A** is $m \times n$, the order of its transpose $\mathbf{A}^T$ is $n \times m$. It should also be noted that if a matrix is transposed twice, it reverts to its original form, that is,

$$(\mathbf{A}^T)^T = \mathbf{A} \tag{A.31}$$

If the elements of the matrix **A** are matrices themselves, i.e., submatrices, the transpose of **A** is formed by interchanging rows of submatrices for the corresponding columns and transposing elements in each submatrix. Thus if the elements of the matrix Eq. (A.29) denote submatrices, then

$$\mathbf{A}^T = \begin{bmatrix} \mathbf{a}_{11}{}^T & \mathbf{a}_{21}{}^T \\ \mathbf{a}_{12}{}^T & \mathbf{a}_{22}{}^T \\ \mathbf{a}_{13}{}^T & \mathbf{a}_{23}{}^T \end{bmatrix} \tag{A.30a}$$

For square matrices which are either symmetric or skew-symmetric there are two important relations:

$$\mathbf{A}^T = \mathbf{A} \qquad \text{for symmetric matrices} \tag{A.32}$$

and $$\mathbf{A}^T = -\mathbf{A} \text{ for skew-symmetric matrices} \tag{A.33}$$

A.6 MATRIX MULTIPLICATION

Multiplication of the matrix **A** by a scalar c is defined as the multiplication of every element of the matrix by the scalar c. Thus the elements of the product $c\mathbf{A}$ are ca_{ij}.

Two matrices **A** and **B** can be multiplied together in order **AB** only *when the number of columns in* **A** *is equal to the number of rows in* **B**. When this condition is fulfilled, the matrices **A** and **B** are said to be conformable for multiplication. Otherwise, matrix multiplication is not defined. The product of two conformable matrices **A** and **B** of order $m \times p$ and $p \times n$, respectively, is defined as matrix **C** of order $m \times n$ whose elements are calculated from

$$c_{ij} = \sum_{r=1}^{p} a_{ir} b_{rj} \qquad i = 1, 2, \ldots, m, \qquad j = 1, 2, \ldots, n \tag{A.34}$$

where a_{ir} and b_{rj} are the elements of **A** and **B**, respectively.

The multiplication process can be extended to products of more than two matrices, provided the adjacent matrices in the product are conformable. It is therefore important that the multiplication sequence be preserved. A useful rule for testing whether matrix products are conformable follows. If the matrix product is $\mathbf{ABC} = \mathbf{D}$, where the orders of **A**, **B**, and **C** are $m \times n$, $n \times o$, and $o \times p$, respectively, the matrix orders are written in order of multiplication, that is,

$$(m \times n)(n \times o)(o \times p)$$

and a check is made whether the second number in the resulting product is the same as third, the fourth number is the same as fifth, etc. In the example quoted here the matrix **D** is of order $m \times p$. However, this checking is not very helpful if the matrices are all square matrices.

To show the application of Eq. (A.34) two simple examples will be considered. As the first example the following matrix product will be evaluated:

$$\begin{bmatrix} 1 & 2 & 3 \\ 4 & 5 & 6 \\ 3 & 1 & 2 \end{bmatrix} \begin{bmatrix} 1 & 3 \\ 2 & -1 \\ 7 & 1 \end{bmatrix} = \begin{bmatrix} 1 \times 1 + 2 \times 2 + 3 \times 7 & 1 \times 3 - 2 \times 1 + 3 \times 1 \\ 4 \times 1 + 5 \times 2 + 6 \times 7 & 4 \times 3 - 5 \times 1 + 6 \times 1 \\ 3 \times 1 + 1 \times 2 + 2 \times 7 & 3 \times 3 - 1 \times 1 + 2 \times 1 \end{bmatrix} = \begin{bmatrix} 26 & 4 \\ 56 & 13 \\ 19 & 10 \end{bmatrix} \tag{A.35}$$

For the second example the matrix product of two matrices whose elements are denoted by symbols will be evaluated:

$$\begin{bmatrix} a_{11} & a_{12} & a_{13} \\ a_{21} & a_{22} & a_{23} \end{bmatrix} \begin{bmatrix} b_{11} & b_{12} \\ b_{21} & b_{22} \\ b_{31} & b_{32} \end{bmatrix} = \begin{bmatrix} a_{11}b_{11} + a_{12}b_{21} + a_{13}b_{31} & a_{11}b_{12} + a_{12}b_{22} + a_{13}b_{32} \\ a_{21}b_{11} + a_{22}b_{21} + a_{23}b_{31} & a_{21}b_{12} + a_{22}b_{22} + a_{23}b_{32} \end{bmatrix} \tag{A.36}$$

The associative and distributive laws apply to matrix multiplication, provided that the multiplication sequence is preserved. For example,

$$\mathbf{A(BC)} = \mathbf{(AB)C} = \mathbf{ABC} \tag{A.37}$$

and $$\mathbf{A(B + C)} = \mathbf{AB} + \mathbf{AC} \tag{A.38}$$

The commutative property, however, does not in general apply to multiplication and

$$\mathbf{AB} \neq \mathbf{BA} \tag{A.39}$$

There may be cases when $\mathbf{AB} = \mathbf{BA}$, and the two matrices $\mathbf{A}$ and $\mathbf{B}$ are then said to be commutable. For example, the unit matrix $\mathbf{I}$ commutes with any square matrix of the same order, that is,

$$\mathbf{AI} = \mathbf{IA} = \mathbf{A} \tag{A.40}$$

When two matrices $\mathbf{A}$ and $\mathbf{B}$ are multiplied, the product $\mathbf{AB}$ is referred to either as $\mathbf{B}$ premultiplied by $\mathbf{A}$ or as $\mathbf{A}$ postmultiplied by $\mathbf{B}$.

The process of matrix multiplication can also be extended to partitioned matrices, provided the individual products of submatrices are conformable for multiplication. For example, the multiplication

$$\mathbf{AB} = \begin{bmatrix} \mathbf{A}_{11} & \mathbf{A}_{12} \\ \mathbf{A}_{21} & \mathbf{A}_{22} \end{bmatrix} \begin{bmatrix} \mathbf{B}_{11} & \mathbf{B}_{12} \\ \mathbf{B}_{21} & \mathbf{B}_{22} \end{bmatrix} = \begin{bmatrix} \mathbf{A}_{11}\mathbf{B}_{11} + \mathbf{A}_{12}\mathbf{B}_{21} & \mathbf{A}_{11}\mathbf{B}_{12} + \mathbf{A}_{12}\mathbf{B}_{22} \\ \mathbf{A}_{21}\mathbf{B}_{11} + \mathbf{A}_{22}\mathbf{B}_{21} & \mathbf{A}_{21}\mathbf{B}_{12} + \mathbf{A}_{22}\mathbf{B}_{22} \end{bmatrix} \tag{A.41}$$

is possible provided the products $\mathbf{A}_{11}\mathbf{B}_{11}$, $\mathbf{A}_{12}\mathbf{B}_{21}$, etc., are conformable. For this condition to be fulfilled, it is only necessary for the vertical partitions in $\mathbf{A}$ to encompass the number of columns equal to the number of rows in the corresponding horizontal partitions in $\mathbf{B}$. The partitioning of $\mathbf{A}$ by horizontal lines and of $\mathbf{B}$ by vertical lines is arbitrary, and it does not affect the conformability of the submatrices.

A.7 CRACOVIANS

Except for the rule of multiplication, cracovians are the same as matrices. The product of cracovians $\mathbf{A}$ and $\mathbf{B}$, taken in that order, is defined as the matrix product $\mathbf{A}^T\mathbf{B}$ of the matrices $\mathbf{A}$ and $\mathbf{B}$. The algebra of cracovians was introduced by the Polish mathematician Banachiewicz, and it is still in favor among some European mathematicians. Since any cracovian can be regarded as a matrix, it is clear that the preference for one or the other can be based only on tradition.

A.8 DETERMINANTS

BASIC DEFINITIONS

The determinant of a square matrix $\mathbf{A}$ is denoted by

$$|\mathbf{A}| = \begin{vmatrix} a_{11} & a_{12} & \cdots & a_{1n} \\ a_{21} & a_{22} & \cdots & a_{2n} \\ \cdots & \cdots & \cdots & \cdots \\ a_{n1} & a_{n2} & \cdots & a_{nn} \end{vmatrix} \tag{A.42}$$

and is defined formally as the following summation:

$$|\mathbf{A}| = \sum \pm (a_{1i} \quad a_{2j} \quad a_{3k} \quad \cdots) \tag{A.43}$$

where the row suffixes of the elements appear in the normal order $1, 2, \ldots, n$, while the column suffixes $i, j, k, \ldots$ appear as some permutation of the normal order. The positive or negative sign associated with a particular product in the summation depends on whether when deriving the required permutation sequence an even or odd number of interchanges of adjacent suffixes from the normal order is required. The summation extends over $n!$ permutations, half of which are even and half odd.

PROPERTIES OF DETERMINANTS

The various properties of determinants are summarized here without proof.

1. The determinants of a matrix and its transpose are equal.
2. Interchanging any two rows or columns changes the sign of the determinant.
3. If two rows, or two columns, in a determinant are identical, the value of the determinant is zero.
4. If all the elements in a row, or in a column, are zero, the determinant is zero.
5. Multiplication by a constant c of all the elements in a row, or in a column, of a determinant $|\mathbf{A}|$ results in a determinant of value $c\,|\mathbf{A}|$.
6. The addition of a constant multiple of a row (or column) to the corresponding elements of any other row (or column) leaves the value of the determinant unchanged.

MINORS AND COFACTORS

The first minor of a determinant $|\mathbf{A}|$, corresponding to the element a_{ij}, is defined as the determinant obtained by omission of the ith row and the jth column of $|\mathbf{A}|$. Therefore, if $|\mathbf{A}|$ is of order n, any first minor is of order $n - 1$. This definition can be extended to second, third, etc., minors, and thus, in general, the determinant obtained by omission of any s rows and s columns from $|\mathbf{A}|$ is defined as an sth minor or as the minor of order $n - s$. The first minor corresponding to the element a_{ij} is denoted by M_{ij}.

If the first minor M_{ij} is multiplied by $(-1)^{i+j}$, it becomes the cofactor of a_{ij}. The cofactors are usually denoted by A_{ij}. Hence

$$A_{ij} = (-1)^{i+j} M_{ij} \tag{A.44}$$

EXPANSION OF DETERMINANTS

It can be demonstrated, using the elements of the ith row, that the determinant $|\mathbf{A}|$ can be expanded in terms of the cofactors of the ith row. Thus the expansion by the ith row is

$$|\mathbf{A}| = a_{i1}A_{i1} + a_{i2}A_{i2} + \cdots + a_{in}A_{in} = \sum_{r=1}^{n} a_{ir}A_{ir} \tag{A.45}$$

Similarly, since all rows and columns can be interchanged without changing the value of the determinant, the expansion by the jth column is

$$|\mathbf{A}| = a_{1j}A_{1j} + a_{2j}A_{2j} + \cdots + a_{nj}A_{nj} = \sum_{r=1}^{n} a_{rj}A_{rj} \tag{A.46}$$

Equations (A.45) and (A.46) are called *Laplace expansion formulas*, and they are particular cases of more general expansions of the cofactors of a determinant. These general expansions are expressed as

$$\sum_{r=1}^{n} a_{ir}A_{kr} = \begin{cases} |\mathbf{A}| & \text{for } k = i \\ 0 & \text{for } k \neq i \end{cases} \tag{A.47}$$

when expanding the determinant by rows and

$$\sum_{r=1}^{n} a_{rj}A_{rk} = \begin{cases} |\mathbf{A}| & \text{for } k = j \\ 0 & \text{for } k \neq j \end{cases} \tag{A.48}$$

when expanding the determinant by columns. When $k \neq i$ in Eq. (A.47) and $k \neq j$ in Eq. (A.48) the summations are then called *expansions in alien cofactors*; these expansions are of some importance in the derivation of reciprocals of matrices.

A.9 MATRIX INVERSION (RECIPROCAL OF A SQUARE MATRIX)

Consider a square matrix of order $n \times n$ whose cofactors are given by A_{ij}. The square matrix

$$\hat{\mathbf{A}} = \begin{bmatrix} A_{11} & A_{21} & A_{31} & \cdots & A_{n1} \\ A_{12} & A_{22} & A_{32} & \cdots & A_{n2} \\ A_{13} & A_{23} & A_{33} & \cdots & A_{n3} \\ \cdots & \cdots & \cdots & \cdots & \cdots \\ A_{1n} & A_{2n} & A_{3n} & \cdots & A_{nn} \end{bmatrix} \tag{A.49}$$

formed from the cofactors A_{ij} is defined as the adjoint of the matrix **A** and is denoted by $\hat{\mathbf{A}}$. From this definition it follows that the adjoint matrix is the transpose of the matrix of cofactors. Upon forming the product

$$\mathbf{A}\hat{\mathbf{A}} = \mathbf{P} \tag{A.50}$$

and using the cofactor expansion form Eq. (A.47) it follows that a typical element in **P** is given by

$$\begin{aligned} p_{ij} &= a_{i1}A_{j1} + a_{i2}A_{j2} + \cdots + a_{in}A_{jn} = \sum_{r=1}^{n} a_{ir}A_{jr} \\ &= \begin{cases} |\mathbf{A}| & \text{if } i = j \\ 0 & \text{if } i \neq j \end{cases} \end{aligned} \tag{A.51}$$

Thus only the elements on the principal diagonal are nonzero and equal to $|\mathbf{A}|$, while all other elements are zero. Hence **P** is a scalar matrix, and

$$\mathbf{A}\hat{\mathbf{A}} = |\mathbf{A}|\mathbf{I} \tag{A.52}$$

where **I** is the unit matrix of the same order as **A**. Dividing throughout by $|\mathbf{A}|$, which is admissible only provided $|\mathbf{A}| \neq 0$, gives

$$\frac{\mathbf{A}\hat{\mathbf{A}}}{|\mathbf{A}|} = \mathbf{A}\mathbf{A}^{-1} = \mathbf{I} \tag{A.53}$$

where the matrix

$$\mathbf{A}^{-1} = \frac{\hat{\mathbf{A}}}{|\mathbf{A}|} \tag{A.54}$$

is defined as the reciprocal or inverse of **A**. Only matrices for which $|\mathbf{A}| \neq 0$ have an inverse. Such matrices are called *nonsingular*. This implies that reciprocals of rectangular matrices do not exist. Starting with the product $\hat{\mathbf{A}}\mathbf{A}$, it can be shown that

$$\mathbf{A}^{-1}\mathbf{A} = \mathbf{I} \tag{A.55}$$

It is therefore apparent from the above discussion that the inverse of a square matrix performs a function in matrix algebra analogous to division in ordinary algebra.

To illustrate the inversion process, the inverse of

$$\mathbf{A} = \begin{bmatrix} a_{11} & a_{12} \\ a_{21} & a_{22} \end{bmatrix} \tag{A.56}$$

will be evaluated. The inverse will be assumed to be given by

$$\mathbf{A}^{-1} = \begin{bmatrix} b_{11} & b_{12} \\ b_{21} & b_{22} \end{bmatrix} \tag{A.57}$$

Now from the definition of matrix inversion $\mathbf{A}^{-1}\mathbf{A} = \mathbf{I}$ it follows that

$$\begin{bmatrix} b_{11} & b_{12} \\ b_{21} & b_{22} \end{bmatrix}\begin{bmatrix} a_{11} & a_{12} \\ a_{21} & a_{22} \end{bmatrix} = \begin{bmatrix} b_{11}a_{11} + b_{12}a_{21} & b_{11}a_{12} + b_{12}a_{22} \\ b_{21}a_{11} + b_{22}a_{21} & b_{21}a_{12} + b_{22}a_{22} \end{bmatrix} = \begin{bmatrix} 1 & 0 \\ 0 & 1 \end{bmatrix} \tag{A.58}$$

Equating the corresponding elements results in

$$\begin{aligned} b_{11}a_{11} + b_{12}a_{21} &= 1 \\ b_{11}a_{12} + b_{12}a_{22} &= 0 \\ b_{21}a_{11} + b_{22}a_{21} &= 0 \\ b_{21}a_{12} + b_{22}a_{22} &= 1 \end{aligned} \tag{A.59}$$

Solving Eqs. (A.59) for b_{11}, b_{12}, b_{21}, and b_{22} and then substituting into Eq. (A.57) gives

$$\mathbf{A}^{-1} = \frac{1}{a_{11}a_{22} - a_{21}a_{12}}\begin{bmatrix} a_{22} & -a_{12} \\ -a_{21} & a_{11} \end{bmatrix} \tag{A.60}$$

The above equation can be recognized as $\mathbf{A}^{-1} = \hat{\mathbf{A}}/|\mathbf{A}|$. This equation gives an explicit formula for the calculation of the inverse of 2×2 matrices. For matrices of order 3×3 the cofactors A_{ij} and the value of the determinant $|\mathbf{A}|$ can be easily evaluated and the matrix inverse found from Eq. (A.54). For larger matrices, however, special techniques suitable for automatic computation must be used. Some of these techniques are described in Secs. A.17 and A.18.

If the matrix to be inverted is a diagonal matrix

$$\mathbf{A} = \lceil a_{11} \quad a_{22} \quad \cdots \quad a_{nn} \rfloor \tag{A.61}$$

then $$\mathbf{A}^{-1} = \left\lceil \frac{1}{a_{11}} \quad \frac{1}{a_{22}} \quad \cdots \quad \frac{1}{a_{nn}} \right\rfloor \tag{A.62}$$

Hence the inverse of a diagonal matrix is also a diagonal matrix whose elements are the reciprocals of the elements in the original matrix.

If a matrix is inverted twice, it reverts to its original form. Hence

$$(\mathbf{A}^{-1})^{-1} = \mathbf{A} \tag{A.63}$$

A.10 RANK AND DEGENERACY

If the rows of a square matrix $\mathbf{A}$ of order $n \times n$ are not linearly independent, the determinant $|\mathbf{A}| = 0$, and the matrix is said to be singular. If the rows of a singular matrix are linearly connected by a single relation, the matrix is defined as simply degenerate or is said to have a degeneracy of one. Naturally, there may be s such relations, in which case the matrix is multiply degenerate or is said to have degeneracy s. The rank of the matrix is then defined as

$r = n - s$. The preceding remarks are, of course, true for the rows as well as the columns.

Somewhat different definitions of degeneracy and rank of a square matrix are as follows: when at least one of the minors of order r of an $n \times n$ matrix does not vanish, whereas all its minors of order $r + 1$ do vanish, the matrix is said to have degeneracy $n - r$, and its rank is r.

The concept of a rank can also be extended to rectangular matrices. Thus a matrix of order $m \times n$ has rank r when not all its minors of order r vanish while all of order $r + 1$ do so. It is evident that r is always less than or equal to either m or n.

A rectangular (or square) matrix of rank r has r linearly independent rows and also r such columns.

A.11 TRANSPOSITION AND INVERSION OF MATRIX PRODUCTS (REVERSAL RULE)

When a matrix product is transposed, the sequence of matrices in the product must be reversed. This rule holds true for any number of matrices. For example, if

$$\mathbf{F} = \mathbf{ABCD} \tag{A.64}$$

then $$\mathbf{F}^T = \mathbf{D}^T\mathbf{C}^T\mathbf{B}^T\mathbf{A}^T \tag{A.65}$$

In structural problems, a matrix product of the form

$$\mathbf{A}^T\mathbf{F}\mathbf{A} = \mathbf{C} \tag{A.66}$$

is often found, where $\mathbf{F}$ is a symmetric matrix and $\mathbf{A}$ is a rectangular matrix. Applying the general rule for matrix transposition, it therefore follows that

$$\mathbf{C}^T = \mathbf{A}^T\mathbf{F}^T(\mathbf{A}^T)^T = \mathbf{A}^T\mathbf{F}\mathbf{A} = \mathbf{C} \tag{A.67}$$

which implies that $\mathbf{C}$ is also a symmetric matrix.

The inversion of a matrix product requires reversal of the matrix sequence, as in the transposition of a matrix product. For example, if

$$\mathbf{F} = \mathbf{ABCD} \tag{A.68}$$

then $$\mathbf{F}^{-1} = \mathbf{D}^{-1}\mathbf{C}^{-1}\mathbf{B}^{-1}\mathbf{A}^{-1} \tag{A.69}$$

A.12 SOLUTION OF SIMULTANEOUS EQUATIONS

N linear simultaneous equations in the unknowns $x_1, x_2, \ldots, x_n$

$$\begin{aligned} a_{11}x_1 + a_{12}x_2 + \cdots + a_{1n}x_n &= b_1 \\ a_{21}x_1 + a_{22}x_2 + \cdots + a_{2n}x_n &= b_2 \\ \cdots\cdots\cdots\cdots\cdots\cdots\cdots\cdots \\ a_{n1}x_1 + a_{n2}x_2 + \cdots + a_{nn}x_n &= b_n \end{aligned} \tag{A.70}$$

can be arranged in matrix form as

$$\begin{bmatrix} a_{11} & a_{12} & \cdots & a_{1n} \\ a_{21} & a_{22} & \cdots & a_{2n} \\ \cdots & \cdots & \cdots & \cdots \\ a_{n1} & a_{n2} & \cdots & a_{nn} \end{bmatrix} \begin{bmatrix} x_1 \\ x_2 \\ \cdot \\ x_n \end{bmatrix} = \begin{bmatrix} b_1 \\ b_2 \\ \cdot \\ b_n \end{bmatrix} \tag{A.70a}$$

or simply $\mathbf{AX} = \mathbf{B}$ (A.70*b*)

Provided $|\mathbf{A}| \neq 0$, both sides of Eq. (A.70*b*) can be premultiplied by $\mathbf{A}^{-1}$, so that

$$\mathbf{A}^{-1}\mathbf{AX} = \mathbf{A}^{-1}\mathbf{B}$$

and hence $\mathbf{X} = \mathbf{A}^{-1}\mathbf{B}$ (A.71)

Thus once the inverse of the matrix $\mathbf{A}$ has been found, the solution for the unknowns $\mathbf{X} = \{x_1 \quad x_2 \quad \cdots \quad x_n\}$ can be obtained from the matrix multiplication of $\mathbf{A}^{-1}\mathbf{B}$.

A.13 INVERSION BY MATRIX PARTITIONING

Given a partitioned square nonsingular matrix, is it possible to find its inverse by a process involving only inversion and multiplication of submatrices. Consider a square matrix $\mathbf{A}$ and its inverse $\mathbf{A}^{-1}$, each partitioned into four submatrices in such a way that the submatrices on the principal diagonals are square. Thus if the matrix $\mathbf{A}$ is partitioned into

$$\mathbf{A} = \begin{bmatrix} \mathbf{A}_{11} & \mathbf{A}_{12} \\ \mathbf{A}_{21} & \mathbf{A}_{22} \end{bmatrix} \tag{A.72}$$

and

$$\mathbf{A}^{-1} = \begin{bmatrix} \mathbf{B}_{11} & \mathbf{B}_{12} \\ \mathbf{B}_{21} & \mathbf{B}_{22} \end{bmatrix} \tag{A.73}$$

then from $\mathbf{A}^{-1}\mathbf{A} = \mathbf{I}$ it follows that

$$\begin{bmatrix} \mathbf{B}_{11} & \mathbf{B}_{12} \\ \mathbf{B}_{21} & \mathbf{B}_{22} \end{bmatrix} \begin{bmatrix} \mathbf{A}_{11} & \mathbf{A}_{12} \\ \mathbf{A}_{21} & \mathbf{A}_{22} \end{bmatrix} = \begin{bmatrix} \mathbf{I} & \mathbf{0} \\ \mathbf{0} & \mathbf{I} \end{bmatrix} \tag{A.74}$$

The diagonal unit submatrices on the right-hand side of Eq. (A.74) are of the same order as $\mathbf{A}_{11}$ and $\mathbf{A}_{22}$, respectively, while the remaining submatrices are null. Multiplying the matrix product on the left-hand side of Eq. (A.74) and then equating its submatrices to those of the unit matrix on the right-hand side gives the following four matrix equations:

$$\mathbf{B}_{11}\mathbf{A}_{11} + \mathbf{B}_{12}\mathbf{A}_{21} = \mathbf{I} \tag{A.75a}$$

$$\mathbf{B}_{11}\mathbf{A}_{12} + \mathbf{B}_{12}\mathbf{A}_{22} = \mathbf{0} \tag{A.75b}$$

$$\mathbf{B}_{21}\mathbf{A}_{11} + \mathbf{B}_{22}\mathbf{A}_{21} = \mathbf{0} \tag{A.75c}$$

$$\mathbf{B}_{21}\mathbf{A}_{12} + \mathbf{B}_{22}\mathbf{A}_{22} = \mathbf{I} \tag{A.75d}$$

Equations (A.75) can be solved for the unknown submatrices $\mathbf{B}_{11}$, $\mathbf{B}_{12}$, $\mathbf{B}_{21}$, and $\mathbf{B}_{22}$. Since the solution of simultaneous matrix equations differs from the solution of similar equations in ordinary algebra, the solution of Eqs. (A.75) will be demonstrated in detail. Postmultiplying Eq. (A.75*a*) by $\mathbf{A}_{11}^{-1}$ gives

$$\mathbf{B}_{11}\mathbf{A}_{11}\mathbf{A}_{11}^{-1} + \mathbf{B}_{12}\mathbf{A}_{21}\mathbf{A}_{11}^{-1} = \mathbf{A}_{11}^{-1}$$

Hence $$\mathbf{B}_{11} = \mathbf{A}_{11}^{-1} - \mathbf{B}_{12}\mathbf{A}_{21}\mathbf{A}_{11}^{-1} \tag{A.76}$$

which, when substituted into Eq. (A.75*b*), gives

$$\mathbf{A}_{11}^{-1}\mathbf{A}_{12} - \mathbf{B}_{12}\mathbf{A}_{21}\mathbf{A}_{11}^{-1}\mathbf{A}_{12} + \mathbf{B}_{12}\mathbf{A}_{22} = \mathbf{0}$$

or $$\mathbf{B}_{12}(\mathbf{A}_{22} - \mathbf{A}_{21}\mathbf{A}_{11}^{-1}\mathbf{A}_{12}) = -\mathbf{A}_{11}^{-1}\mathbf{A}_{12} \tag{A.77}$$

Postmultiplying Eq. (A.77) by $(\mathbf{A}_{22} - \mathbf{A}_{21}\mathbf{A}_{11}^{-1}\mathbf{A}_{12})^{-1}$ leads to

$$\mathbf{B}_{12} = -\mathbf{A}_{11}^{-1}\mathbf{A}_{12}(\mathbf{A}_{22} - \mathbf{A}_{21}\mathbf{A}_{11}^{-1}\mathbf{A}_{12})^{-1} \tag{A.78}$$

Substituting now Eq. (A.78) into (A.76) gives

$$\mathbf{B}_{11} = \mathbf{A}_{11}^{-1} + \mathbf{A}_{11}^{-1}\mathbf{A}_{12}(\mathbf{A}_{22} - \mathbf{A}_{21}\mathbf{A}_{11}^{-1}\mathbf{A}_{12})^{-1}\mathbf{A}_{21}\mathbf{A}_{11}^{-1} \tag{A.79}$$

Postmultiplication of Eq. (A.75*c*) by $\mathbf{A}_{11}^{-1}$ gives

$$\mathbf{B}_{21} = -\mathbf{B}_{22}\mathbf{A}_{21}\mathbf{A}_{11}^{-1} \tag{A.80}$$

which, when substituted into Eq. (A.75*d*), leads to

$$\mathbf{B}_{22} = (\mathbf{A}_{22} - \mathbf{A}_{21}\mathbf{A}_{11}^{-1}\mathbf{A}_{12})^{-1} \tag{A.81}$$

Therefore from Eqs. (A.80) and (A.81) it follows that

$$\mathbf{B}_{21} = -(\mathbf{A}_{22} - \mathbf{A}_{21}\mathbf{A}_{11}^{-1}\mathbf{A}_{12})^{-1}\mathbf{A}_{21}\mathbf{A}_{11}^{-1} \tag{A.82}$$

Examination of Eqs. (A.78) to (A.82) shows that the calculation of the matrix inverse of the partitioned matrix **A** involves only inversions of matrices of the same order as the order of submatrices in the matrix **A**, and thus appreciable saving of computing time may be achieved, since matrix inversions of much smaller order than of **A** are required. Although these equations appear to be rather lengthy, it must be remembered that the matrix multiplication, addition, or subtraction can be carried out much more rapidly than matrix inversion. Furthermore, these equations can be simplified if new matrices are introduced such that

$$\mathbf{X} = \mathbf{A}_{11}^{-1}\mathbf{A}_{12} \tag{A.83}$$

$$\mathbf{Y} = \mathbf{A}_{21}\mathbf{A}_{11}^{-1} \tag{A.84}$$

$$\mathbf{Z} = \mathbf{A}_{22} - \mathbf{Y}\mathbf{A}_{12} = \mathbf{A}_{22} - \mathbf{A}_{21}\mathbf{X} \tag{A.85}$$

It can then be shown that

$$\mathbf{B}_{11} = \mathbf{A}_{11}^{-1} + \mathbf{X}\mathbf{Z}^{-1}\mathbf{Y} \tag{A.86}$$

$$\mathbf{B}_{12} = -\mathbf{X}\mathbf{Z}^{-1} \tag{A.87}$$

$$\mathbf{B}_{21} = -\mathbf{Z}^{-1}\mathbf{Y} \qquad \mathbf{B}_{22} = \mathbf{Z}^{-1} \tag{A.88}$$

A.14 EXTRACTION OF THE INVERSE OF A REDUCED MATRIX

Consider a partitioned square matrix

$$\mathbf{M} = \begin{bmatrix} \mathbf{A} & \mathbf{B} \\ \mathbf{C} & \mathbf{D} \end{bmatrix} \tag{A.89}$$

where $\mathbf{A}$ and $\mathbf{D}$ are square submatrices. The inverse of this matrix will be assumed to exist and is given by

$$\mathbf{M}^{-1} = \begin{bmatrix} \mathbf{a} & \mathbf{b} \\ \mathbf{c} & \mathbf{d} \end{bmatrix} \tag{A.90}$$

It is required to find the inverse of the reduced matrix

$$\mathbf{M}_r = \mathbf{A} \tag{A.91}$$

The inverse can be found most conveniently from the inverse of the original matrix $\mathbf{M}$, assuming of course that this has already been found. From Eqs. (A.89) to (A.91) it can be shown that

$$\mathbf{M}_r^{-1} = \mathbf{a} - \mathbf{b}\mathbf{d}^{-1}\mathbf{c} \tag{A.92}$$

The method is advantageous only where the size of the matrix $\mathbf{d}$ is considerably smaller than that of $\mathbf{M}_r$; otherwise it would be more expedient to determine the inverse of $\mathbf{M}_r$ directly.

A.15 INVERSION OF A MODIFIED MATRIX

Consider a square matrix

$$\mathbf{M}_0 = \begin{bmatrix} \mathbf{A}_0 & \mathbf{B} \\ \mathbf{C} & \mathbf{D} \end{bmatrix} \tag{A.93}$$

for which the inverse has been calculated and is given by

$$\mathbf{M}_0^{-1} = \begin{bmatrix} \mathbf{a}_0 & \mathbf{b} \\ \mathbf{c} & \mathbf{d} \end{bmatrix} \tag{A.94}$$

where the submatrices on the principal diagonals are square. If now only a portion of $\mathbf{M}_0$ is to be modified, a problem which occurs very frequently in structural analysis when part of the structure is modified, such that the modified matrix can be expressed as

$$\mathbf{M}_m = \mathbf{M}_0 + \begin{bmatrix} \Delta\mathbf{A} & \mathbf{0} \\ \mathbf{0} & \mathbf{0} \end{bmatrix} = \begin{bmatrix} \mathbf{A}_m & \mathbf{B} \\ \mathbf{C} & \mathbf{D} \end{bmatrix} \tag{A.95}$$

where $\quad \mathbf{A}_m = \mathbf{A}_0 + \Delta\mathbf{A} \quad$ (A.96)

then it can be demonstrated that the inverse of the modified matrix $\mathbf{M}_m$ can be calculated from

$$\mathbf{M}_m^{-1} = \mathbf{M}_0^{-1} - \begin{bmatrix} \mathbf{a}_0 \\ \mathbf{c} \end{bmatrix} \mathbf{q}^{-1}\, \Delta\mathbf{A}\, [\mathbf{a}_0 \quad \mathbf{b}] \tag{A.97}$$

where $\quad \mathbf{q} = \mathbf{I} + \Delta\mathbf{A}\,\mathbf{a}_0 \quad$ (A.98)

Here again the method is advantageous only if the size of the matrix $\mathbf{q}$ is small in comparison with the size of the modified matrix $\mathbf{M}_m$.

A.16 INVERSION OF A TRIPLE-BAND MATRIX

The stiffness matrix $\mathbf{K}$ used in the matrix displacement method of analysis can be arranged into a triple-band matrix provided the columns in $\mathbf{K}$ refer to structural partitions in which each partition has common boundaries with not more than two adjacent structural partitions. In this way the only coupling submatrices will be those representing adjacent structural partitions.

If the triple-band matrix is denoted by $\mathbf{K}$ and its inverse by $\mathbf{F}$, then from $\mathbf{KF} = \mathbf{I}$

$$\begin{bmatrix} \mathbf{K}_{11} & \mathbf{K}_{12} & & & \\ \mathbf{K}_{21} & \mathbf{K}_{22} & \mathbf{K}_{23} & & \mathbf{0} \\ & \mathbf{K}_{32} & \mathbf{K}_{33} & \mathbf{K}_{34} & \\ \cdots & \cdots & \cdots & \cdots & \cdots \\ & \mathbf{0} & & & \\ & & & \mathbf{K}_{n-1,n} & \mathbf{K}_{nn} \end{bmatrix} \begin{bmatrix} \mathbf{F}_{11} & \mathbf{F}_{12} & \mathbf{F}_{13} & \cdots & \mathbf{F}_{1n} \\ \mathbf{F}_{21} & \mathbf{F}_{22} & \mathbf{F}_{23} & \cdots & \mathbf{F}_{2n} \\ \mathbf{F}_{31} & \mathbf{F}_{32} & \mathbf{F}_{33} & \cdots & \mathbf{F}_{3n} \\ \cdots & \cdots & \cdots & \cdots & \cdots \\ \mathbf{F}_{n1} & \mathbf{F}_{n2} & \mathbf{F}_{n3} & \cdots & \mathbf{F}_{nn} \end{bmatrix} = \mathbf{I} \tag{A.99}$$

It will be assumed that the matrix $\mathbf{K}$ is symmetric ($\mathbf{K}_{ij} = \mathbf{K}_{ji}{}^T$), and therefore the inverse $\mathbf{F} = \mathbf{K}^{-1}$ is also symmetric. By multiplying out the product $\mathbf{FK}$ in Eq. (A.99) a sufficient number of equations is obtained from which the unknown submatrices $\mathbf{F}_{ij}$ can be determined. Since $\mathbf{F}$ is symmetric, only the diagonal submatrices and those above the principal diagonal need be calculated. The results are summarized below.

For the first row in **F**

$$\begin{aligned}
\mathbf{F}_{11} &= \mathbf{K}_{11}{}^{-1} + \mathbf{T}_{12}\mathbf{F}_{22}\mathbf{F}_{12}{}^{T} \\
\mathbf{F}_{12} &= \mathbf{T}_{12}\mathbf{F}_{22} \\
\mathbf{F}_{13} &= \mathbf{T}_{12}\mathbf{F}_{23} \\
&\cdots\cdots \\
\mathbf{F}_{1n} &= \mathbf{T}_{12}\mathbf{F}_{2n}
\end{aligned} \tag{A.100}$$

where $\mathbf{T}_{12} = -\mathbf{K}_{11}{}^{-1}\mathbf{K}_{12}$ (A.101)

For the second row

$$\begin{aligned}
\mathbf{F}_{22} &= (\mathbf{K}_{22} + \mathbf{K}_{21}\mathbf{T}_{12})^{-1} + \mathbf{T}_{23}\mathbf{F}_{33}\mathbf{T}_{23}{}^{T} \\
\mathbf{F}_{23} &= \mathbf{T}_{23}\mathbf{F}_{33} \\
\mathbf{F}_{24} &= \mathbf{T}_{23}\mathbf{F}_{34} \\
&\cdots\cdots \\
\mathbf{F}_{2n} &= \mathbf{T}_{23}\mathbf{F}_{3n}
\end{aligned} \tag{A.102}$$

where $\mathbf{T}_{23} = -(\mathbf{K}_{22} + \mathbf{K}_{21}\mathbf{T}_{12})^{-1}\mathbf{K}_{23}$ (A.103)

Similarly, for the $(n - 1)$st row

$$\begin{aligned}
\mathbf{F}_{n-1,n-1} &= (\mathbf{K}_{n-1,n-1} + \mathbf{K}_{n-1,n-2}\mathbf{T}_{n-2,n-1})^{-1} + \mathbf{T}_{n-1,n}\mathbf{F}_{n,n}\mathbf{T}^{T}_{n-1,n} \\
\mathbf{F}_{n-1,n} &= \mathbf{T}_{n-1,n}\mathbf{F}_{n,n}
\end{aligned} \tag{A.104}$$

where $\mathbf{T}_{n-1,n} = -(\mathbf{K}_{n-1,n-1} + \mathbf{K}_{n-1,n-2}\mathbf{T}_{n-2,n-1})^{-1}\mathbf{K}_{n-1,n}$ (A.105)

while for the nth row only one submatrix is required

$$\mathbf{F}_{n,n} = (\mathbf{K}_{n,n} + \mathbf{K}_{n,n-1}\mathbf{T}_{n-1,n})^{-1} \tag{A.106}$$

Once the **T** matrices have been calculated, the back substitution of Eq. (A.106) into (A.104), etc., produces the required matrix inverse. It should be noted that in any given row in **F** the submatrices located to the right of the principal diagonal submatrices are expressed simply as matrix multiples of the submatrices in the row below, and thus the computational effort is greatly simplified.

A.17 INVERSION BY SUCCESSIVE TRANSFORMATIONS (JORDAN TECHNIQUE)

The matrix inversion of a nonsingular matrix **A** can also be achieved by a series of transformations which gradually change the matrix **A** into the unit matrix **I**.

This method is usually referred to as the *Jordan technique*, and its main steps will now be outlined. If the matrix $\mathbf{A}$ to be inverted is given by

$$\mathbf{A} = \begin{bmatrix} a_{11} & a_{12} & \cdots & a_{1n} \\ a_{21} & a_{22} & \cdots & a_{2n} \\ \cdots & \cdots & \cdots & \cdots \\ a_{n1} & a_{n2} & \cdots & a_{nn} \end{bmatrix} \tag{A.107}$$

it is possible to determine a series of transformation matrices $\mathbf{T}_1, \mathbf{T}_2, \ldots, \mathbf{T}_n$ such that

$$\mathbf{T}_n\mathbf{T}_{n-1} \cdots \mathbf{T}_2\mathbf{T}_1\mathbf{A} = \mathbf{I}_n \tag{A.108}$$

from which it follows that

$$\mathbf{T}_n\mathbf{T}_{n-1} \cdots \mathbf{T}_2\mathbf{T}_1 = \mathbf{A}^{-1} \tag{A.109}$$

The above matrix multiplication can be carried out by starting with a unit matrix and premultiplying it successively by $\mathbf{T}_1, \mathbf{T}_2, \ldots,$ and $\mathbf{T}_n$; thus while the given matrix $\mathbf{A}$ is systematically reduced to the unit matrix $\mathbf{I}_n$, the transformation matrices $\mathbf{T}$ produce, in stages, the required matrix inverse $\mathbf{A}^{-1}$. The premultiplication of $\mathbf{A}$ by $\mathbf{T}_1$ produces a matrix in which the first column is the same as the first column in the unit matrix; i.e., the first element is unity while the remaining elements are zero. The premultiplication of the product $\mathbf{T}_1\mathbf{A}$ by $\mathbf{T}_2$ produces a matrix in which the first two columns are the same as in the unit matrix, and so on for subsequent premultiplications till a complete unit matrix is generated, as given by Eq. (A.108).

To eliminate the elements of the first column of $\mathbf{A}$ except for a_{11}, while reducing the latter to unity, $\mathbf{A}$ is premultiplied by the matrix $\mathbf{T}_1$, given by

$$\mathbf{T}_1 = \left[\begin{array}{c|c} \dfrac{1}{a_{11}} & \mathbf{0} \\ \hline \dfrac{-a_{21}}{a_{11}} & \\ \dfrac{-a_{31}}{a_{11}} & \\ \cdot & \mathbf{I}_{n-1} \\ \cdot & \\ \cdot & \\ \dfrac{-a_{n1}}{a_{11}} & \end{array}\right] \tag{A.110}$$

Hence
$$\mathbf{B} = \mathbf{T}_1\mathbf{A} = \left[\begin{array}{c|cccc} 1 & b_{12} & b_{13} & \cdots & b_{1n} \\ \hline & b_{22} & b_{23} & \cdots & b_{2n} \\ \mathbf{0} & b_{32} & b_{33} & \cdots & b_{3n} \\ & \cdots & \cdots & \cdots & \cdots \\ & b_{n2} & b_{n3} & \cdots & b_{nn} \end{array}\right] \tag{A.111}$$

where
$$b_{1j} = \frac{a_{1j}}{a_{11}} \tag{A.112}$$

and
$$b_{ij} = a_{ij} - \frac{a_{i1}a_{1j}}{a_{11}} \qquad \text{for } i \neq 1 \tag{A.113}$$

To eliminate the elements of the second column of **B** except for b_{22}, while reducing the latter to unity, **B** is premultiplied by the matrix $\mathbf{T}_2$, given by

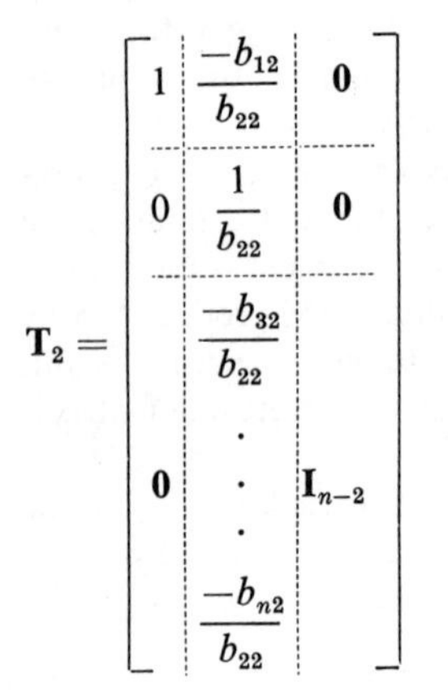

(A.114)

The premultiplication of $\mathbf{T}_1\mathbf{A}$ by $\mathbf{T}_2$ leads to

$$\mathbf{C} = \mathbf{T}_2\mathbf{B} = \mathbf{T}_2\mathbf{T}_1\mathbf{A} = \left[\begin{array}{c|cccc} \mathbf{I}_2 & c_{13} & c_{14} & \cdots & c_{1n} \\ & c_{23} & c_{24} & \cdots & c_{2n} \\ \hline & c_{33} & c_{34} & \cdots & c_{3n} \\ \mathbf{0} & c_{43} & c_{44} & \cdots & c_{4n} \\ & \cdots & \cdots & \cdots & \cdots \\ & c_{n3} & c_{n4} & \cdots & c_{nn} \end{array}\right] \tag{A.115}$$

where $$c_{2j} = \frac{b_{2j}}{b_{22}} \tag{A.116}$$

and $$c_{ij} = b_{ij} - \frac{b_{i2}b_{2j}}{b_{22}} \qquad \text{for } i \neq 2 \tag{A.117}$$

This procedure is followed on until the last premultiplication with

$$\mathbf{T}_n = \left[\begin{array}{c:c} \mathbf{I}_{n-1} & \begin{matrix} \dfrac{-n_{1n}}{n_{nn}} \\ \cdot \\ \cdot \\ \cdot \\ \dfrac{-n_{n-1,n}}{n_{nn}} \end{matrix} \\ \hdashline \mathbf{0} & \dfrac{1}{n_{nn}} \end{array}\right] \tag{A.118}$$

reduces the matrix $\mathbf{A}$ to the unit matrix $\mathbf{I}_n$ [see Eq. (A.108)].

If in some stage of the calculation of the matrices $\mathbf{T}_1, \mathbf{T}_2, \mathbf{T}_3, \ldots$ one of the diagonal elements $a_{11}, b_{22}, c_{33}, \ldots$ is found to be equal to zero, the elimination process cannot be continued. It is then necessary to interchange the affected column with some subsequent column, and this implies that the rows in the matrix inverse will be taken in a different order. The method proceeds in the usual way until a further zero diagonal element is found, and then a further column interchange is introduced. It should be observed that these interchanges do not affect the previously calculated transformation matrices $\mathbf{T}$, but the matrix inverse will have the affected rows interchanged. To preserve the original row sequence the affected rows in the matrix inverse can be interchanged back to their original positions.

If instead of the matrix inverse only the solution to a set of simultaneous equations

$$\mathbf{AX} = \mathbf{Z} \tag{A.119}$$

is required, namely,

$$\mathbf{X} = \mathbf{A}^{-1}\mathbf{Z} \tag{A.120}$$

where $\mathbf{Z}$ is a column matrix, then in order to find $\mathbf{X}$ it is preferable to perform first the multiplication $\mathbf{T}_1\mathbf{Z}$ followed by the premultiplications by $\mathbf{T}_2, \mathbf{T}_3, \ldots, \mathbf{T}_n$. The advantages of this approach are obvious, since the premultiplications would be performed on matrices of order $n \times 1$ as compared with $n \times n$ order when the matrix inverse is calculated first.

To illustrate the Jordan technique consider the nonsingular matrix

$$\mathbf{A} = \begin{bmatrix} 4 & 2 & 1 \\ 2 & 3 & 2 \\ 1 & 2 & 3 \end{bmatrix} \tag{A.121}$$

for which the transformation matrix $\mathbf{T}_1$ given by Eq. (A.110) becomes

$$\mathbf{T}_1 = \begin{bmatrix} \frac{1}{4} & 0 & 0 \\ -\frac{1}{2} & 1 & 0 \\ -\frac{1}{4} & 0 & 1 \end{bmatrix} \tag{A.122}$$

Premultiplying Eq. (A.121) by (A.122) gives

$$\mathbf{B} = \mathbf{T}_1\mathbf{A} = \begin{bmatrix} \frac{1}{4} & 0 & 0 \\ -\frac{1}{2} & 1 & 0 \\ -\frac{1}{4} & 0 & 1 \end{bmatrix} \begin{bmatrix} 4 & 2 & 1 \\ 2 & 3 & 2 \\ 1 & 2 & 3 \end{bmatrix} = \begin{bmatrix} 1 & \frac{1}{2} & \frac{1}{4} \\ 0 & 2 & \frac{3}{2} \\ 0 & \frac{3}{2} & \frac{11}{4} \end{bmatrix} \tag{A.123}$$

Following now the steps outlined in the method leads to

$$\mathbf{C} = \mathbf{T}_2\mathbf{B} = \begin{bmatrix} 1 & -\frac{1}{4} & 0 \\ 0 & \frac{1}{2} & 0 \\ 0 & -\frac{3}{4} & 1 \end{bmatrix} \begin{bmatrix} 1 & \frac{1}{2} & \frac{1}{4} \\ 0 & 2 & \frac{3}{2} \\ 0 & \frac{3}{2} & \frac{11}{4} \end{bmatrix} = \begin{bmatrix} 1 & 0 & -\frac{1}{8} \\ 0 & 1 & \frac{3}{4} \\ 0 & 0 & \frac{13}{8} \end{bmatrix} \tag{A.124}$$

$$\mathbf{D} = \mathbf{T}_3\mathbf{C} = \begin{bmatrix} 1 & 0 & \frac{1}{13} \\ 0 & 1 & -\frac{6}{13} \\ 0 & 0 & \frac{8}{13} \end{bmatrix} \begin{bmatrix} 1 & 0 & -\frac{1}{8} \\ 0 & 1 & \frac{3}{4} \\ 0 & 0 & \frac{13}{8} \end{bmatrix} = \begin{bmatrix} 1 & 0 & 0 \\ 0 & 1 & 0 \\ 0 & 0 & 1 \end{bmatrix} \tag{A.125}$$

The matrix inverse can now be calculated from Eq. (A.109), and hence

$$\mathbf{A}^{-1} = \mathbf{T}_3\mathbf{T}_2\mathbf{T}_1$$

$$= \begin{bmatrix} 1 & 0 & \frac{1}{13} \\ 0 & 1 & \frac{6}{13} \\ 0 & 0 & \frac{8}{13} \end{bmatrix} \begin{bmatrix} 1 & -\frac{1}{4} & 0 \\ 0 & \frac{1}{2} & 0 \\ 0 & -\frac{3}{4} & 1 \end{bmatrix} \begin{bmatrix} \frac{1}{4} & 0 & 0 \\ -\frac{1}{2} & 1 & 0 \\ -\frac{1}{4} & 0 & 1 \end{bmatrix}$$

$$= \begin{bmatrix} 1 & 0 & \frac{1}{13} \\ 0 & 1 & -\frac{6}{13} \\ 0 & 0 & \frac{8}{13} \end{bmatrix} \begin{bmatrix} \frac{3}{8} & -\frac{1}{4} & 0 \\ -\frac{1}{4} & \frac{1}{2} & 0 \\ \frac{1}{8} & -\frac{3}{4} & 1 \end{bmatrix} = \frac{1}{13} \begin{bmatrix} 5 & -4 & 1 \\ -4 & 11 & -6 \\ 1 & -6 & 8 \end{bmatrix} \tag{A.126}$$

Once the matrix inverse $\mathbf{A}^{-1}$ has been determined, it is advisable to calculate either $\mathbf{A}^{-1}\mathbf{A}$ or $\mathbf{A}\mathbf{A}^{-1}$ as a check on the arithmetic; either product must, of course, be equal to a unit matrix.

From the computational point of view it is advantageous to combine Eqs. (A.108) and (A.109) into one equation

$$\mathbf{T}_n\mathbf{T}_{n-1}\cdots\mathbf{T}_2\mathbf{T}_1[\mathbf{A} \mid \mathbf{I}] = [\mathbf{I} \mid \mathbf{A}^{-1}] \tag{A.127}$$

The successive premultiplications by the $\mathbf{T}$ matrices are here performed on the rectangular matrix $[\mathbf{A} \mid \mathbf{I}]$ instead on the matrix $\mathbf{A}$ alone. The rectangular matrix $[\mathbf{A} \mid \mathbf{I}]$ is referred to as the *augmented matrix*. If only the solution to a set of simultaneous equations is required, the $\mathbf{T}$ matrices can be operated on an augmented matrix $[\mathbf{A} \mid \mathbf{Z}]$, where $\mathbf{Z}$ is defined by Eq. (A.119). The solution for the unknowns $\mathbf{X}$ is then obtained from

$$\mathbf{T}_n\mathbf{T}_{n-1}\cdots\mathbf{T}_2\mathbf{T}_1[\mathbf{A} \mid \mathbf{Z}] = [\mathbf{I} \mid \mathbf{X}] \tag{A.128}$$

The Jordan transformation technique has been found very useful for the determination of redundant forces using the equilibrium equations. This particular application is discussed in Sec. 8.3.

A.18 INVERSION BY THE CHOLESKI METHOD

The Choleski method of matrix inversion depends upon the properties of triangular matrices. The basis of this method is that any square matrix can be expressed as the product of an upper and a lower triangular matrix. The actual inversion of the matrix is then reduced to the inversion of the two triangular matrices, which is a very simple procedure. In practically all structural problems the matrices to be inverted are symmetric, and for these cases only one triangular matrix need be inverted.

We assume first that the nonsingular matrix $\mathbf{A}$ can be expressed as the product of a lower triangular matrix $\mathbf{L}$ and a unit upper triangular matrix $\mathbf{U}(1)$, so that

$$\mathbf{A} = \mathbf{L}\mathbf{U}(1) \tag{A.129}$$

Writing the above equation in full, we have

$$\begin{bmatrix} a_{11} & a_{12} & a_{13} & \cdots & a_{1n} \\ a_{21} & a_{22} & a_{23} & \cdots & a_{2n} \\ a_{31} & a_{32} & a_{33} & \cdots & a_{3n} \\ \cdots & \cdots & \cdots & \cdots & \cdots \\ a_{n1} & a_{n2} & a_{n3} & \cdots & a_{nn} \end{bmatrix} = \begin{bmatrix} l_{11} & 0 & 0 & \cdots & 0 \\ l_{21} & l_{22} & 0 & \cdots & 0 \\ l_{31} & l_{32} & l_{33} & \cdots & 0 \\ \cdots & \cdots & \cdots & \cdots & \cdots \\ l_{n1} & l_{n2} & l_{n3} & \cdots & l_{nn} \end{bmatrix} \begin{bmatrix} 1 & u_{12} & u_{13} & \cdots & u_{1n} \\ 0 & 1 & u_{23} & \cdots & u_{2n} \\ 0 & 0 & 1 & \cdots & u_{3n} \\ \cdots & \cdots & \cdots & \cdots & \cdots \\ 0 & 0 & 0 & \cdots & 1 \end{bmatrix} \tag{A.130}$$

Multiplying out the triangular matrices in Eq. (A.130) and equating the corresponding matrix elements gives

$$l_{11} = a_{11}$$

$$l_{11}u_{12} = a_{12} \qquad u_{12} = \frac{a_{12}}{l_{11}}$$

$$\cdots\cdots \qquad \cdots\cdots$$

$$l_{11}u_{1n} = a_{1n} \qquad u_{1n} = \frac{a_{1n}}{l_{11}}$$

$$l_{21} = a_{21}$$

$$l_{21}u_{12} + l_{22} = a_{22} \qquad l_{22} = a_{22} - l_{21}u_{12}$$

$$l_{21}u_{13} + l_{22}u_{23} = a_{23} \qquad u_{23} = \frac{a_{23} - l_{21}u_{13}}{l_{22}}$$

$$\cdots\cdots \qquad \cdots\cdots$$

$$l_{21}u_{1n} + l_{22}u_{2n} = a_{2n} \qquad u_{2n} = \frac{a_{2n} - l_{21}u_{1n}}{l_{22}}$$

$$l_{31} = a_{31}$$

$$l_{31}u_{12} + l_{32} = a_{32} \qquad l_{32} = a_{32} - l_{31}u_{12}$$

$$l_{31}u_{13} + l_{32}u_{23} + l_{33} = a_{33} \qquad l_{33} = a_{33} - l_{31}u_{13} - l_{32}u_{23}$$

$$l_{31}u_{14} + l_{32}u_{24} + l_{33}u_{34} = a_{34} \qquad u_{34} = \frac{a_{34} - l_{31}u_{14} - l_{32}u_{24}}{l_{33}}$$

$$\cdots\cdots \qquad \cdots\cdots$$

$$l_{31}u_{1n} + l_{32}u_{2n} + l_{33}u_{3n} = a_{3n} \qquad u_{3n} = \frac{a_{3n} - l_{31}u_{1n} - l_{32}u_{2n}}{l_{33}}$$

and so on. The above equations can be written in a more compact form as

$$l_{ij} = a_{ij} - \sum_{r=1}^{j-1} l_{ir}u_{rj} \qquad i \geqslant j \qquad \text{lower triangular matrix}$$

$$u_{ij} = \frac{a_{ij} - \sum_{r=1}^{i-1} l_{ir}u_{rj}}{l_{ii}} \qquad i < j \qquad \text{upper triangular matrix}$$

$$u_{ii} = 1 \tag{A.131}$$

Once the two triangular matrices $\mathbf{L}$ and $\mathbf{U}(1)$ have been calculated, the inverse of the matrix $\mathbf{A}$ is determined from

$$\mathbf{A}^{-1} = [\mathbf{LU}(1)]^{-1} = [\mathbf{U}(1)]^{-1}\mathbf{L}^{-1} \tag{A.132}$$

To determine the inverse of the lower triangular matrix we shall use the relationship

$$\mathbf{L}\mathbf{L}^{-1} = \mathbf{L}\mathbf{M} = \mathbf{I} \tag{A.133}$$

where $\quad \mathbf{M} = \mathbf{L}^{-1}$ (A.134)

Eq. (A.133) can be written fully as

$$\begin{bmatrix} l_{11} & 0 & 0 & \cdots & 0 \\ l_{21} & l_{22} & 0 & \cdots & 0 \\ l_{31} & l_{32} & l_{33} & \cdots & 0 \\ \cdots & \cdots & \cdots & \cdots & \cdots \\ l_{n1} & l_{n2} & l_{n3} & \cdots & l_{nn} \end{bmatrix} \begin{bmatrix} m_{11} & m_{12} & m_{13} & \cdots & m_{1n} \\ m_{21} & m_{22} & m_{23} & \cdots & m_{2n} \\ m_{31} & m_{32} & m_{33} & \cdots & m_{3n} \\ \cdots & \cdots & \cdots & \cdots & \cdots \\ m_{n1} & m_{n2} & m_{n3} & \cdots & m_{nn} \end{bmatrix} = \begin{bmatrix} 1 & 0 & 0 & \cdots & 0 \\ 0 & 1 & 0 & \cdots & 0 \\ 0 & 0 & 1 & \cdots & 0 \\ \cdots & \cdots & \cdots & \cdots & \cdots \\ 0 & 0 & 0 & \cdots & 1 \end{bmatrix} \tag{A.135}$$

Multiplying out the matrices **L** and **M** and equating the corresponding matrix elements, we have that

$$\begin{aligned}
&l_{11}m_{11} = 1 && m_{11} = \frac{1}{l_{11}} \\
&l_{11}m_{12} = 0 && m_{12} = 0 \\
&\cdots\cdots && \cdots\cdots \\
&l_{11}m_{1n} = 0 && m_{1n} = 0 \\
&l_{21}m_{11} + l_{22}m_{21} = 0 && m_{21} = \frac{-l_{21}m_{11}}{l_{22}} \\
&l_{21}m_{12} + l_{22}m_{22} = 1 && m_{22} = \frac{1}{l_{22}} \\
&l_{21}m_{13} + l_{22}m_{23} = 0 && m_{23} = 0 \\
&\cdots\cdots && \cdots\cdots \\
&l_{21}m_{1n} + l_{22}m_{2n} = 0 && m_{2n} = 0 \\
&l_{31}m_{11} + l_{32}m_{21} + l_{33}m_{31} = 0 && m_{31} = \frac{-(l_{31}m_{11} + l_{32}m_{21})}{l_{33}} \\
&l_{31}m_{12} + l_{32}m_{22} + l_{33}m_{32} = 0 && m_{32} = \frac{-l_{32}m_{22}}{l_{33}} \\
&l_{31}m_{13} + l_{32}m_{23} + l_{33}m_{33} = 1 && m_{33} = \frac{1}{l_{33}} \\
&l_{31}m_{14} + l_{32}m_{24} + l_{33}m_{34} = 0 && m_{34} = 0 \\
&\cdots\cdots && \cdots\cdots \\
&l_{31}m_{1n} + l_{32}m_{2n} + l_{33}m_{3n} = 0 && m_{3n} = 0
\end{aligned}$$

and so on. The above derived expressions for the elements of **M** can be put into a more compact form as

$$m_{ii} = \frac{1}{l_{ii}} \qquad \text{diagonal elements}$$

$$m_{ij} = -\frac{1}{l_{ii}} \sum_{r=j}^{i-1} l_{ir} m_{rj} \qquad i > j \qquad \text{elements below the diagonal}$$

$$m_{ij} = 0 \qquad i < j \qquad \text{elements above the diagonal} \tag{A.136}$$

Thus the inverse of the lower triangular matrix **L** is also a lower triangular matrix. Similarly, it can be shown that the inverse of the unit upper triangular matrix **U**(1) will be another upper triangular matrix, so that if

$$[\mathbf{U}(1)]^{-1} = \mathbf{N} \tag{A.137}$$

then the elements of **N** are given by

$$n_{ii} = 1 \qquad \text{diagonal elements}$$

$$n_{ij} = -\sum_{r=i}^{j-1} n_{ir} u_{rj} \qquad i < j \qquad \text{elements above the diagonal} \tag{A.138}$$

$$n_{ij} = 0 \qquad i > j \qquad \text{elements below the diagonal}$$

Hence having found the elements of the triangular matrices **M** and **N**, we can calculate the matrix inverse $\mathbf{A}^{-1}$ from

$$\mathbf{A}^{-1} = \mathbf{U}(1)^{-1}\mathbf{L}^{-1} = \mathbf{NM} \tag{A.139}$$

If in the calculation of the lower triangular matrix **L** a diagonal element $l_{ii} = 0$ is found, thereby indicating that the method appears to fail, it is necessary to interchange the ith row with some subsequent row. This implies that the equations associated with the matrix inverse will be taken in a different order. The method proceeds in the usual way until a further diagonal element, say $l_{kk} = 0$, is found, and then the kth row is interchanged with a subsequent row. It should be noted that because of these interchanges of rows, the matrix inverse will have the corresponding columns interchanged. If necessary, the original column sequence can be restored by interchanging all the affected columns.

When the matrix **A** is symmetric, the process of finding its inverse can be shortened. This is due to the fact that it is then possible to express **A** in the form

$$\mathbf{A} = \mathbf{\Lambda}\mathbf{\Lambda}^T \tag{A.140}$$

where $\mathbf{\Lambda}$ is a lower triangular matrix given by

$$\mathbf{\Lambda} = \begin{bmatrix} \lambda_{11} & 0 & 0 & \cdots & 0 \\ \lambda_{21} & \lambda_{22} & 0 & \cdots & 0 \\ \lambda_{31} & \lambda_{32} & \lambda_{33} & \cdots & 0 \\ \cdots & \cdots & \cdots & \cdots & \cdots \\ \lambda_{n1} & \lambda_{n2} & \lambda_{n3} & \cdots & \lambda_{nn} \end{bmatrix} \tag{A.141}$$

Applying now the method used for finding the elements of **L** and **U**(1), we can show that the elements of $\boldsymbol{\Lambda}$ are given by

$$\lambda_{ii} = \left(a_{ii} - \sum_{r=1}^{i-1} \lambda_{ir}^{2}\right)^{\frac{1}{2}}$$

$$\lambda_{ij} = \frac{a_{ij} - \sum_{r=1}^{j-1} \lambda_{ir}\lambda_{jr}}{\lambda_{jj}} \qquad i > j \tag{A.142}$$

$$\lambda_{ij} = 0 \qquad i < j$$

The elements μ_{ij} of the inverse of $\boldsymbol{\Lambda}$ can be determined from $\boldsymbol{\Lambda}\boldsymbol{\Lambda}^{-1} = \mathbf{I}_r$, which leads to

$$\mu_{ii} = \frac{1}{\lambda_{ii}}$$

$$\mu_{ij} = -\frac{\sum_{r=j}^{i-1} \lambda_{ir}\mu_{rj}}{\lambda_{ii}} \qquad i > j \tag{A.143}$$

$$\mu_{ij} = 0 \qquad i < j$$

Hence the inverse of $\boldsymbol{\Lambda}$ is a lower triangular matrix. To determine the elements of $\boldsymbol{\Lambda}^{-1}$ we note that for any nonsingular matrix **A**

$$(\mathbf{A}^{T})^{-1} = \frac{\hat{\mathbf{A}}^{T}}{|\mathbf{A}|} = (\mathbf{A}^{-1})^{T} \tag{A.144}$$

and therefore $\quad (\boldsymbol{\Lambda}^{T})^{-1} = (\boldsymbol{\Lambda}^{-1})^{T} \qquad$ (A.145)

The matrix inverse for the symmetric matrix **A** is then calculated from

$$\mathbf{A}^{-1} = (\boldsymbol{\Lambda}\boldsymbol{\Lambda}^{T})^{-1} = (\boldsymbol{\Lambda}^{-1})^{T}\boldsymbol{\Lambda}^{-1} \tag{A.146}$$

A.19 PROCEDURE FOR IMPROVING THE ACCURACY OF A MATRIX INVERSE

Since the inverse of a large matrix is obtained by a succession of matrix operations, its accuracy is affected by the rounding-off errors, and therefore it is desirable to have a simple procedure whereby the accuracy of an approximate matrix inverse can be improved. Suppose that the inverse of a matrix **A** has been determined approximately by any of the standard numerical methods and it is required to find a closer approximation to the true inverse. If this approximate inverse is given by the matrix $\mathbf{A}_1^{-1}$, then

$$\mathbf{A}^{-1} = \mathbf{A}_1^{-1} + \Delta\mathbf{A}_1 \tag{A.147}$$

where $\Delta\mathbf{A}_1$ represents a matrix of small corrections to $\mathbf{A}_1^{-1}$. The matrix $\mathbf{A}_1^{-1}$ can also be interpreted here as the true inverse of $\mathbf{A}_1$, which differs not too greatly from $\mathbf{A}$. Premultiplying Eq. (A.147) by the product $\mathbf{A}_1^{-1}\mathbf{A}$, we have

$$\mathbf{A}_1^{-1} = \mathbf{A}_1^{-1}\mathbf{A}\mathbf{A}_1^{-1} + \mathbf{A}_1^{-1}\mathbf{A}\,\Delta\mathbf{A}_1 \tag{A.148}$$

Assuming now that the product $\mathbf{A}_1^{-1}\mathbf{A}$ is approximately equal to $\mathbf{I}$, it follows from Eq. (A.148) that

$$\Delta\mathbf{A}_1 \approx \mathbf{A}_1^{-1}(\mathbf{I} - \mathbf{A}\mathbf{A}_1^{-1}) \tag{A.149}$$

which, when substituted into Eq. (A.147), leads to

$$\mathbf{A}^{-1} \approx \mathbf{A}_1^{-1}(2\mathbf{I} - \mathbf{A}\mathbf{A}_1^{-1}) \tag{A.150}$$

Equation (A.150) provides a means of determining a closer approximation to the true inverse $\mathbf{A}^{-1}$ from the approximate inverse $\mathbf{A}_1^{-1}$. If still better accuracy is required, Eq. (A.150) can be used for subsequent iteration. Denoting the nth approximation to the inverse by $\mathbf{A}_n^{-1}$, we have, from Eq. (A.150),

$$\mathbf{A}_n^{-1} \approx \mathbf{A}_{n-1}^{-1}(2\mathbf{I} - \mathbf{A}\mathbf{A}_{n-1}^{-1}) \tag{A.151}$$

A.20 TEST MATRIX FOR INVERSION COMPUTER PROGRAMS

Very often it is necessary to check out a computer program for inversion of large matrices. This can be done most conveniently using the following $n \times n$ nonsingular matrix

$$\mathbf{A} = \begin{bmatrix} \frac{n+2}{2n+2} & -\frac{1}{2} & 0 & 0 & \cdots & 0 & \frac{1}{2n+2} \\ -\frac{1}{2} & 1 & -\frac{1}{2} & 0 & \cdots & 0 & 0 \\ 0 & -\frac{1}{2} & 1 & -\frac{1}{2} & \cdots & 0 & 0 \\ \cdots & \cdots & \cdots & \cdots & \cdots & \cdots & \cdots \\ 0 & 0 & \cdots & \cdots & -\frac{1}{2} & 1 & -\frac{1}{2} \\ \frac{1}{2n+2} & 0 & \cdots & \cdots & 0 & -\frac{1}{2} & \frac{n+2}{2n+2} \end{bmatrix} \tag{A.152}$$

whose inverse is given by

$$
\mathbf{A}^{-1} = \begin{bmatrix}
n & n-1 & n-2 & \cdots & 2 & 1 \\
n-1 & n & n-1 & \cdots & 3 & 2 \\
n-2 & n-1 & n & \cdots & 4 & 3 \\
\cdots & \cdots & \cdots & \cdots & \cdots & \cdots \\
2 & 3 & 4 & \cdots & n & n-1 \\
1 & 2 & 3 & \cdots & n-1 & n
\end{bmatrix} \tag{A.153}
$$

A.21 DIFFERENTIATION OF MATRICES

If the elements a_{ij} of a matrix $\mathbf{A}$ are functions of a parameter t, we define matrix differentiation as

$$
\frac{d\mathbf{A}}{dt} = \dot{\mathbf{A}} = \begin{bmatrix}
\frac{da_{11}}{dt} & \frac{da_{12}}{dt} & \cdots & \frac{da_{1n}}{dt} \\
\frac{da_{21}}{dt} & \frac{da_{22}}{dt} & \cdots & \frac{da_{2n}}{dt} \\
\cdots & \cdots & \cdots & \cdots \\
\frac{da_{m1}}{dt} & \frac{da_{m2}}{dt} & \cdots & \frac{da_{mn}}{dt}
\end{bmatrix} \tag{A.154}
$$

That is, the matrix $\mathbf{A}$ is differentiated by differentiating every element in the conventional manner. In the same way we define higher derivatives of matrices. For example, the second derivative of a matrix with respect to time requires that every element be differentiated twice.

In matrix structural analysis we also find need for differentiation with respect to every element in a column matrix $\mathbf{X} = \{X_1 \quad X_2 \quad \cdots \quad X_n\}$. For example, we may require derivatives of some energy function U represented by a 1×1 matrix. Symbolically this may be written as

$$
\begin{bmatrix}
\frac{\partial U}{\partial X_1} \\
\frac{\partial U}{\partial X_2} \\
\cdot \\
\cdot \\
\cdot \\
\frac{\partial U}{\partial X_n}
\end{bmatrix} = \frac{\partial U}{\partial \mathbf{X}} \tag{A.155}
$$

Very often the matrix U is obtained from matrix multiplication represented symbolically as

$$U = \tfrac{1}{2}\mathbf{X}^T\mathbf{A}\mathbf{X} + \mathbf{X}^T\mathbf{B} + C \tag{A.156}$$

For this expression we obtain

$$\frac{\partial U}{\partial \mathbf{X}} = \mathbf{A}\mathbf{X} + \mathbf{B} \tag{A.157}$$

Equation (A.157) can be proved formally by multiplying out the matrix products in Eq. (A.156) and then performing differentiation, as defined above.

A.22 INTEGRATION OF MATRICES

If the elements a_{ij} of a matrix $\mathbf{A}$ are functions of a parameter t, we define matrix integration as

$$\int \mathbf{A}\,dt = \begin{bmatrix} \int a_{11}\,dt & \int a_{12}\,dt & \cdots & \int a_{1n}\,dt \\ \int a_{21}\,dt & \int a_{22}\,dt & \cdots & \int a_{2n}\,dt \\ \cdots & \cdots & \cdots & \cdots \\ \int a_{m1}\,dt & \int a_{m2}\,dt & \cdots & \int a_{mn}\,dt \end{bmatrix} \tag{A.158}$$

Multiple integrals are defined in an analogous manner. If we want to find an integral of a matrix product such as, for example,

$$\mathbf{M} = \iiint \mathbf{A}^T\mathbf{C}\mathbf{A}\,dx\,dy\,dz \tag{A.159}$$

in which the elements of $\mathbf{A}$ are functions of x, y, and z while the elements of $\mathbf{C}$ are constants, we must first carry out matrix multiplication and then integrate each element in the resulting matrix.

A.23 EIGENVALUES AND EIGENVECTORS

Consider a matrix equation

$$\mathbf{A}\mathbf{x} = \lambda\mathbf{x}$$

or

$$(\mathbf{A} - \lambda\mathbf{I})\mathbf{x} = \mathbf{0} \tag{A.160}$$

in which λ is some undetermined multiplier and $\mathbf{A}$ is a square matrix. When this equation is written in full, we have

$$\begin{aligned} (a_{11} - \lambda)x_1 + a_{12}x_2 + \cdots + a_{1n}x_n &= 0 \\ a_{21}x_1 + (a_{22} - \lambda)x_2 + \cdots + a_{2n}x_n &= 0 \\ \cdots\cdots\cdots\cdots\cdots\cdots\cdots\cdots\cdots & \\ a_{n1}x_1 + a_{n2}x_2 + \cdots + (a_{nn} - \lambda)x_n &= 0 \end{aligned} \tag{A.160a}$$

A nontrivial solution of these equations is possible provided that

$$\begin{vmatrix} (a_{11} - \lambda) & a_{12} & \cdots & a_{1n} \\ a_{21} & (a_{22} - \lambda) & \cdots & a_{2n} \\ \cdots & \cdots & \cdots & \cdots \\ a_{n1} & a_{n2} & \cdots & (a_{nn} - \lambda) \end{vmatrix} = 0 \tag{A.161}$$

or, in matrix notation,

$$|\mathbf{A} - \lambda \mathbf{I}| = 0 \tag{A.161a}$$

When the determinant in (A.161) is expanded, we obtain a polynomial in λ of the form

$$\lambda^n + a_1 \lambda^{n-1} + a_2 \lambda^{n-2} + \cdots + a_n = 0 \tag{A.162}$$

which is known as the characteristic equation of the matrix **A**. The roots $\lambda_1, \lambda_2, \ldots, \lambda_n$ of the characteristic equation are known as the *characteristic values* of the matrix **A**. An alternative name used is the *eigenvalues* of the matrix **A**.

The eigenvalues $\lambda_1, \lambda_2, \ldots, \lambda_n$ fulfill two conditions:

$$\sum_{i=1}^{n} \lambda_i = \text{tr } \mathbf{A} \tag{A.163}$$

and $$\prod_i \lambda_i = |\mathbf{A}| \tag{A.164}$$

where tr **A** is the trace of the matrix **A** defined as $\sum_{i=1}^{n} a_{ii}$ and $\prod_i \lambda_i$ is the product of all eigenvalues. These two conditions can serve as valuable checks on the calculated eigenvalues.

As an example consider the equation

$$\left\{ \begin{bmatrix} 2 & 1 \\ 2 & 3 \end{bmatrix} - \lambda \begin{bmatrix} 1 & 0 \\ 0 & 1 \end{bmatrix} \right\} \begin{bmatrix} x_1 \\ x_2 \end{bmatrix} = \begin{bmatrix} 0 \\ 0 \end{bmatrix} \tag{A.165}$$

The characteristic equation is given by the determinant

$$\begin{vmatrix} 2 - \lambda & 1 \\ 2 & 3 - \lambda \end{vmatrix} = 0 \tag{A.166}$$

which when expanded yields

$$\lambda^2 - 5\lambda + 4 = 0 \tag{A.167}$$

The roots of this equation are $\lambda_1 = 1$ and $\lambda_2 = 4$. Substituting $\lambda = 1$ into Eq. (A.165) and assuming that $x_1 = 1$, we find that $x_2 = -1$. Similarly for $\lambda = 4$ when $x_1 = 1$ is assumed, we find that $x_2 = 2$. The vectors

$$\mathbf{x}^{(1)} = \{1 \quad -1\} \tag{A.168}$$

$$\mathbf{x}^{(2)} = \{1 \quad 2\} \tag{A.169}$$

are described as the *eigenvectors* corresponding to the eigenvalues $\lambda = 1$ and 4, respectively. Since Eq. (A.165) is homogeneous, only relative values of the eigenvectors $\mathbf{x}$ can be determined for each eigenvalue. In vibration analysis the eigenvectors are referred to as *eigenmodes*.

The more general eigenvalue problem is of the form

$$(\mathbf{A} - \lambda\mathbf{B})\mathbf{X} = \mathbf{0} \tag{A.170}$$

where both $\mathbf{A}$ and $\mathbf{B}$ are square matrices. The characteristic equation in this case is

$$|\mathbf{A} - \lambda\mathbf{B}| = 0 \tag{A.171}$$

Equations of this type are found in the vibration analysis of elastic systems and also in buckling analysis. If either $\mathbf{A}$ or $\mathbf{B}$ is nonsingular, it is possible by premultiplying Eq. (A.170) by either $(1/\lambda)\mathbf{A}^{-1}$ or $\mathbf{B}^{-1}$ to reduce it to the standard form (A.160).

A.24 EIGENVALUES AND EIGENVECTORS OF THE QUADRATIC MATRIX EQUATION

In Chap. 12 we saw how the frequency-dependent mass and stiffness matrices lead to an equation of motion in the form of a matrix series with ascending powers of the frequency ω. By retaining terms up to and including ω^4 we obtain an equation of the form

$$(\mathbf{A} - \omega^2\mathbf{B} - \omega^4\mathbf{C})\mathbf{q} = \mathbf{0} \tag{A.172}$$

where $\mathbf{A}$, $\mathbf{B}$, and $\mathbf{C}$ are square matrices. The condition for nonzero solution for $\mathbf{q}$ is

$$|\mathbf{A} - \omega^2\mathbf{B} - \omega^4\mathbf{C}| = 0 \tag{A.173}$$

This determinant can be expanded, and the roots ω^2 of the resulting polynomial can then be substituted into Eq. (A.172) in order to calculate the corresponding eigenvectors (eigenmodes). Two other methods for obtaining the eigenvalues and eigenvectors of Eq. (A.172) can also be used. These methods are described below.

ITERATIVE SOLUTION

It will be assumed that Eq. (A.172) has eigenvectors $\mathbf{p}_1, \mathbf{p}_2, \ldots, \mathbf{p}_n$ corresponding to positive (including zero) eigenvalues $\omega_1, \omega_2, \ldots, \omega_n$. After introducing

$$\mathbf{\Omega}^2 = [\omega_1^2 \quad \omega_2^2 \quad \cdots \quad \omega_n^2] \tag{A.174}$$

and $$\mathbf{p} = [\mathbf{p}_1 \quad \mathbf{p}_2 \quad \cdots \quad \mathbf{p}_n] \tag{A.175}$$

Eq. (A.172) can be written as

$$\mathbf{Ap} - \mathbf{Bp\Omega}^2 - \mathbf{Cp\Omega}^4 = \mathbf{0} \tag{A.176}$$

Postmultiplying Eq. (A.176) by $\mathbf{p}^{-1}$, we obtain

$$\mathbf{A} - \mathbf{Bp\Omega}^2\mathbf{p}^{-1} - \mathbf{Cp\Omega}^4\mathbf{p}^{-1} = \mathbf{0} \tag{A.177}$$

We now introduce a new matrix defined by

$$\mathbf{E} = \mathbf{p\Omega}^2\mathbf{p}^{-1} \tag{A.178}$$

so that Eq. (A.177) becomes, after premultiplying by $\mathbf{C}^{-1}$,

$$\mathbf{E}^2 = \mathbf{C}^{-1}\mathbf{A} - \mathbf{C}^{-1}\mathbf{BE} \tag{A.179}$$

Equation (A.179) forms the basis for the iterative solution. We define next the matrix $\mathbf{E}$ as

$$\mathbf{E} = \mathbf{E}_0 + \Delta\mathbf{E} \tag{A.180}$$

where $\quad \mathbf{E}_0 = \mathbf{p}_0\mathbf{\Omega}_0{}^2\mathbf{p}_0{}^{-1} \quad$ (A.181)

is obtained from the conventional solution eigenvectors $\mathbf{p}_0$ and eigenvalues $\mathbf{\Omega}_0$ satisfying the equation

$$\mathbf{Ap}_0 - \mathbf{Bp}_0\mathbf{\Omega}_0{}^2 = \mathbf{0} \tag{A.182}$$

and $\Delta\mathbf{E}$ is a correction matrix. Substituting Eq. (A.180) into (A.179), we have

$$(\mathbf{E}_0 + \Delta\mathbf{E})\mathbf{E} = \mathbf{C}^{-1}\mathbf{A} - \mathbf{C}^{-1}\mathbf{BE} \tag{A.183}$$

Hence $\quad \mathbf{E} \approx (\mathbf{E}_0 + \mathbf{C}^{-1}\mathbf{B})^{-1}\mathbf{C}^{-1}\mathbf{A} \quad$ (A.184)

which is used to establish the iterative loop

$$\mathbf{E}_{n+1} \approx (\mathbf{E}_n + \mathbf{C}^{-1}\mathbf{B})^{-1}\mathbf{C}^{-1}\mathbf{A} \tag{A.185}$$

where n denotes the nth step. Once satisfactory convergence has been achieved for $\mathbf{E}$, the quadratic-equation eigenvalues and eigenvectors are determined from the conventional equation, obtained from Eq. (A.178),

$$\mathbf{Ep} - \mathbf{p}\Omega^2 = \mathbf{0} \tag{A.186}$$

The above method has been used by Przemieniecki[275] for systems with up to 10 degrees of freedom. One drawback of this method is that occasionally, since the quadratic equation has both positive and negative eigenvalues, the iteration procedure will produce some negative eigenvalues in place of a few highest eigenvalues (frequencies). If these frequencies are required, alternative methods must be used for these cases. This usually occurs when $\mathbf{E}_0$ differs greatly from the final value of $\mathbf{E}$.

DIRECT SOLUTION

A direct method of solving for the eigenvalues and eigenvectors of Eq. (A.172) is described by Buckingham. In this method we introduce

$$\dot{\mathbf{q}} = \omega^2 \mathbf{q} \tag{A.187}$$

The fundamental equation (A.172) is premultiplied by $\mathbf{C}^{-1}$ so that

$$\mathbf{C}^{-1}\mathbf{A}\mathbf{q} - \mathbf{C}^{-1}\mathbf{B}\dot{\mathbf{q}} - \omega^2\dot{\mathbf{q}} = \mathbf{0} \tag{A.188}$$

which is then combined with Eq. (A.187) to form

$$\begin{bmatrix} -\omega^2\mathbf{I} & \mathbf{I} \\ \mathbf{C}^{-1}\mathbf{A} & -(\mathbf{C}^{-1}\mathbf{B} + \omega^2\mathbf{I}) \end{bmatrix} \begin{bmatrix} \mathbf{q} \\ \dot{\mathbf{q}} \end{bmatrix} = \mathbf{0} \tag{A.189}$$

or

$$\left\{ \begin{bmatrix} \mathbf{0} & \mathbf{I}_n \\ \mathbf{C}^{-1}\mathbf{A} & -\mathbf{C}^{-1}\mathbf{B} \end{bmatrix} - \omega^2\mathbf{I}_{2n} \right\} \begin{bmatrix} \mathbf{q} \\ \dot{\mathbf{q}} \end{bmatrix} = \mathbf{0} \tag{A.189a}$$

Since Eq. (A.189*a*) is of a standard form, any conventional eigenvalue computer programs can be used. The only disadvantage of this method is that it involves $2n$ unknown components in each eigenvector compared with n components in the original system.

APPENDIX B
BIBLIOGRAPHY

GENERAL REFERENCES

BIEZENO, C. B., and R. GRAMMEL: "Engineering Dynamics," vol. 1, 2d ed., D. Van Nostrand Company, Inc., Princeton, N.J., 1954.

BISPLINGHOFF, R. L., J. W. MAR, and T. H. H. PIAN: "Statics of Deformable Solids," Addison-Wesley Publishing Company, Inc., Reading, Mass., 1965.

BODEWIG, E.: "Matrix Calculus," North-Holland Publishing Company, Amsterdam, 1959.

BOLEY, B. A., and J. H. WEINER: "Theory of Thermal Stresses," John Wiley & Sons, Inc., New York, 1960.

BUCKINGHAM, R. A.: "Numerical Methods," Sir Isaac Pitman & Sons, Ltd., London, 1957.

CASTIGLIANO, A.: Engineering thesis, University of Turin, 1873; *Trans. Acad. Sci. Turin*, **10**:380 (1875), **11**:127–286 (1876); "Theory of Equilibrium of Elastic Systems and its Applications," Paris, 1879, translated by E. S. Andrews as "Elastic Stresses in Structures," Scott, Greenwood and Son, London, 1919.

ENGESSER, F.: On the Statically Indeterminate Girders with Arbitrary Stress–strain Law . . . , *Z. Architekten Ing.-Ver. Hannover*, **35**:733–744 (1889).

ERINGEN, A. C.: "Nonlinear Theory of Continuous Media," McGraw-Hill Book Company, New York, 1962.

FRAZER, R. A., W. J. DUNCAN, and A. R. COLLAR: "Elementary Matrices," Cambridge University Press, New York, 1946.

FREDERICK, D., and T. S. CHANG: "Continuum Mechanics," Allyn and Bacon, Inc., Boston, 1965.

HOFF, N. J.: "Analysis of Structures," John Wiley & Sons, Inc., New York, 1956.

NAVIER, M. H.: *Mem. Acad. Sci.*, **7** (1827).

Proc. Conf. Matrix Methods Struct. Mech., Wright-Patterson Air Force Base, Ohio, Oct. 26–28, 1965, AFFDL-TR 66-80, 1966.

TIMOSHENKO, S., and J. M. GERE: "Theory of Elastic Stability," 2d ed., McGraw-Hill Book Company, New York, 1961.

TIMOSHENKO, S., and J. N. GOODIER: "Theory of Elasticity," 2d ed., McGraw-Hill Book Company, New York, 1951.

TIMOSHENKO, S., and S. WOINOWSKY-KRIEGER: "Theory of Plates and Shells," 2d ed., McGraw-Hill Book Company, New York, 1959.

WANG, C.-T.: "Applied Elasticity," McGraw-Hill Book Company, New York, 1953.

WESTERGAARD, H. M.: On the Method of Complementary Energy, *Proc. ASCE*, **67**:199–227 (1941); *Trans. ASCE*, **107**:765–803 (1942).

REFERENCES ON MATRIX STRUCTURAL ANALYSIS

1. ABBOTT, M. B.: Structural Analysis by Digital Computer, *Engineering*, **187**:666–667 (1959).

2. ADINI, A: "Analysis of Shell Structures by the Finite Element Method," Ph.D. dissertation, Civil Eng. Dept., University of California, Berkeley, 1961.

3. ALLEY, V. L., and A. H. GERRINGER: A Matrix Method for the Determination of the Natural Frequencies of Free-Free Unsymmetrical Beams with Applications to Launch Vehicles, *NASA Tech. Note* D-1247, 1962.

4. ANG, A. H.-S.: Numerical Approach for Wave Motions in Nonlinear Solid Media, *Proc. Conf. Matrix Methods Struct. Mech., Wright-Patterson Air Force Base, Ohio*, Oct. 26–28, 1965, AFFDL TR 66-80, 1966.

5. ARCHER, J. S.: Digital Computation for Stiffness Matrix Analysis, *Proc. ASCE*, **84** (ST6): paper 1814, (1958).

6. ARCHER, J. S.: Consistent Mass Matrix for Distributed Mass Systems, *J. Struct. Div. Proc. ASCE*, **89**:161–178 (1963).

7. ARCHER, J. S.: Consistent Matrix Formulation for Structural Analysis Using Finite-element Techniques, *J. Am. Inst. Aeron. Astron.*, **3**:1910–1918 (1965).

8. ARCHER, J. S., and C. P. RUBIN: Improved Linear Axisymmetric Shell-fluid Model for Launch Vehicle Longitudinal Response Analysis, *Proc. Conf. Matrix Methods Struct. Mech., Wright-Patterson Air Force Base, Ohio*, Oct. 26–28, 1965, AFFDL TR 66-80, 1966.

9. ARCHER, J. S., and C. H. SAMSON: Structural Idealization for Digital Computer Analysis, *Proc. 2d Conf. Electron. Computation ASCE, Pittsburgh, Pa.*, pp. 283–325, Sept. 8–9, 1960.

10. ARGYRIS, J. H.: Energy Theorems and Structural Analysis, *Aircraft Eng.*, **26**:347–356, 383–387, 394 (1954); **27**:42–58, 80–94, 125–134, 145–158 (1955); see also "Energy Theorems and Structural Analysis," Butterworth Scientific Publications, London, 1960.

11. ARGYRIS, J. H.: The Matrix Analysis of Structures with Cut-outs and Modifications, *Proc. 9th Intern. Congr. Appl. Mech., Sect.* II, *Mech. Solids*, September, 1956.

12. ARGYRIS, J. H.: The Matrix Theory of Statics (in German), *Ingr. Arch.*, **25**:174–192 (1957).

13. ARGYRIS, J. H.: On the Analysis of Complex Elastic Structures, *Appl. Mech. Rev.*, **11**:331–338 (1958).

14. ARGYRIS, J. H.: "Recent Advances in Matrix Methods of Structural Analysis," Progress in Aeronautical Sciences, vol. IV, The Macmillan Company, New York, 1964; also published in B. Fraeijs de Veubeke (ed.), Matrix Methods of Structural Analysis, *AGARDograph*, **72**:1–164 (1964).

15. ARGYRIS, J. H.: "The Trondheim Lectures on the Matrix Theory of Structures," Oslo University Press, Oslo, 1964, and John Wiley & Sons, Inc., New York, 1965.

16. ARGYRIS, J. H.: Matrix Analysis of Three-dimensional Elastic Media Small and Large Deflections, *J. Am. Inst. Aeron. Astron.*, **3**:45–51 (1965).

17. ARGYRIS, J. H.: Three-dimensional Anisotropic and Inhomogeneous Elastic Media Matrix Analysis for Small and Large Displacements, *Ingen. Arch.*, **34**:33–35 (1965).

18. ARGYRIS, J. H.: Continua and Discontinua, *Proc. Conf. Matrix Methods Struct. Mech., Wright-Patterson Air Force Base, Ohio*, Oct. 26-28, 1965, AFFDL TR 66-80, 1966.

19. ARGYRIS, J. H.: Elasto-plastic Matrix Displacement Analysis of Three-dimensional Continua, *J. Roy. Aeron. Soc.*, **69**:633–636 (1965).

20. ARGYRIS, J. H.: Triangular Elements with Linearly Varying Strain for the Matrix Displacement Method, *J. Roy. Aeron. Soc.*, **69**:711–713 (1965).

21. ARGYRIS, J. H.: Reinforced Fields of Triangular Elements with Linearly Varying Strain: Effect of Initial Strains, *J. Roy. Aeron. Soc.*, **69**:799–801 (1965).

22. ARGYRIS, J. H.: Matrix Displacement Analysis of Anisotropic Shells by Triangular Elements, *J. Roy. Aeron. Soc.*, **69**:801–805 (1965).

23. ARGYRIS, J. H.: Tetrahedron Elements with Linearly Varying Strain for the Matrix Displacement Method, *J. Roy. Aeron. Soc.*, **69**:877–880 (1965).

24. ARGYRIS, J. H.: Arbitrary Quadrilateral Spar Webs for the Matrix Displacement Method, *J. Roy. Aeron. Soc.*, **70**:359–362 (1966).

25. ARGYRIS, J. H.: Membrane Parallelogram Element with Linearly Varying Edge Strain for Matrix Displacement Method, *J. Roy. Aeron. Soc.*, **70**:599–604 (1966).

26. ARGYRIS, J. H.: Matrix Analysis of Plates and Shells: Prolegomena to a General Theory, *Ingen. Arch.*, **35**:102–142 (1966).

27. ARGYRIS, J. H., and S. KELSEY: The Matrix Force Method of Structural Analysis and Some New Applications, *Aeron. Res. Council (London) R & M* 3034, February, 1956.

28. ARGYRIS, J. H., and S. KELSEY: Structural Analysis by the Matrix Force Method with Applications to Aircraft Wings, *Wiss. Ges. Luftfahrt Jahrb.*, **1956**:78–98.

29. ARGYRIS, J. H., and S. KELSEY: The Analysis of Fuselages of Arbitrary Cross-section and Taper, *Aircraft Eng.*, **31**:62–74, 101–112, 133–143, 169–180, 192–203, 244–256, 272–283 (1959); **33**:71–83, 103–113, 164–174, 193–200, 227–238 (1961); see also "Modern Fuselage Analysis and the Elastic Aircraft," Butterworth Scientific Publications, London, 1963.

30. ARGYRIS, J. H., and S. KELSEY: Initial Strains in the Matrix Force Method of Structural Analysis, *J. Roy. Aeron. Soc.*, **64**:493–495 (1960).

31. ARGYRIS, J. H., and S. KELSEY: The Validity of the Initial Strain Concept, *J. Roy. Aeron. Soc.*, **65**:129–138 (1961).

32. ASPLUND, S. O.: Inversion of Band Matrices, *Proc. 2d Conf. Electron. Computation, Pittsburgh, Pa., ASCE*, pp. 513–522, Sept. 8–9, 1960.

33. BARTA, J.: On the Minimum Weight of Certain Redundant Structures, *Acta Tech. Acad. Sci. Hung. Budapest*, **18**: 67–76 (1957).

34. BATEMAN, E. H.: Elastic Stress Analysis of Multi-bay Single-storey Frameworks, *Engineering*, **172**:772–774, 804–806 (1951).

35. BAZELEY, G. P., Y. K. CHEUNG, B. M. IRONS, and O. C. ZIENKIEWICZ: Triangular Elements in Plate Bending: Conforming and Nonconforming Solutions, *Proc. Conf. Matrix Methods in Structural Mech., Wright-Patterson Air Force Base, Ohio*, Oct. 26–28, 1965, AFFDL TR 66-80, 1966.

36. BEITCH, L.: Shell Structures Solved Numerically by Using a Network of Partial Panels, *Proc. 7th Struct. Mater. Conf. Am. Inst. Aero. Astro., Cocoa Beach, Fla.*, pp. 35–44, April 18–20, 1966; also published in *J. Am. Inst. Aeron. Astro.*, **5**:418–424 (1967).

37. BENSCOTER, S. U.: The Partitioning of Matrices in Structural Analysis, *J. Appl. Mech.*, **15**:303–307 (1948).

38. BERGER, W. I., and E. SAIBEL: On the Inversion of Continuant Matrices, *J. Franklin Inst.*, **256**:249–253 (1953).

39. BERMAN, F. R.: The Use of a Transformation Chain in Matrix Structural Analysis *Proc. 1st Conf. Electron. Computation ASCE, Kansas City, Mo.*, November, 1958.

40. BERMAN, F. R.: Some Basic Concepts in Matrix Structural Analysis, *Proc. 2d Conf. Electron. Computation ASCE, Pittsburgh, Pa.*, pp. 165–194, Sept. 8–9, 1960.

41. BERMAN, J. H., and J. SKLEROV: Calculation of Natural Modes of Vibration for Free-Free Structures in Three-Dimensional Space, *J. Am. Inst. Aeron. Astron.*, **3**:158–160 (1965).

42. BERMAN, R. R.: Introduction to Matrix Algebra and Its Use in Medium Sized Electronic Digital Computers, *Proc. 1st Conf. Electron. Computation ASCE, Kansas City, Mo.*, pp. 73–87, November, 1958.

43. BESSELING, J. F.: The Complete Analogy Between the Matrix Equations and the Continuous Field Equations of Structural Analysis, *Intern. Symp. Analogue Digital Techn. Appl. Aeron., Liege*, September 9–12, 1963.

44. BESSELING, J. F.: Matrix Analysis of Creep and Plasticity Problems, *Proc. Conf. Matrix Methods Struct. Mech., Wright-Patterson Air Force Base, Ohio*, Oct. 26-28, 1965, AFFDL TR 66-80, 1966.

45. BEST, G. C.: A Formula for Certain Types of Stiffness Matrices of Structural Elements, *J. Am. Inst. Aeron. Astron.*, **1**:212–213 (1963).

46. BEST, G. C.: A Method of Structural Weight Minimization Suitable for High-speed Digital Computers, *J. Am. Inst. Aeron. Astron.*, **1**:478–479 (1963).

47. BEST, G.C.: A General Formula for Stiffness Matrices of Structural Elements, *J. Am. Inst. Aeron. Astron.*, **1**:1920–1921 (1963).

48. BEST. G. C.: Completely Automatic Weight-minimization Method for High-speed Digital Computers, *J. Aircraft*, **1**:129–133 (1964).

49. BEST, G. C.: Comment on "Derivation of Element Stiffness Matrices" by T. H. H. Pian, *J. Am. Inst. Aeron. Astron.*, **2**:1515–1516 (1964).

50. BEST, G. C.: Vibration Analysis of a Cantilevered Square Plate by the Stiffness Matrix Method, *Proc. Conf. Matrix Methods Struct. Mech., Wright-Patterson Air Force Base, Ohio*, Oct. 26–28, 1965, AFFDL TR 66-80, 1966.

51. BIGGS, J. M.: "Introduction to Structural Dynamics," McGraw-Hill Book Company, New York, 1964.

52. BISHOP, R. E. D., G. M. L. GLADWELL, and S. MICHAELSON: "The Matrix Analysis of Vibration," Cambridge University Press, New York, 1965.

53. BISPLINGHOFF, R. L., H. ASHLEY, and R. L. HALFMAN: "Aeroelasticity," Addison-Wesley Publishing Company, Inc., Reading, Mass., 1955.

54. BOGNER, F. K., R. L. FOX, and L. A. SCHMIT: The Generation of Interelement-compatible Stiffness and Mass Matrices by the Use of Interpolation Formulas, *Proc. Conf. Matrix Methods Struct. Mech., Wright-Patterson Air Force Base, Ohio*, Oct. 26–28, 1965, AFFDL TR 66-80, 1966.

55. BOGNER, F. K., R. H. MALLETT, M. D. MINICH, and L. A. SCHMIT: Development and Evaluation of Energy Search Methods of Nonlinear Structural Analysis, *Flight Dynamics Lab. Rept.*, AFFDL TR 65-113, 1965.

56. BORGES, J. F.: Computer Analysis of Structures, *Proc. 2d Conf. Electron. Computation ASCE, Pittsburgh, Pa.*, pp. 195–212, Sept. 8–9, 1960.

57. BROCK, J. E.: Matrix Analysis of Flexible Filaments, *Proc. 1st U.S. Nat. Congr. Appl. Mech., Chicago, Ill.*, pp. 285–289, June 11–16, 1951.

58. BROCK, J. E.: Matrix Method for Flexibility Analysis of Piping Systems, *J. Appl. Mech., Trans. ASME*, **74**:501–516 (1952).

59. BRUSH, D. O.: Strain-energy Expression in Nonlinear Shell Analysis, *J. Aerospace Sci.*, **27**:555–556 (1960).

60. CHEN, L. H.: Piping Flexibility Analysis by Stiffness Matrix, *J. Appl. Mech.*, **26**:608–612 (1959).

61. CHEUNG, Y. K., and O. C. ZIENKIEWICZ: Plates and Tanks on Elastic Foundations: An Application of Finite Element Method, *Intern. J. Solids Struct.*, **1**:451–461 (1965).

62. CLOUGH, R. W.: Use of Modern Computers in Structural Analysis, *Proc. ASCE*, **84** (ST 3): paper 1636 (1958).

63. CLOUGH, R. W.: Structural Analysis by Means of a Matrix Algebra Program, *Proc. 1st. Conf. Electron. Computation ASCE, Kansas City, Mo.*, November, 1958.

64. CLOUGH, R. W.: The Finite Element Method in Plane Stress Analysis, *Proc. 2d Conf. Electron. Computation ASCE, Pittsburgh, Pa.*, pp. 345–378, Sept. 8–9, 1960.

65. CLOUGH, R. W.: The Finite Element Method in Structural Mechanics, Chap. 7, pp. 85–119, in "Stress Analysis," John Wiley & Sons, Inc., New York, 1965.

66. CLOUGH, R. W., and Y. RASHID: Finite Element Analysis of Axisymmetric Solids, *J. Eng. Mech. Div. ASCE*, **91**:71–85 (1965).

67. CLOUGH, R. W., and J. L. TOCHER: Analysis of Thin Arch Dams by the Finite Element Method, *Proc. Intern. Symp. Theory Arch Dams, Southampton University, England*, April, 1964, Pergamon Press, New York, 1964.

68. CLOUGH, R. W., and J. L. TOCHER: Finite Element Stiffness Matrices for Analysis of Plate Bending, *Proc. Conf. Matrix Methods Struct. Mech., Wright-Patterson Air Force Base, Ohio*, Oct. 26–28, 1965, AFFDL TR 66-80, 1966.

69. CLOUGH, R. W., E. L. WILSON, and I. P. KING: Large Capacity Multi-story Frame Analysis Program, *J. Struct. Div. ASCE*, **89**(ST 4):179–204 (1963).

70. COX, H. L.: Vibration of Missiles, I. Vibration of Missiles Accelerating along Initial Trajectories, *Aircraft Eng.*, **33**:2–7 (1961).

71. COX, H. L.: Vibration of Missiles, II. Vibration of Missiles on Launch Stands, *Aircraft Eng.*, **33**:48–55 (1961).

72. CRICHLOW, W. J., and G. W. HAGGENMACHER: The Analysis of Redundant Structures by the Use of High-speed Digital Computers, *J. Aerospace Sci.*, **27**:595–606, 614 (1960).

73. ČERNIAK, E.: Rigid Frame Analysis with the Aid of Digital Computers, *Proc. ASCE*, **84**(EM 1): paper 1494 (1958).

74. DAWE, D. J.: A Finite Element Approach to Plate Vibration Problems, *J. Mech. Eng. Sci.*, **7**:28–32 (1965).

75. DAWE, D. J.: Parallelogrammic Elements in the Solution of Rhombic Cantilever Plate Problems, *J. Strain Anal.*, **1**:223–230 (1966).

76. DENKE, P. H.: A Matric Method of Structural Analysis, *Proc. 2d U.S. Natl. Congr. Appl. Mech. ASME*, pp. 445–451, June, 1954.

77. DENKE, P. H.: The Matrix Solution of Certain Nonlinear Problems in Structural Analysis, *J. Aeron. Sci.*, **23**:231–236 (1956).

78. DENKE, P. H.: A General Digital Computer Analysis of Statically Indeterminate Structures, *NASA Tech. Note* D-1666, 1962.

79. DENKE, P. H.: Digital Analysis of Nonlinear Structures by the Force Method, in B. Fraeijs de Veubeke (ed.), Matrix Methods of Structural Analysis, *AGARDograph*, **72**:317–342 (1964).

80. DENKE, P. H.: A Computerized Static and Dynamic Structural Analysis System; pt. III, Engineering Aspects and Mathematical Formulation of the Problem, Douglas Aircraft Company, paper 3213, *Soc. Automotive Eng. Intern. Congr. Exposition, Detroit*, Jan. 11–15, 1965.

81. D'EWART, B. B., and R. F. FARRELL: The Computation of Dynamic Response Time Histories of a System Using a Matrix Iterative Technique, *Proc. Conf. Matrix Methods Struct. Mech., Wright-Patterson Air Force Base, Ohio*, Oct. 26–28, 1965, AFFDL TR 66-80, 1966.

82. DORN, W. S., R. E. GOMORY, and H. J. GREENBERG: Automatic Design of Optimal Structures, *J. Mechanique*, **3**:25–52 (1964).

83. DUNCAN, W. J.: Reciprocation of Triply Partitioned Matrices, *J. Roy. Aeron. Soc.*, **60**:131–132 (1956).

84. DURRETT, J. C.: "Nonlinear Matrix Analysis of Discrete Element Structures," M.S. thesis, Air Force Institute of Technology, Wright-Patterson Air Force Base, Ohio, 1963.

85. EHRICH, F. F.: A Matrix Solution for the Vibration of Nonuniform Discs, *J. Appl. Mech.*, **23**:109–115 (1956).

86. EISEMANN, K., and S. NAMYET: Space Frame Analysis by Matrices and Computer, *J. Struct. Div. ASCE*, (ST 6): 245–277 (1962).

87. ESHLEMAN, A. L., and J. D. VAN DYKE: A Rational Method of Analysis by Matrix Methods of Acoustically Loaded Structure for Prediction of Sonic Fatigue Strength, *Proc. 2d Intern. Conf. Acoust. Fatigue Aerospace Structures, Dayton, Ohio*, pp. 721–746, April 29–May 1, 1964, Syracuse University Press, Syracuse, N.Y., 1965.

88. FALCONER, R. W., and R. D. LANE: A Regularization Method for Calculation of Stresses, Deformations, and Elastic Modes, *Proc. Conf. Matrix Methods Struct. Mech., Wright-Patterson Air Force Base, Ohio*, Oct. 26–28, 1965, AFFDL TR 66-80, 1966.

89. FALKENHEINER, H.: Systematic Analysis of Redundant Elastic Systems (in French), *Rech. Aeron.*, **17**:17–31 (1950).

90. FALKENHEINER, H.: Systematic Analysis of Redundant Elastic Structures by Means of Matrix Calculus, *J. Aeron. Sci.*, **20**:293–294 (1953).

91. FARBRIDGE, J. E. F., F. A. WOODWARD, and G. E. A. THOMANN: Aeroelastic Problems of Low Aspect Ratio Wings, V. Application of the Structural and Aeroelastic Matrices to the Solution of Steady State Aeroelastic Problem, *Aircraft Eng.*, **28**:196–198 (1956).

92. FENVES, S. J.: Structural Analysis by Networks, Matrices and Computers, *J. Struct. Div. ASCE*, **92**:199–221 (1966).

93. FENVES, S. J., and F. H. BRANIN: Network-topological Formulation of Structural Analysis, *J. Struct. Div. ASCE*, **89**(ST4):483–514 (1963).

94. FENVES, S. J., R. D. LOGCHER, and S. P. MAUCH: "Stress: A Reference Manual," The M.I.T. Press, Cambridge, Mass., 1965.

95. FENVES, S. J., R. D. LOGCHER, S. P. MAUCH, and K. F. REINSCHMIDT: "Stress: A User's Manual," The M.I.T. Press, Cambridge, Mass., 1964.

96. FILHO, F. V.: Groups of Unknowns in Structural Analysis, *J. Roy. Aeron. Soc.*, **66**:322–323 (1962).

97. FILHO, F. V.: The Basic Redundant Systems in the Analysis of Complex Structures, *J. Aerospace Sci.*, **29**:1006–1007 (1962).

98. FILHO, F. V.: Unification of Matrix Methods of Structural Analysis, *J. Am. Inst. Aeron. Astron.*, **1**:916–917 (1963).

99. FILHO, F. V.: Orthogonalization of Internal Force and Strain Systems, *Proc. Conf. Matrix Methods Struct. Mech., Wright-Patterson Air Force Base, Ohio*, Oct. 26–28, 1965, AFFDL TR 66-80, 1966.

100. FOX, R. L., and L. A. SCHMIT: Advances in the Integrated Approach to Structural Synthesis, *J. Spacecraft Rockets*, **3**:858–866 (1966).

101. FRAEIJS DE VEUBEKE, B. M.: Upper and Lower Bounds in Matrix Structural Analysis, in B. Fraeijs de Veubeke (ed.), Matrix Methods of Structural Analysis, *AGARDograph*, **72**:165–201 (1964).

102. FRAEIJS DE VEUBEKE, B. M.: Displacement and Equilibrium Models in the Finite Element Method, chap. 9, pp. 145–197, in "Stress Analysis," John Wiley & Sons, Inc., New York, 1965.

103. FRAEIJS DE VEUBEKE, B. M.: Bending and Stretching of Plates: Special Models for Upper and Lower Bounds, *Proc. Conf. Matrix Methods Struct. Mech., Wright-Patterson Air Force Base, Ohio*, Oct. 26–28, 1965, AFFDL TR 66-80, 1966.

104. GALLAGHER, R. H.: Matrix Structural Analysis of Heated Airframes, *Proc. Symp. Aerothermoelasticity, Dayton, Ohio*, Oct. 30–Nov. 7, 1961, Rept. ASD TR 61-645, pp. 879–917, 1961.

105. GALLAGHER, R. H.: Techniques for the Derivation of Element Stiffness Matrices, *J. Am. Inst. Aeron. Astron.*, **1**:1431–1432 (1963).

106. GALLAGHER, R. H.: Comments on "Derivation of Element Stiffness Matrices by Assumed Stress Distributions" by T. H. H. Pian, *J. Am. Inst. Aeron. Astron.*, **3**:186–187 (1965).

107. GALLAGHER, R. H.: "Development and Evaluation of Matrix Methods for Thin Shell Structural Analysis," Ph.D. dissertation, State University of New York, Buffalo, N.Y., June, 1966.

108. GALLAGHER, R. H., and J. PADLOG: Discrete Element Approach to Structural Instability Analysis, *J. Am. Inst. Aeron. Astron.*, **1**:1437–1439 (1963).

109. GALLAGHER, R. H., J. PADLOG, and P. P. BIJLAARD: Stress Analysis of Heated Complex Shapes, *J. Am. Rocket Soc.*, **32**:700–707 (1962).

110. GALLAGHER, R. H., and I. RATTINGER: The Theoretical and Experimental Determination of the Elastic Characteristics of Modern Airframes, *AGARD Rept.*, **400** (1960).

111. GALLAGHER, R. H., and I. RATTINGER: The Deformational Behavior of Low Aspect Ratio Multi-web Wings: I. Experimental Data, II. Elementary and Plate Bending Theories, III. Discrete Element Idealizations, *Aeron. Quart.*, **12**:361–371 (1961); **13**:71–87, 143–166 (1962).

112. GALLAGHER, R. H., I. RATTINGER, and J. S. ARCHER: A Correlation Study of Methods of Matrix Structural Analysis, *AGARDograph* **69**, The Macmillan Company, New York, 1964.

113. GARVEY, S. J.: The Quadrilateral Shear Panel, *Aircraft Eng.*, **23**:134–135, 144 (1951).

114. GATEWOOD. B. E., and N. OHANIAN: Tri-diagonal Matrix Method for Complex Structures, *J. Struct. Div., ASCE*, **91**:27–41 (1965).

115. GATEWOOD, B. E., and N. OHANIAN: Examples of Solution Accuracy in Certain Large Simultaneous Equation Systems, *Proc. Conf. Matrix Methods Struct. Mech., Wright-Patterson Air Force Base, Ohio*, Oct. 26–28, 1965, AFFDL TR 66-80, 1966.

116. GAUZY, H.: Vibration Testing by Harmonic Excitation, *AGARD Manual on Aeroelasticity*, **1**: chap. 4, 1–21 (1961).

117. GELLATLY, R. A., and R. H. GALLAGHER: Development of Advanced Structural Optimization Programs and Their Application to Large Order Systems, *Proc. Conf. Matrix Methods Struct Mech., Wright-Patterson Air Force Base, Ohio*, Oct. 26–28, 1965, AFFDL TR 66-80, 1966.

118. GELLATLY, R. A., and R. H. GALLAGHER: A Procedure for Automated Minimum Weight Structural Design, I. Theoretical Basis, *Aeronaut. Quart.*, **17**:216–230 (1966).

119. GELLATLY, R. A., R. H. GALLAGHER, and W. A. LUBERACKI: Development of a Procedure for Automated Synthesis of Minimum Weight Structures, *Air Force Flight Dynamics Lab. Rept.* AFFDL TDR 64-141, 1964.

120. GILLIS, P., and K. H. GERSTLE: Analysis of Structures by Combining Redundants, *J. Struct. Div. ASCE*, **87**:41–56 (1961).

121. GOFMAN, S. M.: Matrix Formulas for the Analysis of Some Complicated Plane Frame Systems (in Russian), *Tr. Inst. Sooruzh. Akad. Nauk Uz. SSR*, no. 4, 1954.

122. GOLOLOBOV, M.: Special Cases of the Application of the Deformation Matrix Method to the Solution of Statically Indeterminate Structures (in Czech), *Zpravodaj VZLU*, no. 2, pp. 9–14, 1964.

123. GOLOLOBOV, M.: Application of the Influence Coefficients of the Individual Parts of a Statically Indeterminate Structure to Its Solution by the Deformation Matrix Method (in Czech), *Zpravodaj VZLU*, no. 3, pp. 3–7, 1964.

124. GOODEY, W. J.: Note on a General Method of Treatment of Structural Discontinuities, *J. Roy. Aeron. Soc.*, **59**:695–697 (1955).

125. GRAFTON, P. E., and D. R. STROME: Analysis of Axisymmetrical Shells by the Direct Stiffness Method, *J. Am. Inst. Aeron. Astron.*, **1**:2342–2347 (1963).

126. GREENE, B. E.: Application of Generalized Constraints in the Stiffness Method of Structural Analysis, *J. Am. Inst. Aeron. Astron.*, **4**:1531–1537 (1966).

127. GREENE, B. E., D. R. STROME, and R. C. WEIKEL: Application of the Stiffness Method to the Analysis of Shell Structures, *ASME paper* 61-AV-58, 1961.

128. GRIFFIN, K. H.: Analysis of Pin-jointed Space Frameworks: Nonlinear Effects, *Air Force Inst. Tech. Rept.* AFIT TR 66–16, 1966.

129. GRZEDZIELSKI, A. L. M.: Organization of a Large Computation in Aircraft Stress Analysis, *Natl. Res. Council Can. Rept.* LR-257, 1959.

130. GRZEDZIELSKI, A. L. M.: Note on the Applications of the Matrix Force Method of Structural Analysis, *J. Roy. Aeron. Soc.*, **64**:354–357 (1960).

131. GRZEDZIELSKI, A. L. M.: The Initial Strain Concept, *J. Roy. Aeron. Soc.*, **65**:127–129, 136–137 (1961).

132. GRZEDZIELSKI, A. L. M.: Theory of Multi-spar and Multi-rib Structures, *Natl. Res. Council Can. Rept.* LR-297, 1961.

133. GUYAN, R. J.: Reduction of Stiffness and Mass Matrices, *J. Am. Inst. Aeron. Astron.*, **3**:380 (1965).

134. GUYAN, R. J.: Distributed Mass Matrix for Plate Element Bending, *J. Am. Inst. Aeron. Astron.*, **3**:567–568 (1965).

135. HALL, A. S., and R. W. WOODHEAD: "Frame Analysis," John Wiley & Sons, Inc., New York, 1961.

136. HENDERSON, J. C. DE and W. G. BICKLEY: Statical Indeterminacy of a Structure, *Aircraft Eng.*, **27**:400–402 (1955).

137. HERRESHOFF, J. B.: Flutter Analysis Using Influence Matrices and Steady-state Aerodynamics, *J. Am. Inst. Aeron. Astron.*, **1**:2853–2855 (1963).

138. HERRMANN, L. R.: A Bending Analysis for Plates, *Proc. Conf. Matrix Methods Struct. Mech., Wright-Patterson Air Force Base, Ohio*, Oct. 26–28, 1965, AFFDL TR 66-80, 1966.

139. HESSEL, A.: Analysis of Plates and Shells by Matrix Methods, *SAAB Aircraft Company Linköping, Sweden, Tech. Note* 48, 1961.

140. HESTER, K. L.: "Analytical and Experimental Determination of the Natural Frequencies and Mode Shapes of Skew Plates," M.S. thesis, Air Force Institute of Technology, Wright-Patterson Air Force Base, Ohio, 1965.

141. HOOLEY, R. F., and P. D. HIBBERT: Bounding Plane Stress Solutions by Finite Elements, *J. Struct. Div. ASCE*, **92**:39–48 (1966).

142. HOSKIN, B. C.: A Note on Modifications in Redundant Structures, *Aeron. Res. Lab Melbourne, Australia, Rept.* ARL/SM 280, 1963.

143. HRENNIKOFF, A.: Solution of Problems in Elasticity by the Framework Method, *J. Appl. Mech.*, **8**:A169–A175 (1941).

144. HUFF, R. D., and R. H. GALLAGHER, Thermoelastic Effects on Hypersonic Stability and Control, pt. II, vol. III, Elastic Response Analysis of Fuselage and Combined Wing-fuselage Structures, *Flight Control Lab. Rept.* ASD TR 61-287, 1963.

145. HUNG, F. C. et al.: Dynamics of Shell-like Lifting Bodies, I. The Analytical Investigation, *Flight Dynamics Lab. Rept.* AFFDL TR 65-17, 1965.

146. HUNG, F. C., and D. J. STONE: Prediction of Stiffness and Vibration Characteristics of Trusses, Multi-stage Cylinders, and Clustered Cylinders, *Proc. Symp. Aerothermoelasticity, Dayton, Ohio*, Oct. 30–Nov. 1, 1961, ASD TR 61–645, pp. 219–289, 1961.

147. HUNT, P. M.: The Electronic Digital Computer in Aircraft Structural Analysis, *Aircraft Eng.*, **28**:70–76, 111–118, 155–165 (1956).

148. HURTY, W. C.: Dynamic Analysis of Structural Systems Using Component Modes, *J. Am. Inst. Aeron. Astron.*, **3**:678–685 (1965).

149. HURTY, W. C.: Truncation Errors in Natural Frequencies as Computed by the Method of Component Mode Synthesis, *Proc. Conf. Matrix Methods Struct. Mech., Wright-Patterson Air Force Base, Ohio*, Oct. 26–28, 1965, AFFDL TR 66-80, 1966.

150. HURTY, W. C., and M. F. RUBINSTEIN: "Dynamics of Structures," Prentice-Hall, Inc., Englewood Cliffs, N.J., 1964.

151. IRONS, B. M. R.: Structural Eigenvalue Problems: Elimination of Unwanted Variables, *J. Am. Inst. Aeron. Astron.*, **3**:961–962 (1965).

152. IRONS, B. M. R.: Comment on "Distributed Mass Matrix for Plate Element Bending" by R. J. Guyan, *J. Am. Inst. Aeron. Astron.*, **4**:189 (1966).

153. IRONS, B. M. R., and J. BARLOW: Comment on "Matrices for the Direct Stiffness Method" by R. J. Melosh, *J. Am. Inst. Aeron. Astron.*, **2**:403-404 (1964).

154. IRONS, B. M. R., and K. J. DRAPER: Inadequacy of Nodal Corrections in a Stiffness Solution for Plate Bending, *J. Am. Inst. Aeron. Astron.*, **3**:961 (1965).

155. JENNINGS, A.: Natural Vibrations of a Free Structure, *Aircraft Eng.*, **34**:81–83 (1962).

156. JENSEN, W. R., W. E. FALBY, and N. PRINCE: Matrix Analysis Methods for Anisotropic Inelastic Structures, *Flight Dynamics Lab. Rept.* AFFDL TR 65-220, 1966.

157. JONES, R. E.: A Generalization of the Direct Stiffness Method of Structural Analysis, *J. Am. Inst. Aeron. Astron.*, **2**:821–826 (1964).

158. JONES, R. E., and D. R. STROME: A Survey of the Analysis of Shells by the Displacement Method, *Proc. Conf. Matrix Methods Struct. Mech., Wright-Patterson Air Force Base, Ohio*, Oct. 26–28, 1965, AFFDL TR 66-80, 1966.

159. JONES, R. E., and D. R. STROME: Direct Stiffness Method Analysis of Shells of Revolution Utilizing Curved Elements, *J. Am. Inst. Aeron. Astron.*, **4**:1519–1525 (1966).

160. KAMEL, H.: "Automatic Analysis of Fuselages and Problems of Conditioning," Ph.D. dissertation, University of London, London, 1964.

161. KANDIDOV, V. P.: Approximate Analysis of Nonhomogeneous Plates by Separation into Elements (in Russian), *Mosk. Univ. Vestn. Ser. Mat. Mekhan.*, **19**:67–73 (1964).

162. KAPUR, K. K.: "Buckling of Thin Plates Using the Matrix Stiffness Method," Ph.D. dissertation, Dept. of Civil Engineering, University of Washington, Seattle, Wash., 1965.

163. KAPUR, K. K.: Stability of Plates Using the Finite Element Method, *J. Eng. Mech. Div. ASCE*, **92**:177–195 (1966).

164. KAPUR, K. K.: Prediction of Plate Vibrations Using a Consistent Mass Matrix, *J. Am. Inst. Aeron. Astron.* **4**:565–566 (1966).

165. KAUFMAN, S., D. B. HALL, and J. KAUFMAN: Molecular Vibrations by a Matrix Force Method, *J. Mol. Spectry.*, **16**:264–277 (1965).

166. KEITH, J. S. et al.: Methods in Structural Dynamics for Thin Shell Clustered Launch Vehicles, *Flight Dynamics Lab. Rept.* FDL TDR 64-105, 1965.

167. KHANNA, J.: Criterion for Selecting Stiffness Matrices, *J. Am. Inst. Aeron. Astron.*, **3**:1976 (1965).

168. KLEIN, B.: A Simple Method of Matric Structural Analysis, *J. Aeron. Sci.*, **24**:39-46 (1957).

169. KLEIN, B.: A Simple Method of Matric Structural Analysis, II. Effects of Taper and a Consideration of Curvature, *J. Aeron. Sci.*, **24**:813-820 (1957).

170. KLEIN, B.: Application of the Witte Rearranging Method to a Typical Structural Matrix, *J. Aerospace Sci.*, **25**:342–343 (1958).

171. KLEIN, B.: A Simple Method of Matric Structural Analysis, III. Analysis of Flexible Frames and Stiffened Cylindrical Shells, *J. Aerospace Sci.*, **25**:385–394 (1958).

172. KLEIN, B.: A Simple Method of Matric Structural Analysis, IV. Nonlinear Problems, *J. Aerospace Sci.*, **26**:351–359 (1959).

173. KLEIN, B.: Simultaneous Calculation of Influence Coefficients and Influence Loads for Arbitrary Structures, *J. Aerospace Sci.*, **26**:451–452 (1959).

174. KLEIN, B.: A Simple Method of Matric Structural Analysis, V. Structures Containing Plate Elements of Arbitrary Shape and Thickness, *J. Aerospace Sci.*, **27**:859–866 (1960).

175. KLEIN, B.: Some Comments on the Inversion of Certain Large Matrices, *J. Aerospace Sci.*, **28**:432 (1961).

176. KLEIN, B.: A Simple Method of Matric Structural Analysis, VI. Bending of Plates of Arbitrary Shape and Thickness under Arbitrary Normal Loading, *J. Aerospace Sci.*, **29**:306–310, 322 (1962).

177. KLEIN, B. et al.: Nonlinear Shell Matrix Analysis, *Flight Dynamics Lab. Rept.* RTD TDR 63-4255, 1964.

178. KLEIN, B., and M. CHIRICO: New Methods in Matric Structural Analysis, *Proc. 2d Conf. Electron. Computation ASCE, Pittsburgh, Pa.*, pp. 213–223, Sept. 8–9, 1960.

179. KLEIN, S.: "Matrix Analysis of Shell Structures," M.S. thesis, Dept. Aeronautics and Astronautics, Massachusetts Institute of Technology, Cambridge, Mass., 1964.

180. KLEIN, S.: A Study of the Matrix Displacement Method as Applied to Shells of Revolution, *Proc. Conf. Matrix Methods Struct. Mech., Wright-Patterson Air Force Base, Ohio*, Oct. 26–28, 1965, AFFDL TR 66-80, 1966.

181. KLEIN, S., and R. J. SYLVESTER: The Linear Elastic Dynamic Analysis of Shells of Revolution by the Matrix Displacement Method, *Proc. Conf. Matrix Struct. Mech., Wright-Patterson Air Force Base, Ohio*, Oct. 26–28, 1965, AFFDL TR 66-80, 1966.

182. KOSKO, E.: Effect of Local Modifications in Redundant Structures, *J. Aeron. Sci.*, **21**:206-207 (1954).

183. KOSKO, E.: Reciprocation of Triply Partitioned Matrices, *J. Roy. Aeron. Soc.*, **60**:490–491 (1965).

184. KOSKO, E.: Matrix Inversion by Partitioning, *Aeron. Quart.*, **8**:157–184 (1957).

185. KOSKO, E.: The Equivalence of Force and Displacement Methods in the Matrix Analysis of Elastic Structures, *Proc. Conf. Matrix Methods Struct. Mech., Wright-Patterson Air Force Base, Ohio*, Oct. 26–28, 1965, AFFDL TR 66-80, 1966.

186. KRON, G.: Tensorial Analysis and Equivalent Circuits of Elastic Structures, *J. Franklin Inst.*, **238**:399–442 (1944).

187. KRON, G.: Solving Highly Complex Elastic Structures in Easy Stages, *J. Appl. Mech.*, **22**:235–244 (1955).

188. LANGEFORS, B.: Structural Analysis of Sweptback Wings by Matrix Transformation, *SAAB Aircraft Company, Linköping, Sweden, Tech. Note* 3, 1951.

189. LANGEFORS, B.: Analysis of Elastic Structures by Matrix Transformation with Special Regard to Semi-monocoque Structures, *J. Aeron. Sci.*, **19**:451–458 (1952).

190. LANGEFORS, B.: Matrix Methods for Redundant Structures, *J. Aeron. Sci.*, **20**:292–293 (1953).

191. LANGEFORS, B.: Exact Reduction and Solution by Parts of Equations for Elastic Structures, *SAAB Aircraft Company, Linköping, Sweden, Tech. Note* 24, 1953.

192. LANGEFORS, B.: Algebraic Methods for the Numerical Analysis of Built-up Systems, *SAAB Aircraft Company, Linköping, Sweden, Tech. Note* 38, 1957.

193. LANGEFORS, B.: Theory of Aircraft Structural Analysis, *Z. Flugwiss.*, **6**:281–291 (1958).

194. LANGEFORS, B.: Algebraic Topology for Elastic Networks, *SAAB Aircraft Company, Linköping, Sweden, Tech. Note* 49, 1961.

195. LANGEFORS, B.: Triangular Plates in Structural Analysis, *SAAB Aircraft Company, Linköping, Sweden, Rept.* LNC-140, 1961.

196. LANSING, W., W. R. JENSEN, and W. FALBY: Matrix Analysis Methods for Inelastic Structures, *Proc. Conf. Matrix Methods Struct. Mech., Wright-Patterson Air Force Base, Ohio*, Oct. 26–28, 1965, AFFDL TR 66-80, 1966.

197. LANSING, W., I. W. JONES, and P. RATNER: A Matrix Force Method for Analyzing Heated Wings, *Symp. Struct. Dynamics High Speed Flight, Aerospace Industries Association and ONR, Los Angeles, Calif.*, Apr. 24–26, 1961.

198. LANSING, W., I. W. JONES, and P. RATNER: Nonlinear Analysis of Heated, Cambered Wings by the Matrix Force Method, *J. Am. Inst. Aeron. Astron.*, **1**:1619–1626 (1963).

199. LECKIE, F. A., and G. M. LINDBERG: The Effect of Lumped Parameters on Beam Frequencies, *Aeron. Quart.*, **14**:224–240 (1963).

200. LEVIEN, K. W., and B. J. HARTZ: Dynamic Flexibility Matrix Analysis of Frames, *J. Struct. Div. ASCE*, **89**(ST4):515–536 (1963).

201. LEVY, S: Computation of Influence Coefficients for Aircraft Structures with Discontinuities and Sweepback, *J. Aeron. Sci.*, **14**:547–560 (1947).

202. LEVY, S.: Structural Analysis and Influence Coefficients for Delta Wings, *J. Aeron. Sci.*, **20**:449–454 (1953).

203. LEYDS, J.: Aeroelastic Problems of Low Aspect Ratio Wings, III. Aerodynamic Forces on an Oscillating Delta Wing, IV. Application of the Structural and Aerodynamic Matrices to the Solution of the Flutter Problem, *Aircraft Eng.*, **28**:119–122, 166–167 (1956).

204. LIN, Y. K.: Transfer Matrix Representation of Flexible Airplanes in Gust Response Study, *J. Aircraft*, **2**:116–121 (1965).

205. LIN, Y. K.: A Method of the Determination of the Matrix of Impulse Response Functions with Special Reference to Applications in Random Vibration Problems, *Proc. Conf. Matrix Methods Struct. Mech., Wright-Patterson Air Force Base, Ohio*, Oct. 26–28, 1965, AFFDL TR 66-80, 1966.

206. LINDBERG, G. M.: Vibration of Non-uniform beams, *Aeron. Quart.*, **14**:387–395 (1963).

207. LIVESLEY, R. K.: Analysis of Rigid Frames by an Electronic Computer, *Engineering*, **176**:230–232 (1953).

208. LIVESLEY, R. K.: "Matrix Methods of Structural Analysis," Pergamon Press, New York, 1964.

209. LU, Z. A., J. PENZIEN, and E. P. POPOV: Finite Element Solution for Thin Shells of Revolution, *NASA Tech. Rept.* CR-37, 1964.

210. LUNDER, C. A.: "Derivation of a Stiffness Matrix for a Right Triangular Plate in Bending and Subjected to Initial Stresses," M.S. thesis, Dept. of Aeronautics and Astronautics, University of Washington, Seattle, Wash., 1962.

211. MAC NEAL, R. H.: Application of the Compensation Theorem to the Modification of Redundant Structures, *J. Aeron. Sci.*, **20**:726–727 (1953).

212. MARGUERRE, K.: Vibrations and Stability Problems of Beams Treated by Matrices, *J. Math. Phys.*, **35**:28–43 (1956).

213. MARTIN, H. C.: Truss Analysis by Stiffness Considerations, *Proc. ASCE*, **82**(EM4): paper 1070, (1965); also in *Trans. ASCE*, **123**:1182–1194 (1958).

214. MARTIN, H. C.: Large Deflection and Stability Analysis by the Direct Stiffness Method, *NASA Tech. Rept.* 32-931, 1966.

215. MARTIN, H. C.: "Introduction to Matrix Methods of Structural Analysis," McGraw-Hill Book Company, New York, 1966.

216. MARTIN, H. C.: On the Derivation of Stiffness Matrices for the Analysis of Large Deflection and Stability Problems, *Proc. Conf. Matrix Methods Struct. Mech., Wright-Patterson Air Force Base, Ohio*, Oct. 26–28, 1965, AFFDL TR 66-80, 1966.

217. MAYERJAK, R. J.: On the Weight and Design of a Redundant Truss, *Aeron. Res. Lab. Rept.* 62-338, *Wright-Patterson Air Force Base, Ohio*, 1962.

218. MC CALLEY, R. B.: Error Analysis for Eigenvalue Problems, *Proc. 2d Conf. Electron. Computation ASCE, Pittsburgh, Pa.*, pp. 523–550, Sept. 8–9, 1960.

219. MC CORMICK, C. W.: Plane Stress Analysis, *J. Struct. Div. ASCE*, **89**:37–54 (1963).

220. MC MAHAN, L. L.: "Development and Application of the Direct Stiffness Method for Out-of-plane Bending Using a Triangular Plate Element," M.S. thesis, Dept. of Aeronautics and Astronautics, University of Washington, Seattle, Wash., 1962.

221. MC MINN, S. J.: "Matrices for Structural Analysis," John Wiley & Sons, Inc., New York, 1962.

222. MC MINN, S. J.: The Effect of Axial Loads on the Stiffness of Rigid-jointed Plane Frames, *Proc. Conf. Matrix Methods Struct. Mech., Wright-Patterson Air Force Base, Ohio*, Oct. 26–28, 1965, AFFDL TR 66-80, 1966.

223. MEISSNER, C. J.: Direct Beam Stiffness-matrix Calculations Including Shear Effects, *J. Aerospace Sci.*, **29**:247–248 (1962).

224. MELOSH, R. J.: A Stiffness Matrix for the Analysis of Thin Plates in Bending, *J. Aerospace Sci.*, **28**:34–42, 64 (1961).

225. MELOSH, R. J.: Matrix Methods of Structural Analysis, *J. Aerospace Sci.*, **29**:365–366 (1962).

226. MELOSH, R. J.: Basis for Derivation of Matrices for the Direct Stiffness Method, *J. Am. Inst. Aeron. Astron.*, **1**:1631–1637 (1963).

227. MELOSH, R. J.: Structural Analysis of Solids, *J. Struct. Div. ASCE*, **89**:205–223 (1963).

228. MELOSH, R. J.: A Flat Triangular Shell Element Stiffness Matrix, *Proc. Conf. Matrix*

Methods Struct. Mech., Wright-Patterson Air Force Base, Ohio, Oct. 26–28, 1965, AFFDL TR 66–80, 1966.

229. MELOSH, R. J., and T. E. LANG: Modified Potential Energy Mass Representations for Frequency Prediction, *Proc. Conf. Matrix Methods Structural Mechanics, Wright-Patterson Air Force Base, Ohio*, Oct. 26–28, 1965, AFFDL TR 66-80, 1966.

230. MELOSH, R. J., and R. G. MERRIT: Evaluation of Spar Matrices for Stiffness Analysis, *J. Aerospace Sci.*, **25**:537–543 (1958).

231. MENTEL, T. J.: Study and Development of Simple Matrix Methods for Inelastic Structures, *J. Spacecraft Rockets*, **3**:449–457 (1966).

232. METHERELL, A. F.: Matrix Methods in Nonlinear Damping Analysis, *Air Force Mater. Lab. Rept.* AFML TR 65-128, 1965.

233. MEYETTE, R. J.: "Deflections and Stresses of Heated Truss-type Structures by a Direct Stiffness Method," M.S. thesis, Air Force Institute of Technology, Wright-Patterson Air Force Base, Ohio, 1961.

234. MICHIELSEN, H. F., and A. DIJK: Structural Modifications in Redundant Structures, *J. Aeron. Sci.*, **20**:286-288 (1953).

235. MILLER, R. E.: Structural Analysis Flexible Grid Technique for SST Wing Parametric Studies, *J. Aircraft*, **2**:257–261 (1965).

236. MORICE, P. B.: "Linear Structural Analysis," The Ronald Press Company, New York, 1959.

237. MOROSOW, G., and I. J. JASZLICS: Titan III 20% Dynamic Model Characteristics, Comparison of Theory and Experiment, *Proc. Conf. Matrix Methods Struct. Mech., Wright-Patterson Air Force Base, Ohio*, Oct. 26–28, 1965, AFFDL TR 66-80, 1966.

238. MURTY, A. V. K.: A Lumped Inertia Force Method for Vibration Problems, *Aeron. Quart.*, **17**:127–140 (1966).

239. NEUBERT, V. H.: Computer Methods for Dynamic Structural Response, *Proc. 2d Conf. Electron. Computation ASCE, Pittsburgh, Pa.*, pp. 455–465, Sept. 8–9, 1960.

240. NIELSEN, N. N.: Vibration Tests of a Nine-story Steel Frame Building, *J. Eng. Mech. Div. ASCE*, **92**(EM1):81–110 (1966).

241. ODEN, J. T.: Calculation of Geometric Stiffness Matrices for Complex Structures, *J. Am. Inst. Aeron. Astron.*, **4**:1480–1482 (1966).

242. PADLOG, J., R. D. HUFF, and G. F. HOLLOWAY: Unelastic Behavior of Structures Subjected to Cyclic, Thermal and Mechanical Stressing Conditions, WADD TR 60–271, 1960.

243. PEARSON, C. E.: A Computational Technique for Three-dimensional Pin-jointed Structures, *Proc. 2d Conf. Electron. Computation ASCE, Pittsburgh, Pa.*, pp. 327–343, Sept. 8-9, 1960.

244. PEI, M. L.: Matrix Solution of Beams with Variable Moments of Inertia, *Proc. ASCE*, **85**(ST 8): paper 2218 (1959).

245. PEI, M. L.: Stiffness Method of Rigid Frame Analysis, *Proc. 2d Conf. Electron. Computation ASCE, Pittsburgh, Pa.*, pp. 225–248, Sept. 8–9, 1960.

246. PERCY, J. H., W. A. LODEN, and D. NAVARATNA: A Study of Matrix Analysis Methods for Inelastic Structures, *Flight Dynamic Lab. Rept.* RTD TDR 63-4032, 1963.

247. PERCY, J. H., T. H. H. PIAN, S. KLEIN, and D. R. NAVARATNA: Application of Matrix Displacement Methods to Linear Elastic Analysis of Shells of Revolution, *J. Am. Inst. Aeron. Astron.*, **3**:2138–2145 (1965).

248. PESTEL, E. C.: Dynamics of Structures by Transfer Matrices, *Air Force Office Sci. Res. Rept.* AFOSR 1449, 1961.

249. PESTEL, E. C.: Application of Transfer Matrices to Cylindrical Shells (in German), *Z. Angew. Math. Mech.*, **43**:T89–T92 (1963).

250. PESTEL, E. C.: Dynamic Stiffness Matrix Formulation by Means of Hermitian Polynomials, *Proc. Conf. Matrix Methods Struct. Mech., Wright-Patterson Air Force Base, Ohio*, Oct. 26–28, 1965, AFFDL TR 66-80, 1966.

251. PESTEL, E. C., and F. A. LECKIE: "Matrix Methods in Elastomechanics," McGraw-Hill Book Company, New York, 1963.

252. PETYT, M.: Vibration Analysis of Plates and Shells by the Method of Finite Elements, *Proc. 5th Congr. Intern. Acoust., Liege*, Sept. 7–14, 1965.

253. PIAN, T. H. H.: Derivation of Element Stiffness Matrices, *J. Am. Inst. Aeron. Astron.*, **2**:576–577, 1964.

254. PIAN, T. H. H.: Derivation of Element Stiffness Matrices by Assumed Stress Distribution, *J. Am. Inst. Aeron. Astron.*, **2**:1333–1336 (1964).

255. PIAN, T. H. H.: Element Stiffness Matrices for Boundary Compatibility and Prescribed Boundary Stresses, *Proc. Conf. Matrix Methods Struct. Mech., Wright-Patterson Air Force Base, Ohio*, Oct. 26–28, 1965, AFFDL TR 66-80, 1966.

256. PIPES, L. A.: "Matrix Methods for Engineering," Prentice-Hall, Inc., Englewood Cliffs, N.J., 1963.

257. PLUNKETT, R.: A Matrix Method of Calculating Propeller-blade Moments and Deflections, *J. Appl. Mech.*, **16**:361–369 (1949).

258. POPE, G. G.: A Discrete Method for the Analysis of Plane Elastoplastic Stress Problems, *Aeron. Quart.*, **17**:83–104 (1966).

259. POPE, G. G.: The Application of the Matrix Displacement Method in Plane Elastoplastic Problems, *Proc. Conf. Matrix Methods Struct. Mech., Wright-Patterson Air Force Base, Ohio*, Oct. 26–28, 1965, AFFDL TR 66-80, 1966.

260. POPOV, E. P., J. PENZIEN, and Z.-A. LU: Finite Element Solution for Axisymmetrical Shells, *J. Eng. Mech. Div. ASCE*, **90**:119–145 (1964).

261. POPPLETON, E. D.: A Note on the Design of Redundant Structures, *Univ. Toronto Inst. Aerophysics Tech. Note* 36, July, 1960.

262. POPPLETON, E. D.: The Redesign of Redundant Structures Having Undesirable Stress Distributions, *J. Aerospace Sci.*, **28**:347–348 (1961).

263. POPPLETON, E. D.: Note on the Matrix Analysis of Nonlinear Structures, *Univ. Toronto Inst. Aerophysics, Tech. Note* 46, March, 1961.

264. PÖSCHL, T.: On an Application of Matrix Calculus to the Theory of Frames, *Ingr. Arch.*, **19**:69–74 (1951).

265. PRENTIS, J. M., and F. A. LECKIE: "Mechanical Vibrations: An Introduction to Matrix Methods," Longmans, Green & Co., Ltd., London, 1963.

266. PRZEMIENIECKI, J. S.: Matrix Analysis of Fuselage Structures, *Bristol Aircraft Ltd., England, Tech. Rept.* TOR 104, 1957.

267. PRZEMIENIECKI, J. S.: Matrix Analysis of Shell Structures with Flexible Frames, *Aeron. Quart.*, **9**:361–394 (1958).

268. PRZEMIENIECKI, J. S.: Analysis of Aircraft Structures by the Substructure Relaxation Method, *Bristol Aircraft Ltd., England, Tech. Rept.* TOR 145, 1961.

269. PRZEMIENIECKI, J. S.: Thermal Stresses, pp. 211–257 in "Introduction to Structural Problems in Nuclear Reactor Engineering," Pergamon Press, New York, 1962.

270. PRZEMIENIECKI, J. S.: Matrix Structural Analysis of Substructures, *J. Am. Inst. Aeron. Astron.*, **1**:138–147 (1963).

271. PRZEMIENIECKI, J. S.: Triangular Plate Elements in the Matrix Force Method of Structural Analysis, *J. Am. Inst. Aeron. Astron.*, **1**:1895–1897 (1963).

272. PRZEMIENIECKI, J. S.: Matrix Analysis of Aerospace Structures, *Proc. 5th Intern. Symp. Space Techn. Sci., Tokyo*, pp. 477–500, Sept. 2–7, 1963.

273. PRZEMIENIECKI, J. S.: Tetrahedron Elements in the Matrix Force Method of Structural Analysis, *J. Am. Inst. Aeron. Astron.*, **2**:1152–1154 (1964).

274. PRZEMIENIECKI, J. S.: Generalization of the Unit Displacement Theorem with Applications to Dynamics, *Proc. Conf. Matrix Methods Struct. Anal., Wright-Patterson Air Force Base, Ohio*, Oct. 27–28, 1964.

275. PRZEMIENIECKI, J. S.: Quadratic Matrix Equations for Determining Vibration Modes

and Frequencies of Continuous Elastic Systems, *Proc. Conf. Matrix Methods Struct. Mech., Wright-Patterson Air Force Base, Ohio*, Oct. 26–28, 1965, AFFDL TR 66-80, 1966.

276. PRZEMIENIECKI, J. S.: Equivalent Mass Matrices for Rectangular Plates in Bending, *J. Am. Inst. Aeron. Astron.*, **4**:949–950 (1966).

277. PRZEMIENIECKI, J. S., and L. BERKE: Digital Computer Program for the Analysis of Aerospace Structures by the Matrix Displacement Method, *Flight Dynamics Lab. Rept.* FDL TDR 64-18, 1964.

278. PRZEMIENIECKI, J. S., and P. H. DENKE: Joining of Complex Substructures by the Matrix Force Method, *J. Aircraft*, **3**:236–243 (1966).

279. RAZANI, R.: Behavior of Fully Stressed Design of Structures and Its Relationship to Minimum-weight Design, *J. Am. Inst. Aeron. Astron.*, **3**:2262–2268 (1965).

280. RENTON, J. D.: Stability of Space Frames by Computer Analysis, *J. Struct. Div. ASCE*, **88**(ST4):81–103 (1962).

281. RICHARDSON, J. R.: A More Realistic Method for Routine Flutter Calculations, *Proc. Symp. Struct. Dynamics Aeroelasticity, Am. Inst. Aeron. Astron.*, pp. 10–17, Aug. 30–Sept. 1, 1965.

282. ROBINSON, J.: Automatic Selection of Redundancies in the Matrix Force Method: The Rank Technique, *Can. Aeron. Space J.*, **11**:9–12 (1965).

283. ROBINSON, J.: Dissertation on the "Rank Technique" and its Application, *J. Roy. Aeron. Soc.*, **69**:280–283 (1965).

284. ROBINSON, J.: Computer Utilization for Structural Analysis: World Survey, April 1965, *J. Roy. Aeron. Soc.*, **70**:735–737 (1966).

285. ROBINSON, J., and R. R. REGL: An Automated Matrix Analysis for General Plane Frames (The Rank Technique), *J. Am. Helicopter Soc.*, **8**:16–35 (1963).

286. RODDEN, W. P.: A Matrix Approach to Flutter Analysis, *Inst. Aerospace Sci. Paper* FF-23, 1959.

287. RODDEN, W. P.: Structural Influence Coefficients for a Redundant System Including Beam-Column Effects, *Proc. Conf. Matrix Methods Struct. Mech., Wright-Patterson Air Force Base, Ohio*, Oct. 26–28, 1965, AFFDL TR 66-8C, 1966.

288. RODDEN, W. P., and E. F. FARKAS: On Accelerating Convergence in the Iteration of Complex Matrices, *J. Aerospace Sci.*, **29**:1143 (1962).

289. RODDEN, W. P., E. F. FARKAS, G. L. COMMERFORD, and H. A. MALCOM: Structural Influence Coefficients for a Redundant System Including Beam-Column Effects: Analytical Development and Computational Procedures, *Aerospace Corp. Rept.* SSD TR 65-114, 1965.

290. RODDEN, W. P., E. F. FARKAS, and H. A. MALCOM: Flutter and Vibration Analysis by a Modal Method: Analytical Development and Computational Procedure, *Aerospace Corp. Rept.* SSD TDR 63-158, 1963.

291. RODDEN, W. P., J. P. JONES, and P. G. BHUTA: A Matrix Formulation of the Transverse Structural Influence Coefficients of an Axially Loaded Timoshenko Beam, *J. Am. Inst. Aeron. Astron.*, **1**:225–227 (1963).

292. RODDEN, W. P., and J. D. REVELL: The Status of Unsteady Aerodynamics Influence Coefficients, *Inst. Aerospace Sci. Fairchild Fund paper* FF-33, 1962.

293. RODGERS, G. L.: "Dynamics of Framed Structures," John Wiley & Sons, Inc., New York, 1959.

294. ROSANOFF, R. A., and T. A. GINSBURG: Matrix Error Analysis for Engineers, *Proc. Conf. Matrix Methods Struct. Mech., Wright-Patterson Air Force Base, Ohio*, Oct. 26–28, 1965, AFFDL TR 66-80, 1966.

295. ROTHMAN, H., and P. MASON: Analysis of Large Movable Structures, *J. Struct. Div. ASCE*, **89**(ST4):353–387 (1963).

296. RUBINSTEIN, M. F.: Multi-story Frame Analysis by Digital Computers, *Proc. 2d Conf. Electron. Computation ASCE, Pittsburgh, Pa.*, pp. 261–281, Sept. 8–9, 1960.

297. SAMSON, C. H., and H. W. BERGMANN: Analysis of Low Aspect Ratio Aircraft Structures, *J. Aerospace Sci.*, **27**:679–693, 711 (1960).

298. SANDER, G.: Upper and Lower Limits in the Matrix Analysis of Plates in Flexure and Torsion (in French), *Bull. Soc. Roy. Sci., Liege*, **33**:33–37 (1964).

299. SAUNDERS, H.: Matrix Analysis of a Nonuniform Beam Column on Multisupports, *J. Am. Inst. Aeron. Astron.*, **1**:951–952 (1963).

300. SCHEFFEY, C. F.: Optimization of Structures by Variation of Critical Parameters, *Proc. 2d Conf. Electron. Computation ASCE, Pittsburgh, Pa.*, pp. 133–144, Sept. 8–9, 1960.

301. SCHMIT, L. A.: Structural Design by Systematic Synthesis, *Proc. 2d Conf. Electron. Computation ASCE, Pittsburgh, Pa.*, pp. 105–132, Sept. 8–9, 1960.

302. SCHMIT, L. A.: Comment on "Completely Automatic Weight-minimization Method of High-speed Digital Computers" by G. C. Best, *J. Aircraft*, **1**:375–377 (1964).

303. SCHMIT, L. A., and R. L. FOX: An Integrated Approach to Structural Synthesis and Analysis, *J. Am. Inst. Aeron. Astron.*, **3**:1104–1112 (1965).

304. SCHMIT, L. A., and R. H. MALLET: Structural Synthesis and Design Parameter Hierarchy, *J. Struct. Div. ASCE*, **89**(ST4):269–299 (1963).

305. SCHNELL, W.: Matrix Stability Calculations for Multiply-supported Beam Columns (in German), *Z. Ange. Math. Mech.*, **35**:269-284 (1955).

306. SCIARRA, J. J.: Dynamic Unified Structural Analysis Method Using Stiffness Matrices, *Proc. 7th Struct. Mater. Conf. Am. Inst. Aeron. Astron., Cocoa Beach, Fla*, pp. 94–111, April 18–20, 1966.

307. SHAIKEVICH, V. D.: On Some Problems of the Application of the Theory of Matrices to the Analysis of Statically Indeterminate Systems, *Dnepropet. Met. Inst.*, 1955.

308. SHORE, S.: Analysis of Space Structures, *Proc. 1st Conf. Electron. Computation ASCE, Kansas City, Mo.*, pp. 133–151, November, 1958.

309. SHORE, S.: The Elements of Matrix Structural Analysis, *Proc. 2d Conf. Electron. Computation ASCE, Pittsburgh, Pa.*, pp. 145–164, Sept. 8–9, 1960.

310. SIMPSON, A.: An Improved Displaced Frequency Method for Estimation of Dynamical Characteristics of Mechanical Systems, *J. Roy. Aeron. Soc.*, **70**:661–665 (1966).

311. SMOLLEN, L. E.: Generalized Matrix Method for the Design and Analysis of Vibration-isolation System, *J. Acoust. Soc. Am.*, **40**:195–204 (1966).

312. SPILLERS, W. R.: Applications of Topology in Structural Analysis, *J. Struct. Div. ASCE*, **89** (ST4):301–313 (1963).

313. STRICKLIN, J. A.: Computation of Stress Resultants from the Element Stiffness Matrices, *J. Am. Inst. Aeron. Astron.*, **4**:1095–1096 (1966).

314. SYLVESTER, R. J.: Computer Solutions to Linear Buckling Problems, *Proc. 2d Conf. Electron. Computation ASCE, Pittsburgh, Pa.*, pp. 429–442, Sept. 8–9, 1960.

315. TAIG, I. C.: Structural Analysis by the Displacement Method, *English Electric Aviation Ltd., Warton, England, Rept.* SO 17, 1961.

316. TAIG, I. C.: Computer Applications in the Development of Efficient Aircraft Structures, *J. Roy. Aeron. Soc.*, **67**:706–710, 715 (1963).

317. TAIG, I. C.: Automated Stress Analysis Using Substructures, *Proc. Conf. Matrix Methods Struct. Mech., Wright-Patterson Air Force Base, Ohio*, Oct. 26–28, 1965, AFFDL TR 66-80, 1966.

318. TAIG, I. C., and R. I. KERR: Some Problems in the Discrete Element Representation of Aircraft Structures, in B. Fraeijs de Veubeke (ed.), Matrix Methods of Structural Analysis, *AGARDograph*, **72**:267–315 (1964).

319. TATMAN, D. I.: "Matrix Analysis of Axisymmetrical Shells," M.S. thesis, Air Force Institute of Technology, Wright-Patterson Air Force Base, Ohio, 1965.

320. THOMANN, G. E. A.: Aeroelastic Problems of Low Aspect Ratio Wings, I. Structural Analysis, *Aircraft Eng.*, **28**:36–42 (1956).

321. THOMSON, W. T.: Matrix Solution for the Vibration of Non-uniform Beams, *J. Appl. Mech.*, **72**:337–339 (1950).

322. TOCHER, J. L.: "Analysis of Plate Bending Using Triangular Elements," Ph.D. dissertation, Civil Engineering Dept., University of California, Berkeley, 1962.

323. TOCHER, J. L.: Selective Inversion of Stiffness Matrices, *J. Struct. Div. ASCE*, **92**:75–88 (1966).

324. TOCHER, J. L., and K. K. KAPUR: Comment on "Basis for Derivation of Matrices for the Direct Stiffness Method" by R. J. Melosh, *J. Am. Inst. Aeron. Astron.*, **3**:1215–1216 (1965).

325. TOTTENHAM, H.: The Matrix Progression Method in Structural Analysis, in "Introduction to Structural Problems in Nuclear Reactor Engineering," pp. 189–210, Pergamon Press, New York, 1962.

326. TRAILL-NASH, R. W.: The Symmetric Vibrations of Aircraft, *Aeron. Quart.*, **3**:1–22 (1951).

327. TRAILL-NASH, R. W.: The Antisymmetric Vibrations of Aircraft, *Aeron. Quart.*, **3**:145–160 (1951).

328. TURNER, M. J.: The Direct Stiffness Method of Structural Analysis, Structures, and Materials Panel Paper, *AGARD Meeting*, *Aachen*, *Germany*, Sept. 17, 1959.

329. TURNER, M. J., R. W. CLOUGH, H. C. MARTIN, and L. J. TOPP: Stiffness and Deflection Analysis of Complex Structures, *J. Aeron. Sci.*, **23**:805-823, 854 (1956).

330. TURNER, M. J., E. H. DILL, H. C. MARTIN, and R. J. MELOSH: Large Deflections of Structures Subjected to Heating and External Loads, *J. Aerospace Sci.*, **27**:97–102, 127 (1960).

331. TURNER, M. J., H. C. MARTIN, and R. C. WEIKEL: Further Development and Application of the Stiffness Method, in B. Fraeijs de Veubeke (ed.), Matrix Methods of Structural Analysis, *AGARDograph*, **72**:203–266 (1964).

332. UTKU, S.: Stiffness Matrices for Thin Triangular Elements of Non-zero Gaussian Curvature, *Am. Inst. Aeron. Astron.*, *4th Aerospace Sci. Meeting*, *Los Angeles*, *Calif.*, paper 66-530, June 27–29, 1966.

333. VISSER, W.: A Finite Element Method for the Determination of Non-stationary Temperature Distribution and Thermal Deformations, *Proc. Conf. Matrix Methods Struct. Mech.*, *Wright-Patterson Air Force Base*, *Ohio*, Oct. 26–28, 1965, AFFDL TR 66-80, 1966.

334. WALLACE, C. D.: "Matrix Analysis of Axisymmetric Shells under General Loading," M.S. thesis, Air Force Institute of Technology, Wright-Patterson Air Force Base, Ohio, 1966.

335. WARREN, D. S.: A Matrix Method for the Analysis of the Buckling of Structural Panels Subjected to Creep Environment, *Flight Dynamics Lab. Rept.* ASD TDR 62-740, 1962.

336. WARREN, D. S., R. A. CASTLE, and R. C. GLORIA: An Evaluation of the State-of-the-art of Thermomechanical Analysis of Structures, *Flight Dynamics Lab. Rept.* WADD TR 61-152, 1962.

337. WARREN, D. S., and P. H. DENKE: Matrix Analysis of the Dynamic Behavior of Geometrically Nonlinear Structures, *Proc. 2d Intern. Conf. Acoust. Fatigue Aerospace Structures*, *Dayton*, *Ohio*, pp. 265–277, April 29–May 1, 1964, Syracuse University Press, Syracuse, N.Y., 1965.

338. WEAVER, W.: Dynamics of Discrete Parameter Structures, *Devel. Theoret. Appl. Mech.*, *Proc. 2d Southeastern Conf.*, pp. 629–651, March 5–6, 1964.

339. WEHLE, L. B., and W. LANSING: A Method of Reducing the Analysis of Complex Redundant Structures to a Routine Procedure, *J. Aeron. Sci.*, **19**:677–684 (1952).

340. WEI, B. C. F., and P. NIELSEN: A Computer Analysis of Large Booster Structures for Design Optimization, *SAE Natl. Aeron. Space Eng. Manuf. Meeting*, paper 746A, Sept. 23–27, 1963.

341. WEIKEL, R. C., R. E. JONES, J. A. SEILER, H. C. MARTIN, and B. E. GREENE: Nonlinear and Thermal Effects on Elastic Vibrations, *Flight Dynamics Lab. Rept.* ASD TDR 62-156, 1962.

342. WILLIAMS, D.: Recent Developments in the Structural Approach to Aeroelastic Problems, *J. Roy. Aeron. Soc.*, **58**:403–428 (1954).

343. WILLIAMS, D: A General Method for Deriving the Structural Influence Coefficients of Aeroplane Wings, I, II, *Aeron. Res. Council (London) R & M* 3048, 1956.

344. WILSON, E. L.: Matrix Analysis of Nonlinear Structures, *Proc. 2d Conf. Electron. Computation ASCE, Pittsburgh, Pa.*, pp. 415–428, Sept. 8–9, 1960.

345. WILSON, E. L.: Structural Analysis of Axisymmetric Solids, *J. Am. Inst. Aeron. Astron.*, **3**:2269–2274 (1965).

346. WISSMANN, J. W.: Nonlinear Structural Analysis: Tensor Formulation, *Proc. Conf. Matrix Methods Struct. Mech., Wright-Patterson Air Force Base, Ohio*, Oct. 26–28, 1965, AFFDL TR 66-80, 1966.

347. WOODWARD, F. A.: Aeroelastic Problems of Low Aspect Ratio Wings, II. Aerodynamic Forces on an Elastic Wing in Supersonic Flow, *Aircraft Eng.*, **28**:77–81 (1956).

348. YETTRAM, A. L., and H. M. HUSAIN: Generalized Matrix Force and Displacement Methods of Linear Structural Analysis, *J. Am. Inst. Aeron. Astron.*, **3**:1154–1156 (1965).

349. ZIENKIEWICZ, O. C.: Finite Element Procedures in the Solution of Plate and Shell Problems, chap. 8, pp. 120–144, in "Stress Analysis," John Wiley & Sons, Inc., New York, 1965.

350. ZIENKIEWICZ, O. C.: Solution of Anisotropic Seepage by Finite Elements, *J. Eng. Mech. Div. ASCE*, **92**:111–120 (1966).

351. ZIENKIEWICZ, O. C., and Y. K. CHEUNG: Finite Element Method of Analysis of Arch Dam Shells and Comparison with Finite Difference Problems, *Proc. Intern. Symp. Theory Arch Dams, Southampton University*, April, 1964, Pergamon Press, New York, 1964.

352. ZIENKIEWICZ, O. C., and Y. K. CHEUNG: The Finite Element Method for Analysis of Elastic, Isotropic and Orthotropic slabs, *Proc. Inst. Civil Engrs. (London)*, **28**:471–488 (1964).

ADDITIONAL REFERENCES

353. AKYUZ, F. A.: On the Solution of Two-dimensional Nonlinear Problems of Elastoplasticity by Finite Element and Direct Stiffness Method, *Am. Inst. Aeron. Astron., 5th Aerospace Sciences Meeting*, Jan. 23–26, 1967, AIAA paper 67-144, 1967.

354. ANTEBI, J., I. K. SHAH, and J. M. SHAH: Computer Analysis of the F-111 Wing Assembly, *J. Structural Div. ASCE*, **92**:405–424 (1966).

355. ARGYRIS, J. H.: Some Results on the Free-free Oscillations of Aircraft Type Structures, *Rev. Soc. Franc. Mecan.*, **15**:59–73 (1965).

356. ARGYRIS, J. H.: A Tapered TRIM 6 Element for the Matrix Displacement Method, *J. Roy. Aeron. Soc.*, **70**:1040–1043 (1966).

357. ARGYRIS, J. H.: The TRIAX 6-Element for Axisymmetric Analysis by the Matrix Displacement Method: pt. I: Foundations; pt. II: Elastic Stiffness and Influence of Initial Strains, *J. Roy. Aeron. Soc.*, **70**:1102–1106 (1966).

358. ARGYRIS, J. H., and P. C. PATTON: The Large Computer in a European University Research Institute, *Appl. Mech. Rev.*, **19**:1029–1039 (1966).

359. ARGYRIS, J. H., and P. C. PATTON: A Look into the Future: How Computers Will Influence Engineering, *J. Roy. Aeron. Soc.*, **71**:244–252 (1967).

360. BOGNER, F. K., R. L. FOX, and L. A. SCHMIT: A Cylindrical Shell Discrete Element, *J. Am. Inst. Aeron. Astron.*, **5**:745–750 (1967).

361. CHOW, H. Y., Z. A. LU, J. F. ABEL, E. P. POPOV, and J. PENZIEN: A Computer Program for the Static and Dynamic Finite Element Analysis of Axisymmetric Thin Shells, *Structural Eng. Lab., Univ. of Calif., Berkeley, Calif. Rept.* SESM 66-18, 1966.

362. DEAK, A. L., and T. H. H. PIAN: Application of the Smooth-surface Interpolation to the Finite-Element Analysis, *J. Am. Inst. Aeron. Astron.*, **5**:187–189 (1967).

363. DENKE, P. H.:Nonlinear and Thermal Effects on Elastic Vibrations, *Douglas Aircraft Company, Inc., Tech. Rept.* SM-30426, October, 1960.

364. FRAEIJS DE VEUBEKE, B. M.: Basis of a Well-conditioned Force Program for Equilibrium Models via the Southwell Slab Analogies, *Air Force Flight Dynamics Lab. Rept.* AFFDL TR 67-10, 1967.

365. GALLAGHER, R. H., R. A. GELLATLY, J. PADLOG, and R. H. MALLET: A Discrete Element Procedure for Thin-shell Instability Analysis, *J. Am. Inst. Aeron. Astron.*, **5**:138–145 (1967).

366. GELLATLY, R. A., and R. H. GALLAGHER: A Procedure for Automated Minimum Weight Structural Design; pt. II: Applications, *Aeron. Quart.*, **17**:332–342 (1966).

367. GERE, J. M., and W. WEAVER: "Analysis of Framed Structures," D. Van Nostrand Company, Inc., Princeton, N.J., 1965.

368. IRONS, B. M.: Engineering Applications of Numerical Integration in Stiffness Method, *J. Am. Inst. Aeron. Astron.*, **4**:2035–2037 (1966).

369. KAPUR, K. K.: Vibrations of a Timoshenko Beam Using Finite-element Approach, *J. Acoust. Soc. Am.*, **40**:1058–1063 (1966).

370. KAPUR, K. K., and B. J. HARTZ: Stability of Plates Using the Finite-Element Method, *J. Eng. Mech. Div. ASCE*, **92**:177–195 (1966).

371. KHANNA, J., and R. F. HOOLEY: Comparison and Evaluation of Stiffness Matrices, *J. Am. Inst. Aeron. Astron.*, **4**:2105–2111 (1966).

372. KICHER, T. P.: Optimum Design—Minimum Weight Versus Fully Stressed, *J. Struct. Div. ASCE*, **92**:265–279 (1966).

373. KRAHULA, J. L.: Analysis of Bent and Twisted Bars Using the Finite-Element Method, *J. Am. Inst. Aeron. Astron.*, **5**:1194–1195 (1967).

374. LANG, T. E.: Summary of the Functions and Capabilities of the Structural Analysis and Matrix Interpretive System Computer Program, *Jet Propulsion Lab. Tech. Rept.* 32-1075, 1967.

375. LAURSEN, H. I.: "Matrix Analysis of Structures," McGraw-Hill Book Company, New York, 1966.

376. LOEWY, R. G.: A Matrix Holzer Analysis for Bending Vibrations of Clustered Launch Vehicles, *J. Spacecraft and Rockets*, **3**:1625–1637 (1966).

377. MALLETT, R. H., and L. BERKE: Automated Method for the Large Deflection and Instability Analysis of Three-dimensional Truss and Frame Assemblies, *Air Force Flight Dynamics Lab. Rept.* AFFDL TR 66-102, 1966.

378. MARGUERRE, K.: On the Approximation of Continuous Systems by Systems with a Finite Number of Degrees of Freedom, *Rev. Francaise Mecan.*, **15**:51–57 (1965).

379. MC LAY, R. W.: Completeness and Convergence Properties of Finite-Element Displacement Functions: A General Treatment, *Am. Inst. Aeron. Astron., 5th Aerospace Sciences Meeting*, Jan. 23–26, 1967, AIAA paper 67-143, 1967.

380. NAVARATNA, D. R.: Elastic Stability of Shells of Revolution by the Variational Approach Using Discrete Elements, *Ballistic Systems Div. Rept.* BSD TR 66-261, 1966.

381. NAVARATNA, D. R.: Natural Vibrations of Deep Spherical Shells, *J. Am. Inst. Aeron. Astron.*, **4**:2056–2058 (1966).

382. NAVARATNA, D. R.: Computation of Stress Resultants in Finite-Element Analysis, *J. Am. Inst. Aeron. Astron.*, **4**:2058–2060 (1966).

383. ODEN, J. T.: "Mechanics of Elastic Structures," McGraw-Hill Book Company, New York, 1967.

384. PAPENFUSS, S. W.: "Lateral Plate Deflection by Stiffness Matrix Methods," M.S. thesis, Dept. of Civil Eng., Univ. of Washington, Seattle, Wash., 1959.

385. PERCY, J. H.: Quadrilateral Finite Element in Elastic-Plastic Plane-Stress Analysis, *J. Am. Ins. Aeron. Astron.*, **5**:367 (1967).

386. PICKARD, J.: Format II—Second Version of Fortran Matrix Abstraction Technique, vol. I, Engineering User Rept., *Air Force Flight Dynamics Lab. Rept.* AFFDL TR 66-207, 1966.

387. PRZEMIENIECKI, J. S.: Stability Analysis of Complex Structures Using Discrete Element Techniques, Symposium on Structural Stability and Optimization, *Roy. Aeron. Soc. and Loughborough Univ. of Tech., England*, March 23–24, 1967.

388. PRZEMIENIECKI, J. S.: Large Deflections of Frame Structures, *7th Intern. Symp. of Space Tech. and Sci., Tokyo, Japan*, May 15–20, 1967.

389. RAO, A. K., and S. P. G. RAJU: A Matrix Method for Vibration and Stability Problems, *J. Aeron. Soc. India*, **18**:90–99 (1966).

390. RAO, C. V. J., and A. V. K. MURTHY: Transverse Vibrations of Trusses, *J. Aeron. Soc. India*, **18**:41–46 (1966).

391. ROBINSON, J.: "Structural Matrix Analysis for the Engineer," John Wiley & Sons, Inc., New York, 1966.

392. RODDEN, W. P.: A Method of Deriving Structural Influence Coefficients from Ground Vibration Tests, *J. Am. Inst. Aeron. Astron.*, **5**:991–1000 (1967).

393. ROSANOFF, R. A., and P. RADKOWSKI: Research Direction in Matrix Error Analysis, *Am. Inst. Aeron. Astron., 5th Aerospace Sciences Meeting*, Jan. 23–26, 1967, AIAA paper 67-142 (1967).

394. RUBINSTEIN, M. F.: "Matrix Computer Analysis of Structures," Prentice-Hall, Inc., Englewood Cliffs, N.J., 1966.

395. SANDER, G., and B. M. FRAEIJS DE VEUBEKE: Upper and Lower Bounds to Structural Deformations by Dual Analysis in Finite Elements, *Air Force Flight Dynamics Lab. Rept.* AFFDL TR 66-199, 1967.

396. SHAH, J. M.: Ill-conditioned Stiffness Matrices, *J. Struct. Div. ASCE*, **92**:443–457 (1966).

397. SMITH, R. G., and J. A. WEBSTER: Matrix Analysis of Beam-Columns by the Method of Finite Elements, *Proc. Royal Soc.* (*Edinburgh*), *sec. A*, **67**:156–173 (1964–65).

398. STRICKLIN, J. A., D. R. NAVARATNA, and T. H. H. PIAN: Improvements on the Analysis of Shells of Revolution by the Matrix Displacement Method, *J. Am. Inst. Aeron. Astron.*, **4**:2069–2072 (1966).

399. TAIG, I. C.: The Computer in the Stress and Design Offices, *J. Roy. Aeron. Soc.*, **71**:256–261 (1967).

400. UHRIG, R.: The Transfer Matrix Method Seen as One Method of Structural Analysis Among Others, *J. Sound and Vibration*, **4**:136–155 (1966).

401. UTKU, S., and R. J. MELOSH: Behavior of Triangular Shell Element Stiffness Matrices Associated with Polyhedral Deflection Distributions, *Am. Inst. Aeron. Astron., 5th Aerospace Sciences Meeting*, Jan. 23–26, 1967, AIAA paper 67-114, 1967.

402. WANG, C. K.: "Matrix Methods of Structural Analysis," International Textbook Company, Scranton, Pa., 1966.

403. WANG, P. C.: "Numerical and Matrix Methods in Structural Mechanics," John Wiley & Sons, Inc., New York, 1966.

404. WEAVER, W.: "Computer Programs for Structural Analysis," D. Van Nostrand Company, Princeton, N.J., 1967.

INDEX

A CATALOG OF
SELECTED DOVER BOOKS
IN ALL FIELDS OF INTEREST

A CATALOG OF SELECTED DOVER BOOKS IN ALL FIELDS OF INTEREST

CONCERNING THE SPIRITUAL IN ART, Wassily Kandinsky. Pioneering work by father of abstract art. Thoughts on color theory, nature of art. Analysis of earlier masters. 12 illustrations. 80pp. of text. 5⅜ × 8½. 23411-8 Pa. $2.50

LEONARDO ON THE HUMAN BODY, Leonardo da Vinci. More than 1200 of Leonardo's anatomical drawings on 215 plates. Leonardo's text, which accompanies the drawings, has been translated into English. 506pp. 8⅜ × 11¾. 24483-0 Pa. $10.95

GOBLIN MARKET, Christina Rossetti. Best-known work by poet comparable to Emily Dickinson, Alfred Tennyson. With 46 delightfully grotesque illustrations by Laurence Housman. 64pp. 4 × 6¾. 24516-0 Pa. $2.50

THE HEART OF THOREAU'S JOURNALS, edited by Odell Shepard. Selections from *Journal*, ranging over full gamut of interests. 228pp. 5⅜ × 8½. 20741-2 Pa. $4.50

MR. LINCOLN'S CAMERA MAN: MATHEW B. BRADY, Roy Meredith. Over 300 Brady photos reproduced directly from original negatives, photos. Lively commentary. 368pp. 8⅜ × 11¼. 23021-X Pa. $11.95

PHOTOGRAPHIC VIEWS OF SHERMAN'S CAMPAIGN, George N. Barnard. Reprint of landmark 1866 volume with 61 plates: battlefield of New Hope Church, the Etawah Bridge, the capture of Atlanta, etc. 80pp. 9 × 12. 23445-2 Pa. $6.00

A SHORT HISTORY OF ANATOMY AND PHYSIOLOGY FROM THE GREEKS TO HARVEY, Dr. Charles Singer. Thoroughly engrossing non-technical survey. 270 illustrations. 211pp. 5⅜ × 8½. 20389-1 Pa. $4.50

REDOUTE ROSES IRON-ON TRANSFER PATTERNS, Barbara Christopher. Redouté was botanical painter to the Empress Josephine; transfer his famous roses onto fabric with these 24 transfer patterns. 80pp. 8¼ × 10⅞. 24292-7 Pa. $3.50

THE FIVE BOOKS OF ARCHITECTURE, Sebastiano Serlio. Architectural milestone, first (1611) English translation of Renaissance classic. Unabridged reproduction of original edition includes over 300 woodcut illustrations. 416pp. 9⅜ × 12¼. 24349-4 Pa. $14.95

CARLSON'S GUIDE TO LANDSCAPE PAINTING, John F. Carlson. Authoritative, comprehensive guide covers, every aspect of landscape painting. 34 reproductions of paintings by author; 58 explanatory diagrams. 144pp. 8⅜ × 11. 22927-0 Pa. $4.95

101 PUZZLES IN THOUGHT AND LOGIC, C.R. Wylie, Jr. Solve murders, robberies, see which fishermen are liars—purely by reasoning! 107pp. 5⅜ × 8½. 20367-0 Pa. $2.00

TEST YOUR LOGIC, George J. Summers. 50 more truly new puzzles with new turns of thought, new subtleties of inference. 100pp. 5⅜ × 8½. 22877-0 Pa. $2.25

THE MURDER BOOK OF J.G. REEDER, Edgar Wallace. Eight suspenseful stories by bestselling mystery writer of 20s and 30s. Features the donnish Mr. J.G. Reeder of Public Prosecutor's Office. 128pp. 5⅜ × 8½. (Available in U.S. only)
24374-5 Pa. $3.50

ANNE ORR'S CHARTED DESIGNS, Anne Orr. Best designs by premier needlework designer, all on charts: flowers, borders, birds, children, alphabets, etc. Over 100 charts, 10 in color. Total of 40pp. 8¼ × 11. 23704-4 Pa. $2.25

BASIC CONSTRUCTION TECHNIQUES FOR HOUSES AND SMALL BUILDINGS SIMPLY EXPLAINED, U.S. Bureau of Naval Personnel. Grading, masonry, woodworking, floor and wall framing, roof framing, plastering, tile setting, much more. Over 675 illustrations. 568pp. 6½ × 9¼. 20242-9 Pa. $8.95

MATISSE LINE DRAWINGS AND PRINTS, Henri Matisse. Representative collection of female nudes, faces, still lifes, experimental works, etc., from 1898 to 1948. 50 illustrations. 48pp. 8⅜ × 11¼. 23877-6 Pa. $2.50

HOW TO PLAY THE CHESS OPENINGS, Eugene Znosko-Borovsky. Clear, profound examinations of just what each opening is intended to do and how opponent can counter. Many sample games. 147pp. 5⅜ × 8½. 22795-2 Pa. $2.95

DUPLICATE BRIDGE, Alfred Sheinwold. Clear, thorough, easily followed account: rules, etiquette, scoring, strategy, bidding; Goren's point-count system, Blackwood and Gerber conventions, etc. 158pp. 5⅜ × 8½. 22741-3 Pa. $3.00

SARGENT PORTRAIT DRAWINGS, J.S. Sargent. Collection of 42 portraits reveals technical skill and intuitive eye of noted American portrait painter, John Singer Sargent. 48pp. 8¼ × 11⅛. 24524-1 Pa. $2.95

ENTERTAINING SCIENCE EXPERIMENTS WITH EVERYDAY OBJECTS, Martin Gardner. Over 100 experiments for youngsters. Will amuse, astonish, teach, and entertain. Over 100 illustrations. 127pp. 5⅜ × 8½. 24201-3 Pa. $2.50

TEDDY BEAR PAPER DOLLS IN FULL COLOR: A Family of Four Bears and Their Costumes, Crystal Collins. A family of four Teddy Bear paper dolls and nearly 60 cut-out costumes. Full color, printed one side only. 32pp. 9¼ × 12¼.
24550-0 Pa. $3.50

NEW CALLIGRAPHIC ORNAMENTS AND FLOURISHES, Arthur Baker. Unusual, multi-useable material: arrows, pointing hands, brackets and frames, ovals, swirls, birds, etc. Nearly 700 illustrations. 80pp. 8⅜ × 11¼.
24095-9 Pa. $3.75

DINOSAUR DIORAMAS TO CUT & ASSEMBLE, M. Kalmenoff. Two complete three-dimensional scenes in full color, with 31 cut-out animals and plants. Excellent educational toy for youngsters. Instructions; 2 assembly diagrams. 32pp. 9¼ × 12¼. 24541-1 Pa. $3.95

SILHOUETTES: A PICTORIAL ARCHIVE OF VARIED ILLUSTRATIONS, edited by Carol Belanger Grafton. Over 600 silhouettes from the 18th to 20th centuries. Profiles and full figures of men, women, children, birds, animals, groups and scenes, nature, ships, an alphabet. 144pp. 8⅜ × 11¼. 23781-8 Pa. $4.95

25 KITES THAT FLY, Leslie Hunt. Full, easy-to-follow instructions for kites made from inexpensive materials. Many novelties. 70 illustrations. 110pp. 5⅜ × 8½. 22550-X Pa. $2.25

PIANO TUNING, J. Cree Fischer. Clearest, best book for beginner, amateur. Simple repairs, raising dropped notes, tuning by easy method of flattened fifths. No previous skills needed. 4 illustrations. 201pp. 5⅜ × 8½. 23267-0 Pa. $3.50

EARLY AMERICAN IRON-ON TRANSFER PATTERNS, edited by Rita Weiss. 75 designs, borders, alphabets, from traditional American sources. 48pp. 8¼ × 11. 23162-3 Pa. $1.95

CROCHETING EDGINGS, edited by Rita Weiss. Over 100 of the best designs for these lovely trims for a host of household items. Complete instructions, illustrations. 48pp. 8¼ × 11. 24031-2 Pa. $2.25

FINGER PLAYS FOR NURSERY AND KINDERGARTEN, Emilie Poulsson. 18 finger plays with music (voice and piano); entertaining, instructive. Counting, nature lore, etc. Victorian classic. 53 illustrations. 80pp. 6½ × 9¼. 22588-7 Pa. $1.95

BOSTON THEN AND NOW, Peter Vanderwarker. Here in 59 side-by-side views are photographic documentations of the city's past and present. 119 photographs. Full captions. 122pp. 8¼ × 11. 24312-5 Pa. $6.95

CROCHETING BEDSPREADS, edited by Rita Weiss. 22 patterns, originally published in three instruction books 1939-41. 39 photos, 8 charts. Instructions. 48pp. 8¼ × 11. 23610-2 Pa. $2.00

HAWTHORNE ON PAINTING, Charles W. Hawthorne. Collected from notes taken by students at famous Cape Cod School; hundreds of direct, personal *apercus*, ideas, suggestions. 91pp. 5⅜ × 8½. 20653-X Pa. $2.50

THERMODYNAMICS, Enrico Fermi. A classic of modern science. Clear, organized treatment of systems, first and second laws, entropy, thermodynamic potentials, etc. Calculus required. 160pp. 5⅜ × 8½. 60361-X Pa. $4.00

TEN BOOKS ON ARCHITECTURE, Vitruvius. The most important book ever written on architecture. Early Roman aesthetics, technology, classical orders, site selection, all other aspects. Morgan translation. 331pp. 5⅜ × 8½. 20645-9 Pa. $5.50

THE CORNELL BREAD BOOK, Clive M. McCay and Jeanette B. McCay. Famed high-protein recipe incorporated into breads, rolls, buns, coffee cakes, pizza, pie crusts, more. Nearly 50 illustrations. 48pp. 8¼ × 11. 23995-0 Pa. $2.00

THE CRAFTSMAN'S HANDBOOK, Cennino Cennini. 15th-century handbook, school of Giotto, explains applying gold, silver leaf; gesso; fresco painting, grinding pigments, etc. 142pp. 6⅛ × 9¼. 20054-X Pa. $3.50

FRANK LLOYD WRIGHT'S FALLINGWATER, Donald Hoffmann. Full story of Wright's masterwork at Bear Run, Pa. 100 photographs of site, construction, and details of completed structure. 112pp. 9¼ × 10. 23671-4 Pa. $6.50

OVAL STAINED GLASS PATTERN BOOK, C. Eaton. 60 new designs framed in shape of an oval. Greater complexity, challenge with sinuous cats, birds, mandalas framed in antique shape. 64pp. 8¼ × 11. 24519-5 Pa. $3.50

THE BOOK OF WOOD CARVING, Charles Marshall Sayers. Still finest book for beginning student. Fundamentals, technique; gives 34 designs, over 34 projects for panels, bookends, mirrors, etc. 33 photos. 118pp. 7¾ × 10⅝. 23654-4 Pa. $3.95

CARVING COUNTRY CHARACTERS, Bill Higginbotham. Expert advice for beginning, advanced carvers on materials, techniques for creating 18 projects—mirthful panorama of American characters. 105 illustrations. 80pp. 8⅜ × 11.
24135-1 Pa. $2.50

300 ART NOUVEAU DESIGNS AND MOTIFS IN FULL COLOR, C.B. Grafton. 44 full-page plates display swirling lines and muted colors typical of Art Nouveau. Borders, frames, panels, cartouches, dingbats, etc. 48pp. 9⅜ × 12¼.
24354-0 Pa. $6.00

SELF-WORKING CARD TRICKS, Karl Fulves. Editor of *Pallbearer* offers 72 tricks that work automatically through nature of card deck. No sleight of hand needed. Often spectacular. 42 illustrations. 113pp. 5⅜ × 8½. 23334-0 Pa. $3.50

CUT AND ASSEMBLE A WESTERN FRONTIER TOWN, Edmund V. Gillon, Jr. Ten authentic full-color buildings on heavy cardboard stock in H-O scale. Sheriff's Office and Jail, Saloon, Wells Fargo, Opera House, others. 48pp. 9¼ × 12¼.
23736-2 Pa. $3.95

CUT AND ASSEMBLE AN EARLY NEW ENGLAND VILLAGE, Edmund V. Gillon, Jr. Printed in full color on heavy cardboard stock. 12 authentic buildings in H-O scale: Adams home in Quincy, Mass., Oliver Wight house in Sturbridge, smithy, store, church, others. 48pp. 9¼ × 12¼. 23536-X Pa. $3.95

THE TALE OF TWO BAD MICE, Beatrix Potter. Tom Thumb and Hunca Munca squeeze out of their hole and go exploring. 27 full-color Potter illustrations. 59pp. 4¼ × 5½. (Available in U.S. only) 23065-1 Pa. $1.50

CARVING FIGURE CARICATURES IN THE OZARK STYLE, Harold L. Enlow. Instructions and illustrations for ten delightful projects, plus general carving instructions. 22 drawings and 47 photographs altogether. 39pp. 8⅜ × 11.
23151-8 Pa. $2.50

A TREASURY OF FLOWER DESIGNS FOR ARTISTS, EMBROIDERERS AND CRAFTSMEN, Susan Gaber. 100 garden favorites lushly rendered by artist for artists, craftsmen, needleworkers. Many form frames, borders. 80pp. 8¼ × 11.
24096-7 Pa. $3.50

CUT & ASSEMBLE A TOY THEATER/THE NUTCRACKER BALLET, Tom Tierney. Model of a complete, full-color production of Tchaikovsky's classic. 6 backdrops, dozens of characters, familiar dance sequences. 32pp. 9⅜ × 12¼.
24194-7 Pa. $4.50

ANIMALS: 1,419 COPYRIGHT-FREE ILLUSTRATIONS OF MAMMALS, BIRDS, FISH, INSECTS, ETC., edited by Jim Harter. Clear wood engravings present, in extremely lifelike poses, over 1,000 species of animals. 284pp. 9 × 12.
23766-4 Pa. $9.95

MORE HAND SHADOWS, Henry Bursill. For those at their 'finger ends,'' 16 more effects—Shakespeare, a hare, a squirrel, Mr. Punch, and twelve more—each explained by a full-page illustration. Considerable period charm. 30pp. 6½ × 9¼.
21384-6 Pa. $1.95

SURREAL STICKERS AND UNREAL STAMPS, William Rowe. 224 haunting, hilarious stamps on gummed, perforated stock, with images of elephants, geisha girls, George Washington, etc. 16pp. one side. 8¼ × 11. 24371-0 Pa. $3.50

GOURMET KITCHEN LABELS, Ed Sibbett, Jr. 112 full-color labels (4 copies each of 28 designs). Fruit, bread, other culinary motifs. Gummed and perforated. 16pp. 8¼ × 11. 24087-8 Pa. $2.95

PATTERNS AND INSTRUCTIONS FOR CARVING AUTHENTIC BIRDS, H.D. Green. Detailed instructions, 27 diagrams, 85 photographs for carving 15 species of birds so life-like, they'll seem ready to fly! 8¼ × 11. 24222-6 Pa. $2.75

FLATLAND, E.A. Abbott. Science-fiction classic explores life of 2-D being in 3-D world. 16 illustrations. 103pp. 5⅜ × 8. 20001-9 Pa. $2.00

DRIED FLOWERS, Sarah Whitlock and Martha Rankin. Concise, clear, practical guide to dehydration, glycerinizing, pressing plant material, and more. Covers use of silica gel. 12 drawings. 32pp. 5⅜ × 8½. 21802-3 Pa. $1.00

EASY-TO-MAKE CANDLES, Gary V. Guy. Learn how easy it is to make all kinds of decorative candles. Step-by-step instructions. 82 illustrations. 48pp. 8¼ × 11. 23881-4 Pa. $2.50

SUPER STICKERS FOR KIDS, Carolyn Bracken. 128 gummed and perforated full-color stickers: GIRL WANTED, KEEP OUT, BORED OF EDUCATION, X-RATED, COMBAT ZONE, many others. 16pp. 8¼ × 11. 24092-4 Pa. $2.50

CUT AND COLOR PAPER MASKS, Michael Grater. Clowns, animals, funny faces...simply color them in, cut them out, and put them together, and you have 9 paper masks to play with and enjoy. 32pp. 8¼ × 11. 23171-2 Pa. $2.25

A CHRISTMAS CAROL: THE ORIGINAL MANUSCRIPT, Charles Dickens. Clear facsimile of Dickens manuscript, on facing pages with final printed text. 8 illustrations by John Leech, 4 in color on covers. 144pp. 8⅜ × 11¼. 20980-6 Pa. $5.95

CARVING SHOREBIRDS, Harry V. Shourds & Anthony Hillman. 16 full-size patterns (all double-page spreads) for 19 North American shorebirds with step-by-step instructions. 72pp. 9¼ × 12¼. 24287-0 Pa. $4.95

THE GENTLE ART OF MATHEMATICS, Dan Pedoe. Mathematical games, probability, the question of infinity, topology, how the laws of algebra work, problems of irrational numbers, and more. 42 figures. 143pp. 5⅜ × 8½. (EBE) 22949-1 Pa. $3.50

READY-TO-USE DOLLHOUSE WALLPAPER, Katzenbach & Warren, Inc. Stripe, 2 floral stripes, 2 allover florals, polka dot; all in full color. 4 sheets (350 sq. in.) of each, enough for average room. 48pp. 8¼ × 11. 23495-9 Pa. $2.95

MINIATURE IRON-ON TRANSFER PATTERNS FOR DOLLHOUSES, DOLLS, AND SMALL PROJECTS, Rita Weiss and Frank Fontana. Over 100 miniature patterns: rugs, bedspreads, quilts, chair seats, etc. In standard dollhouse size. 48pp. 8¼ × 11. 23741-9 Pa. $1.95

THE DINOSAUR COLORING BOOK, Anthony Rao. 45 renderings of dinosaurs, fossil birds, turtles, other creatures of Mesozoic Era. Scientifically accurate. Captions. 48pp. 8¼ × 11. 24022-3 Pa. $2.25

JAPANESE DESIGN MOTIFS, Matsuya Co. Mon, or heraldic designs. Over 4000 typical, beautiful designs: birds, animals, flowers, swords, fans, geometrics; all beautifully stylized. 213pp. 11⅜ × 8¼. 22874-6 Pa. $7.95

THE TALE OF BENJAMIN BUNNY, Beatrix Potter. Peter Rabbit's cousin coaxes him back into Mr. McGregor's garden for a whole new set of adventures. All 27 full-color illustrations. 59pp. 4¼ × 5½. (Available in U.S. only) 21102-9 Pa. $1.50

THE TALE OF PETER RABBIT AND OTHER FAVORITE STORIES BOXED SET, Beatrix Potter. Seven of Beatrix Potter's best-loved tales including Peter Rabbit in a specially designed, durable boxed set. 4¼ × 5½. Total of 447pp. 158 color illustrations. (Available in U.S. only) 23903-9 Pa. $10.80

PRACTICAL MENTAL MAGIC, Theodore Annemann. Nearly 200 astonishing feats of mental magic revealed in step-by-step detail. Complete advice on staging, patter, etc. Illustrated. 320pp. 5⅜ × 8½. 24426-1 Pa. $5.95

CELEBRATED CASES OF JUDGE DEE (DEE GOONG AN), translated by Robert Van Gulik. Authentic 18th-century Chinese detective novel; Dee and associates solve three interlocked cases. Led to van Gulik's own stories with same characters. Extensive introduction. 9 illustrations. 237pp. 5⅜ × 8½.
23337-5 Pa. $4.50

CUT & FOLD EXTRATERRESTRIAL INVADERS THAT FLY, M. Grater. Stage your own lilliputian space battles.By following the step-by-step instructions and explanatory diagrams you can launch 22 full-color fliers into space. 36pp. 8¼ × 11. 24478-4 Pa. $2.95

CUT & ASSEMBLE VICTORIAN HOUSES, Edmund V. Gillon, Jr. Printed in full color on heavy cardboard stock, 4 authentic Victorian houses in H-O scale: Italian-style Villa, Octagon, Second Empire, Stick Style. 48pp. 9¼ × 12¼.
23849-0 Pa. $3.95

BEST SCIENCE FICTION STORIES OF H.G. WELLS, H.G. Wells. Full novel *The Invisible Man,* plus 17 short stories: "The Crystal Egg," "Aepyornis Island," "The Strange Orchid," etc. 303pp. 5⅜ × 8½. (Available in U.S. only)
21531-8 Pa. $4.95

TRADEMARK DESIGNS OF THE WORLD, Yusaku Kamekura. A lavish collection of nearly 700 trademarks, the work of Wright, Loewy, Klee, Binder, hundreds of others. 160pp. 8¾ × 8. (Available in U.S. only) 24191-2 Pa. $5.00

THE ARTIST'S AND CRAFTSMAN'S GUIDE TO REDUCING, ENLARGING AND TRANSFERRING DESIGNS, Rita Weiss. Discover, reduce, enlarge, transfer designs from any objects to any craft project. 12pp. plus 16 sheets special graph paper. 8¼ × 11. 24142-4 Pa. $3.25

TREASURY OF JAPANESE DESIGNS AND MOTIFS FOR ARTISTS AND CRAFTSMEN, edited by Carol Belanger Grafton. Indispensable collection of 360 traditional Japanese designs and motifs redrawn in clean, crisp black-and-white, copyright-free illustrations. 96pp. 8¼ × 11. 24435-0 Pa. $3.95

CHANCERY CURSIVE STROKE BY STROKE, Arthur Baker. Instructions and illustrations for each stroke of each letter (upper and lower case) and numerals. 54 full-page plates. 64pp. 8¼ × 11. 24278-1 Pa. $2.50

THE ENJOYMENT AND USE OF COLOR, Walter Sargent. Color relationships, values, intensities; complementary colors, illumination, similar topics. Color in nature and art. 7 color plates, 29 illustrations. 274pp. 5⅜ × 8½. 20944-X Pa. $4.50

SCULPTURE PRINCIPLES AND PRACTICE, Louis Slobodkin. Step-by-step approach to clay, plaster, metals, stone; classical and modern. 253 drawings, photos. 255pp. 8⅛ × 11. 22960-2 Pa. $7.50

VICTORIAN FASHION PAPER DOLLS FROM HARPER'S BAZAR, 1867-1898, Theodore Menten. Four female dolls with 28 elegant high fashion costumes, printed in full color. 32pp. 9¼ × 12¼. 23453-3 Pa. $3.50

FLOPSY, MOPSY AND COTTONTAIL: A Little Book of Paper Dolls in Full Color, Susan LaBelle. Three dolls and 21 costumes (7 for each doll) show Peter Rabbit's siblings dressed for holidays, gardening, hiking, etc. Charming borders, captions. 48pp. 4¼ × 5½. 24376-1 Pa. $2.25

NATIONAL LEAGUE BASEBALL CARD CLASSICS, Bert Randolph Sugar. 83 big-leaguers from 1909-69 on facsimile cards. Hubbell, Dean, Spahn, Brock plus advertising, info, no duplications. Perforated, detachable. 16pp. 8¼ × 11. 24308-7 Pa. $2.95

THE LOGICAL APPROACH TO CHESS, Dr. Max Euwe, et al. First-rate text of comprehensive strategy, tactics, theory for the amateur. No gambits to memorize, just a clear, logical approach. 224pp. 5⅜ × 8½. 24353-2 Pa. $4.50

MAGICK IN THEORY AND PRACTICE, Aleister Crowley. The summation of the thought and practice of the century's most famous necromancer, long hard to find. Crowley's best book. 436pp. 5⅜ × 8½. (Available in U.S. only) 23295-6 Pa. $6.50

THE HAUNTED HOTEL, Wilkie Collins. Collins' last great tale; doom and destiny in a Venetian palace. Praised by T.S. Eliot. 127pp. 5⅜ × 8½. 24333-8 Pa. $3.00

ART DECO DISPLAY ALPHABETS, Dan X. Solo. Wide variety of bold yet elegant lettering in handsome Art Deco styles. 100 complete fonts, with numerals, punctuation, more. 104pp. 8⅛ × 11. 24372-9 Pa. $4.00

CALLIGRAPHIC ALPHABETS, Arthur Baker. Nearly 150 complete alphabets by outstanding contemporary. Stimulating ideas; useful source for unique effects. 154 plates. 157pp. 8⅜ × 11¼. 21045-6 Pa. $4.95

ARTHUR BAKER'S HISTORIC CALLIGRAPHIC ALPHABETS, Arthur Baker. From monumental capitals of first-century Rome to humanistic cursive of 16th century, 33 alphabets in fresh interpretations. 88 plates. 96pp. 9 × 12. 24054-1 Pa. $4.50

LETTIE LANE PAPER DOLLS, Sheila Young. Genteel turn-of-the-century family very popular then and now. 24 paper dolls. 16 plates in full color. 32pp. 9¼ × 12¼. 24089-4 Pa. $3.50

KEYBOARD WORKS FOR SOLO INSTRUMENTS, G.F. Handel. 35 neglected works from Handel's vast oeuvre, originally jotted down as improvisations. Includes Eight Great Suites, others. New sequence. 174pp. 9⅜ × 12¼.
24338-9 Pa. $7.50

AMERICAN LEAGUE BASEBALL CARD CLASSICS, Bert Randolph Sugar. 82 stars from 1900s to 60s on facsimile cards. Ruth, Cobb, Mantle, Williams, plus advertising, info, no duplications. Perforated, detachable. 16pp. 8¼ × 11.
24286-2 Pa. $2.95

A TREASURY OF CHARTED DESIGNS FOR NEEDLEWORKERS, Georgia Gorham and Jeanne Warth. 141 charted designs: owl, cat with yarn, tulips, piano, spinning wheel, covered bridge, Victorian house and many others. 48pp. 8¼ × 11.
23558-0 Pa. $1.95

DANISH FLORAL CHARTED DESIGNS, Gerda Bengtsson. Exquisite collection of over 40 different florals: anemone, Iceland poppy, wild fruit, pansies, many others. 45 illustrations. 48pp. 8¼ × 11. 23957-8 Pa. $1.75

OLD PHILADELPHIA IN EARLY PHOTOGRAPHS 1839-1914, Robert F. Looney. 215 photographs: panoramas, street scenes, landmarks, President-elect Lincoln's visit, 1876 Centennial Exposition, much more. 230pp. 8⅜ × 11¾.
23345-6 Pa. $9.95

PRELUDE TO MATHEMATICS, W.W. Sawyer. Noted mathematician's lively, stimulating account of non-Euclidean geometry, matrices, determinants, group theory, other topics. Emphasis on novel, striking aspects. 224pp. 5⅜ × 8½.
24401-6 Pa. $4.50

ADVENTURES WITH A MICROSCOPE, Richard Headstrom. 59 adventures with clothing fibers, protozoa, ferns and lichens, roots and leaves, much more. 142 illustrations. 232pp. 5⅜ × 8½. 23471-1 Pa. $3.95

IDENTIFYING ANIMAL TRACKS: MAMMALS, BIRDS, AND OTHER ANIMALS OF THE EASTERN UNITED STATES, Richard Headstrom. For hunters, naturalists, scouts, nature-lovers. Diagrams of tracks, tips on identification. 128pp. 5⅜ × 8. 24442-3 Pa. $3.50

VICTORIAN FASHIONS AND COSTUMES FROM HARPER'S BAZAR, 1867-1898, edited by Stella Blum. Day costumes, evening wear, sports clothes, shoes, hats, other accessories in over 1,000 detailed engravings. 320pp. 9⅜ × 12¼.
22990-4 Pa. $9.95

EVERYDAY FASHIONS OF THE TWENTIES AS PICTURED IN SEARS AND OTHER CATALOGS, edited by Stella Blum. Actual dress of the Roaring Twenties, with text by Stella Blum. Over 750 illustrations, captions. 156pp. 9 × 12.
24134-3 Pa. $8.50

HALL OF FAME BASEBALL CARDS, edited by Bert Randolph Sugar. Cy Young, Ted Williams, Lou Gehrig, and many other Hall of Fame greats on 92 full-color, detachable reprints of early baseball cards. No duplication of cards with *Classic Baseball Cards*. 16pp. 8¼ × 11. 23624-2 Pa. $3.50

THE ART OF HAND LETTERING, Helm Wotzkow. Course in hand lettering, Roman, Gothic, Italic, Block, Script. Tools, proportions, optical aspects, individual variation. Very quality conscious. Hundreds of specimens. 320pp. 5⅜ × 8½.
21797-3 Pa. $4.95

HOW THE OTHER HALF LIVES, Jacob A. Riis. Journalistic record of filth, degradation, upward drive in New York immigrant slums, shops, around 1900. New edition includes 100 original Riis photos, monuments of early photography. 233pp. 10 × 7⅞. 22012-5 Pa. $7.95

CHINA AND ITS PEOPLE IN EARLY PHOTOGRAPHS, John Thomson. In 200 black-and-white photographs of exceptional quality photographic pioneer Thomson captures the mountains, dwellings, monuments and people of 19th-century China. 272pp. 9⅜ × 12¼. 24393-1 Pa. $12.95

GODEY COSTUME PLATES IN COLOR FOR DECOUPAGE AND FRAMING, edited by Eleanor Hasbrouk Rawlings. 24 full-color engravings depicting 19th-century Parisian haute couture. Printed on one side only. 56pp. 8¼ × 11. 23879-2 Pa. $3.95

ART NOUVEAU STAINED GLASS PATTERN BOOK, Ed Sibbett, Jr. 104 projects using well-known themes of Art Nouveau: swirling forms, florals, peacocks, and sensuous women. 60pp. 8¼ × 11. 23577-7 Pa. $3.50

QUICK AND EASY PATCHWORK ON THE SEWING MACHINE: Susan Aylsworth Murwin and Suzzy Payne. Instructions, diagrams show exactly how to machine sew 12 quilts. 48pp. of templates. 50 figures. 80pp. 8¼ × 11. 23770-2 Pa. $3.50

THE STANDARD BOOK OF QUILT MAKING AND COLLECTING, Marguerite Ickis. Full information, full-sized patterns for making 46 traditional quilts, also 150 other patterns. 483 illustrations. 273pp. 6⅞ × 9⅝. 20582-7 Pa. $5.95

LETTERING AND ALPHABETS, J. Albert Cavanagh. 85 complete alphabets lettered in various styles; instructions for spacing, roughs, brushwork. 121pp. 8¾ × 8. 20053-1 Pa. $3.75

LETTER FORMS: 110 COMPLETE ALPHABETS, Frederick Lambert. 110 sets of capital letters; 16 lower case alphabets; 70 sets of numbers and other symbols. 110pp. 8⅛ × 11. 22872-X Pa. $4.50

ORCHIDS AS HOUSE PLANTS, Rebecca Tyson Northen. Grow cattleyas and many other kinds of orchids—in a window, in a case, or under artificial light. 63 illustrations. 148pp. 5⅜ × 8½. 23261-1 Pa. $2.95

THE MUSHROOM HANDBOOK, Louis C.C. Krieger. Still the best popular handbook. Full descriptions of 259 species, extremely thorough text, poisons, folklore, etc. 32 color plates; 126 other illustrations. 560pp. 5⅜ × 8½. 21861-9 Pa. $8.50

THE DORÉ BIBLE ILLUSTRATIONS, Gustave Doré. All wonderful, detailed plates: Adam and Eve, Flood, Babylon, life of Jesus, etc. Brief King James text with each plate. 241 plates. 241pp. 9 × 12. 23004-X Pa. $8.95

THE BOOK OF KELLS: Selected Plates in Full Color, edited by Blanche Cirker. 32 full-page plates from greatest manuscript-icon of early Middle Ages. Fantastic, mysterious. Publisher's Note. Captions. 32pp. 9¾ × 12¼. 24345-1 Pa. $4.50

THE PERFECT WAGNERITE, George Bernard Shaw. Brilliant criticism of the Ring Cycle, with provocative interpretation of politics, economic theories behind the Ring. 136pp. 5⅜ × 8½. (Available in U.S. only) 21707-8 Pa. $3.00

THE RIME OF THE ANCIENT MARINER, Gustave Doré, S.T. Coleridge. Doré's finest work, 34 plates capture moods, subtleties of poem. Full text. 77pp. 9¼ × 12. 22305-1 Pa. $4.95

SONGS OF INNOCENCE, William Blake. The first and most popular of Blake's famous "Illuminated Books," in a facsimile edition reproducing all 31 brightly colored plates. Additional printed text of each poem. 64pp. 5¼ × 7. 22764-2 Pa. $3.00

AN INTRODUCTION TO INFORMATION THEORY, J.R. Pierce. Second (1980) edition of most impressive non-technical account available. Encoding, entropy, noisy channel, related areas, etc. 320pp. 5⅜ × 8½. 24061-4 Pa. $4.95

THE DIVINE PROPORTION: A STUDY IN MATHEMATICAL BEAUTY, H.E. Huntley. "Divine proportion" or "golden ratio" in poetry, Pascal's triangle, philosophy, psychology, music, mathematical figures, etc. Excellent bridge between science and art. 58 figures. 185pp. 5⅜ × 8½. 22254-3 Pa. $3.95

THE DOVER NEW YORK WALKING GUIDE: From the Battery to Wall Street, Mary J. Shapiro. Superb inexpensive guide to historic buildings and locales in lower Manhattan: Trinity Church, Bowling Green, more. Complete Text; maps. 36 illustrations. 48pp. 3⅞ × 9¼. 24225-0 Pa. $2.50

NEW YORK THEN AND NOW, Edward B. Watson, Edmund V. Gillon, Jr. 83 important Manhattan sites: on facing pages early photographs (1875-1925) and 1976 photos by Gillon. 172 illustrations. 171pp. 9¼ × 10. 23361-8 Pa. $7.95

HISTORIC COSTUME IN PICTURES, Braun & Schneider. Over 1450 costumed figures from dawn of civilization to end of 19th century. English captions. 125 plates. 256pp. 8⅜ × 11¼. 23150-X Pa. $7.50

VICTORIAN AND EDWARDIAN FASHION: A Photographic Survey, Alison Gernsheim. First fashion history completely illustrated by contemporary photographs. Full text plus 235 photos, 1840-1914, in which many celebrities appear. 240pp. 6½ × 9¼. 24205-6 Pa. $6.00

CHARTED CHRISTMAS DESIGNS FOR COUNTED CROSS-STITCH AND OTHER NEEDLECRAFTS, Lindberg Press. Charted designs for 45 beautiful needlecraft projects with many yuletide and wintertime motifs. 48pp. 8¼ × 11. 24356-7 Pa. $1.95

101 FOLK DESIGNS FOR COUNTED CROSS-STITCH AND OTHER NEEDLECRAFTS, Carter Houck. 101 authentic charted folk designs in a wide array of lovely representations with many suggestions for effective use. 48pp. 8¼ × 11. 24369-9 Pa. $2.25

FIVE ACRES AND INDEPENDENCE, Maurice G. Kains. Great back-to-the-land classic explains basics of self-sufficient farming. The one book to get. 95 illustrations. 397pp. 5⅜ × 8½. 20974-1 Pa. $4.95

A MODERN HERBAL, Margaret Grieve. Much the fullest, most exact, most useful compilation of herbal material. Gigantic alphabetical encyclopedia, from aconite to zedoary, gives botanical information, medical properties, folklore, economic uses, and much else. Indispensable to serious reader. 161 illustrations. 888pp. 6½ × 9¼. (Available in U.S. only) 22798-7, 22799-5 Pa., Two-vol. set $16.45

DECORATIVE NAPKIN FOLDING FOR BEGINNERS, Lillian Oppenheimer and Natalie Epstein. 22 different napkin folds in the shape of a heart, clown's hat, love knot, etc. 63 drawings. 48pp. 8¼ × 11. 23797-4 Pa. $1.95

DECORATIVE LABELS FOR HOME CANNING, PRESERVING, AND OTHER HOUSEHOLD AND GIFT USES, Theodore Menten. 128 gummed, perforated labels, beautifully printed in 2 colors. 12 versions. Adhere to metal, glass, wood, ceramics. 24pp. 8¼ × 11. 23219-0 Pa. $2.95

EARLY AMERICAN STENCILS ON WALLS AND FURNITURE, Janet Waring. Thorough coverage of 19th-century folk art: techniques, artifacts, surviving specimens. 166 illustrations, 7 in color. 147pp. of text. 7⅞ × 10¾. 21906-2 Pa. $9.95

AMERICAN ANTIQUE WEATHERVANES, A.B. & W.T. Westervelt. Extensively illustrated 1883 catalog exhibiting over 550 copper weathervanes and finials. Excellent primary source by one of the principal manufacturers. 104pp. 6⅛ × 9¼. 24396-6 Pa. $3.95

ART STUDENTS' ANATOMY, Edmond J. Farris. Long favorite in art schools. Basic elements, common positions, actions. Full text, 158 illustrations. 159pp. 5⅜ × 8½. 20744-7 Pa. $3.95

BRIDGMAN'S LIFE DRAWING, George B. Bridgman. More than 500 drawings and text teach you to abstract the body into its major masses. Also specific areas of anatomy. 192pp. 6½ × 9¼. (EA) 22710-3 Pa. $4.50

COMPLETE PRELUDES AND ETUDES FOR SOLO PIANO, Frederic Chopin. All 26 Preludes, all 27 Etudes by greatest composer of piano music. Authoritative Paderewski edition. 224pp. 9 × 12. (Available in U.S. only) 24052-5 Pa. $7.50

PIANO MUSIC 1888-1905, Claude Debussy. Deux Arabesques, Suite Bergamesque, Masques, 1st series of Images, etc. 9 others, in corrected editions. 175pp. 9⅜ × 12¼. (ECE) 22771-5 Pa. $5.95

TEDDY BEAR IRON-ON TRANSFER PATTERNS, Ted Menten. 80 iron-on transfer patterns of male and female Teddys in a wide variety of activities, poses, sizes. 48pp. 8¼ × 11. 24596-9 Pa. $2.25

A PICTURE HISTORY OF THE BROOKLYN BRIDGE, M.J. Shapiro. Profusely illustrated account of greatest engineering achievement of 19th century. 167 rare photos & engravings recall construction, human drama. Extensive, detailed text. 122pp. 8¼ × 11. 24403-2 Pa. $7.95

NEW YORK IN THE THIRTIES, Berenice Abbott. Noted photographer's fascinating study shows new buildings that have become famous and old sights that have disappeared forever. 97 photographs. 97pp. 11⅜ × 10. 22967-X Pa. $6.50

MATHEMATICAL TABLES AND FORMULAS, Robert D. Carmichael and Edwin R. Smith. Logarithms, sines, tangents, trig functions, powers, roots, reciprocals, exponential and hyperbolic functions, formulas and theorems. 269pp. 5⅜ × 8½. 60111-0 Pa. $3.75

HANDBOOK OF MATHEMATICAL FUNCTIONS WITH FORMULAS, GRAPHS, AND MATHEMATICAL TABLES, edited by Milton Abramowitz and Irene A. Stegun. Vast compendium: 29 sets of tables, some to as high as 20 places. 1,046pp. 8 × 10½. 61272-4 Pa. $19.95

REASON IN ART, George Santayana. Renowned philosopher's provocative, seminal treatment of basis of art in instinct and experience. Volume Four of *The Life of Reason.* 230pp. 5⅜ × 8. 24358-3 Pa. $4.50

LANGUAGE, TRUTH AND LOGIC, Alfred J. Ayer. Famous, clear introduction to Vienna, Cambridge schools of Logical Positivism. Role of philosophy, elimination of metaphysics, nature of analysis, etc. 160pp. 5⅜ × 8½. (USCO) 20010-8 Pa. $2.75

BASIC ELECTRONICS, U.S. Bureau of Naval Personnel. Electron tubes, circuits, antennas, AM, FM, and CW transmission and receiving, etc. 560 illustrations. 567pp. 6½ × 9¼. 21076-6 Pa. $8.95

THE ART DECO STYLE, edited by Theodore Menten. Furniture, jewelry, metalwork, ceramics, fabrics, lighting fixtures, interior decors, exteriors, graphics from pure French sources. Over 400 photographs. 183pp. 8⅜ × 11¼. 22824-X Pa. $6.95

THE FOUR BOOKS OF ARCHITECTURE, Andrea Palladio. 16th-century classic covers classical architectural remains, Renaissance revivals, classical orders, etc. 1738 Ware English edition. 216 plates. 110pp. of text. 9½ × 12¾. 21308-0 Pa. $11.50

THE WIT AND HUMOR OF OSCAR WILDE, edited by Alvin Redman. More than 1000 ripostes, paradoxes, wisecracks: Work is the curse of the drinking classes, I can resist everything except temptations, etc. 258pp. 5⅜ × 8½. (USCO) 20602-5 Pa. $3.50

THE DEVIL'S DICTIONARY, Ambrose Bierce. Barbed, bitter, brilliant witticisms in the form of a dictionary. Best, most ferocious satire America has produced. 145pp. 5⅜ × 8½. 20487-1 Pa. $2.50

ERTÉ'S FASHION DESIGNS, Erté. 210 black-and-white inventions from *Harper's Bazar,* 1918-32, plus 8pp. full-color covers. Captions. 88pp. 9 × 12. 24203-X Pa. $6.50

ERTÉ GRAPHICS, Erté. Collection of striking color graphics: *Seasons, Alphabet, Numerals, Aces* and *Precious Stones.* 50 plates, including 4 on covers. 48pp. 9⅜ × 12¼. 23580-7 Pa. $6.95

PAPER FOLDING FOR BEGINNERS, William D. Murray and Francis J. Rigney. Clearest book for making origami sail boats, roosters, frogs that move legs, etc. 40 projects. More than 275 illustrations. 94pp. 5⅜ × 8½. 20713-7 Pa. $2.25

ORIGAMI FOR THE ENTHUSIAST, John Montroll. Fish, ostrich, peacock, squirrel, rhinoceros, Pegasus, 19 other intricate subjects. Instructions. Diagrams. 128pp. 9 × 12. 23799-0 Pa. $4.95

CROCHETING NOVELTY POT HOLDERS, edited by Linda Macho. 64 useful, whimsical pot holders feature kitchen themes, animals, flowers, other novelties. Surprisingly easy to crochet. Complete instructions. 48pp. 8¼ × 11. 24296-X Pa. $1.95

CROCHETING DOILIES, edited by Rita Weiss. Irish Crochet, Jewel, Star Wheel, Vanity Fair and more. Also luncheon and console sets, runners and centerpieces. 51 illustrations. 48pp. 8¼ × 11. 23424-X Pa. $2.00

CATALOG OF DOVER BOOKS

YUCATAN BEFORE AND AFTER THE CONQUEST, Diego de Landa. Only significant account of Yucatan written in the early post-Conquest era. Translated by William Gates. Over 120 illustrations. 162pp. 5⅜ × 8½. 23622-6 Pa. $3.50

ORNATE PICTORIAL CALLIGRAPHY, E.A. Lupfer. Complete instructions, over 150 examples help you create magnificent "flourishes" from which beautiful animals and objects gracefully emerge. 8⅛ × 11. 21957-7 Pa. $2.95

DOLLY DINGLE PAPER DOLLS, Grace Drayton. Cute chubby children by same artist who did Campbell Kids. Rare plates from 1910s. 30 paper dolls and over 100 outfits reproduced in full color. 32pp. 9¼ × 12¼. 23711-7 Pa. $3.50

CURIOUS GEORGE PAPER DOLLS IN FULL COLOR, H. A. Rey, Kathy Allert. Naughty little monkey-hero of children's books in two doll figures, plus 48 full-color costumes: pirate, Indian chief, fireman, more. 32pp. 9¼ × 12¼. 24386-9 Pa. $3.50

GERMAN: HOW TO SPEAK AND WRITE IT, Joseph Rosenberg. Like *French, How to Speak and Write It.* Very rich modern course, with a wealth of pictorial material. 330 illustrations. 384pp. 5⅜ × 8½. (USUKO) 20271-2 Pa. $4.75

CATS AND KITTENS: 24 Ready-to-Mail Color Photo Postcards, D. Holby. Handsome collection; feline in a variety of adorable poses. Identifications. 12pp. on postcard stock. 8¼ × 11. 24469-5 Pa. $2.95

MARILYN MONROE PAPER DOLLS, Tom Tierney. 31 full-color designs on heavy stock, from *The Asphalt Jungle, Gentlemen Prefer Blondes,* 22 others. 1 doll. 16 plates. 32pp. 9⅜ × 12¼. 23769-9 Pa. $3.50

FUNDAMENTALS OF LAYOUT, F.H. Wills. All phases of layout design discussed and illustrated in 121 illustrations. Indispensable as student's text or handbook for professional. 124pp. 8⅛.× 11. 21279-3 Pa. $4.50

FANTASTIC SUPER STICKERS, Ed Sibbett, Jr. 75 colorful pressure-sensitive stickers. Peel off and place for a touch of pizzazz: clowns, penguins, teddy bears, etc. Full color. 16pp. 8¼ × 11. 24471-7 Pa. $2.95

LABELS FOR ALL OCCASIONS, Ed Sibbett, Jr. 6 labels each of 16 different designs—baroque, art nouveau, art deco, Pennsylvania Dutch, etc.—in full color. 24pp. 8¼ × 11. 23688-9 Pa. $2.95

HOW TO CALCULATE QUICKLY: RAPID METHODS IN BASIC MATHEMATICS, Henry Sticker. Addition, subtraction, multiplication, division, checks, etc. More than 8000 problems, solutions. 185pp. 5 × 7¼. 20295-X Pa. $2.95

THE CAT COLORING BOOK, Karen Baldauski. Handsome, realistic renderings of 40 splendid felines, from American shorthair to exotic types. 44 plates. Captions. 48pp. 8¼ × 11. 24011-8 Pa. $2.25

THE TALE OF PETER RABBIT, Beatrix Potter. The inimitable Peter's terrifying adventure in Mr. McGregor's garden, with all 27 wonderful, full-color Potter illustrations. 55pp. 4¼ × 5½. (Available in U.S. only) 22827-4 Pa. $1.60

BASIC ELECTRICITY, U.S. Bureau of Naval Personnel. Batteries, circuits, conductors, AC and DC, inductance and capacitance, generators, motors, transformers, amplifiers, etc. 349 illustrations. 448pp. 6½ × 9¼. 20973-3 Pa. $7.95

CATALOG OF DOVER BOOKS

SOURCE BOOK OF MEDICAL HISTORY, edited by Logan Clendening, M.D. Original accounts ranging from Ancient Egypt and Greece to discovery of X-rays: Galen, Pasteur, Lavoisier, Harvey, Parkinson, others. 685pp. 5⅜ × 8½.
20621-1 Pa. $10.95

THE ROSE AND THE KEY, J.S. Lefanu. Superb mystery novel from Irish master. Dark doings among an ancient and aristocratic English family. Well-drawn characters; capital suspense. Introduction by N. Donaldson. 448pp. 5⅜ × 8½.
24377-X Pa. $6.95

SOUTH WIND, Norman Douglas. Witty, elegant novel of ideas set on languorous Meditterranean island of Nepenthe. Elegant prose, glittering epigrams, mordant satire. 1917 masterpiece. 416pp. 5⅜ × 8½. (Available in U.S. only)
24361-3 Pa. $5.95

RUSSELL'S CIVIL WAR PHOTOGRAPHS, Capt. A.J. Russell. 116 rare Civil War Photos: Bull Run, Virginia campaigns, bridges, railroads, Richmond, Lincoln's funeral car. Many never seen before. Captions. 128pp. 9⅜ × 12¼.
24283-8 Pa. $6.95

PHOTOGRAPHS BY MAN RAY: 105 Works, 1920-1934. Nudes, still lifes, landscapes, women's faces, celebrity portraits (Dali, Matisse, Picasso, others), rayographs. Reprinted from rare gravure edition. 128pp. 9⅜ × 12¼. (Available in U.S. only) 23842-3 Pa. $6.95

STAR NAMES: THEIR LORE AND MEANING, Richard H. Allen. Star names, the zodiac, constellations: folklore and literature associated with heavens. The basic book of its field, fascinating reading. 563pp. 5⅜ × 8½. 21079-0 Pa. $7.95

BURNHAM'S CELESTIAL HANDBOOK, Robert Burnham, Jr. Thorough guide to the stars beyond our solar system. Exhaustive treatment. Alphabetical by constellation: Andromeda to Cetus in Vol. 1; Chamaeleon to Orion in Vol. 2; and Pavo to Vulpecula in Vol. 3. Hundreds of illustrations. Index in Vol. 3. 2000pp. 6⅛ × 9¼. 23567-X, 23568-8, 23673-0 Pa. Three-vol. set $36.85

THE ART NOUVEAU STYLE BOOK OF ALPHONSE MUCHA, Alphonse Mucha. All 72 plates from *Documents Decoratifs* in original color. Stunning, essential work of Art Nouveau. 80pp. 9⅜ × 12¼. 24044-4 Pa. $7.95

DESIGNS BY ERTE; FASHION DRAWINGS AND ILLUSTRATIONS FROM "HARPER'S BAZAR," Erte. 310 fabulous line drawings and 14 *Harper's Bazar* covers, 8 in full color. Erte's exotic temptresses with tassels, fur muffs, long trains, coifs, more. 129pp. 9⅜ × 12¼. 23397-9 Pa. $6.95

HISTORY OF STRENGTH OF MATERIALS, Stephen P. Timoshenko. Excellent historical survey of the strength of materials with many references to the theories of elasticity and structure. 245 figures. 452pp. 5⅜ × 8½. 61187-6 Pa. $8.95

Prices subject to change without notice.

Available at your book dealer or write for free catalog to Dept. GI, Dover Publications, Inc., 31 East 2nd St. Mineola, N.Y. 11501. Dover publishes more than 175 books each year on science, elementary and advanced mathematics, biology, music, art, literary history, social sciences and other areas.